The Protective Arm of the Renin–Angiotensin System

Functional Aspects and Therapeutic Implications

The Protective Arm of the Renin–Angiotensin System

Functional Aspects and Therapeutic Implications

First Edition

Thomas Unger
Ulrike M. Steckelings
Robson A.S. dos Santos

AMSTERDAM • BOSTON • HEIDELBERG • LONDON
NEW YORK • OXFORD • PARIS • SAN DIEGO
SAN FRANCISCO • SINGAPORE • SYDNEY • TOKYO

Academic Press is an imprint of Elsevier

Academic Press is an imprint of Elsevier
125 London Wall, London, EC2Y 5AS, UK
525 B Street, Suite 1800, San Diego, CA 92101-4495, USA
225 Wyman Street, Waltham, MA 02451, USA
The Boulevard, Langford Lane, Kidlington, Oxford OX5 1GB, UK

Library of Congress Cataloging-in-Publication Data
A catalog record for this book is available from the Library of Congress

British Library Cataloguing in Publication Data
A catalogue record for this book is available from the British Library.

ISBN: 978-0-12-801364-9

For information on all Academic Press publications
visit our website at http://store.elsevier.com/

Printed and bound in the United Kingdom

Working together
to grow libraries in
developing countries

www.elsevier.com • www.bookaid.org

Contents

37. ACE2/Ang-(1-7)/Mas Axis and Physical Exercise

Daisy Motta-Santos and Lenice Becker

38. Angiotensin-Converting Enzyme 2/ Angiotensin-(1-7)/Mas Receptor Axis: Emerging Pharmacological Target for Pulmonary Diseases

Vinayak Shenoy, Anderson J. Ferreira, Michael Katovich and Mohan K. Raizada

39. Nanocarriers for Improved Delivery of Angiotensin-(1-7)

Frédéric Frézard, Robson Augusto Souza dos Santos and Ana Paula Correa Oliveira Bahia

40. Mas Agonists

Rodrigo A. Fraga-Silva, Walkyria Oliveira Sampaio and Robson Augusto Souza dos Santos

41. Preclinical and Clinical Development of Angiotensin Peptides (Mas/Ang(1-7)/ ACE2): Future Clinical Application

Kathleen E. Rodgers and Gere S. diZerega

Contributors

Marie-Isabel Aguilar Department of Biochemistry and Molecular Biology, Monash University, Clayton Victoria 3800, Australia

Natalia Alenina Max-Delbrueck-Center for Moleculare Medicine (MDC), Berlin-Buch, Germany; Federal University of Minas Gerais, Belo Horizonte, Brazil, and Instituto Nacional de Ciência e Tecnologia em, NanoBiofarmacêutica (NanoBIOFAR), Belo Horizonte, Brazil

Peter W. Angus Department of Medicine, The University of Melbourne, Austin Health, Heidelberg, Victoria, Australia

Michael Bader Max Delbrück Center for Molecular Medicine (MDC), Berlin, Germany; Charité Medical Faculty, Berlin, Germany; Universidade Federal de Minas Gerais, Belo Horizonte, Brazil; and Institute for Biology, University of Lübeck, Germany

Ana Paula Correa Oliveira Bahia Departamento de Fisiologia e Biofísica—ICB, Universidade Federal de Minas Gerais, Belo Horizonte, Minas Gerais, Brazil

Dhaniel Baraldi Department of Pharmacology, Monash University, Clayton Victoria 3800, Australia

Lívia Corrêa Barroso Departamento de Bioquimica e Imunologia, UFMG, Belo Horizonte, Minas Gerais, Brazil

Lenice Becker Sport Center of UFOP, Federal University of Ouro Preto, Ouro Preto, Minas Gerais, Brazil

Douglas M. Bennion Department of Physiology and Functional Genomics, University of Florida, Gainesville, Florida, USA

Serge P. Bottari Laboratory for Fundamental and Applied Bioenergetics, INSERM U1055, Université Grenoble—Alpes, and Radioanalysis Unit, Institute for Biology and Pathology, CHU de Grenoble, Grenoble, France

Peter Buehlmayer Hangstrasse 18, CH-4144 Arlesheim, Switzerland

Maria J. Campagnole-Santos Department of Physiology and Biophysics, INCT-Nanobiofar, Institute of Biological Sciences, Federal University of Minas Gerais-UFMG, Brazil

Robert M. Carey Department of Medicine, University of Virginia, Charlottesville, Virginia, USA

Harshita Chodavarapu Department of Pharmacology, Louisiana State University Health Sciences Center, New Orleans, Louisiana, USA

Leoluca Criscione MangiaSano Consulting, Rümelinbachweg 10, CH-4054 Basel, Switzerland

Juraj Culman Institute of Experimental and Clinical Pharmacology, University Hospitals of Schleswig-Holstein, Campus Kiel, Kiel, Germany

Björn Dahlöf Department of Molecular and Clinical Medicine, Institute of Medicine, Sahlgrenska Academy, Gothenburg, Sweden

Leon Alexander Danyel Center for Cardiovascular Research (CCRI), Institute of Pharmacology, Charité-Universitätsmedizin Berlin, Germany

Marc de Gasparo Cardiovascular & Metabolic Syndrome Adviser, Rossemaison, Switzerland

Mark Del Borgo Department of Biochemistry and Molecular Biology, Monash University, Clayton Victoria 3800, Australia

Kate M. Denton Department of Physiology, Monash University, Clayton Victoria 3800, Australia

Isha S. Dhande University of Hosuton, Houston, Texas, USA

Gere S. diZerega US Biotest, San Luis Obispo, and Keck School of Medicine, Los Angeles, CA, USA

Fernando Pablo Dominici Department of Biological Chemistry, IQUIFIB, University of Buenos Aires, Buenos Aires, Argentina

Adelina M. dos Reis Department of Physiology & Biophysics, Universidade Federal de Minas Gerais, Belo Horizonte, Brazil

Victor J. Dzau Institute of Medicine of the National Academies, Washington, DC, USA

Veronica Valero Esquitino Center for Cardiovascular Research (CCRI), Institute of Pharmacology, Charité-Universitätsmedizin Berlin, Germany

Anderson J. Ferreira Department of Morphology, Federal University of Minas Gerais, Belo Horizonte, MG 31270-901, Brazil

Anna Foryst-Ludwig Center for Cardiovascular Research (CCR), Institute of Pharmacology, Charité-Universitätsmedizin Berlin, Berlin, Germany

Sébastien Foulquier CARIM School for Cardiovascular Diseases, Maastricht University, Maastricht, The Netherlands

Rodrigo A. Fraga-Silva Institute of Bioengineering, Ecole Polytechnique Federale de Lausanne, Lausanne, Vaud, Switzerland

Frédéric Frézard Departamento de Fisiologia e Biofísica—ICB, Universidade Federal de Minas Gerais, Belo Horizonte, Minas Gerais, Brazil

Pascal Furet Novartis Institute for Biochemical Research, CH-4002 Basel, Switzerland

Patricia E. Gallagher Hypertension and Vascular Research Center, Wake Forest School of Medicine, Winston-Salem, North Carolina, USA

Lie Gao Department of Cellular and Integrative Physiology, University of Nebraska Medical Center, Omaha, Nebraska, USA

Tracey A. Gaspari Department of Pharmacology, Monash University, Clayton Victoria 3800, Australia

Mariela M. Gironacci Department of Biological Chemistry, IQUIFIB-CONICET, University of Buenos Aires, Buenos Aires, Argentina

Anders Hallberg Department of Medicinal Chemistry, Division of Organic Pharmaceutical Chemistry, Uppsala University, Uppsala, Sweden

Chandana B. Herath Department of Medicine, The University of Melbourne, Austin Health, Heidelberg, Victoria, Australia

Lucinda M. Hilliard Department of Physiology, Monash University, Clayton Victoria 3800, Australia

Masatsugu Horiuchi Molecular Cardiovascular Biology and Pharmacology, Ehime University Graduate School of Medicine, Tohon, Shitsukawa, Ehime, Japan

Tahir Hussain University of Hosuton, Houston, Texas, USA

Rebecka Isaksson Department of Medicinal Chemistry, Division of Organic Pharmaceutical Chemistry, Uppsala University, Uppsala, Sweden

Emma S. Jones Department of Pharmacology, Monash University, Clayton Victoria 3800, Australia

Elena Kaschina Center for Cardiovascular Research (CCR), Institute of Pharmacology, Charité-Universitätsmedizin Berlin, Berlin, Germany

Michael Katovich Department of Pharmacodynamics, College of Pharmacy, University of Florida, Gainesville, FL 32610, USA

Ulrich Kintscher Center for Cardiovascular Research (CCR), Institute of Pharmacology, Charité-Universitätsmedizin Berlin, Berlin, Germany

Mats Larhed Department of Medicinal Chemistry, Division of Organic Pharmaceutical Chemistry, Uppsala University, Uppsala, Sweden

Eric Lazartigues Department of Pharmacology, Louisiana State University Health Sciences Center, New Orleans, Louisiana, USA

Kai Y. Mak Department of Medicine, The University of Melbourne, Austin Health, Heidelberg, Victoria, Australia

Nephtali Marina Department of Clinical Pharmacology and Experimental Therapeutics, University College London, London, UK

Claudia A. McCarthy Department of Pharmacology, Monash University, Clayton Victoria 3800, Australia

Adam P. Mecca Department of Physiology and Functional Genomics, University of Florida, Gainesville, Florida, USA

Marcos Barrouin Melo Department of Physiology and Pharmacology, UFMG (Federal University of Minas Gerais), Belo Horizonte, Minas Gerais, Brazil

Katrina M. Mirabito Department of Physiology, Monash University, Clayton Victoria 3800, Australia

Masaki Mogi Molecular Cardiovascular Biology and Pharmacology, Ehime University Graduate School of Medicine, Tohon, Shitsukawa, Ehime, Japan

Angie Molina Department of Molecular Medicine, Institut Gustave Roussy, INSERM U981, Université Paris Sud, 94800 Villejuif, France

Augusto C. Montezano Institute of Cardiovascular & Medical Sciences, University of Glasgow, Glasgow, UK

Daisy Motta-Santos Department of Physiology and Biophysics, Federal University of Minas Gerais, Belo Horizonte, Minas Gerais, Brazil

Clara Nahmias Department of Molecular Medicine, Institut Gustave Roussy, INSERM U981, Université Paris Sud, 94800 Villejuif, France

Pawel Namsolleck CARIM School for Cardiovascular Diseases, Maastricht University, Maastricht, The Netherlands

Anne Nehlig Department of Molecular Medicine, Institut Gustave Roussy, INSERM U981, Université Paris Sud, 94800 Villejuif, France

Aurelie Nguyen Institute of Cardiovascular & Medical Sciences, University of Glasgow, Glasgow, UK

Valéria Nunes-Souza Max Delbrück Center for Molecular Medicine, Berlin, Germany; Laboratório de Reatividade Cardiovascular (LRC), Setor de Fisiologia e Farmacologia, ICBS, Universidade Federal de Alagoas, Maceió, Al, Brazil, and National Institute of Science and Technology in NanoBiopharmaceutics (N-BIOFAR), Brazil

Ludovit Paulis Institute of Pathophysiology, Faculty of Medicine, Comenius University, Bratislava, Slovakia

Sérgio Veloso Brant Pinheiro Department of Pediatrics, Federal University of Minas Gerais—UFMG, Belo Horizonte, Minas Gerais, Brazil

Luiza A. Rabelo Max Delbrück Center for Molecular Medicine, Berlin, Germany; Laboratório de Reatividade Cardiovascular (LRC), Setor de Fisiologia e Farmacologia, ICBS, Universidade Federal de Alagoas, Maceió, Al, Brazil, and National Institute of Science and Technology in NanoBiopharmaceutics (N-BIOFAR), Brazil

Mohan K. Raizada Department of Physiology and Functional Genomics, College of Medicine, University of Florida, Gainesville, FL 32610, USA

Chiara Recarti CARIM School for Cardiovascular Diseases, Maastricht University, Maastricht, The Netherlands

Robert W. Regenhardt Department of Physiology and Functional Genomics, University of Florida, Gainesville, Florida, USA

Fernando M. Reis Department of Obstetrics & Gynecology and Division of Human Reproduction, Universidade Federal de Minas Gerais, Belo Horizonte, Brazil

Kathleen E. Rodgers School of Pharmacy, University of Southern California, Los Angeles, California, USA

Sylvie Rodrigues-Ferreira Department of Molecular Medicine, Institut Gustave Roussy, INSERM U981, Université Paris Sud, 94800 Villejuif, France

Walkyria Oliveira Sampaio Department of Biological Sciences, Universidade de Itaúna, Itaúna, and Department of Physiology, Universidade Federal de Minas Gerais, Belo Horizonte, Minas Gerais, Brazil

Robson Augusto Souza dos Santos Departamento de Fisiologia e Biofísica—ICB, Universidade Federal de Minas Gerais, Belo Horizonte, Minas Gerais, Brazil

Sérgio Henrique Sousa Santos Department of Pharmacology, Universidade Federal de Minas Gerais, Belo Horizonte, Minas Gerais, Brazil

Carmine Savoia Clinical and Molecular Medicine Department, Sapienza University of Rome, Rome, Italy

Vinayak Shenoy Department of Pharmacodynamics, College of Pharmacy, University of Florida, Gainesville, FL 32610, USA

Kátia Daniella Silveira Interdisciplinary Laboratory of Medical Investigation—LIIM, Federal University of Minas Gerais—UFMG, Belo Horizonte, Minas Gerais, Brazil

Ana Cristina Simões Silva Department of Pediatrics, and Interdisciplinary Laboratory of Medical Investigation—LIIM, Federal University of Minas Gerais—UFMG, Belo Horizonte, Minas Gerais, Brazil

Ulrike M. Steckelings Department of Cardiovascular and Renal Research, Institute of Molecular Medicine, University of Southern Denmark, Odense, Denmark, and IMM-Department of Cardiovascular and Renal Research, University of Southern Denmark, Odense, Denmark

Colin Sumners Department of Physiology and Functional Genomics, University of Florida, Gainesville, Florida, USA

E. Ann Tallant Hypertension and Vascular Research Center, Wake Forest School of Medicine, Winston-Salem, North Carolina, USA

Mauro Martins Teixeira Departamento de Bioquímica e Imunologia, Instituto de Ciências Biológicas, Belo Horizonte, Minas Gerais, Brazil

Rhian M. Touyz Institute of Cardiovascular & Medical Sciences, University of Glasgow, Glasgow, UK

Anthony J. Turner School of Molecular & Cellular Biology, Faculty of Biological Sciences, University of Leeds, Leeds, UK

Thomas Unger CARIM School for Cardiovascular Diseases, Maastricht University, Maastricht, The Netherlands

Daniel C. Villela Department of Basic Science - Faculty of Medicine, Federal University of Jequitinhonha and Mucuri Valleys, Diamantina, Brazil

Antony Vinh Department of Pharmacology, Monash University, Clayton Victoria 3800, Australia

Massimo Volpe Clinical and Molecular Medicine Department, Sapienza University of Rome, Rome, Italy

Yan Wang Department of Pharmacology, Monash University, Clayton Victoria 3800, Australia

Iresha Welungoda Department of Pharmacology, Monash University, Clayton Victoria 3800, Australia

Steven Whitebread Novartis Institute for Biochemical Research, Cambridge, Massachusetts 02139, USA

Robert E. Widdop Department of Pharmacology, Monash University, Clayton Victoria 3800, Australia

Bryan Williams Institute of Cardiovascular Science, University College London, London, UK

Irving H. Zucker Department of Cellular and Integrative Physiology, University of Nebraska Medical Center, Omaha, Nebraska, USA

Chapter 1

The Angiotensin AT$_2$ Receptor: From Enigma to Therapeutic Target

Thomas Unger,* Ulrike M. Steckelings,† Victor J. Dzau‡

*CARIM School for Cardiovascular Diseases, Maastricht University, Maastricht, The Netherlands, †Department of Cardiovascular and Renal Research, Institute of Molecular Medicine, University of Southern Denmark, Odense, Denmark, ‡Institute of Medicine of the National Academies, Washington, DC, USA

INTRODUCTION

The angiotensin AT$_2$ receptor (AT$_2$R) has mutated from an "enigmatic receptor",[1–3] with its designation from the very beginning of its discovery 25 years ago in the late 1980s. It has long been enigmatic, because of its constitutive action, atypical intracellular signaling, and "hidden" (patho)physiological functions that took years, almost decades, to unveil. Even today, its various actions, some of them still controversial, its numerous intracellular signaling pathways, its interaction with other membrane receptors, and its role in the context of the renin–angiotensin system (RAS) are far from being fully elucidated. We know much more than in the early days about the AT$_2$R at least, we can confidently classify it as an important member of the "protective arm" of the RAS (Figure 1), and we have even identified this receptor as a therapeutic target.

Until about 1989, the scientific community was convinced that angiotensin II (Ang II), the main effector peptide of the RAS, utilized only one single receptor to exert its various actions in the cardiovascular system and beyond: the angiotensin receptor. Several peptidergic angiotensin derivatives, such as saralasin, had been developed to antagonize those Ang II actions thought to be harmful or just to serve as pharmaceutical tools to gain more insight into the role of the RAS. Indeed, these angiotensin derivatives worked quite well to attenuate or block Ang II-mediated effects, for instance, vasoconstriction of isolated blood vessels. In some cases, they could also be used to lower blood pressure in hypertensive individuals, especially when the RAS was activated. However, expectations as to developing these compounds into clinically useful antihypertensives could not be fulfilled, since they were acting only under specific clinical conditions of an activated RAS and, moreover, as peptides, they were of low oral bioavailability and were rapidly degraded in the organism.

Around this time, three pharmaceutical companies, searching for new tools and better drugs interfering with the RAS, came up, independently of each other, with compounds that were differentially binding to angiotensin receptors in tissues such as the uterus, adrenal gland, and blood vessels, pointing to distinct angiotensin receptor populations in these tissues.[4,5] Losartan (DuP 753), developed by DuPont, was selectively binding to the "classical" angiotensin receptor in blood vessels and other tissues, later designated as the angiotensin AT$_1$ receptor (AT$_1$R), while CGP 42112A, EXP 655, PD 123177, and PD 123319, developed by Ciba-Geigy (Novartis), DuPont, and Parke-Davis, respectively, were binding to a "new" angiotensin receptor in the uterus and adrenal gland, later designed as the angiotensin AT$_2$R.[5]

The scientific RAS community was confused. Intensive, sometimes quite controversial, discussions followed, first about the nomenclature (as described by M. de Gasparo et al. in Chapter 2) and then about the function of the different angiotensin receptors.

In retrospective, the surprise about two or even more receptor subtypes in one biological system such as the RAS seems to be somewhat out of place, since many of these systems operate on several receptors, for instance, dopamine, adrenaline, and noradrenaline as the effectors of the sympathetic nervous system or bradykinin, endothelin, vasopressin, and serotonin, and the effects mediated by different receptors in these systems are often different from or even opposing each other. However, at the time, 25 years ago, the established picture of the RAS, carefully painted by a great number of authors and publications, received some severe scratches but, as we learned step by step later on, became also much more colorful.

Proof of the existence of the AT$_2$R was first provided by the molecular cloning of cDNA of this receptor by Victor Dzau's laboratory at Stanford[6,7] and Tadashi Inagami's group at Vanderbilt.[8–10] With further characterization of the genomic

The Protective Arm of the Renin–Angiotensin System (RAS). http://dx.doi.org/10.1016/B978-0-12-801364-9.00001-8

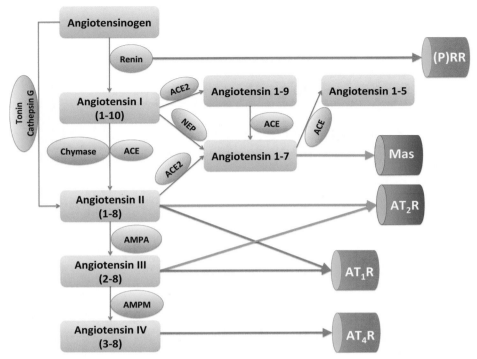

FIGURE 1 The renin–angiotensin system with its "classical" arms (in red) and its "new" protective arms (in green). Abbreviations: ACE: angiotensin-converting enzyme; ACE2: angiotensin-converting enzyme 2; AMPA: aminopeptidase A; AMPM: aminopeptidase M; NEP: neutral endopeptidase; (P)RR: (pro)renin receptor; Mas: Mas receptor; AT$_1$R: angiotensin AT$_1$ receptor; AT$_2$R: angiotensin AT$_2$ receptor; AT$_4$R: angiotensin AT$_4$ receptor.

structure[11,12] and the documentation of mouse phenotype by gene deletion experiments by the Dzau-Kobilka[13] and Inagami[14] laboratories, the AT$_2$R was no longer a pharmacologic binding phenomenon but a real biologic entity.

Back in 1991, two new members, Monika Stoll and Kaj Metsärinne, joined Thomas Unger's laboratory in Heidelberg, Germany. The Heidelberg group was interested in the effects of Ang II on cardiac hypertrophy and, in particular, on the effect of angiotensin on heart capillary endothelial cells, because recent findings from this laboratory had shown that long-term treatment with a converting enzyme inhibitor increased capillary length density in the hearts of spontaneously hypertensive rats (SHR) *in vivo*.[15] These results implied that a reduction of Ang II by ACE inhibition would preserve the capillary growth potential and that, in turn, Ang II, despite inducing hypertrophy or hyperplasia in vascular smooth muscle cells, might have a growth-inhibitory effect on capillary endothelial cells.

The two new members of the group started a project on the cellular growth effects of angiotensin peptides and came up with the finding that Ang II had a profound antiproliferative effect on coronary endothelial cells from SHR (but not from normotensive Wistar-Kyoto controls), which was, however, not clearly concentration-dependent and, therefore, and for other reasons (see below), difficult to publish.[16] The authors speculated about the AT$_2$R, the "new" angiotensin receptor, to be responsible for their findings but, at this time, there was no evidence for the AT$_2$R as a potential mediator of antiproliferative Ang II effects. Thus, they stated the following: "…The molecular structure and the function of the AT$_2$ receptor are unknown, but recently it was reported that it may signal through activation of a membrane-associated phosphotyrosine phosphatase (PTP)[17] …If AT$_2$ receptors signal through activation of a PTP, …decreased angiogenesis after treatment with Ang II may possibly occur."[16]

In those days, the research community was not ready to accept findings against a mantra of the time stating that Ang II was exclusively acting as a proliferative/hypertrophic agent in cells of the cardiovascular system as first described by Schelling et al.[18] It took some more years until it was generally accepted that Ang II, via its AT$_2$R, could act indeed as an antiproliferative principle. The breakthrough was mainly brought about by two independent publications in 1995. In the first study, coming from Victor Dzau's laboratory, overexpression of the AT$_2$R in an intimal proliferation model in rats attenuated carotid neointima formation, and in cultured smooth muscle cells, AT$_2$R transfection reduced proliferation and inhibited mitogen-activated protein kinase activity.[19] In the second study from the Heidelberg group, selective stimulation of AT$_2$R (by Ang II in the presence of losartan to block AT$_1$Rs or directly by CGP 42112) inhibited bFGF-stimulated proliferation of microvascular endothelial cells from SHR.[20] These two studies assigned for the first time a clear biological function to the "enigmatic" AT$_2$R: antiproliferation.

At this time, the AT$_2$R had already been cloned in rodents and in man by the groups of Victor Dzau[6,7] and Tadashi Inagami,[8–10] aspects of its regulation had been described,[21] and its predominant localization in, among others, fetal tissue, the adrenal gland, and some areas of the brain had been confirmed (for more extensive and recent reviews on AT$_2$R tissue distribution and functional aspects, see Horiuchi et al.[21,22]; Csikos et al.[23]; De Gasparo et al.[5]; Volpe et al.[24]; Utsunomiya et al.[25]; Steckelings et al.[26]; Padia and Carey[27]).

Very early on, it became clear that AT$_2$R signaling differed markedly from AT$_1$R signaling. While both receptors, though having only 33–34% homology,[5,6] could be assigned to the seven-transmembrane domain family of receptors, which usually bind G proteins, the AT$_2$R exhibited some atypical features. First of all, binding to a G protein was not easy to establish but, after all, seems to occur in some cases involving Giα2 and Gα3 proteins[2,28–31] (see also Recarti et al., Chapter 4). Second, the AT$_2$R does not follow the classical cellular internalization and reconstitution patterns as does the AT$_1$R.[32–34] Third, the AT$_2$R undergoes extensive homo- and heterodimerization with other receptors like the AT$_1$R, the bradykinin B$_2$ receptor, or the Mas receptor,[35–38] and fourth, the AT$_2$R uses several direct physical protein interaction partners (binding proteins). These are, as known today, (a) the above-mentioned G proteins, (b) an SHP-1 phosphatase linked to an atypical signaling initiation with a Gαs alone,[39] (c) a binding protein family called ATIP or ATBP,[40–42] and (d) a transcription factor called PLZF,[43] which incidentally also represents a major binding protein of the (pro)renin receptor.[44] Binding to SHP-1 and to ATIP/ATBP has been shown to be related to antiproliferation, intracellular receptor transport, and receptor activation, whereas binding to PLZF, in the presence of additional growth factors, may link AT$_2$R stimulation to hypertrophic effects.[31]

Based on these physical interaction partners, the AT$_2$R engages a complex intracellular signaling network linked to distinct physiological functions: antiproliferation/proliferation,[19,20,45] anti-inflammation,[46,47] cellular differentiation,[48] antifibrosis,[49] neuroprotection or neuroregeneration,[50,51] natriuresis,[52] and induction or inhibition of apoptosis[53–58] (Figure 2). These various actions have been reviewed in detail elsewhere[2,5,26,59–62] and will be addressed in further chapters of this book.

Returning to the early years of AT$_2$R research, there were two major conceptions concerning AT$_2$R function emerging right from the beginning of the discovery of the second angiotensin receptor.

One concept saw this receptor as "the little brother" of the classical Ang II receptor, basically having the same, maybe somewhat modified or weaker, effects as its big brother AT$_1$R. The second concept circled around the role of the AT$_2$R in the therapeutic mechanisms of AT$_1$R receptor blockers (ARBs). According to this concept, part of the beneficial effects of the ARBs in cardiovascular diseases was thought to be due to stimulation of the unopposed AT$_2$R by increased Ang II concentrations during AT$_1$R blockade.

The "little brother" concept was fueled over the years by experimental findings that, for instance, stimulation of the AT$_2$R might inhibit a protein tyrosine phosphatase and thus a growth-inhibiting process[8] that the AT$_2$R would cause constitutive growth of cardiomyocytes,[63] or that the AT$_2$R would contribute to vascular and cardiac hypertrophy and fibrosis.[64–67]

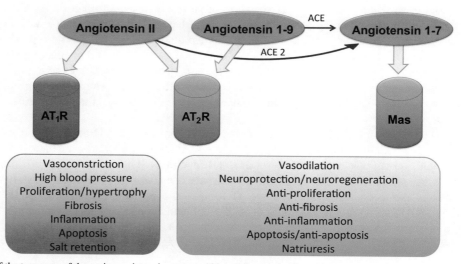

FIGURE 2 Effects of the two arms of the renin–angiotensin system. Abbreviations as in Figure 1.

These findings were contrasted by the above-mentioned studies by Nakajima, et al.[19] and Stoll et al.[20] as well as those by Masaki et al.[68] on cardiac-specific overexpression of the AT_2R and, in addition, also by the majority of investigations into various tissues and cell types demonstrating antiproliferative or antifibrotic effects in the heart or blood vessels following AT_2R stimulation (for a review, see Horiuchi et al.;[22] Funke-Kaiser et al.;[31] Namsolleck et al.;[61] Steckelings et al.[69]). But still, the discrepancies concerning the functional role of the AT_2R have long been a matter of concern or of "love and hate"[26] to researchers.

More recently, a model to clear up these contradictions has been proposed in which the type of adapter protein (ATIP/ATBP or SHP-1 versus PLZF) recruited to the receptor, as well as the presence or absence of growth factors, determines the cellular effects assuming a multiple-state receptor model with several activated states. Such a "biased agonism" model could also be the basis for our mechanistic understanding of drug actions, for instance, those of the selective AT_2R agonist, compound 21.[31] In addition, different patterns of homo- or heterodimerization (with the AT_1R, the Mas receptor, or the bradykinin B_2 receptor) have to be considered as functional determinants of AT_2R action.

The above considerations, together with further experimental findings pointing to novel actions of the AT_2R, different from all what was hitherto known about the RAS, for instance, neuroprotection and neuroregeneration (see below), clarified the picture: The AT_2R is by no means just the "little brother" of the AT_1R; on the contrary, it is quite often an opponent of the AT_1R as an independent biological modulator and an important part of the "protective arm of the RAS." It became also clear that the AT_2R system harbors some unique features, which had not been described before with regard to the classical RAS. Apart from specific cellular signaling pathways clearly distinct from those of the AT_1R, the AT_2R was found to be constitutively active[35] so that the regulation of its expression gains importance against the regulation of agonist concentrations. It also became apparent that this receptor is distinctly regulated: while being highly expressed in fetal tissues but suppressed in many tissues in the adult organism, its expression can be drastically upregulated not only under the condition of ischemic or traumatic tissue injury but also in atherosclerotic lesions.[2,5,70–73]

The other concept, closely related to the drug development of the AT_1R blockers, was more therapeutically oriented: The idea was that during AT_1R blockade with the new class of selective AT_1R antagonists, the sartans or ARBs, Ang II levels would rise and that an increased number of Ang II molecules would then bind to the unopposed AT_2R. Enhanced stimulation of the AT_2R would be responsible for some of the beneficial effects of AT_1R blockade.[5,74]

This idea gained support from a number of experimental studies, which in principle all followed a similar pattern. The effects of AT_2R-inhibiting compounds were studied under various conditions after blockade of the AT_1R. The following three out of numerous examples may suffice: (i) A marked increase of the vascular cGMP content, indicative of nitric oxide (NO) generation, was found in the aortic tissue of SHR, but not in normotensive WKY controls, after treatment with losartan, and this effect was sensitive to AT_2R blockade by PD 123319[75,76]; (ii) both the increase of NO synthase expression induced by candesartan in bovine pulmonary endothelial cells and candesartan-induced coronary vasodilation in rat hearts were eliminated by an AT_2R antagonist[77]; and (iii) experiments in a rat model of ischemic stroke showed that blockade of brain AT_2R by intracerebroventricular administration of the AT_2R antagonist, PD 123177, abolished the neuroprotective effects of central AT_1R blockade with irbesartan on infarct size and neurological outcome.[72]

The AT_2R-mediated increase of the NO/cGMP system seen under various conditions and involving bradykinin and its BK1 and BK2 receptors in vascular tissue[60,75,78] was shown to be linked to vascular repair or vasodilation, sometimes to blood pressure lowering. From these findings, the idea was derived that the AT_2R represents a counterregulatory system to AT_1R-mediated vasoconstriction, which may be inoperative in certain disease states such as hypertension.[60,79] Based on this concept, it was also tempting to speculate that blockade of AT_1R by sartans would engender additional mechanisms including the bradykinin and NO systems, reminiscent of the well-established mechanism of action of the ACE inhibitors. In the case of the latter, bradykinin-mediated NO generation and other effects had been made responsible not only for part of their antihypertensive actions[80] but also for additional tissue-protective effects[81] despite the decreased number of Ang II molecules reaching the AT_2R.

In the case of the AT_1R blockers, AT_2R stimulation acting via the bradykinin–NO pathway could now be assigned a protective role, similar to bradykinin potentiation in the case of the ACE inhibitors.[82] However, there are some drawbacks to this concept: While it holds for various, but not all,[83] experimental conditions, it has never been confirmed in humans due to the fact that neither specific AT_2R agonists nor AT_2R antagonists have been clinically approved so far. In addition, concerning the vasodilator effects, the recent use of the selective AT_2R agonist, compound 21, has revealed that antihypertensive actions following direct AT_2R stimulation are weak, if at all present, and depend very much on the conditions of vascular precontraction.[79,84] This does not generally exclude the beneficial actions of ARBs to be mediated via the AT_2R axis, but it appears that the AT_2R contributes little if at all to their antihypertensive actions.

A further aspect of AT$_2$R signaling and function deserves particular attention, a feature quite novel and unique for the RAS: neuroprotection and neuroregeneration, modulation of sympathoexcitation, and possibly improvement of cognitive function. These characteristics of the AT$_2$R have been discussed recently[61,85–87] and will be dealt with in separate chapters of this book. However, the way that different groups involved in AT$_2$R research came about this feature of the "enigmatic receptor" is quite exemplary for early AT$_2$R research and shall therefore be reported here in more detail.

One of the first publications on neuronal differentiation via the AT$_2$R in NG 108-15 cells[88] describes the issue as follows: "Considering the abundance of T-type Ca2 + channels (previously shown to be inhibited via AT$_2$R by the same authors[89]) in neurons from fetal brain, the crucial role of Ca2+ in neuronal differentiation and the abundance of AT$_2$Rs in this developmental period, it could be postulated that Ang II, via the AT$_2$R could afford some aspects of neuronal differentiation."

At the same time, another group published a study in PC12W cells, a pheochromocytoma-derived cell line, which, like the NG 108-15 cells, only harbors AT$_2$R but no AT$_1$R in the undifferentiated state. Stimulation of AT$_2$R induced growth arrest and neuronal differentiation in these cells. Still, under the impression of the antiproliferative actions of the AT$_2$R not only in endothelial cells[20] but also in view of the above-cited involvement of calcium in fetal development and the AT$_2$R-mediated inhibitory action on some calcium channels, the authors postulated that growth inhibition might be part of a biological program of cell differentiation involved in development and some (patho)physiological conditions.[90]

A third group tackled the effects of AT$_2$R stimulation on neuronal activity by investigating membrane K+ and Ca+ currents and channels.[59] Based on previous findings that neonate animals express higher levels of AT$_2$R than AT$_1$R in the brain,[91] on AT$_2$R-induced neuronal cell differentiation (see above) and apoptosis,[53] on inhibitory effects of AT$_2$R stimulation on central AT$_1$R-mediated Ang II effects like vasopressin release and drinking,[92] and on the fact that mutant mice lacking the gene encoding the AT$_2$R demonstrated deficits in brain-controlled behavior,[13,14] they reported on distinct signaling pathways modulated by neuronal AT$_1$R and AT$_2$R, in line with central anti-excitatory effects of the AT$_2$R in general and the inhibitory action of the AT$_2$R on (AT$_1$R-induced) sympathoexcitation (as reviewed more recently[93]).

Taken together, different experimental approaches used by three independent groups revealed a fairly consistent picture of the AT$_2$R in the nervous system, revealing that its signaling and actions were clearly distinct from those of the much better described AT$_1$R, and that the AT$_2$R-induced neuronal differentiation and inhibition of neuroexcitation could be not only part of neuronal development but also the basis for therapeutic considerations about drug-induced AT$_2$R stimulation.[94] This idea was supported later on with the help of the first nonpeptide AT$_2$R agonist by a wealth of results in different experimental disease models as detailed in recent reviews (see below) and in several other chapters of this book.

Although the search for potential drug targets is always in the background of researchers' minds when trying to reveal expression pattern, signaling pathways and biological functions of receptors like the AT$_1$R and the AT$_2$R, initial thoughts focused much more on mechanism of action and clinical potential of inhibitors of the RAS, such as the ACE inhibitors, the ARBs, and the renin inhibitors, than on the possibility of making the AT$_2$R itself a drug target.[74,95] A possible contribution of the unopposed AT$_2$R to the antihypertensive and tissue-protective actions of the ARBs was taken into consideration and was also observed in some experimental studies, but due to the fact that the experimental tools of the time were not sufficient to delineate clear beneficial effects of AT$_2$R stimulation that would carry in a translational way from bench to bedside, many of these ideas remained speculative.

The decisive step forward came with the publication of the first highly selective, orally active AT$_2$R agonist by the group of Hallberg and Alterman in Uppsala, Sweden[96], as reviewed in detail elsewhere[97] and in this book. Until now, compound 21, the main representative of this family of AT$_2$R agonists, not only has helped to unravel many of the secrets of AT$_2$R signaling and function but also, moreover and most importantly, has opened the door to numerous potential clinical indications of AT$_2$R stimulation not only in the cardiovascular field but also in the areas of renal,[27,62] metabolic,[98] and neuronal[61,85,93] diseases as well as in many other indication areas where anti-inflammatory, antifibrotic, or antiproliferative actions are therapeutically required[61] (Figure 3). It should be noted that compound 21, currently approaching Phase I study, is not the only representative of agents interfering with the AT$_2$R. Other investigators have also followed this track of drug development as shown in Table 1.

The future will tell whether or not the "not-anymore-so-enigmatic" AT$_2$R will hold its promises as a clinically successful drug target.[99] However, what AT$_2$R research has certainly taught us over the years is to look at both sides of the RAS with respect to not only scientific but also therapeutic aspects: The therapeutic goal of the future is not only to block the "harmful arm" of the RAS with ACE inhibitors, ARBs, and renin inhibitors as we have done in the past but also to stimulate the AT$_2$R as a major part of the "beneficial arm" of the RAS, a potent endogenous protective system.

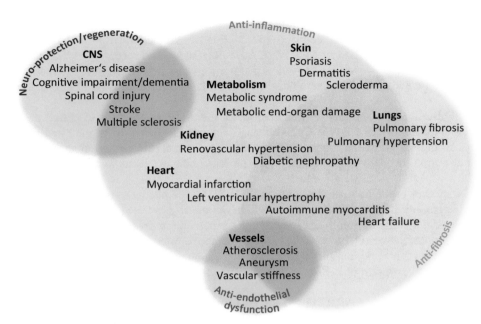

FIGURE 3 Potential clinical indications of compounds stimulating the angiotensin AT_2 receptor, in particular compound 21.

TABLE 1 Drug development based on compounds interfering with the angiotensin AT_2 receptor

Name	Company	Mechanism	Primary indication	Note
Compound 21	Vicore Pharma	AT_2 receptor agonism	Idiopathic pulmonary fibrosis	Small molecule compound; Phase I/II starts in 2014
LP2	Lanthio Pharma	AT_2 receptor agonism	Idiopathic pulmonary fibrosis	Peptide-based compound; Phase I starts in 2014
MP-157	Mitsubishi Tanabe Pharma	AT_2 receptor agonism	Hypertension	Phase I in Europe
EMA401	Spinifex Pharmaceuticals	AT_2 receptor antagonism	Neuropathic pain	Phase II in postherpetic neuralgia finished

REFERENCES

1. Unger T. The angiotensin type 2 receptor: variations on an enigmatic theme. *J Hypertens* 1999;**17**:1775–86.
2. Horiuchi M, Akishita M, Dzau VJ. Recent progress in angiotensin II type 2 receptor research in the cardiovascular system. *Hypertension* 1999;**33**:613–21.
3. Gallinat S, Busche S, Raizada MK, Sumners C. The angiotensin II type 2 receptor: an enigma with multiple variations. *Am J Physiol Endocrinol Metab* 2000;**278**:E357–74.
4. Whitebread S, Mele M, Kamber B, de Gasparo M. Preliminary biochemical characterization of two angiotensin II receptor subtypes. *Biochem Biophys Res Commun* 1989;**163**:284–91.
5. De Gasparo M, Catt KJ, Inagami T, Wright JW, Unger T. International union of pharmacology. XXIII. The angiotensin II receptors. *Pharmacol Rev* 2000;**52**:415–72.
6. Mukoyama M, Nakajima M, Horiuchi M, Sasamura H, Pratt RE, Dzau VJ. Expression cloning of type 2 angiotensin II receptor reveals a unique class of seven-transmembrane receptors. *J Biol Chem* 1993;**268**:24539–42.
7. Nakajima M, Mukoyama M, Pratt RE, Horiuchi M, Dzau VJ. Cloning of cDNA and analysis of the gene for mouse angiotensin II type 2 receptor. *Biochem Biophys Res Commun* 1993;**197**:393–9.
8. Kambayashi Y, Bardhan S, Takahashi K, Tsuzuki S, Inui H, Hamakubo T, et al. Molecular cloning of a novel angiotensin II receptor isoform involved in phosphotyrosine phosphatase inhibition. *J Biol Chem* 1993;**268**:24543–6.
9. Tsuzuki S, Ichiki T, Nakakubo H, Kitami Y, Guo DF, Shirai H, et al. Molecular cloning and expression of the gene encoding human angiotensin II type 2 receptor. *Biochem Biophys Res Commun* 1994;**200**:1449–54.

10. Ichiki T, Herold CL, Kambayashi Y, Bardhan S, Inagami T. Cloning of the cDNA and the genomic DNA of the mouse angiotensin II type 2 receptor. *Biochim Biophys Acta* 1994;**1189**:247–50.

11. Koike G, Horiuchi M, Yamada T, Szpirer C, Jacob HJ, Dzau VJ. Human type 2 angiotensin II receptor gene: cloned, mapped to the X chromosome, and its mRNA is expressed in the human lung. *Biochem Biophys Res Commun* 1994;**203**:1842–50.

12. Katsuya T, Horiuchi M, Minami S, Koike G, Santoro NF, Hsueh AJ, et al. Genomic organization and polymorphism of human angiotensin II type receptor: no evidence for its gene mutation in two families of human premature ovarian failure syndrome. *Mol Cell Endocrinol* 1997;**127**(2):221–8.

13. Hein L, Barsh GS, Pratt RE, Dzau VJ, Kobilka BK. Behavioural and cardiovascular effects of disrupting the angiotensin II type-2 receptor in mice. *Nature* 1995;**377**:744–7.

14. Ichiki T, Labosky PA, Shiota C, Okuyama S, Imagawa Y, Fogo A, et al. Effects on blood pressure and exploratory behaviour of mice lacking angiotensin II type-2 receptor. *Nature* 1995;**377**:748–50.

15. Unger T, Mattfeldt T, Lamberty V, Bock P, Mall G, Linz W, et al. Effect of early onset angiotensin converting enzyme inhibition on myocardial capillaries. *Hypertension* 1992;**20**:478–82.

16. Metsärinne KP, Stoll M, Gohlke P, Paul M, Unger T. Angiotensin II is antiproliferative for coronary endothelial cells in vitro. *Pharm Pharmacol Lett* 1992;**2**:150–2.

17. Bottari SP, King IN, Reichlin S, Dahlstroem I, Lydon N, de Gasparo M. The angiotensin AT2 receptor stimulates protein tyrosine phosphatase activity and mediates inhibition of particulate guanylate cyclase. *Biochem Biophys Res Commun* 1992;**183**:206–11.

18. Schelling P, Ganten D, Speck G, Fischer H. Effects of angiotensin II and angiotensin II antagonist saralasin on cell growth and renin in 3T3 and SV3T3 cells. *J Cell Physiol* 1979;**98**:503–13.

19. Nakajima M, Hutchinson HG, Fujinaga M, Hayashida W, Morishita R, Zhang L, et al. The angiotensin II type 2 (AT2) receptor antagonizes the growth effects of the AT1 receptor: gain-of-function study using gene transfer. *Proc Natl Acad Sci U S A* 1995;**92**:10663–7.

20. Stoll M, Steckelings UM, Paul M, Bottari SP, Metzger R, Unger T. The angiotensin AT2-receptor mediates inhibition of cell proliferation in coronary endothelial cells. *J Clin Invest* 1995;**95**:651–7.

21. Horiuchi M, Koike G, Yamada T, Mukoyama M, Nakajima M, Dzau VJ. The growth-dependent expression of angiotensin II type 2 receptor is regulated by transcription factors interferon regulatory factor-1 and 2. *J Biol Chem* 1995;**270**:20225–30.

22. Horiuchi M, Iwanami J, Mogi M. Regulation of angiotensin II receptors beyond the classical pathway. *Clin Sci Lond Eng 1979* 2012;**123**:193–203.

23. Csikós T, Chung O, Unger T. Receptors and their classification: focus on angiotensin II and the AT2 receptor. *J Hum Hypertens* 1998;**12**:311–18.

24. Volpe M, Musumeci B, De Paolis P, Savoia C, Morganti A. Angiotensin II AT2 receptor subtype: an uprising frontier in cardiovascular disease? *J Hypertens* 2003;**21**:1429–43.

25. Utsunomiya H, Nakamura M, Kakudo K, Inagami T, Tamura M. Angiotensin II AT2 receptor localization in cardiovascular tissues by its antibody developed in AT2 gene-deleted mice. *Regul Pept* 2005;**126**:155–61.

26. Steckelings UM, Kaschina E, Unger T. The AT2 receptor—a matter of love and hate. *Peptides* 2005;**26**:1401–9.

27. Padia SH, Carey RM. AT2 receptors: beneficial counter-regulatory role in cardiovascular and renal function. *Pflüg Arch Eur J Physiol* 2013;**465**:99–110.

28. Hayashida W, Horiuchi M, Dzau VJ. Intracellular third loop domain of angiotensin II type-2 receptor: role in mediating signal transduction and cellular function. *J Biol Chem* 1996;**271**:21985–92.

29. Kang J, Posner P, Sumners C. Angiotensin II type 2 receptor stimulation of neuronal K+ currents involves an inhibitory GTP binding protein. *Am J Physiol* 1994;**267**:C1389–97.

30. Zhang J, Pratt RE. The AT2 receptor selectively associates with Gialpha2 and Gialpha3 in the rat fetus. *J Biol Chem* 1996;**271**:15026–33.

31. Funke-Kaiser H, Reinemund J, Steckelings UM, Unger T. Adapter proteins and promoter regulation of the angiotensin AT2 receptor–implications for cardiac pathophysiology. *J Renin-Angiotensin-Aldosterone Syst* 2010;**11**:7–17.

32. Mukoyama M, Horiuchi M, Nakajima M, Pratt RE, Dzau VJ. Characterization of rat type 2 angiotensin II receptor stably expressed in 293 cells. *Mol Cell Endocrinol* 1995;**112**:61–8.

33. Hein L, Meinel L, Pratt RE, Dzau VJ, Kobilka BK. Intracellular trafficking of angiotensin II and its AT1 and AT2 receptors: evidence for selective sorting of receptor and ligand. *Mol Endocrinol* 1997;**11**:1266–77.

34. Csikós T, Balmforth AJ, Grojec M, Gohlke P, Culman J, Unger T. Angiotensin AT2 receptor degradation is prevented by ligand occupation. *Biochem Biophys Res Commun* 1998;**243**:142–7.

35. Miura S-I, Karnik SS, Saku K. Constitutively active homo-oligomeric angiotensin II type 2 receptor induces cell signaling independent of receptor conformation and ligand stimulation. *J Biol Chem* 2005;**280**:18237–44.

36. AbdAlla S, Lother H, Abdel-tawab AM, Quitterer U. The angiotensin II AT2 receptor is an AT1 receptor antagonist. *J Biol Chem* 2001;**276**:39721–6.

37. Abadir PM, Periasamy A, Carey RM, Siragy HM. Angiotensin II type 2 receptor-bradykinin B2 receptor functional heterodimerization. *Hypertension* 2006;**48**:316–22.

38. Villela DC, Munter L-M, Multhaup G, Mayer M, Benz V, Namsolleck P, et al. Evidence of a direct MAS-AT2 receptor dimerization. *J Hypertens* 2012;**30**:e117.

39. Feng Y-H, Sun Y, Douglas JG. Gbeta gamma-independent constitutive association of Galpha s with SHP-1 and angiotensin II receptor AT2 is essential in AT2-mediated ITIM-independent activation of SHP-1. *Proc Natl Acad Sci U S A* 2002;**99**:12049–54.

40. Nouet S, Amzallag N, Li J-M, Louis S, Seitz I, Cui T-X, et al. Trans-inactivation of receptor tyrosine kinases by novel angiotensin II AT2 receptor-interacting protein, ATIP. *J Biol Chem* 2004;**279**:28989–97.

41. Wruck CJ, Funke-Kaiser H, Pufe T, Kusserow H, Menk M, Schefe JH, et al. Regulation of transport of the angiotensin AT2 receptor by a novel membrane-associated Golgi protein. *Arterioscler Thromb Vasc Biol* 2005;**25**:57–64.

42. Rodrigues-Ferreira S, Nahmias C. An ATIPical family of angiotensin II AT2 receptor-interacting proteins. *Trends Endocrinol Metab* 2010;**21**:684–90.

43. Senbonmatsu T, Saito T, Landon EJ, Watanabe O, Price Jr. E, Roberts RL, et al. A novel angiotensin II type 2 receptor signaling pathway: possible role in cardiac hypertrophy. *EMBO J* 2003;**22**:6471–82.

44. Schefe JH, Menk M, Reinemund J, Effertz K, Hobbs RM, Pandolfi PP, et al. A novel signal transduction cascade involving direct physical interaction of the renin/prorenin receptor with the transcription factor promyelocytic zinc finger protein. *Circ Res* 2006;**99**:1355–66.

45. Horiuchi M, Hayashida W, Akishita M, Tamura K, Daviet L, Lehtonen JY, et al. Stimulation of different subtypes of angiotensin II receptors, AT1 and AT2 receptors, regulates STAT activation by negative crosstalk. *Circ Res* 1999;**84**:876–82.

46. Akishita M, Horiuchi M, Yamada H, Zhang L, Shirakami G, Tamura K, et al. Inflammation influences vascular remodeling through AT2 receptor expression and signaling. *Physiol Genomics* 2000;**2**(1):13–20.

47. Rompe F, Artuc M, Hallberg A, Alterman M, Ströder K, Thöne-Reineke C, et al. Direct angiotensin II type 2 receptor stimulation acts anti-inflammatory through epoxyeicosatrienoic acid and inhibition of nuclear factor kappa B. *Hypertension* 2010;**55**:924–31.

48. Yamada H, Akishita M, Ito M, Tamura K, Daviet L, Lehtonen JY, et al. AT2 receptor and vascular smooth muscle cell differentiation in vascular development. *Hypertension* 1999;**33**(6):1414–19.

49. Ohkubo N, Matsubara H, Nozawa Y, Mori Y, Murasawa S, Kijima K, et al. Angiotensin type 2 receptors are reexpressed by cardiac fibroblasts from failing myopathic hamster hearts and inhibit cell growth and fibrillar collagen metabolism. *Circulation* 1997;**96**(11):3954–62.

50. Lucius R, Gallinat S, Rosenstiel P, Herdegen T, Sievers J, Unger T. The angiotensin II type 2 (AT2) receptor promotes axonal regeneration in the optic nerve of adult rats. *J Exp Med* 1998;**188**(4):661–70.

51. Reinecke K, Lucius R, Reinecke A, Rickert U, Herdegen T, Unger T. Angiotensin II accelerates functional recovery in the rat sciatic nerve in vivo: role of the AT2 receptor and the transcription factor NF-kappaB. *FASEB J* 2003;**17**(14):2094–6, Epub 2003 Sep 18.

52. Kemp BA, Howell NL, Gildea JJ, Keller SR, Padia SH, Carey RM. AT2 receptor activation induces natriuresis and lowers blood pressure. *Circ Res* 2014;**115**(3):388–99, Jun 5. pii: CIRC RESAHA.114.304110. [Epub ahead of print].

53. Yamada T, Horiuchi M, Dzau VJ. Angiotensin II type 2 receptor mediates programmed cell death. *Proc Natl Acad Sci U S A* 1996;**93**:156–60.

54. Horiuchi M, Hayashida W, Kambe T, Yamada T, Dzau VJ. Angiotensin type 2 receptor dephosphorylates Bcl-2 by activating mitogen-activated protein kinase phosphatase-1 and induces apoptosis. *J Biol Chem* 1997;**272**:19022–6.

55. Horiuchi M, Yamada T, Hayashida W, Dzau VJ. Interferon regulatory factor-1 up-regulates angiotensin II type 2 receptor and induces apoptosis. *J Biol Chem* 1997;**272**(18):11952–8.

56. Yamada T, Akishita M, Pollman M, Gibbons GH, Dzau VJ, Horiuchi M. Angiotensin II type 2 receptor mediates vascular muscle cell apoptosis and antagonizes angiotensin II type 1 receptor action: an in vitro gene transfer study. *Life Sci* 1998;**63**(19):289–95.

57. Lehtonen JY, Horiuchi M, Daviet L, Akishita M, Dzau VJ. Activation of the de novo biosynthesis of sphingolipids mediates angiotensin II type 2 receptor-induced apoptosis. *J Biol Chem* 1999;**274**(24):16901–6.

58. Gallinat S, Busche S, Schütze S, Krönke M, Unger T. AT2 receptor stimulation induces generation of ceramides in PC12W cells. *FEBS Lett* 1999;**443**(1):75–9.

59. Sumners C, Gelband CH. Neuronal ion channel signalling pathways: modulation by angiotensin II. *Cell Signal* 1998;**10**:303–11.

60. Carey RM. Cardiovascular and renal regulation by the angiotensin type 2 receptor: the AT2 receptor comes of age. *Hypertension* 2005;**45**:840–4.

61. Namsolleck P, Recarti C, Foulquier S, Steckelings UM, Unger T. AT(2) receptor and tissue injury: therapeutic implications. *Curr Hypertens Rep* 2014;**16**:416.

62. Danyel LA, Schmerler P, Paulis L, Unger T, Steckelings UM. Impact of AT2-receptor stimulation on vascular biology, kidney function, and blood pressure. *Integr Blood Press Control* 2013;**6**:153–61.

63. D'Amore A, Black MJ, Thomas WG. The angiotensin II type 2 receptor causes constitutive growth of cardiomyocytes and does not antagonize angiotensin II type 1 receptor-mediated hypertrophy. *Hypertension* 2005;**46**:1347–54.

64. Levy BI, Benessiano J, Henrion D, Caputo L, Heymes C, Duriez M, et al. Chronic blockade of AT2-subtype receptors prevents the effect of angiotensin II on the rat vascular structure. *J Clin Invest* 1996;**98**:418–25.

65. Senbonmatsu T, Ichihara S, Price Jr. E, Gaffney FA, Inagami T. Evidence for angiotensin II type 2 receptor-mediated cardiac myocyte enlargement during in vivo pressure overload. *J Clin Invest* 2000;**106**:R25–9.

66. Mifune M, Sasamura H, Shimizu-Hirota R, Miyazaki H, Saruta T. Angiotensin II type 2 receptors stimulate collagen synthesis in cultured vascular smooth muscle cells. *Hypertension* 2000;**36**:845–50.

67. Ichihara S, Senbonmatsu T, Price Jr. E, Ichiki T, Gaffney FA, Inagami T. Angiotensin II type 2 receptor is essential for left ventricular hypertrophy and cardiac fibrosis in chronic angiotensin II-induced hypertension. *Circulation* 2001;**104**:346–51.

68. Masaki H, Kurihara T, Yamaki A, Norio I, Nozawa Y, Mori Y, et al. Cardiac-specific overexpression of angiotensin II AT2 receptor causes attenuated response to AT1 receptor-mediated pressor and chronotropic effects. *J Clin Invest* 1998;**101**(3):527–35.

69. Steckelings UM, Widdop RE, Paulis L, Unger T. The angiotensin AT2 receptor in left ventricular hypertrophy. *J Hypertens* 2010;**28**:S50–5.

70. Akishita M, Ito M, Lehtonen JY, Daviet L, Dzau VJ, Horiuchi M. Expression of the AT2 receptor developmentally programs extracellular signal-regulated kinase activity and influences fetal vascular growth. *J Clin Invest* 1999;**103**(1):63–71.

71. Gallinat S, Yu M, Dorst A, Unger T, Herdegen T. Sciatic nerve transection evokes lasting up-regulation of angiotensin AT2 and AT1 receptor mRNA in adult rat dorsal root ganglia and sciatic nerves. *Brain Res Mol Brain Res* 1998;**57**:111–22.

72. Li J, Culman J, Hörtnagl H, Zhao Y, Gerova N, Timm M, et al. Angiotensin AT2 receptor protects against cerebral ischemia-induced neuronal injury. *J Off Publ Fed Am Soc Exp Biol* 2005;**19**:617–19.

73. Sales VL, Sukhova GK, Lopez-Ilasaca MA, Libby P, Dzau VJ, Pratt RE. Angiotensin type 2 receptor is expressed in murine atherosclerotic lesions and modulates lesion evolution. *Circulation* 2005;**112**:3328–36.

74. Dzau VJ, Mukoyama M, Pratt RE. Molecular biology of angiotensin receptors: target for drug research? *J Hypertens* 1994;**12**(2):S1–5.

75. Gohlke P, Pees C, Unger T. AT2 receptor stimulation increases aortic cyclic GMP in SHRSP by a kinin-dependent mechanism. *Hypertension* 1998;**31**:349–55.

76. Pees C, Unger T, Gohlke P. Effect of angiotensin AT2 receptor stimulation on vascular cyclic GMP production in normotensive Wistar Kyoto rats. *Int J Biochem Cell Biol* 2003;**35**:963–72.

77. Thai H, Wollmuth J, Goldman S, Gaballa M. Angiotensin subtype 1 receptor (AT1) blockade improves vasorelaxation in heart failure by up-regulation of endothelial nitric-oxide synthase via activation of the AT2 receptor. *J Pharmacol Exp Ther* 2003;**307**:1171–8.

78. Barker TA, Massett MP, Korshunov VA, Mohan AM, Kennedy AJ, Berk BC. Angiotensin II type 2 receptor expression after vascular injury: differing effects of angiotensin-converting enzyme inhibition and angiotensin receptor blockade. *Hypertension* 2006;**48**:942–9.

79. Jones ES, Vinh A, McCarthy CA, Gaspari TA, Widdop RE. AT2 receptors: functional relevance in cardiovascular disease. *Pharmacol Ther* 2008;**120**:292–316.

80. Danckwardt L, Shimizu I, Bönner G, Rettig R, Unger T. Converting enzyme inhibition in kinin-deficient brown Norway rats. *Hypertension* 1990;**16**:429–35.

81. Linz W, Wiemer G, Gohlke P, Unger T, Schölkens BA. Contribution of kinins to the cardiovascular actions of angiotensin-converting enzyme inhibitors. *Pharmacol Rev* 1995;**47**:25–49.

82. Yayama K, Okamoto H. Angiotensin II-induced vasodilation via type 2 receptor: role of bradykinin and nitric oxide. *Int Immunopharmacol* 2008;**8**:312–18.

83. Zhao Y, Biermann T, Luther C, Unger T, Culman J, Gohlke P. Contribution of bradykinin and nitric oxide to AT2 receptor-mediated differentiation in PC12 W cells. *J Neurochem* 2003;**85**:759–67.

84. Steckelings UM, Rompe F, Kaschina E, Namsolleck P, Grzesiak A, Funke-Kaiser H, et al. The past, present and future of angiotensin II type 2 receptor stimulation. *J Renin-Angiotensin-Aldosterone Syst* 2010;**11**:67–73.

85. Guimond M-O, Gallo-Payet N. How does angiotensin AT(2) receptor activation help neuronal differentiation and improve neuronal pathological situations? *Front Endocrinol* 2012;**3**:164.

86. Sumners C, Horiuchi M, Widdop RE, McCarthy C, Unger T, Steckelings UM. Protective arms of the renin-angiotensin-system in neurological disease. *Clin Exp Pharmacol Physiol* 2013;**40**:580–8.

87. Jing F, Mogi M, Sakata A, Iwanami J, Tsukuda K, Ohshima K, et al. Direct stimulation of angiotensin II type 2 receptor enhances spatial memory. *J Cereb Blood Flow Metab* 2012;**32**:248–55.

88. Laflamme L, Gasparo M, Gallo JM, Payet MD, Gallo-Payet N. Angiotensin II induction of neurite outgrowth by AT2 receptors in NG108-15 cells. Effect counteracted by the AT1 receptors. *J Biol Chem* 1996;**271**:22729–35.

89. Buisson B, Laflamme L, Bottari SP, de Gasparo M, Gallo-Payet N, Payet MD. A G protein is involved in the angiotensin AT2 receptor inhibition of the T-type calcium current in non-differentiated NG108-15 cells. *J Biol Chem* 1995;**270**:1670–4.

90. Meffert S, Stoll M, Steckelings UM, Bottari SP, Unger T. The angiotensin II AT2 receptor inhibits proliferation and promotes differentiation in PC12W cells. *Mol Cell Endocrinol* 1996;**122**:59–67.

91. Tsutsumi K, Saavedra JM. Characterization and development of angiotensin II receptor subtypes (AT1 and AT2) in rat brain. *Am J Physiol* 1991;**261**:R209–16.

92. Höhle S, Spitznagel H, Rascher W, Culman J, Unger T. Angiotensin AT1 receptor-mediated vasopressin release and drinking are potentiated by an AT2 receptor antagonist. *Eur J Pharmacol* 1995;**275**:277–82.

93. Gao L, Zucker IH. AT2 receptor signaling and sympathetic regulation. *Curr Opin Pharmacol* 2011;**11**:124–30.

94. Rosenstiel P, Gallinat S, Arlt A, Unger T, Sievers J, Lucius R. Angiotensin AT2 receptor ligands: do they have potential as future treatments for neurological disease? *CNS Drugs* 2002;**16**:145–53.

95. Unger T, Chung O, Csikos T, Culman J, Gallinat S, Gohlke P, et al. Angiotensin receptors. *J Hypertens* 1996;**14**:S95–103.

96. Wan Y, Wallinder C, Plouffe B, Beaudry H, Mahalingam AK, Wu X, et al. Design, synthesis, and biological evaluation of the first selective nonpeptide AT2 receptor agonist. *J Med Chem* 2004;**47**:5995–6008.

97. Steckelings UM, Larhed M, Hallberg A, Widdop RE, Jones ES, Wallinder C, et al. Non-peptide AT2-receptor agonists. *Curr Opin Pharmacol* 2011;**11**:187–92.

98. Ohshima K, Mogi M, Jing F, Iwanami J, Tsukuda K, Min L-J, et al. Direct angiotensin II type 2 receptor stimulation ameliorates insulin resistance in type 2 diabetes mice with PPARγ activation. *PLoS One* 2012;**7**:e48387.

99. Willyard C. As drug target reemerges, the question is to block or stimulate it. *Nat Med* 2014;**20**:222.

Chapter 2

The AT$_2$ Receptor: Historical Perspective

Marc de Gasparo,* Steven Whitebread,† Leoluca Criscione,‡ Peter Buehlmayer,¶ Pascal Furet§

*Cardiovascular & Metabolic Syndrome Adviser, Rossemaison, Switzerland, †Novartis Institute for Biochemical Research, Cambridge, Massachusetts 02139, USA, ‡MangiaSano Consulting, Rümelinbachweg 10, CH-4054 Basel, Switzerland, ¶Hangstrasse 18, CH-4144 Arlesheim, Switzerland, §Novartis Institute for Biochemical Research, CH-4002 Basel, Switzerland

BACKGROUND TO THE EARLY INVOLVEMENT OF CIBA-GEIGY IN THE ANGIOTENSIN FIELD

The octapeptide angiotensin II (Ang II), isolated simultaneously in 1940 by Braun-Menéndez in Argentina and Page in Indianapolis, was characterized and synthesized at Ciba by Schwyzer et al. and by Schwarz et al. at the Cleveland Clinic in 1957. Early angiotensin binding studies showed differences between tissues and suggested receptor heterogeneity, but it was not until the late 1980s that tools became available to prove the existence of different angiotensin receptor subtypes.

As the aldosterone receptor antagonist project was successful in Ciba-Geigy with the characterization of epoxymexrenone, now called eplerenone, a compound with 3–10-fold decrease in the androgenic and progestogenic effect without disturbance of the ovulatory cycle compared with spironolactone,[1] it was decided to put more weight on another target of the RAS, that is, the angiotensin receptor. Initial experiments were started in 1984.

In the aldosterone project, we were using the Kagawa test[2] with adrenalectomized rats to evaluate the antimineralocorticoid activity of the compounds. It was therefore easy to collect these adrenals to develop an angiotensin receptor binding assay.

Biological activity and binding of Ang II to its receptor depend essentially on the aromatic amino acids Tyr^4, His^6, and Phe^8 as well as the carboxyl group of the Phe^8. Replacing Arg^1 by sarcosine increases biological activity, but substituting aliphatic amino acids for Phe^8 turns the peptide into an antagonist.[3] Modification of the peptidic structure was therefore an interesting approach to study the postulated angiotensin receptor and various peptide analogs with changes in the 4–8 core of Ang II as well as with substitution at the N- and C-terminal ends were synthesized by Bruno Kamber.[4] His^6 was found to be important for binding, as was the N-terminal portion. Among the peptides synthesized, CGP 42112, a modified pentapeptide analog of Ang II with the structure nicotinic acid-Tyr-(N$^\alpha$-benzyloxycarbonyl-Arg)-Lys-His-Pro-Ile-OH (Figure 1), had the highest affinity. CGP 42112 A and B are the hydrochloride and trifluoroacetate salts, respectively.

The first nonpeptidic compounds, which inhibited the effects of angiotensin *in vivo*, were reported in two patents by Furukawa et al.[5] and Chiu et al.,[6] and Wong et al.[7] from DuPont characterized these imidazole-containing molecules further and demonstrated micromolar activity in a rat adrenal [^{3}H]Ang II binding assay. A patent was published shortly before by Carini and Duncia.[8] A key compound covered by this patent was Ex 89 (DuP 753, losartan) (Figure 1). Other imidazole derivatives were described in a 1987 patent by Warner-Lambert,[9] for example, PD 123177 (Ex 13) and PD 123319 (Ex 15) (Figure 1). Ciba-Geigy chemists were quick to synthesize some of these compounds and therefore, three tools were available to us in 1988: DuP 753, PD 123177, and CGP 42112.

In 1980, Franz Weber started a Swiss federal initiative against vivisection, which was rejected by the Swiss people in 1985. As a Swiss citizen, speaking French, M. de Gasparo was requested by Ciba-Geigy to be deeply involved in the campaign against the initiative in the French-speaking part of the country and to defend the work done in the pharma industry. Refine, replace, and reduce were the leading words to decrease the use of animals in research. It was indicated, however, to really translate these words into laboratory practice. One possibility was to substitute tissues from human origin for the rat adrenals used in the binding assay. What kind of human (contractile) tissue, easy to collect and in sufficient amount, could be available for screening? The human uterus fulfilled these criteria. With the help of a gynecologist in July 1986, the first sample was collected after hysterectomy and immediately prepared for binding assay.

In early experiments with rat adrenal membranes, we had noticed that if we did not add dithiothreitol (DTT), a sulfhydryl-reducing agent to prevent ligand degradation, some Ang II analogs gave biphasic curves, whereas Ang II itself

The Protective Arm of the Renin–Angiotensin System (RAS). http://dx.doi.org/10.1016/B978-0-12-801364-9.00002-X

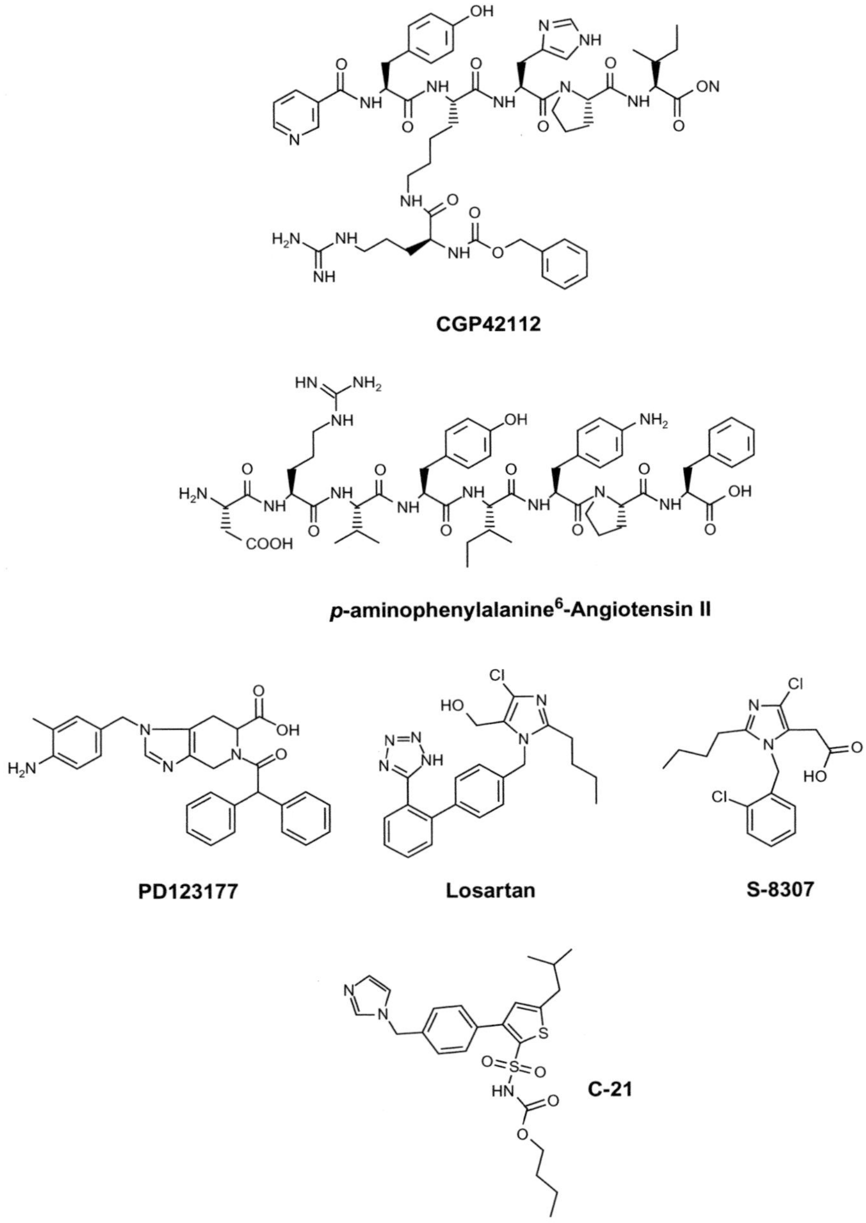

CGP42112

p-aminophenylalanine⁶-Angiotensin II

PD123177   **Losartan**   **S-8307**

C-21

FIGURE 1   Structures of key compounds.

always gave monophasic curves, although this depended on the preparation used.[10] At the end of 1986, a student working in the lab was given a project to devise a cocktail of peptidase inhibitors to reduce degradation of the radioligand. He was able to reduce the level of radioligand degradation from 30% to 10% in the absence of DTT. We soon discovered that in the absence of DTT, we always obtained monophasic curves when using human uterus membranes but often biphasic curves in rat adrenal preparations[10,11] (Figure 2). This situation was highly suggestive of the presence of at least two different binding sites in rat adrenal. Such a role of DTT to discriminate Ang II receptor subtypes was confirmed by Chiu et al.[12] but, in the absence of suitable tool compounds, it was difficult to prove this receptor heterogeneity.

TWO ANG II RECEPTOR SUBTYPES

In the absence of DTT, both CGP 42112 and Ang II had similar Ki's in rat adrenal glomerulosa and human uterus (0.73 and 0.45 nM), whereas DuP 753 had a higher Ki (16.4 nM) in rat adrenal but no affinity at all for human uterus receptors. The competition curves of radiolabeled Ang II with CGP 421122 and DuP 753 were clearly biphasic in the adrenal, strongly

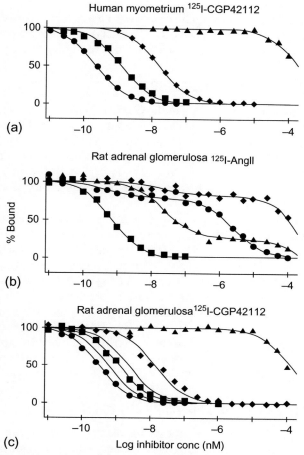

FIGURE 2 Binding of ^{125}I-CGP 42112 or ^{125}I-Ang II to human myometrium or rat adrenal glomerulosa membranes: competition curves with Ang II (■), CGP 42112 (●), PD 123319 (◆), and losartan (▲). *From Whitebread et al.*[11]

suggesting receptor heterogeneity.[11] Indeed, [^{125}I]-Ang II binding in the presence of a sufficiently high concentration of CGP 42112 to block its specific receptor gave a monophasic displacement curve with DuP 753 clearly showing that two distinct populations are present in rat adrenal. Also, DTT almost totally inhibited the [^{125}I]-Ang II binding in the adrenal, whereas it enhanced it in the uterus.[10,13] This difference is probably linked to a reduction of disulfide bridges between the cysteine residues in the extracellular loops of the AT$_1$ receptor. The human uterus assay continued to be used for Ang II screening, although we now also had access to rat vascular smooth muscle cells. Ang II was equally active in the two preparations in the subnanomolar range, but CGP 42112 was only very weakly active in rat smooth muscle cells, whereas it was highly active in the uterus. Based on these results, it was clear that the Ang II receptors in the human uterus were not the same as those in smooth muscle cells[14,15] and that both receptors appeared to be present in the adrenal.

In isolated rabbit aortic rings, DuP 753 inhibited Ang II-induced contractions with an IC$_{50}$ of 23 nM, whereas much higher concentrations of CGP 42112 were required to see the effect (IC$_{50}$ 1850 nM).[14] Clearly, there was no correlation between human uterus binding data and functional results in the aortic ring. This clearly indicated that it was the DuP 753-sensitive receptors in vascular smooth muscle cells that were the ones mediating the pressor responses in the aortic ring assay, not the CGP 42112-sensitive receptors, and that inhibition of the former would have the potential of lowering blood pressure *in vivo*.[14] At this point, we realized we had been following the wrong Ang II receptor (for the indication hypertension), but at the same time, we were uniquely positioned to move fast towards the discovery of a new antihypertensive drug: valsartan (Diovan®). Random and focused screens were quickly initiated using rat smooth muscle cells, while research on the role of the AT$_2$ subtype continued. This situation was considered by Ciba-Geigy to be a real advantage over the competition and should not be published. However, in 1988, there was a publication from the group of DuPont reporting the effect of a nonpeptide Ang II antagonist and suggesting that mechanism(s) other than AII receptor blockade may contribute partly to the hypotensive response to S-8307 (i.e., 2-n-butyl-4-chloro-1-(2-chlorobenzyl)imidazole-5-acetic acid).[7] Also, in May 1989, Wong et al.[16] wrote, "Since different subtypes of Ang II receptors have been reported in different tissues, this series of imidazole Ang II antagonists would be useful tools to explore the properties of these Ang II receptors." Also, studying Sar[1]-Ala[8] Ang I, the same group already observed in 1979 that this molecule "was found to be a good inhibitor *in vivo* in the rat blood pressure assay, somewhat less active in guinea pig ileum and a relatively weak antagonist in rat uterus."[17] Without doubt, from our own data and those reported in these competitor's publications, other Ang II receptors exist and we were allowed to publish our results rapidly.[10]

ANGIOTENSIN II RECEPTOR NOMENCLATURE

The topic of the angiotensin receptor was at the agenda of the Gordon Conference on angiotensin in Oxnard in 1990. The atmosphere was relatively heavy as the colleagues of DuPont were suspecting some leakage from their site to Basel, which was not the case. In fact, using two dissimilar nonpeptidic compounds DuP 753 and Exp 655, Chiu et al. (1989) identified two different subtypes in rat adrenal cortex and medulla.[12] Also, Speth and Kim detected two angiotensin II receptor binding sites in rat liver and PC12 cells, which differ using DuP 753 and p-amino-phe[6]-angiotensin II and in their susceptibility to a sulfhydryl-reducing agent.[18] Thus, three research groups came independently to the discovery and distinction between two major angiotensin receptor subtypes.

The different presentations at the Gordon Conference were sometimes difficult to follow as the nomenclature used in the three groups was different. The receptor with high affinity for losartan was named Ang II-1, Ang II-B, or Ang II-α, whereas the other with the high affinity for CGP 42112 or PD 123177 was called Ang II-2, Ang II-A, or Ang II β.[10,12,18] Bumpus immediately reacted, "Gentleman, we have to speak the same language" and to clarify this issue, a nomenclature committee was established by the American Heart Association Council for High Blood Pressure Research. The committee met in Baltimore in September 1990 to make recommendations and provide guidelines for creating the nomenclature for the angiotensin receptor subtypes. The Ciba-Geigy suggestion was that the DuPont numbering should be followed to recognize their considerable contribution to the identification of the subtypes. It was furthermore proposed to use "AT" instead of Ang for describing the angiotensin receptor (analogous to the previously described endothelin receptor "ET") to avoid confusion with angiotensin and its fragments. The prototypical ligand for the AT_1 receptor was losartan, whereas both CGP 42112 and PD 123177 characterized the AT_2 receptor, both having at least a 2000-fold greater affinity for the AT_2 than for the AT_1 receptor.[19]

EARLY RESEARCH ON THE ROLE OF THE AT_2 RECEPTOR

Comparing PD 123177 and CGP 42112 (Figure 3), it appears that the PD diphenyl system is superposable to the lipophilic pocket of isoleucine in CGP 42112. The imidazole ring of PD overlaps with the peptide bond located between the last two C-terminal residues, Pro and Ile, and the para-aminophenylalanine group of PD is in proximity with the histidine of Ang II.[4] The AT_2 receptor initially did not appear to be coupled to a classical G protein.[20] Indeed, a nonhydrolyzable analog of GTP (GTPγS) affected neither the binding parameters nor the dissociation rate of Ang II from the AT_2 receptor. Moreover, Ang II did not stimulate labeled GTPγS incorporation into tissues expressing only the AT_2 receptor. In fact, the AT_2 receptor cDNA encodes a 363-amino-acid protein with a typical seven-transmembrane domain. It is 33% identical to the AT_1 receptor.[21,22] The gene encoding the AT_2 receptor is located on the long arm of the X chromosome.[23]

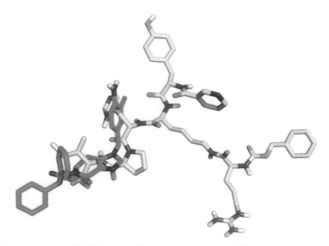

FIGURE 3 Superposition of CGP 42112 and PD 123177 (green). The diphenyl system of PD superimposes on the lipophilic pocket of isoleucine in CGP. *Modified from de Gasparo et al.*[4]

Stimulation of the AT$_2$ receptor triggers G-protein coupling through the third intracellular loop of the receptor. This activates Tyr phosphatases (MKP-1, SHP-1, and PEA)[24] and inhibits MAP kinase, inducing AT$_1$ receptor inactivation. It also stimulates growth factor-activated kinases, eNOS and phospholipase A2, and arachidonic acid production. The AT$_2$ receptor also causes vasodilation through an extracellular bradykinin pathway and the cGMP cascade. AT$_2$ receptor stimulation inhibits proliferation, promotes apoptosis, opens potassium channels, and closes the T-type calcium channel-inducing vasodilation.[25,26] It stimulates neurite outgrowth and its effect is abolished by PD 123319.[27]

CGP 42112 usually acts as a full agonist at the AT$_2$ receptor (Ki 0.5 nM). Indeed, in PC12W cells expressing only the AT$_2$ receptor, CGP 42112, like Ang II itself, decreases atrial natriuretic peptide-stimulated cGMP concentration. This effect is not affected by losartan but is inhibited by the AT$_2$ antagonist PD 123319.[28,29] In contrast, at high concentration, CGP 42112 behaves like an antagonist of the AT$_1$ receptor: It inhibits Ang II-induced contraction in rabbit aortic rings with an IC$_{50}$ of 1.85 μM.[14] Above 1 μM, CGP 42112 increased renal blood flow and urinary sodium excretion suggesting an antagonistic property when the RAS is activated,[30] whereas it acts as an AT$_1$ agonist when the RAS is blocked.[31] To avoid any misinterpretation in animal studies or in systems, which express both receptor subtypes, CGP 42112 can be used as an AT$_2$-selective ligand provided its concentration does not exceed 25 nM (50-fold its Ki for the AT$_2$ receptor).

As has been demonstrated for the alpha- and beta-adrenergic receptors, it was proposed that there is a balance between the two receptors, in that they mediate opposite cellular functions as if they were mutually antagonistic.[28,32,33] This is clinically interesting. Blockade of the AT$_1$ receptor is accompanied by an increased plasma renin activity and thus of plasma Ang II, which could stimulate the unblocked AT$_2$ receptor.

Although the role of the AT$_1$ receptor is well understood, there is still a lot of ambiguity regarding the AT$_2$ receptor.[33] Although CGP 42112 was an excellent tool to start the study of the AT$_2$ receptor properties, it is a peptide and therefore not useful for oral treatment. The development of a selective nonpeptidic AT$_2$ agonist appears therefore to be a clear requirement to better understand the physiological and pathophysiological role of the AT$_2$ receptor, to determine its potential side effects and to consider various possible clinical indications. Such compounds, modeled on the C-terminal pentapeptide structure of Ang II, have been synthesized[34] and preliminary results with C21 appear promising not only in a number of cardiovascular diseases but also for its anti-inflammatory and neuroprotective properties.[35–38]

CONCLUSION

Ang II, one of the active components of the renin–angiotensin cascade,[39] binds to two major receptors, AT$_1$ and AT$_2$. Selective tool compounds discovered by different groups in the 1980s were critical to the identification and initial characterization of these two major Ang II receptor subtypes. This finding set the stage for the discovery of a new class of antihypertensive agents, the angiotensin receptor blockers (ARBs), and initiated the ongoing research into the role of the AT$_2$ receptor subtype. The AT$_2$ receptor appears to be a key mediator of the beneficial action of the RAS on cardiovascular and noncardiovascular indications and often opposes the effects of the AT$_1$ receptor. New selective AT$_2$ receptor nonpeptidic agonists such as C21 are required to better understand the suggested important pathophysiological role of the AT$_2$ receptor and clarify its potential clinical interest for cardiovascular and noncardiovascular diseases.

REFERENCES

1. de Gasparo M, Joss U, Ramjoué HP, Whitebread SE, Haenni H, Schenkel L, et al. Three new epoxy-spirolactone derivatives: characterization *in vivo* and *in vitro*. *J Pharmacol Exp Ther* 1987;**240**:650–6.
2. Kagawa CM. Blocking the renal electrolyte effects of mineralocorticoids with an orally active steroidal spirolactone. *Endocrinology* 1960;**67**:125–32.
3. Khosla MC, Hall MM, Smeby RR, Bumpus FM. Agonist and antagonist relationships in 1- and 8-substituted analogs of angiotensin II. *J Med Chem* 1974;**17**:1156–60.
4. de Gasparo M, Levens NR, Kamber B, Furet P, Whitebread S, Brechler V, et al. The angiotensin II AT$_2$ receptor subtype. In: Saavedra JM, Timmermans PB, editors. *Angiotensin receptor*. New York: Plenum Press; 1994. p. 95–117.
5. Furukawa Y, Kishimoto S, Nishikawa K. Hypotensive imidazole derivatives US patent 4340598, 20 July 1982 and Hypotensive imidazole-5-acetic acid derivatives. US Patent 4355040, 19 October 1982.
6. Chiu AT, Carini DJ, Johnson AL, McCall DE, Price WA, Thoolen M, et al. Non-peptide angiotensin II receptor antagonists. II. Pharmacology of S-8308. *Eur J Pharmacol* 1988;**157**:13–21.
7. Wong PC, Chiu AT, Price WA, Thoolen MJ, Carini DJ, Johnson AL, et al. Nonpeptide angiotensin II receptor antagonists. I. Pharmacological characterization of 2-n-butyl-4-chloro-1-(2-chlorobenzyl)imidazole-5-acetic acid, sodium salt (S-8307). *J Pharmacol Exp Ther* 1988;**247**:1–7.
8. Carini DJ, Duncia JJV. Angiotensin II receptor blocking imidazoles. European Patent EP0253310, 20 January 1988.
9. Blankley J, Hodges JC, Kiely JS, Klutchko SR. 4,5,6,7-Tetrahydro-1H-imidazo[4,5-c]pyridine derivatives and analogs having antihypertensive activity. European Patent EP245637, 19 November 1987.

10. Whitebread S, Mele M, Kamber B, de Gasparo M. Preliminary biochemical characterization of two angiotensin II receptor subtypes. *Biochem Biophys Res Commun* 1989;**163**:284–91.

11. Whitebread SE, Taylor V, Bottari SP, Kamber B, de Gasparo M, Radioiodinated CGP. 42112A: a novel high affinity and highly selective ligand for the characterization of angiotensin AT₂ receptors. *Biochem Biophys Res Commun* 1991;**181**:1365–71.

12. Chiu AT, Herblin WF, McCall DE, Ardecky RJ, Carini DJ, Duncia JV, et al. Identification of angiotensin II receptor subtypes. *Biochem Biophys Res Commun* 1989;**165**:196–203.

13. de Gasparo M, Whitebread S, Bottari SP, Levens SR. Heterogeneity of angiotensin receptor subtypes. In: Timmermans PB, Wexler RR, editors. *Medicinal chemistry of the renin angiotensin system.* Amsterdam: Elsevier; 1994. p. 269–94.

14. Criscione L, Thomann H, Whitebread S, de Gasparo M, Bühlmayer P, Herold P, et al. Binding characteristics and vascular effects of various angiotensin II antagonists. *J Cardiovasc Pharmacol* 1990;**16**(Suppl 4):S56–9.

15. de Gasparo M, Whitebread S, Mele M, Motani AS, Whitcombe PJ, Ramjoué HP, et al. Biochemical characterization of two angiotensin II receptor subtypes in the rat. *J Cardiovasc Pharmacol* 1990;**16**(Suppl 4):S31–5.

16. Wong PC, Price WA, Chiu AT, Thoolen MJ, Duncia JV, Johnson AL, et al. Nonpeptide angiotensin II receptor antagonists. IV. EXP6155 and EXP6803. *Hypertension* 1989;**13**:489–97.

17. Chiu AT, Sutherland Jr JC, Day AR, Freer RJ. Synthesis and pharmacology of a prohormone angiotensin antagonist. *Eur J Pharmacol* 1979;**54**:177–80.

18. Speth RC, Kim KH. Discrimination of two angiotensin II receptor subtypes with a selective agonist analogue of angiotensin II, p-aminophenylalanine⁶ angiotensin II. *Biochem Biophys Res Commun* 1990;**169**:997–1006.

19. Bumpus FM, Catt KJ, Chiu AT, de Gasparo M, Goodfriend T, Husain A, et al. Nomenclature for angiotensin receptors. A report of the Nomenclature Committee of the Council for High Blood Pressure Research. *Hypertension* 1991;**17**:720–1.

20. Bottari SP, Taylor V, King IN, Bogdal Y, Whitebread S, de Gasparo M, et al. AT₂ receptors do not interact with guanine nucleotide binding proteins. *Eur J Pharmacol* 1991;**207**:157–63.

21. Nakajima M, Mukoyama M, Pratt RE, Horiuchi M, Dzau VJ. Cloning of cDNA and analysis of the gene for mouse angiotensin II type 2 receptor. *Biochem Biophys Res Commun* 1993;**197**:393–9.

22. Kambayashi Y, Bardhan S, Takahashi K, Tsuzuki S, Inui H, Hamakubo T, et al. Molecular cloning of a novel angiotensin II receptor isoform involved in phosphotyrosine phosphatase inhibition. *J Biol Chem* 1993;**268**:24543–6.

23. Koike G, Horiuchi M, Yamada T, Szpirer C, Jacob HJ, Dzau VJ. Human type 2 angiotensin II receptor gene: cloned, mapped to the X chromosome, and its mRNA is expressed in the human lung. *Biochem Biophys Res Commun* 1994;**203**:1842–50.

24. Bottari SP, King IN, Reichlin S, Dahlstroem I, Lydon N, de Gasparo M. The angiotensin AT2 receptor stimulates protein tyrosine phosphatase activity and mediates inhibition of particulate guanylate cyclase. *Biochem Biophys Res Commun* 1992;**183**:206–11.

25. Buisson B, Bottari SP, de Gasparo M, Gallo-Payet N, Payet MD. The angiotensin AT₂ receptor modulates T-type calcium current in non-differentiated NG108-15 cells. *FEBS Lett* 1992;**309**:161–4.

26. Buisson B, Laflamme L, Bottari SP, de Gasparo M, Gallo-Payet N, Payet MD. A G protein is involved in the angiotensin AT2 receptor inhibition of the T-type calcium current in non-differentiated NG108-15 cells. *J Biol Chem* 1995;**270**:1670–4.

27. Laflamme L, de Gasparo M, Gallo JM, Payet MD, Gallo-Payet N. Angiotensin II induction of neurite outgrowth by AT₂ receptors in NG108-15 cells. Effect counteracted by the AT₁ receptors. *J Biol Chem* 1996;**271**:227–9.

28. Brechler V, Jones PW, Levens NR, de Gasparo M, Bottari SP. Agonistic and antagonistic properties of angiotensin analogs at the AT₂ receptor in PC12W cells. *Regul Pept* 1993;**44**:207–13.

29. Brechler V, Reichlin S, de Gasparo M, Bottari SP. Angiotensin II stimulates protein tyrosine phosphatase activity through a G-protein independent mechanism. *Receptor Channels* 1994;**2**:89–98.

30. Macari D, Bottari S, Whitebread S, de Gasparo M, Levens N. Renal actions of the selective angiotensin AT₂ receptor ligands CGP 42112B and PD 123319 in the sodium-depleted rat. *Eur J Pharmacol* 1993;**249**:85–93.

31. Macari D, Whitebread S, Cumin F, de Gasparo M, Levens N. Renal actions of the angiotensin AT₂ receptor ligands CGP 42112 and PD 123319 after blockade of the renin-angiotensin system. *Eur J Pharmacol* 1994;**259**:27–36.

32. de Gasparo M, Whitebread S, Levens N, Ramjoue HP, Criscione L, Rogg R, et al. Pharmacology of angiotensin II receptor subtypes. In: Saez JM, Brownie AC, Capponi A, Chambaz EM, Mantero F, editors. *Cellular and molecular biology of the adrenal cortex*, Colloque INSERM, France: John Libbey Eurotext Ltd; 1992. **222**:3–17.

33. de Gasparo M, Siragy HM. The AT2 receptor: fact, fancy and fantasy. *Regul Pept* 1999;**81**:11–24.

34. Murugaiah AM, Wu X, Wallinder C, Mahalingam AK, Wan Y, Sköld C, et al. From the first selective non-peptide AT₂ receptor agonist to structurally related antagonists. *J Med Chem* 2012;**55**:2265–78.

35. Danyel LA, Schmerler P, Paulis L, Unger T, Steckelings UM. Impact of AT₂-receptor stimulation on vascular biology, kidney function, and blood pressure. *Integr Blood Press Control* 2013;**6**:153–61.

36. Namsolleck P, Recarti C, Foulquier S, Steckelings UM, Unger T. AT₂ receptor and tissue injury: therapeutic implications. *Curr Hypertens Rep* 2014;**16**:416.

37. Foulquier S, Steckelings UM, Unger T. Impact of the AT₂ receptor agonist C21 on blood pressure and beyond. *Curr Hypertens Rep* 2012;**14**:403–9.

38. Lauer D, Slavic S, Sommerfeld M, Thöne-Reineke C, Sharkovska Y, Hallberg A, et al. Angiotensin Type 2 receptor stimulation ameliorates left ventricular fibrosis and dysfunction via regulation of tissue inhibitor of matrix metalloproteinase 1/matrix metalloproteinase 9 axis and transforming growth factor β1 in the rat heart. *Hypertension* 2014;**63**:60–7.

39. Ramírez-Sánchez M, Prieto I, Wangensteen R, Banegas I, Segarra AB, Villarejo AB, et al. The renin-angiotensin system: new insight into old therapies. *Curr Med Chem* 2013;**20**:1313–22.

Chapter 3

Animal Models with a Genetic Alteration of AT$_2$ Expression

Masaki Mogi, Masatsugu Horiuchi

Molecular Cardiovascular Biology and Pharmacology, Ehime University Graduate School of Medicine, Tohon, Shitsukawa, Ehime, Japan

ABBREVIATIONS

α-SMA	α-smooth muscle actin
Ang II	angiotensin II
ApoE	apolipoprotein E
AT$_1$ receptor	angiotensin II type 1 receptor
AT$_2$ receptor	angiotensin II type 2 receptor
B$_2$R	bradykinin subtype 2 receptor
DKO	double-knockout
MAP	mean arterial pressure
MI	myocardial infarction
NO	nitric oxide
NOS	nitric oxide synthase

INTRODUCTION

Following cloning of the cDNA for the mouse angiotensin II type 2 receptor (AT$_2$ receptor), genetically modified animals were expected to be generated for functional analysis of AT$_2$ receptor functions. Two kinds of mice with disruption of the AT$_2$ receptor gene (*Agtr2*) were reported simultaneously by Hein et al.[1] and Ichiki et al.[2] In 1995, Hein et al. first demonstrated that AT$_2$ receptor-disrupted (*Agtr2*$^-$) mice, which were generated by disruption of the coding sequence of the *Agtr2* gene via insertion of the neomycin resistance gene (*neo*^{r}) into exon 3, develop normally but have an impaired drinking response to water deprivation as well as a reduction in spontaneous movements.[1] Morphological changes in the heart, lung, kidney, adrenal, spleen, brain, aorta, ovary, uterus, pancreas, eye, skeletal muscle, and blood are not observed in nonmutant, hemizygous and homozygous mutant mice. Ichiki et al. also generated *Agtr2*$^-$ mice by disruption of the *Agtr2* gene via insertion of the neomycin resistance gene (*neo*^{r}) into exon 3.[2] Homozygous female mutant mice are healthy and fertile. Likewise, no obvious morphological changes are observed in the brain (inferior olive and locus coeruleus), heart, lung, kidney, and adrenal of hemizygous male mutant mice. Generation of *Agtr2*$^-$ mice reveals the functional roles of the AT$_2$ receptor clearly and contributes to progress in research on AT$_2$ receptor functions.

BLOOD PRESSURE

Hein et al. reported that an increased vasopressor response to injection of angiotensin II (Ang II) is found in *Agtr2*$^-$ mice, although baseline blood pressure is normal.[1] In contrast, Ichiki et al. reported that mice lacking the AT$_2$ receptor show a significant increase in blood pressure and increased sensitivity to the pressor action of Ang II.[2] Thus, there is

The Protective Arm of the Renin–Angiotensin System (RAS). http://dx.doi.org/10.1016/B978-0-12-801364-9.00003-1

a difference in baseline blood pressure and the vasopressor response to injection of Ang II between these two kinds of $Agtr2^-$ mice. Tanaka et al. reported that the increased vascular reactivity to Ang II in $Agtr2^-$ mice reported by Ichiki et al. is at least partly due to increased vascular angiotensin II type 1 receptor (AT_1 receptor) expression.[3] Therefore, triple gene-knockout mice deficient in all three Ang II receptors (AT_{1a}, AT_{1b}, and AT_2) were generated by intercrossing and breeding of single-knockout mice by Walther's group.[4] Investigation of triple-knockout mice showed that AT_2 receptor deletion increased baseline mean arterial pressure (MAP), whereas mice lacking the AT_{1a} receptor were hypotensive and displayed a reduced heart weight/body weight ratio. Because Ang II failed to alter MAP in triple-knockout mice, it is likely that no other receptors than AT_{1a}, AT_{1b}, and/or AT_2 mediate the pressor effects of Ang II. Central roles of Ang II receptors in blood pressure regulation have been highlighted. Intracerebroventricular Ang II injection increased systolic blood pressure in a dose-dependent manner in $Agtr2^-$ mice; however, this increase was significantly greater in $Agtr2^-$ mice than in wild-type (C57BL/6) mice.[5] Moreover, we generated mice with deletion of both the AT_{1a} and AT_2 receptors (AT_{1a}/AT_2-double-knockout (DKO)) by crossing AT_{1a} and AT_2-KO mice.[5] Intracerebroventricular Ang II injection-induced blood pressure elevation and water intake were significantly greater in $Agtr2^-$ mice, but not in AT_{1a}/AT_2-DKO, indicating that the functional balance of central AT_{1a} and AT_2 receptors plays an important role in the mechanism of hypertension.

CARDIAC DISEASE

Akishita et al. clearly demonstrated that the vasoconstrictor response is exaggerated in $Agtr2^-$ mice by evaluating contraction of aortic rings by Ang II *in vitro*[6] and that coronary arterial thickening and perivascular fibrosis after aortic banding are 50% greater in $Agtr2^-$ mice than in wild-type mice.[7] We reported the potential beneficial effects of AT_2 receptor signaling on cardiac remodeling[8,9] and heart failure after myocardial infarction (MI)[10] in experiments using $Agtr2^-$ mice. Furthermore, chronic loss of the AT_2 receptor prevented collagen deposition and caused cardiac rupture in post-MI models.[11] To evaluate the function of the AT_2 receptor in local tissues, tissue-specific $Agtr2$ gene alterations have been reported since 1998. Masaki et al. demonstrated that mice with cardiac-specific overexpression of the AT_2 receptor showed decreased sensitivity to the AT_1 receptor-mediated pressor action without any effect on cardiac contractility, which was mainly due to the inhibitory effect of the AT_2 receptor on the AT_1 receptor-mediated chronotropic effect,[12] and that this mouse strain also shows attenuated perivascular fibrosis by a kinin/NO-dependent mechanism without the development of cardiomyocyte hypertrophy by Ang II infusion.[13] These transgenic mice were generated by microinjection of α-myosin heavy chain (α-MHC) or α-smooth muscle actin (α-SMA) promoter and AT_2 receptor cDNA into the pronuclei of fertilized embryos. On the other hand, transgenic mice overexpressing AT_2 receptors selectively in ventricular myocytes were generated by Lorell's group, where the AT_2 receptor was driven by the myosin light chain-2v promoter.[14] They reported that overexpression of AT_2 receptors promotes the development of dilated cardiomyopathy and heart failure.[14] They showed that chronic overexpression of the AT_2 receptor has the potential to cause Ca^{2+}- and pH-dependent contractile dysfunction in ventricular myocytes, as well as loss of the inotropic response to Ang II,[15] and that AT_2 receptor signaling modifies the pathological hypertrophic response to aortic banding resulting in chronic pressure overload-induced cardiac hypertrophy.[16] In post-MI remodeling, Voros et al. also showed a beneficial effect of the AT_2 receptor using double-gene-alteration mice. They generated these mice using mice with cardiac overexpression of the AT_2 receptor and AT_{1a}-deficient mice and assessed the intricately intertwined connection of these receptors.[17] They showed that genetic deletion of the AT_{1a} receptor is additive to AT_2 receptor overexpression, due, at least in part, to blood pressure lowering. Moreover, AT_2 receptor and guanylyl cyclase-A DKO (AT_2/GCA-KO) mice were generated by Dr. Nakao's group.[18] Using this model, they showed that GCA inhibits AT_2 receptor-mediated pro-hypertrophic signaling in the heart.

VASCULAR DISEASE

We reported the potential beneficial effects of AT_2 receptor signaling on vascular remodeling in a cuff placement model,[19,20] a wire injury model,[21] and vascular senescence, in experiments using $Agtr2^-$ mice.[22] Mice with smooth muscle cell (VSMC)-specific overexpression of the AT_2 receptor ($smAT_2$-Tg) showed that the production of bradykinin is stimulated and the nitric oxide (NO)/cGMP system is enhanced in a paracrine manner to promote vasodilation.[23] On the other hand, we also generated AT_2 receptor and apolipoprotein E (ApoE)-DKO (AT_2/ApoE-DKO) mice.[24] Because the AT_2 receptor is located on the X chromosome, ApoE-KO mice and $Agtr2^-$ mice were bred to yield mice heterozygous at the ApoE locus and heterozygous (female) or hemizygous (male) at the AT$_2$ receptor locus. Using these DKO mice, it was clearly proved that AT$_2$ receptor stimulation plays an important role in atherosclerotic lesion formation through regulation of oxidative stress,[24] adipocyte differentiation, and adipose tissue dysfunction.[25] Moreover, Takata et al. generated smAT$_2$-Tg and ApoE-knockout (smAT$_2$-Tg/ApoE-KO) mice and demonstrated that vascular AT$_2$ receptor stimulation exerts antiatherogenic actions in an endothelial kinin/NO-dependent manner. Interestingly, such an antioxidative effect is likely to be exerted by inhibition of the accumulation of superoxide-producing mononuclear leukocytes.[26] Moreover, Isbell et al. compared post-MI left ventricular remodeling among mice to overexpression of the AT$_2$ receptor (AT$_2$-Tg), deletion of the bradykinin subtype 2 receptor (B$_2$R) (B$_2$-KO), AT$_2$-Tg with B$_2$R deletion (AT$_2$-Tg/B$_2$R-KO), and wild-type mice and found that attenuation of post-MI remodeling by overexpression of the AT$_2$ receptor is not directly mediated via the B$_2$R pathway.[27] Furthermore, double-gene-alteration mice generated by crossing *Agtr2$^-$* mice and overexpression of adipose tissue angiotensinogen (aAGT-Tg) were reported by Yvan-Charvet et al. These mice exhibited higher blood pressure than that in aAGT-Tg mice. Interestingly, despite a reduction of adipose mass in double-gene-alteration mice, AT$_2$ receptor deficiency caused increased renin production.[28] On the other hand, *Agtr2$^-$* mice bred with an established model of the Marfan syndrome, which are heterozygous for a cysteine substitution in an epidermal growth factor-like domain of fibrillin-1 (*Fbn1$^{C1039G/+}$*), exhibit exacerbated aortic diseases compared with that in Marfan syndrome mice.[29]

RENAL DISEASE

Agtr2$^-$ mice exhibited a reduction in urinary sodium excretion via bradykinin and NO, against the antinatriuretic action after infusion of Ang II.[30] Obst et al. focused on renal NO synthase (NOS) isoforms. *Agtr2$^-$* mice showed upregulation of neuronal NOS (nNOS) and inducible NOS (iNOS).[31] They also investigated the effect of deoxycorticosterone acetate (DOCA)-salt on renal function using *Agtr2$^-$* mice. Under DOCA-salt, renal iNOS was further increased and could contribute to renal damage. *Agtr2$^-$* mice showed increased renal fibrosis and fibrocyte infiltration with concomitant upregulation of renal transcripts of procollagen type Iα1 (COL1A1) compared with wild-type mice, together with an increased number of fibrocytes in the bone marrow.[32] In the urinary tract, some *Agtr2$^-$* mice show anomalies that mimic human congenital anomalies of the kidney and urinary tract as a result of a defect of *in utero* organogenesis of the excretory system.[33] Moreover, the receptor for advanced glycation end products (RAGE) expression in podocytes was moderately higher in *Agtr2$^-$* mice than in control animals after Ang II treatment, indicating that Ang II-mediated RAGE induction in podocytes occurs via the AT$_2$ receptor.[34] In contrast, Hashimoto et al. demonstrated protective roles of the AT$_2$ receptor against glomerular injury under the control of the α-SMA promoter, using AT$_2$-Tg mice subjected to 5/6 nephrectomy.[35]

NEURAL DISEASE

In brain development, von Bohlen und Halbach et al. reported an increase of cell number in the piriform cortex and hippocampus CA1, CA2, and CA3 regions of *Agtr2$^-$* mice compared with wild-type mice.[36] By whole-genome microarray analysis of RNA isolated from the brain of *Agtr2$^-$* mice, it was shown that differentially expressed genes encode molecules involved in multiple cellular processes including microtubule functions associated with dendritic spine morphology, suggesting that AT$_2$ receptor-modulated genes in the brain influence learning and memory.[37] *Agtr2$^-$* mice show anxiety-like behavior involving the noradrenergic system. Okuyama et al. speculated that the amygdala appears to play an important role in responses to fear and anxiety.[38] Moreover, Watanabe et al. reported that *Agtr2$^-$* mice exhibit a lower body temperature than controls, possibly due to a decrease in immunological stress-induced hyperthermia.[39] On the other hand, the pain threshold was significantly lower in *Agtr2$^-$* mice, with a decreased level of brain beta-endorphins.[40] These results indicate that the AT$_2$ receptor has various roles in behavior and brain functions.

CONCLUSION

Animal models with a genetic alteration of AT$_2$ receptor expression have been discussed and are summarized in Table 1. Genetic alteration animals contribute to functional analysis of not only the AT$_2$ receptor but also interactions between other receptors and renin–angiotensin system components.

TABLE 1 Mouse Models with a Genetic Alteration of AT_2 Receptor Expression

AT_2 Receptor Modification	Conventional or Conditional	Target Tissues	Genetic Alteration	Other Gene Alteration	References
Deletion	Conventional	Whole body	AT_2R only		[1,5–10,19–22,32,34]
Deletion	Conventional	Whole body	AT_2R only		[2,3,11,30,31,33,36–40]
Overexpression	Conditional	Cardiac myocytes	AT_2R only		[12,13]
Overexpression	Conditional	Smooth muscle cells	AT_2R only		[23,35]
Overexpression	Conditional	Ventricular myocytes	AT_2R only		[14–16]
Deletion	Conventional	Whole body	Double	Deletion of AT_{1a} receptor (conventional)	[5]
Deletion	Conventional	Whole body	Double	Deletion of ApoE (conventional)	[24,25]
Overexpression	Conditional	Smooth muscle cells	Double	Deletion of ApoE (conventional)	[26]
Overexpression	Conventional	Whole body	Double	Deletion of B_2 receptor (conventional)	[27]
Overexpression	Conditional	Cardiac myocytes	Double	Deletion of AT_{1a} receptor (conventional)	[17]
Deletion	Conventional	Whole body	Double	Overexpression of angiotensinogen (conditional; adipose tissue)	[28]
Deletion	Conventional	Whole body	Double	Deletion of guanylyl cyclase-A	[18]
Deletion	Conventional	Whole body	Double	Heterozygous for a cysteine substitution in an epidermal growth factor-like domain of fibrillin-1 (conventional)	[29]
Deletion	Conventional	Whole body	Triple	Deletion of AT_{1a} and AT_{1b} receptors (conventional)	[4]

REFERENCES

1. Hein L, Barsh GS, Pratt RE, Dzau VJ, Kobilka BK. Behavioural and cardiovascular effects of disrupting the angiotensin II type-2 receptor in mice. *Nature* 1995;**377**:744–7.
2. Ichiki T, Labosky PA, Shiota C, Okuyama S, Imagawa Y, Fogo A, et al. Effects on blood pressure and exploratory behaviour of mice lacking angiotensin II type-2 receptor. *Nature* 1995;**377**:748–50.
3. Tanaka M, Tsuchida S, Imai T, Fujii N, Miyazaki H, Ichiki T, et al. Vascular response to angiotensin II is exaggerated through an upregulation of AT$_1$ receptor in AT$_2$ knockout mice. *Biochem Biophys Res Commun* 1999;**258**:194–8.
4. Gembardt F, Heringer-Walther S, van Esch JH, Sterner-Kock A, van Veghel R, Le TH, et al. Cardiovascular phenotype of mice lacking all three subtypes of angiotensin II receptors. *FASEB J* 2008;**22**:3068–77.
5. Li Z, Iwai M, Wu L, Shiuchi T, Jinno T, Cui TX, et al. Role of AT$_2$ receptor in the brain in regulation of blood pressure and water intake. *Am J Physiol Heart Circ Physiol* 2003;**284**:H116–21.
6. Akishita M, Yamada H, Dzau VJ, Horiuchi M. Increased vasoconstrictor response of the mouse lacking angiotensin II type 2 receptor. *Biochem Biophys Res Commun* 1999;**261**:345–9.
7. Akishita M, Iwai M, Wu L, Zhang L, Ouchi Y, Dzau VJ, et al. Inhibitory effect of angiotensin II type 2 receptor on coronary arterial remodeling after aortic banding in mice. *Circulation* 2000;**102**:1684–9.
8. Wu L, Iwai M, Nakagami H, Chen R, Suzuki J, Akishita M, et al. Effect of angiotensin II type 1 receptor blockade on cardiac remodeling in angiotensin II type 2 receptor null mice. *Arterioscler Thromb Vasc Biol* 2002;**22**:49–54.
9. Oishi Y, Ozono R, Yano Y, Teranishi Y, Akishita M, Horiuchi M, et al. Cardioprotective role of AT$_2$ receptor in postinfarction left ventricular remodeling. *Hypertension* 2003;**41**:814–18.
10. Adachi Y, Saito Y, Kishimoto I, Harada M, Kuwahara K, Takahashi N, et al. Angiotensin II type 2 receptor deficiency exacerbates heart failure and reduces survival after acute myocardial infarction in mice. *Circulation* 2003;**107**:2406–8.
11. Ichihara S, Senbonmatsu T, Price Jr. E, Ichiki T, Gaffney FA, Inagami T. Targeted deletion of angiotensin II type 2 receptor caused cardiac rupture after acute myocardial infarction. *Circulation* 2002;**106**:2244–9.
12. Masaki H, Kurihara T, Yamaki A, Inomata N, Nozawa Y, Mori Y, et al. Cardiac-specific overexpression of angiotensin II AT$_2$ receptor causes attenuated response to AT$_1$ receptor-mediated pressor and chronotropic effects. *J Clin Invest* 1998;**101**:527–35.
13. Kurisu S, Ozono R, Oshima T, Kambe M, Ishida T, Sugino H, et al. Cardiac angiotensin II type 2 receptor activates the kinin/NO system and inhibits fibrosis. *Hypertension* 2003;**41**:99–107.
14. Yan X, Price RL, Nakayama M, Ito K, Schuldt AJ, Manning W, et al. Ventricular-specific expression of angiotensin II type 2 receptors causes dilated cardiomyopathy and heart failure in transgenic mice. *Am J Physiol Heart Circ Physiol* 2003;**285**:H2179–87.
15. Nakayama M, Yan X, Price RL, Borg TK, Ito K, Sanbe A, et al. Chronic ventricular myocyte-specific overexpression of angiotensin II type 2 receptor results in intrinsic myocyte contractile dysfunction. *Am J Physiol Heart Circ Physiol* 2005;**288**:H317–27.
16. Yan X, Schuldt AJ, Price RL, Amende I, Liu FF, Okoshi K, et al. Pressure overload-induced hypertrophy in transgenic mice selectively overexpressing AT$_2$ receptors in ventricular myocytes. *Am J Physiol Heart Circ Physiol* 2008;**294**:H1274–81.
17. Voros S, Yang Z, Bove CM, Gilson WD, Epstein FH, French BA, et al. Interaction between AT1 and AT$_2$ receptors during postinfarction left ventricular remodeling. *Am J Physiol Heart Circ Physiol* 2006;**290**:H1004–10.
18. Li Y, Saito Y, Kuwahara K, Rong X, Kishimoto I, Harada M, et al. Guanylyl cyclase-A inhibits angiotensin II type 2 receptor-mediated pro-hypertrophic signaling in the heart. *Endocrinology* 2009;**150**:3759–65.
19. Wu L, Iwai M, Nakagami H, Li Z, Chen R, Suzuki J, et al. Roles of angiotensin II type 2 receptor stimulation associated with selective angiotensin II type 1 receptor blockade with valsartan in the improvement of inflammation-induced vascular injury. *Circulation* 2001;**104**:2716–21.
20. Suzuki J, Iwai M, Nakagami H, Wu L, Chen R, Sugaya T, et al. Role of angiotensin II-regulated apoptosis through distinct AT$_1$ and AT$_2$ receptors in neointimal formation. *Circulation* 2002;**106**:847–53.
21. Yamamoto Y, Watari Y, Brydun A, Yoshizumi M, Akishita M, Horiuchi M, et al. Role of the angiotensin II type 2 receptor in arterial remodeling after wire injury in mice. *Hypertens Res* 2008;**31**:1241–9.
22. Min LJ, Mogi M, Iwanami J, Jing F, Tsukuda K, Ohshima K, et al. Angiotensin II type 2 receptor-interacting protein prevents vascular senescence. *J Am Soc Hypertens* 2012;**6**:179–84.
23. Tsutsumi Y, Matsubara H, Masaki H, Kurihara H, Murasawa S, Takai S, et al. Angiotensin II type 2 receptor overexpression activates the vascular kinin system and causes vasodilation. *J Clin Invest* 1999;**104**:925–35.
24. Iwai M, Chen R, Li Z, Shiuchi T, Suzuki J, Ide A, et al. Deletion of angiotensin II type 2 receptor exaggerated atherosclerosis in apolipoprotein E-null mice. *Circulation* 2005;**112**:1636–43.
25. Iwai M, Tomono Y, Inaba S, Kanno H, Senba I, Mogi M, et al. AT$_2$ receptor deficiency attenuates adipocyte differentiation and decreases adipocyte number in atherosclerotic mice. *Am J Hypertens* 2009;**22**:784–91.
26. Takata H, Yamada H, Kawahito H, Kishida S, Irie D, Kato T, et al. Vascular angiotensin II type 2 receptor attenuates atherosclerosis via a kinin/NO-dependent mechanism. *J Renin Angiotensin Aldosterone Syst* 2013, in press.
27. Isbell DC, Voros S, Yang Z, DiMaria JM, Berr SS, French BA, et al. Interaction between bradykinin subtype 2 and angiotensin II type 2 receptors during post-MI left ventricular remodeling. *Am J Physiol Heart Circ Physiol* 2007;**293**:H3372–8.
28. Yvan-Charvet L, Massiera F, Lamande N, Ailhaud G, Teboul M, Moustaid-Moussa N, et al. Deficiency of angiotensin type 2 receptor rescues obesity but not hypertension induced by overexpression of angiotensinogen in adipose tissue. *Endocrinology* 2009;**150**:1421–8.

29. Habashi JP, Doyle JJ, Holm TM, Aziz H, Schoenhoff F, Bedja D, et al. Angiotensin II type 2 receptor signaling attenuates aortic aneurysm in mice through ERK antagonism. *Science* 2011;**332**:361–5.

30. Siragy HM, Inagami T, Ichiki T, Carey RM. Sustained hypersensitivity to angiotensin II and its mechanism in mice lacking the subtype-2 (AT$_2$) angiotensin receptor. *Proc Natl Acad Sci U S A* 1999;**96**:6506–10.

31. Obst M, Gross V, Bonartsev A, Janke J, Muller DN, Park JK, et al. Nitric oxide synthase expression in AT$_2$ receptor-deficient mice after DOCA-salt. *Kidney Int* 2004;**65**:2268–78.

32. Sakai N, Wada T, Matsushima K, Bucala R, Iwai M, Horiuchi M, et al. The renin-angiotensin system contributes to renal fibrosis through regulation of fibrocytes. *J Hypertens* 2008;**26**:780–90.

33. Miyazaki Y, Ichikawa I. Role of the angiotensin receptor in the development of the mammalian kidney and urinary tract. *Comp Biochem Physiol A Mol Integr Physiol* 2001;**128**:89–97.

34. Ruster C, Franke S, Wenzel U, Schmidthaupt R, Fraune C, Krebs C, et al. Podocytes of AT$_2$ receptor knockout mice are protected from angiotensin II-mediated RAGE induction. *Am J Nephrol* 2011;**34**:309–17.

35. Hashimoto N, Maeshima Y, Satoh M, Odawara M, Sugiyama H, Kashihara N, et al. Overexpression of angiotensin type 2 receptor ameliorates glomerular injury in a mouse remnant kidney model. *Am J Physiol Renal Physiol* 2004;**286**:F516–25.

36. von Bohlen und Halbach O, Walther T, Bader M, Albrecht D. Genetic deletion of angiotensin AT$_2$ receptor leads to increased cell numbers in different brain structures of mice. *Regul Pept* 2001;**99**:209–16.

37. Pawlowski TL, Heringer-Walther S, Cheng CH, Archie JG, Chen CF, Walther T, et al. Candidate *Agtr2* influenced genes and pathways identified by expression profiling in the developing brain of *Agtr2(−/y)* mice. *Genomics* 2009;**94**:188–95.

38. Okuyama S, Sakagawa T, Chaki S, Imagawa Y, Ichiki T, Inagami T. Anxiety-like behavior in mice lacking the angiotensin II type-2 receptor. *Brain Res* 1999;**821**:150–9.

39. Watanabe T, Hashimoto M, Okuyama S, Inagami T, Nakamura S. Effects of targeted disruption of the mouse angiotensin II type 2 receptor gene on stress-induced hyperthermia. *J Physiol* 1999;**515**(Pt 3):881–5.

40. Sakagawa T, Okuyama S, Kawashima N, Hozumi S, Nakagawasai O, Tadano T, et al. Pain threshold, learning and formation of brain edema in mice lacking the angiotensin II type 2 receptor. *Life Sci* 2000;**67**:2577–85.

Chapter 4

AT$_2$ Receptor Signaling: Solved and Unsolved

Chiara Recarti,* Serge P. Bottari[†,‡]

*CARIM School for Cardiovascular Diseases, Maastricht University, Maastricht, The Netherlands, [†]Laboratory for Fundamental and Applied Bioenergetics, INSERM U1055, Université Grenoble—Alpes, Grenoble, France, [‡]Radioanalysis Unit, Institute for Biology and Pathology, CHU de Grenoble, Grenoble, France

INTRODUCTION

Thanks to the availability of radiolabeled angiotensin II (Ang II), its receptors were identified as early as 1971.[1] Although there had been several indications for the occurrence of angiotensin receptor subtypes early on, it was not until 1989 that two receptor subtypes, AT$_1$ and AT$_2$, were identified and characterized[2] thanks to the availability of novel selective nonpeptidic analogs. Interestingly, in adult organisms, the AT$_2$ receptor (AT$_2$R) expression was found to be limited to a few tissues and organs and the more ubiquitous AT$_1$ receptor (AT$_1$R) was initially found to mediate most if not all known effects of Ang II.[3–6]

The first hint that AT$_2$R may elicit distinct biological responses came from the observation that, as opposed to AT$_1$Rs, their AT$_2$ counterparts appeared to signal through G protein-independent pathways in certain tissues.[7] Some of these pathways were subsequently identified as the activation of phosphotyrosine phosphatase (PTP) activity,[8–10] the inhibition of the ANP/BNP receptor guanylate cyclase (GC-A/NPRA),[8] and modulation of T-type calcium currents.[11] Activation of PTP appeared to be involved in the inhibition of GC-A/NPRA catalytic activity,[12] inactivation of the insulin receptor,[13] and most probably the inhibition of cell proliferation[14–16] through the inhibition of the MAP kinases ERK1/2.[17,18]

The major PTP involved in AT$_2$R-mediated kinase inhibition appears to be SHP-1, which has been reported to inhibit SIRPα1 and IRS-1, EGFR, Pyk2, JNK, PI3K, and Akt kinases,[13,19–21] NF-κB through stabilization and phosphorylation of IκB[22] and cyclin D1 expression and cyclin D1-dependent cdk activity.[16] SHP-1 also appears to be responsible for AT$_2$R-mediated inhibition of the small G protein RhoA and thereby of smooth muscle relaxation.[23]

Interestingly, SHP-1 is recruited by the AT$_2$R through an adaptor protein ATIP1,[24] a member of a new family of seven-transmembrane receptor adaptor proteins encoded by the tumor suppressor gene *MTUS1*.[25] These AT$_2$R binding proteins (ATBP/ATIP)[26] appear to mediate a large number of AT$_2$R-mediated responses including metabolic, antiproliferative, proapoptotic, anti-inflammatory, vasodilatory, and cognitive effects.[23,25] Interestingly, AT$_2$R mutations resulting in defective interaction with ATIP have been reported in cases of mental retardation[27] and ATIP3 appears to inhibit tumor metastasis and is a prognostic marker in breast[28] and bladder cancer.[29]

These observations further strengthen the original hypothesis that the AT$_2$R mostly signals through G protein-independent pathways[7] involving PTP activation.[8] Indeed, ATIP proteins bypass the G proteins and serve as docking proteins allowing the recruitment of SHP-1 to the C-terminal tail of the AT$_2$R as well as its interaction with the microtubules involved in AT$_2$R translocation from the Golgi to the plasma membrane[26] and potentially its oligomerization, which has been reported to allow ligand-independent signaling.[30]

Other proteins have also been reported to directly interact with AT$_2$R, including ErbB3,[31] PLZF,[32] CNK1,[33] NHE6,[34] and TIMP3.[35] PLZF, promyelocytic leukemia zinc finger protein, is another unexpected AT$_2$R interacting protein.[32] Interestingly, the transcription factor PLZF has also been described as a direct binding partner of the recently discovered (pro)renin receptor.[36] In the initial study, PLZF was found to bind and to activate p85α, the regulatory subunit of phosphatidylinositol-3 kinase (PI3K). p85α further activates downstream pathways, including Akt/PKB and p70[S6K] leading to growth and protein synthesis in cardiomyocytes.[32,37] In a recent study, the same group showed that this pathway involves the transcription of the cardiac-specific transcription factor GATA4. They also observed that PLFZ/AT$_2$R mediates activation of p70[S6K] and inhibition of GSK3β, which favors nuclear accumulation of GATA4.[38] Together, these data suggest that in this particular model, the AT_2R mediates and contributes to hypertrophy (see also the introductory chapter).

Interaction of multiple proteins other than heterotrimeric G proteins, named GIPs,[39] has since been reported for many G protein-coupled receptors (GPCRs) and has been shown to play a major role in GPCR trafficking, endocytosis, and recycling as well as signaling.[40] As such, the AT_2R has been among the very first examples of GPCR able to bypass G proteins for signaling.

The apparent contradictory reports regarding the G protein dependency of AT_2R signal transduction therefore comply with the current general concept of GPCR signaling and contribute to the understanding of the diversity of biological responses which have been reported to be elicited by this receptor and which will be briefly described in the next paragraphs.

Another factor that has been responsible for much of the confusion regarding AT_2R signaling was the idea that the two first AT_2R-selective ligands CGP42112 and PD123319 were both antagonists. Whereas PD123319 is indeed an antagonist, though not highly selective, CGP42112 is a partial agonist.[5,41–45] The use of CGP42112 as an antagonist, especially in competition with Ang II, has therefore led to confusing results.

The recent development of well-characterized stable, orally active high-affinity and highly selective nonpeptidic AT_2R agonists, among which most notably compound 21,[46] has greatly facilitated the detailed investigation of AT_2R-mediated responses and opens new perspectives for the therapeutic targeting of AT_2R.[47] Similarly, novel AT_2R antagonists, like compound 38 derived from the same lead structure and displaying a much higher affinity than PD123319,[48] will contribute to further confirm AT_2R-induced responses.

It should nevertheless be kept in mind that the novel agonists may not necessarily trigger all the pathways that are activated by Ang II through AT_2R. Indeed, it has, for example, been reported that compound 21 can induce vasorelaxation through AT_2R-independent mechanisms.[49] This apparently puzzling observation is in fact not that surprising as several synthetic "biased" GPCR ligands have recently been found to display "functional selectivity",[50] a property that is probably due to their ability to induce "asymmetrical dimerization".[51] More apparently conflicting data can therefore be expected with the advent of novel classes of synthetic ligands.

AT_2R DIMERIZATION

Further complexity to AT_2R signaling has been added with the discovery of homo- and heterodimerization of the AT_2R (Figure 1). Miura and colleagues demonstrated the ability of AT_2R to form homodimers. The authors showed that constitutive active homooligomerization is due to disulfide bounding between two conserved cysteine residues independent of ligand stimulation and receptor conformation and that it induces constitutive cell signaling.[30] In 2001, AbdAlla and colleagues reported, for the first time, a direct binding of the AT_2R to the AT_1R.[52] The authors demonstrated that AT_2R and AT_1R form heterodimers in PC12 cells, in rat fetal fibroblasts and human myometrial biopsies. This study showed that this heterodimerization leads to inhibition of AT_1R signaling independently of Ang II binding to the AT_2R. Moreover, the authors demonstrated functional relevance of heterodimerization, reporting decreased AT_1R-AT_2R dimerization in myometrial tissue from pregnant women and increased Ang II responsiveness and showing that direct binding of AT_2R to the AT_1R inhibits AT_1R activation. Interestingly, it was recently reported that Ang(3-4 Val-Tyr), an Ang II-derived peptide, induces the

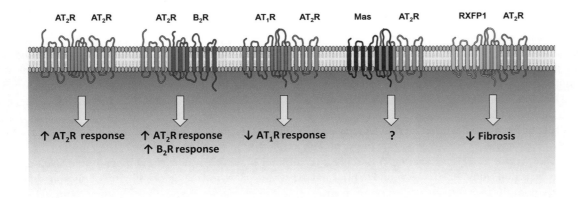

FIGURE 1 AT_2R homodimerization and heterodimerization. Abbreviations: AT_2R, angiotensin AT_2 receptor; B_2R, B_2 bradykinin receptor; AT_1R, angiotensin AT_1 receptor; Mas, Mas receptor; RXFP1, relaxin family peptide receptor 1.

dissociation of AT$_1$R-AT$_2$R dimers in renal proximal tubular cells in SHR, thereby blocking the Ang II-mediated activation of ouabain-resistant Na$^+$-ATPase.[53] These data indicate that some ligands may act by affecting receptor heterodimerization.

In addition, as shown in 2006 by Abadir and colleagues, the AT$_2$R can also form heterodimers with the B$_2$ bradykinin receptor (B$_2$R). This dimerization was reported to occur in function of AT$_2$R and B$_2$R expression levels and independently of ligand binding. Furthermore, AT$_2$R-B$_2$R dimerization induces phosphorylation/dephosphorylation signaling, which leads to dephosphorylation of p38 and extracellular signal-regulated kinase (ERK)1/2 (p42/p44) MAP kinases as well as increased NO and cGMP production.[54] Similarly, AT$_2$R has been postulated to transinactivate the insulin receptor by the formation of heterodimers with it through ATIP.[24]

Recently, Chow and colleagues studying the interaction between relaxin and AT$_2$R, reported the formation of heterodimers of AT$_2$R and relaxin family peptide receptor 1 (RXFP1), which mediates relaxin antifibrotic effects.[55]

Finally, Villela and colleagues also studied the interaction between AT$_2$R and Mas receptor, which is usually linked to the Ang (1-7) peptide but has been reported to mediate Ang II responses as well (see Chapter 24). Using fluorescence resonance energy transfer in transfected HEK 293 cells, the authors showed the ability of AT$_2$R to form heterodimers with the Mas receptor.[56] The functional significance of these dimers remains however elusive.

PROLIFERATION AND DIFFERENTIATION

The observation that the AT$_2$R was able to activate PTP[8] raised the question of its potential role in regulating cell proliferation and differentiation.[5] This hypothesis was further strengthened by the transient expression of AT$_2$R in healing skin[57,58] and embryogenesis.[59,60] The first evidence of the antiproliferative effects of the AT$_2$R was reported by Unger's group in 1995. They showed that AT$_2$R inhibits the proliferation of coronary endothelial cells from SHR as opposed to their AT$_1$R counterpart.[14] These observations were rapidly confirmed by Dzau's group using AT$_2$R gene transfer in balloon-injured rat carotid arteries.[61] Furthermore, the same group showed that AT$_2$R exerts an inhibitory effect on cell growth *via* inhibition of ERK 1/2 activation and consequent inhibition of signal transducers and activators of transcription (STAT), suggesting a negative cross talk with growth receptors.[62] The antiproliferative properties were subsequently confirmed and shown to be associated with differentiating properties in the neuronal cell line PC12W.[15] The mechanisms involved were subsequently shown to involve not only PTP activation[17] as hypothesized previously, but also activation of the zinc finger homeodomain enhancer protein (Zfhep) gene that encodes for a transcription factor inducing differentiation.[63] These observations raised the issue of the AT$_2$R as a potential therapeutic target in cardiovascular disease.[64,65] Other potential therapeutic issues are obviously cancer in AT$_2$R-expressing tumor cells. A few preliminary studies suggest that AT$_2$R knockout increases cancer cell proliferation[66] and treatment of nontumoral prostate cells with compound 21 significantly reduces their number, suggesting that AT$_2$R stimulation might be an approach for prostate cancer prevention.[67]

The AT$_2$R has also been reported to play a role in cell differentiation in nonneuronal cells.[68] An interesting example is that of human pancreatic progenitor cells where the AT$_2$R plays a crucial role in their differentiation into islet-like cell clusters and, more particularly, into β-cells.[69] This effect, which has been confirmed in mouse embryos,[70] opens the way for the use of AT$_2$R agonists for the generation of β-cells for transplantation in diabetic patients. Interestingly, the AT$_2$R also appears to mediate insulinotropic effects of Ang II[71] in agreement with previous reports on the insulin-sensitizing properties of AT$_1$R antagonists.[72,73] With regard to insulin sensitivity, recent reports also point towards the beneficial role of AT$_2$R in adipose tissue where they have been reported to mediate preadipocyte differentiation to well-differentiated small adipocytes expressing high levels of PPARγ, adiponectin, and prostacyclin. Conversely, AT$_2$R appears to inhibit bone marrow-derived mesenchymal stem cell differentiation to adipocytes.[74] On the other hand, the AT$_2$R has been reported to restore normal adipocyte morphology and to improve insulin sensitivity in mature adipocytes in diet-induced insulin-resistant rats.[75] Taken together, these observations indicate that the AT$_2$R plays a beneficial role in glucose and lipid metabolism as opposed to the AT$_1$R and that selective AT$_2$R agonists may therefore become new candidate drugs for the treatment of metabolic syndrome.

As mentioned earlier, AT$_2$R activation leads to neurite outgrowth in neuronal cells.[15] Unger's group showed that AT$_2$R affects neuronal differentiation and nerve regeneration *via* regulation of the cytoskeleton. Thus, in PC12W cells, AT$_2$R stimulation induces an increased expression of polymerized beta-tubulin and MAP2 but a reduction in MAP1B[76] and neurofilament M.[77] This process shares some common pathways with other AT$_2$R-mediated effects, e.g. interaction with ATIP resulting in the activation of SHP-1 and of methyl methanesulfonate-sensitive 2 enzyme (MMS2).[78] In NG108-15 cells, AT$_2$R-mediated neuronal differentiation involves inhibition of PKCα and of p21[RAS79] and sustained ERK 1/2 activation that is crucial for neurite outgrowth.[80] It has been shown later that ERK 1/2 activation and the subsequent neurite outgrowth are mediated by phosphorylation of tyrosine kinase receptor TrkA in NG108-15 cells[81] and in primary neurons.[82] Another pathway that can be activated in parallel to ERK 1/2 phosphorylation[83] is the NO/cGMP signaling: it has been reported that

AT$_2$R-mediated activation of nNOS can lead to increased cGMP production resulting in neurite outgrowth in NG108-15 cells.[84] This does not exclude that other pathways and players could be involved in this mechanism like PLZF and PPARγ.[85]

Apart from these *in vitro* effects, the AT$_2$R has been reported to exert neurotrophic effects *in vivo* as well in optic and sciatic nerve lesion[86,87] and striatal ischemia-induced injury models.

Some of these neurotrophic effects may in fact be mediated by T-type calcium channels. These channels are expressed not only during fetal development but also after birth[88] and have been postulated to promote axonal and dendritic outgrowth.[89,90] Apart from its neurotrophic properties, T-type channels have also been postulated to be involved in cardiac hypertrophy[91] and in skeletal muscle regeneration and repair.[92] Interestingly, the AT$_2$R has been reported to be expressed in healing tissues and may therefore affect remodeling and regeneration at least partly through the modulation of these channels.

Finally, T-type channels are known to play a critical role in neuronal excitability, thereby triggering among others long-term potentiation, which is one of the phenomena underlying synaptic plasticity and a major mechanism in memory and learning.[11,41]

These observations suggest that Ang II also plays a role in neuronal differentiation and regeneration, which may find therapeutic applications through the development of novel, potentially "biased" AT$_2$R agonists devoid of peripheral side effects.

APOPTOSIS

As indicated above, the AT$_2$R has been reported to be involved in apoptosis, an effect that appears to strongly depend on the experimental conditions.

AT$_2$R regulation of apoptosis can occur via different pathways. AT$_2$R stimulation can induce tyrosine phosphatase activation, such as mitogen-activated protein kinase phosphatase-1 (MKP-1) and ERK 1/2 inactivation, causing dephosphorylation of Bcl-2 and increased Bax-mediated proapoptotic actions.[93–96]

An alternative pathway for AT$_2$R-mediated regulation of apoptosis includes the activation of caspases. Thus, AT$_2$R stimulation has been reported to upregulate ceramide synthesis causing activation of stress kinases and caspases leading to apoptosis.[95,97–101]

Other studies reported an antiapoptotic role of AT$_2$R especially in pathological conditions. Recently, Namsolleck and colleagues investigated the role of AT$_2$R on neuroprotection and neurite outgrowth in a model of spinal cord injury in mice. This study demonstrated that stimulation of AT$_2$R with compound 21 not only induces increased expression of brain-derived neurotrophic factor (BDNF) and of the neurotrophin receptors TrkA and TrkB but also upregulates the expression of the antiapoptotic Bcl-2 in primary mouse neurons and astrocytes.[82] This study is in agreement with previous reports on the effects of AT$_2$R on neurite outgrowth in NG108-15 cells.[80] Interestingly, T-type channels also appear to be involved in motoneuron apoptosis[102] and may thus participate in AT$_2$R-dependent regulation of cell survival.

Kaschina and colleagues demonstrated antiapoptotic and anti-inflammatory effects of AT$_2$R stimulation with compound 21 leading to protective effects after myocardial infarction.[103] The authors showed that AT$_2$R-mediated antiapoptosis after myocardial infarction was achieved by rescuing the expression of p38 and ERK 1/2 MAPK and decreasing Fas ligand (FasL) and caspase-3 expression.

Conversely, Tan et al. described a proapoptotic effect of AT$_2$R in vascular smooth muscle. This study reported that AT$_2$R mediates inducible GATA-6 expression via activation of mitogen-activated protein kinase kinase (MEK)—ERK1/2 and c-Jun N-terminal kinase (JNK). GATA-6 in turn activates FasL promoter and FasL expression leading to apoptosis via caspase 8.[104] Similar effects have been reported in response to Ang II stimulation of the intestinal epithelial cell line Caco-2. The proapoptotic effect of AT$_2$R appeared to involve GATA-6 and Bax.[105] This again indicates that responses to AT$_2$R stimulation appear to depend on cell type or tissue and experimental conditions.

INFLAMMATION

The first evidence regarding an anti-inflammatory effect of the AT$_2$R was probably provided by Wu et al. who reported its ability to inhibit MCP-1 expression, NF-κB binding to DNA and IκB degradation.[22] This early observation was followed by reports from Victorino's group who described the role of AT$_2$R in reducing inflammation-induced fluid leak.[106,107]

The two main mechanisms through which the AT$_2$R mediates its anti-inflammatory effects are inhibition of NF-κB activity[22,108] and inhibition of oxidative stress.[109,110]

Stimulation of the AT$_2$R with compound 21 induces anti-inflammatory effects via reduction of cytokine levels *in vitro* and *in vivo*. Rompe and colleagues showed that direct AT$_2$R stimulation reduces TNF-α-induced IL-6 levels in a dose-dependent manner by activation of protein phosphatases, CYP-dependent epoxidation of arachidonic acid (AA) to epoxyeicosatrienoic acids (EET) and inhibition of NF-κB activity and translocation.[108] Through this mechanism, AT$_2$R counteracts AT$_1$R proinflammatory effects, which consist in CYP-dependent hydroxylation of arachidonic acid to 20-hydroxyeicosatetraenoic acid and NF-κB activation.

The second main mechanism of AT$_2$R-mediated inflammation involves the inhibition of oxidative stress. Dandapat et al. hypothesized that AT$_2$R-mediated anti-inflammation counteracts AT$_1$R-mediated oxidative stress,[111–114] reducing prooxidant signals *via* inhibition of NADPH oxidase expression and ROS generation, leading to downregulation of p38 and ERK 1/2 MAP kinase phosphorylation.[115] It is known that during oxidative stress, increased ROS concentration reduces the amount of NO by oxidizing it to ONOO⁻.[116,117] In the two-kidney/one-clip rat model, it has been reported that AT$_2$R stimulation reverses the early renal inflammation, which is characterized by increased TNF-α, IL-6, and TGF-β and decreased NO and cGMP concentrations.[118] Recently, Dhande et al. reported that AT$_2$R-mediated anti-inflammatory action in the kidney is characterized by an increased production of the anti-inflammatory cytokine IL-10 through NO signaling.[119]

Further to these main mechanisms, there is increasing evidence of AT$_2$R anti-inflammatory effects via cellular immunity mechanisms. Curato et al. demonstrated an immune regulatory and cardioprotective action of AT$_2$R in the context of ischemic heart injury. This mechanism involves the downregulation of the expression of proinflammatory cytokines and sustained IL-10 production mediated, at least in part, via a CD8⁺AT$_2$R⁺ T-cell population. The authors characterized the cardioprotective CD8⁺AT$_2$R⁺ T-cell population which displayed upregulated IL-10 and downregulated IL-2 and INF-γ expression, as compared to CD8⁺AT$_2$R⁻ T cells which increased in response to ischemic cardiac injury.[120] Another recent study supports the immune regulatory role of the AT$_2$R. The authors showed that AT$_2$R stimulation leads to inhibition of T-cell recruitment and modulation of the differentiation of naive T cells into proinflammatory T helper Th1 and Th17 subsets while promoting differentiation into anti-inflammatory T regulatory cells.[121]

FIBROSIS

As early as 1997, Carretero's group hypothesized that the cardioprotective effects of AT$_1$R antagonists were at least in part mediated by the AT$_2$R.[122] They later confirmed this and reported AT$_2$R to play an important role in the inhibition of fibrosis in a postmyocardial infarction model.[123]

Subsequently, other authors reported the involvement of AT$_2$R in the inhibition of fibrosis in the liver,[124] in the pancreas,[125] and in the kidney.[126]

Several studies concomitantly reported antifibrotic and anti-inflammatory effects suggesting, as expected, a link between the two mechanisms. Recent studies indicate that AT$_2$R-mediated antifibrotic effects occur through the regulation of metalloproteinases (MMPs) and of their inhibitors (TIMP), which are key elements of the metabolic balance of the extracellular matrix.[115,127–129] Dandapat et al., investigating the role of AT$_2$R in the atherosclerotic plaque, found that AT$_2$R overexpression reduces MMP2 and MMP9 expression and activity as well as collagen accumulation. The authors hypothesized that the upregulation of AT$_2$R leads to anti-inflammatory effects, decrease in collagen formation, and downregulation of MMP expression through the inhibition of NADPH oxidase with consequent decrease of ROS generation and negative regulation of redox signaling events, e.g. p38MAPK phosphorylation and redox-sensitive transcription factors.[115] Brassard et al. showed opposite roles of AT$_1$R and AT$_2$R in vascular remodeling of resistance arteries, demonstrating an AT$_2$R-mediated downregulation of MMP2 and increased elastin, counteracting the AT$_1$R-mediated upregulation of MMP2 and decrease in TIMP2 activity.[127] In agreement with these findings, Jing et al. confirmed that AT$_2$R opposes the effects elicited by AT$_1$R signaling and causes a marked reduction in MMP2 levels in engineered rat smooth muscle cells expressing AT$_2$R in a conditional expression system.[128] In 2013, a study of Kljajic et al. supported a regulatory role of AT$_2$R on MMP expression, demonstrating that stimulation of the receptor with the peptide agonist CGP42112 causes a reduction of MMP2 and MMP9 in a setting of atherosclerosis,[130] and recently, Lauer et al. showed protective effects of long-term AT$_2$R-selective stimulation with compound 21, in a rat model of myocardial infarction.[129] This study demonstrated for the first time that AT$_2$R stimulation drastically downregulates TGF-β1 expression after myocardial infarction and showed an inhibition of IL-1α-induced TGF-β1 expression in primary cardiac fibroblasts. In addition, the proteolytic activity of MMP9 induced by myocardial infarction was attenuated in a dose-dependent fashion by compound 21 in cardiac fibroblasts. This study also showed that AT$_2$R stimulation leads to the regulation of the expression of the MMP9 inhibitor, TIMP1. Moreover, the authors suggest the existence of a feedforward loop of MMPs and TGF-β1 in cardiac remodeling, characterized by a TGF-β1-mediated upregulation of MMP's activity, MMP9- and MMP2-mediated cleavage and activation of TGF-β1 with subsequent activation of TIMP1 transcription.[129]

Recently, Samuel's group reported that the AT$_2$R is required for the effects of relaxin on renal interstitial fibrosis, which involves the disruption of the profibrotic activity of TGF-β1 (Figure 2). This observation appears to be due to the formation of heterodimers comprising the AT$_2$R and the relaxin RXFP1 receptor.[55]

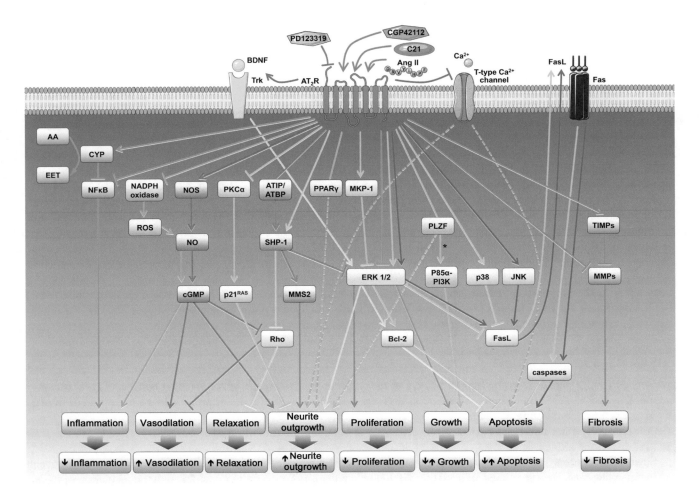

FIGURE 2   Main AT$_2$R-mediated signaling pathways. Stimulation of AT$_2$R results in the activation or inhibition of different signaling pathways leading, among others, to anti-inflammation, vasodilation, relaxation, neurite outgrowth, antifibrosis, regulation of growth, and modulation of apoptosis. This figure does not take into account the impact of homo- and heterodimerization on signaling processed (see Figure 1). Please notice that arrows' ends represent activation, "T-shaped" edges mean inhibition, and green boxes represent final cellular responses. *Growth factors are necessary for p85αPI3K activation. Abbreviations: AT$_2$R, angiotensin AT$_2$ receptor; Ang II, angiotensin II; C21, compound 21 (nonpeptide AT$_2$R agonist); CGP42112, peptide AT$_2$R agonist; PD123319, nonpeptide AT$_2$R antagonist; Trk, tyrosine kinase receptor; BDNF, brain-derived neurotrophic factor; Fas, Fas cell surface death receptor; FasL, Fas ligand; CYP, cytochrome P450; AA, arachidonic acid; EET, epoxyeicosatrienoic acid; NF-κB, nuclear factor-kappa-B; NADPH oxidase, nicotinamide adenine dinucleotide phosphate oxidase; ROS, reactive oxygen species; NOS, nitric oxide synthase; NO, nitric oxide; cGMP, cyclic guanosine monophosphate; PKCα, protein kinase C, alpha; p21^{RAS}, Ras family of small GTP-binding proteins; ATIP/ATBP, AT$_2$ receptor-interacting protein (or AT$_2$ receptor-binding protein); SHP-1, Src homology region 2 domain-containing phosphatase-1; MMS2, methyl methanesulfonate-sensitive 2; Rho, small G protein; PPARγ, peroxisome proliferator-activated receptor gamma; MKP-1, mitogen-activated protein kinase phosphatase-1; ERK1/2 (or p42/p44), extracellular signal-regulated kinase; Bcl-2, apoptosis regulator; PLZF, promyelocytic leukemia zinc finger protein; p85αPI3K, phosphatidylinositol 3-kinase regulatory subunit alpha; p38, mitogen-activated protein kinase; JNK, JUN N-terminal kinase; TIMPs, tissue inhibitor of metalloproteinases; MMPs, matrix metalloproteinases.

SOLVED AND UNSOLVED ISSUES

As it appears from this short overview, major progress has been made in the understanding of the AT$_2$R's pathophysiological significance since its first description 25 years ago.[2] After having been considered to be merely a "binding site" essentially due to the fact that it did not mediate obvious acute responses of Ang II, the AT$_2$R turned out to mediate some unexpected actions of this hormone. It was also one of the first examples of the seven-transmembrane domain receptor family found to mediate many of its actions independently of G proteins[7] through novel docking proteins[24] which had recently been described for two metabotropic glutamate and for the 5-HT$_2$ receptor.[39] This signaling pathway involving GIPs has now been shown to be essential to many GPCRs[40] and has opened new perspectives regarding the pathophysiological actions of Ang II as well as its potential interest as a therapeutic target in unexpected areas such as cancer, metabolism, and neuroprotection. The issue whether or not AT$_2$R signals through G proteins can thus be considered as solved as well as that of the AT$_2$R's functional relevance.

Among the remaining and challenging issues are the discovery and unraveling of the GIP-dependent signaling pathways and of their functional significance. This may take time as the AT$_2$R can be expected to trigger different pathways involving different GIPs and different GIP partners according to the tissue and cell type. Understanding of these pathways will allow further determination of the pathophysiological functions of Ang II and pave the way for novel therapeutic strategies.

Another challenge which is directly linked to the previous one, is the search for novel "biased" AT$_2$R ligands such as compound 21, which could selectively activate or block specific pathways.[50] Such compounds are the obvious successors of the subtype-selective ligands and should constitute truly innovative drugs with a high target response rather than target protein specificity.

The work on the AT$_2$R during recent years has thus opened a new era in the field of Ang II research, which appears to progressively drift away from its initial cardiovascular field into a multitude of research areas where proliferation, differentiation, apoptosis, inflammation, and fibrosis are involved paving the way for the various potential novel therapeutic indications of selective AT$_2$R agonists.

REFERENCES

1. Baudouin M, Meyer P, Worcel M. Specific binding of 3H-angiotensin II in rabbit aorta. *Biochem Biophys Res Commun* 1971;**42**(3):434–40.
2. Whitebread S, Mele M, Kamber B, de Gasparo M. Preliminary biochemical characterization of two angiotensin II receptor subtypes. *Biochem Biophys Res Commun* 1989;**163**(1):284–91.
3. Wong PC, Hart SD, Zaspel AM, Chiu AT, Ardecky RJ, Smith RD, et al. Functional studies of nonpeptide angiotensin II receptor subtype-specific ligands: DuP 753 (AII-1) and PD123177 (AII-2). *J Pharmacol Exp Ther* 1990;**255**(2):584–92.
4. Balla T, Baukal AJ, Eng S, Catt KJ. Angiotensin II receptor subtypes and biological responses in the adrenal cortex and medulla. *Mol Pharmacol* 1991;**40**(3):401–6.
5. Bottari SP, de Gasparo M, Steckelings UM, Levens NR. Angiotensin II receptor subtypes: characterization, signalling mechanisms, and possible physiological implications. *Front Neuroendocrinol* 1993;**14**(2):123–71.
6. De Gasparo M, Catt KJ, Inagami T, Wright JW, Unger T. International union of pharmacology. XXIII. The angiotensin II receptors. *Pharmacol Rev* 2000;**52**(3):415–72.
7. Bottari SP, Taylor V, King IN, Bogdal Y, Whitebread S, de Gasparo M. Angiotensin II AT2 receptors do not interact with guanine nucleotide binding proteins. *Eur J Pharmacol* 1991;**207**(2):157–63.
8. Bottari SP, King IN, Reichlin S, Dahlstroem I, Lydon N, de Gasparo M. The angiotensin AT2 receptor stimulates protein tyrosine phosphatase activity and mediates inhibition of particulate guanylate cyclase. *Biochem Biophys Res Commun* 1992;**183**(1):206–11.
9. Nahmias C, Cazaubon SM, Briend-Sutren MM, Lazard D, Villageois P, Strosberg AD. Angiotensin II AT2 receptors are functionally coupled to protein tyrosine dephosphorylation in N1E-115 neuroblastoma cells. *Biochem J* 1995;**306**(Pt 1):87–92.
10. Brechler V, Reichlin S, De Gasparo M, Bottari SP. Angiotensin II stimulates protein tyrosine phosphatase activity through a G-protein independent mechanism. *Recept Channels* 1994;**2**(2):89–98.
11. Buisson B, Bottari SP, de Gasparo M, Gallo-Payet N, Payet MD. The angiotensin AT2 receptor modulates T-type calcium current in non-differentiated NG108-15 cells. *FEBS Lett* 1992;**309**(2):161–4.
12. Brechler V, Levens NR, De Gasparo M, Bottari SP. Angiotensin AT2 receptor mediated inhibition of particulate guanylate cyclase: a link with protein tyrosine phosphatase stimulation? *Recept Channels* 1994;**2**(2):79–87.
13. Elbaz N, Bedecs K, Masson M, Sutren M, Strosberg AD, Nahmias C. Functional trans-inactivation of insulin receptor kinase by growth-inhibitory angiotensin II AT2 receptor. *Mol Endocrinol* 2000;**14**(6):795–804.
14. Stoll M, Steckelings UM, Paul M, Bottari SP, Metzger R, Unger T. The angiotensin AT2-receptor mediates inhibition of cell proliferation in coronary endothelial cells. *J Clin Invest* 1995;**95**(2):651–7.
15. Meffert S, Stoll M, Steckelings UM, Bottari SP, Unger T. The angiotensin II AT2 receptor inhibits proliferation and promotes differentiation in PC12W cells. *Mol Cell Endocrinol* 1996;**122**(1):59–67.
16. Liakos P, Bourmeyster N, Defaye G, Chambaz EM, Bottari SP. ANG II AT1 and AT2 receptors both inhibit bFGF-induced proliferation of bovine adrenocortical cells. *Am J Physiol* 1997;**273**(4 Pt 1):C1324–34.
17. Bedecs K, Elbaz N, Sutren M, Masson M, Susini C, Strosberg AD, et al. Angiotensin II type 2 receptors mediate inhibition of mitogen-activated protein kinase cascade and functional activation of SHP-1 tyrosine phosphatase. *Biochem J* 1997;**325**(Pt 2):449–54.
18. Cui T, Nakagami H, Iwai M, Takeda Y, Shiuchi T, Daviet L, et al. Pivotal role of tyrosine phosphatase SHP-1 in AT2 receptor-mediated apoptosis in rat fetal vascular smooth muscle cell. *Cardiovasc Res* 2001;**49**(4):863–71.
19. Cui T-X, Nakagami H, Nahmias C, Shiuchi T, Takeda-Matsubara Y, Li J-M, et al. Angiotensin II subtype 2 receptor activation inhibits insulin-induced phosphoinositide 3-kinase and Akt and induces apoptosis in PC12W cells. *Mol Endocrinol* 2002;**16**(9):2113–23.
20. Matsubara H, Shibasaki Y, Okigaki M, Mori Y, Masaki H, Kosaki A, et al. Effect of angiotensin II type 2 receptor on tyrosine kinase Pyk2 and c-Jun NH2-terminal kinase via SHP-1 tyrosine phosphatase activity: evidence from vascular-targeted transgenic mice of AT2 receptor. *Biochem Biophys Res Commun* 2001;**282**(5):1085–91.

21. Shibasaki Y, Matsubara H, Nozawa Y, Mori Y, Masaki H, Kosaki A, et al. Angiotensin II type 2 receptor inhibits epidermal growth factor receptor transactivation by increasing association of SHP-1 tyrosine phosphatase. *Hypertension* 2001;**38**(3):367–72.

22. Wu L, Iwai M, Li Z, Shiuchi T, Min L-J, Cui T-X, et al. Regulation of inhibitory protein-kappaB and monocyte chemoattractant protein-1 by angiotensin II type 2 receptor-activated Src homology protein tyrosine phosphatase-1 in fetal vascular smooth muscle cells. *Mol Endocrinol* 2004;**18**(3):666–78.

23. Guilluy C, Rolli-Derkinderen M, Loufrani L, Bourgé A, Henrion D, Sabourin L, et al. Ste20-related kinase SLK phosphorylates Ser188 of RhoA to induce vasodilation in response to angiotensin II Type 2 receptor activation. *Circ Res* 2008;**102**(10):1265–74.

24. Nouet S, Amzallag N, Li J-M, Louis S, Seitz I, Cui T-X, et al. Trans-inactivation of receptor tyrosine kinases by novel angiotensin II AT2 receptor-interacting protein. ATIP. *J Biol Chem* 2004;**279**(28):28989–97.

25. Rodrigues-Ferreira S, Nahmias C. An ATIPical family of angiotensin II AT2 receptor-interacting proteins. *Trends Endocrinol Metab* 2010;**21**(11):684–90.

26. Wruck CJ, Funke-Kaiser H, Pufe T, Kusserow H, Menk M, Schefe JH, et al. Regulation of transport of the angiotensin AT2 receptor by a novel membrane-associated Golgi protein. *Arterioscler Thromb Vasc Biol* 2005;**25**(1):57–64.

27. Rodrigues-Ferreira S, le Rouzic E, Pawlowski T, Srivastava A, Margottin-Goguet F, Nahmias C. AT2 receptor-interacting proteins ATIPs in the Brain. *Int J Hypertens* 2013;**2013**:513047.

28. Molina A, Velot L, Ghouinem L, Abdelkarim M, Bouchet BP, Luissint A-C, et al. ATIP3, a novel prognostic marker of breast cancer patient survival, limits cancer cell migration and slows metastatic progression by regulating microtubule dynamics. *Cancer Res* 2013;**73**(9):2905–15.

29. Rogler A, Hoja S, Giedl J, Ekici AB, Wach S, Taubert H, et al. Loss of MTUS1/ATIP expression is associated with adverse outcome in advanced bladder carcinomas: data from a retrospective study. *BMC Cancer* 2014;**14**:214.

30. Miura S-I, Karnik SS, Saku K. Constitutively active homo-oligomeric angiotensin II type 2 receptor induces cell signaling independent of receptor conformation and ligand stimulation. *J Biol Chem* 2005;**280**(18):18237–44.

31. Knowle D, Ahmed S, Pulakat L. Identification of an interaction between the angiotensin II receptor sub-type AT2 and the ErbB3 receptor, a member of the epidermal growth factor receptor family. *Regul Pept* 2000;**87**(1–3):73–82.

32. Senbonmatsu T, Saito T, Landon EJ, Watanabe O, Price Jr E, Roberts RL, et al. A novel angiotensin II type 2 receptor signaling pathway: possible role in cardiac hypertrophy. *EMBO J* 2003;**22**(24):6471–82.

33. Fritz RD, Radziwill G. The scaffold protein CNK1 interacts with the angiotensin II type 2 receptor. *Biochem Biophys Res Commun* 2005;**338**(4):1906–12.

34. Pulakat L, Cooper S, Knowle D, Mandavia C, Bruhl S, Hetrick M, et al. Ligand-dependent complex formation between the Angiotensin II receptor subtype AT2 and Na+/H+ exchanger NHE6 in mammalian cells. *Peptides* 2005;**26**(5):863–73.

35. Kang K-H, Park S-Y, Rho SB, Lee J-H. Tissue inhibitor of metalloproteinases-3 interacts with angiotensin II type 2 receptor and additively inhibits angiogenesis. *Cardiovasc Res* 2008;**79**(1):150–60.

36. Schefe JH, Menk M, Reinemund J, Effertz K, Hobbs RM, Pandolfi PP, et al. A novel signal transduction cascade involving direct physical interaction of the renin/prorenin receptor with the transcription factor promyelocytic zinc finger protein. *Circ Res* 2006;**99**(12):1355–66.

37. Funke-Kaiser H, Reinemund J, Steckelings UM, Unger T. Adapter proteins and promoter regulation of the angiotensin AT2 receptor–implications for cardiac pathophysiology. *J Renin-Angiotensin-Aldosterone Syst* 2010;**11**(1):7–17.

38. Wang N, Frank GD, Ding R, Tan Z, Rachakonda A, Pandolfi PP, et al. Promyelocytic leukemia zinc finger protein activates GATA4 transcription and mediates cardiac hypertrophic signaling from angiotensin II receptor 2. *PLoS One* 2012;**7**(4):e35632.

39. Bockaert J, Fagni L, Dumuis A, Marin P. GPCR interacting proteins (GIP). *Pharmacol Ther* 2004;**103**(3):203–21.

40. Maurice P, Guillaume J-L, Benleulmi-Chaachoua A, Daulat AM, Kamal M, Jockers R. GPCR-interacting proteins, major players of GPCR function. *Adv Pharmacol* 2011;**62**:349–80.

41. Buisson B, Laflamme L, Bottari SP, de Gasparo M, Gallo-Payet N, Payet MD. A G protein is involved in the angiotensin AT2 receptor inhibition of the T-type calcium current in non-differentiated NG108-15 cells. *J Biol Chem* 1995;**270**(4):1670–4.

42. Macari D, Bottari S, Whitebread S, De Gasparo M, Levens N. Renal actions of the selective angiotensin AT2 receptor ligands CGP 42112B and PD 123319 in the sodium-depleted rat. *Eur J Pharmacol* 1993;**249**(1):85–93.

43. Gao S, Park BM, Cha SA, Park WH, Park BH, Kim SH. Angiotensin AT2 receptor agonist stimulates high stretch induced-ANP secretion via PI3K/NO/sGC/PKG/pathway. *Peptides* 2013;**47**:36–44.

44. Brechler V, Jones PW, Levens NR, de Gasparo M, Bottari SP. Agonistic and antagonistic properties of angiotensin analogs at the AT2 receptor in PC12W cells. *Regul Pept* 1993;**44**(2):207–13.

45. Laredo J, Shah JR, Lu ZR, Hamilton BP, Hamlyn JM. Angiotensin II stimulates secretion of endogenous ouabain from bovine adrenocortical cells via angiotensin type 2 receptors. *Hypertension* 1997;**29**(1 Pt 2):401–7.

46. Wan Y, Wallinder C, Plouffe B, Beaudry H, Mahalingam AK, Wu X, et al. Design, synthesis, and biological evaluation of the first selective non-peptide AT2 receptor agonist. *J Med Chem* 2004;**47**(24):5995–6008.

47. Steckelings UM, Larhed M, Hallberg A, Widdop RE, Jones ES, Wallinder C, et al. Non-peptide AT2-receptor agonists. *Curr Opin Pharmacol* 2011;**11**(2):187–92.

48. Murugaiah AMS, Wu X, Wallinder C, Mahalingam AK, Wan Y, Sköld C, et al. From the first selective non-peptide AT(2) receptor agonist to structurally related antagonists. *J Med Chem* 2012;**55**(5):2265–78.

49. Verdonk K, Durik M, Abd-Alla N, Batenburg WW, van den Bogaerdt AJ, van Veghel R, et al. Compound 21 induces vasorelaxation via an endothelium- and angiotensin II type 2 receptor-independent mechanism. *Hypertension* 2012;**60**(3):722–9.

50. Luttrell LM. Minireview: more than just a hammer: ligand "bias" and pharmaceutical discovery. *Mol Endocrinol* 2014;**28**(3):281–94.

51. Maurice P, Kamal M, Jockers R. Asymmetry of GPCR oligomers supports their functional relevance. *Trends Pharmacol Sci* 2011;**32**(9):514–20.

52. AbdAlla S, Lother H, Abdel-tawab AM, Quitterer U. The angiotensin II AT2 receptor is an AT1 receptor antagonist. *J Biol Chem* 2001;**276**(43):39721–6.

53. Dias J, Ferrão FM, Axelband F, Carmona AK, Lara LS, Vieyra A. ANG-(3–4) inhibits renal Na+−ATPase in hypertensive rats through a mechanism that involves dissociation of ANG II receptors, heterodimers, and PKA. *Am J Physiol Renal Physiol* 2014;**306**(8):F855–63.

54. Abadir PM, Periasamy A, Carey RM, Siragy HM. Angiotensin II type 2 receptor-bradykinin B2 receptor functional heterodimerization. *Hypertension* 2006;**48**(2):316–22.

55. Chow BSM, Kocan M, Bosnyak S, Sarwar M, Wigg B, Jones ES, et al. Relaxin requires the angiotensin II type 2 receptor to abrogate renal interstitial fibrosis. *Kidney Int* 2014;**86**(1):75–85.

56. Villela DC, Munter L-M, Multhaup G, Mayer M, Benz V, Namsolleck P, et al. Evidence of a direct MAS-AT2 receptor dimerization. *J Hypertens* 2012;**30**:e117.

57. Viswanathan M, Saavedra JM. Expression of angiotensin II AT2 receptors in the rat skin during experimental wound healing. *Peptides* 1992;**13**(4):783–6.

58. Kimura B, Sumners C, Phillips MI. Changes in skin angiotensin II receptors in rats during wound healing. *Biochem Biophys Res Commun* 1992;**187**(2):1083–90.

59. Tsutsumi K, Strömberg C, Viswanathan M, Saavedra JM. Angiotensin-II receptor subtypes in fetal tissue of the rat: autoradiography, guanine nucleotide sensitivity, and association with phosphoinositide hydrolysis. *Endocrinology* 1991;**129**(2):1075–82.

60. Grady EF, Sechi LA, Griffin CA, Schambelan M, Kalinyak JE. Expression of AT2 receptors in the developing rat fetus. *J Clin Invest* 1991;**88**(3):921–33.

61. Nakajima M, Hutchinson HG, Fujinaga M, Hayashida W, Morishita R, Zhang L, et al. The angiotensin II type 2 (AT2) receptor antagonizes the growth effects of the AT1 receptor: gain-of-function study using gene transfer. *Proc Natl Acad Sci U S A* 1995;**92**(23):10663–7.

62. Horiuchi M, Hayashida W, Akishita M, Tamura K, Daviet L, Lehtonen JY, et al. Stimulation of different subtypes of angiotensin II receptors, AT1 and AT2 receptors, regulates STAT activation by negative crosstalk. *Circ Res* 1999;**84**(8):876–82.

63. Stoll M, Hahn AWA, Jonas U, Zhao Y, Schieffer B, Fischer JW, et al. Identification of a zinc finger homoeodomain enhancer protein after AT(2) receptor stimulation by differential mRNA display. *Arterioscler Thromb Vasc Biol* 2002;**22**(2):231–7.

64. Unger T, Sandmann S. Angiotensin receptor blocker selectivity at the AT1- and AT2-receptors: conceptual and clinical effects. *J Renin-Angiotensin-Aldosterone Syst* 2000;**1**(2 Suppl):S6–9.

65. Volpe M, Musumeci B, De Paolis P, Savoia C, Morganti A. Angiotensin II AT2 receptor subtype: an uprising frontier in cardiovascular disease? *J Hypertens* 2003;**21**(8):1429–43.

66. Doi C, Egashira N, Kawabata A, Maurya DK, Ohta N, Uppalapati D, et al. Angiotensin II type 2 receptor signaling significantly attenuates growth of murine pancreatic carcinoma grafts in syngeneic mice. *BMC Cancer* 2010;**10**:67.

67. Guimond M-O, Battista M-C, Nikjouitavabi F, Carmel M, Barres V, Doueik AA, et al. Expression and role of the angiotensin II AT2 receptor in human prostate tissue: in search of a new therapeutic option for prostate cancer. *Prostate* 2013;**73**(10):1057–68.

68. Yamada H, Akishita M, Ito M, Tamura K, Daviet L, Lehtonen JY, et al. AT2 receptor and vascular smooth muscle cell differentiation in vascular development. *Hypertension* 1999;**33**(6):1414–19.

69. Leung KK, Liang J, Ma MT, Leung PS. Angiotensin II type 2 receptor is critical for the development of human fetal pancreatic progenitor cells into islet-like cell clusters and their potential for transplantation. *Stem Cells Dayt* 2012;**30**(3):525–36.

70. Leung KK, Liang J, Zhao S, Chan WY, Leung PS. Angiotensin II type 2 receptor regulates the development of pancreatic endocrine cells in mouse embryos. *Dev Dyn Off Publ Am Assoc Anat* 2014;**243**(3):415–27.

71. Shao C, Zucker IH, Gao L. Angiotensin type 2 receptor in pancreatic islets of adult rats: a novel insulinotropic mediator. *Am J Physiol Endocrinol Metab* 2013;**305**(10):E1281–91.

72. Kintscher U, Bramlage P, Paar WD, Thoenes M, Unger T. Irbesartan for the treatment of hypertension in patients with the metabolic syndrome: a sub analysis of the Treat to Target post authorization survey. Prospective observational, two armed study in 14,200 patients. *Cardiovasc Diabetol* 2007;**6**:12.

73. Saitoh S, Takeishi Y. Pleiotropic effects of ARB in diabetes mellitus. *Curr Vasc Pharmacol* 2011;**9**(2):136–44.

74. Jing F, Mogi M, Horiuchi M. Role of renin-angiotensin-aldosterone system in adipose tissue dysfunction. *Mol Cell Endocrinol* 2013;**378**(1–2):23–8.

75. Shum M, Pinard S, Guimond M-O, Labbé SM, Roberge C, Baillargeon J-P, et al. Angiotensin II type 2 receptor promotes adipocyte differentiation and restores adipocyte size in high-fat/high-fructose diet-induced insulin resistance in rats. *Am J Physiol Endocrinol Metab* 2013;**304**(2):E197–210.

76. Stroth U, Meffert S, Gallinat S, Unger T. Angiotensin II and NGF differentially influence microtubule proteins in PC12W cells: role of the AT2 receptor. *Brain Res Mol Brain Res* 1998;**53**(1–2):187–95.

77. Gallinat S, Csikos T, Meffert S, Herdegen T, Stoll M, Unger T. The angiotensin AT2 receptor down-regulates neurofilament M in PC12W cells. *Neurosci Lett* 1997;**227**(1):29–32.

78. Li J-M, Mogi M, Tsukuda K, Tomochika H, Iwanami J, Min L-J, et al. Angiotensin II-induced neural differentiation via angiotensin II type 2 (AT2) receptor-MMS2 cascade involving interaction between AT2 receptor-interacting protein and Src homology 2 domain-containing protein-tyrosine phosphatase 1. *Mol Endocrinol* 2007;**21**(2):499–511.

79. Beaudry H, Gendron L, Guimond M-O, Payet MD, Gallo-Payet N. Involvement of protein kinase C alpha (PKC alpha) in the early action of angiotensin II type 2 (AT2) effects on neurite outgrowth in NG108-15 cells: AT2-receptor inhibits PKC alpha and p21ras activity. *Endocrinology* 2006;**147**(9):4263–72.

80. Gendron L, Laflamme L, Rivard N, Asselin C, Payet MD, Gallo-Payet N. Signals from the AT2 (angiotensin type 2) receptor of angiotensin II inhibit p21ras and activate MAPK (mitogen-activated protein kinase) to induce morphological neuronal differentiation in NG108-15 cells. *Mol Endocrinol* 1999;**13**(9):1615–26.

81. Plouffe B, Guimond M-O, Beaudry H, Gallo-Payet N. Role of tyrosine kinase receptors in angiotensin II AT2 receptor signaling: involvement in neurite outgrowth and in p42/p44mapk activation in NG108-15 cells. *Endocrinology* 2006;**147**(10):4646–54.

82. Namsolleck P, Boato F, Schwengel K, Paulis L, Matho KS, Geurts N, et al. AT2-receptor stimulation enhances axonal plasticity after spinal cord injury by upregulating BDNF expression. *Neurobiol Dis* 2013;**51**:177–91.

83. Stroth U, Blume A, Mielke K, Unger T. Angiotensin AT(2) receptor stimulates ERK1 and ERK2 in quiescent but inhibits ERK in NGF-stimulated PC12W cells. *Brain Res Mol Brain Res* 2000;**78**(1-2):175–80.

84. Gendron L, Côté F, Payet MD, Gallo-Payet N. Nitric oxide and cyclic GMP are involved in angiotensin II AT(2) receptor effects on neurite outgrowth in NG108-15 cells. *Neuroendocrinology* 2002;**75**(1):70–81.

85. Namsolleck P, Recarti C, Foulquier S, Steckelings UM, Unger T. AT(2) receptor and tissue injury: therapeutic implications. *Curr Hypertens Rep* 2014;**16**(2):416.

86. Lucius R, Gallinat S, Rosenstiel P, Herdegen T, Sievers J, Unger T. The angiotensin II type 2 (AT2) receptor promotes axonal regeneration in the optic nerve of adult rats. *J Exp Med* 1998;**188**(4):661–70.

87. Reinecke K, Lucius R, Reinecke A, Rickert U, Herdegen T, Unger T. Angiotensin II accelerates functional recovery in the rat sciatic nerve in vivo: role of the AT2 receptor and the transcription factor NF-kappaB. *J Off Publ Fed Am Soc Exp Biol* 2003;**17**(14):2094–6.

88. Lory P, Bidaud I, Chemin J. T-type calcium channels in differentiation and proliferation. *Cell Calcium* 2006;**40**(2):135–46.

89. Schmidt-Hieber C, Jonas P, Bischofberger J. Enhanced synaptic plasticity in newly generated granule cells of the adult hippocampus. *Nature* 2004;**429**(6988):184–7.

90. McCobb DP, Kater SB. Membrane voltage and neurotransmitter regulation of neuronal growth cone motility. *Dev Biol* 1988;**130**(2):599–609.

91. Ahmmed GU, Dong PH, Song G, Ball NA, Xu Y, Walsh RA, et al. Changes in Ca(2+) cycling proteins underlie cardiac action potential prolongation in a pressure-overloaded guinea pig model with cardiac hypertrophy and failure. *Circ Res* 2000;**86**(5):558–70.

92. Bijlenga P, Liu JH, Espinos E, Haenggeli CA, Fischer-Lougheed J, Bader CR, et al. T-type alpha 1H Ca2+ channels are involved in Ca2+ signaling during terminal differentiation (fusion) of human myoblasts. *Proc Natl Acad Sci U S A* 2000;**97**(13):7627–32.

93. Horiuchi M, Hayashida W, Kambe T, Yamada T, Dzau VJ. Angiotensin type 2 receptor dephosphorylates Bcl-2 by activating mitogen-activated protein kinase phosphatase-1 and induces apoptosis. *J Biol Chem* 1997;**272**(30):19022–6.

94. Horiuchi M, Akishita M, Dzau VJ. Molecular and cellular mechanism of angiotensin II-mediated apoptosis. *Endocr Res* 1998;**24**(3–4):307–14.

95. Savoia C, D'Agostino M, Lauri F, Volpe M. Angiotensin type 2 receptor in hypertensive cardiovascular disease. *Curr Opin Nephrol Hypertens* 2011;**20**(2):125–32.

96. Yamada T, Akishita M, Pollman MJ, Gibbons GH, Dzau VJ, Horiuchi M. Angiotensin II type 2 receptor mediates vascular smooth muscle cell apoptosis and antagonizes angiotensin II type 1 receptor action: an in vitro gene transfer study. *Life Sci* 1998;**63**(19):L289–95.

97. Gallinat S, Busche S, Schütze S, Krönke M, Unger T. AT2 receptor stimulation induces generation of ceramides in PC12W cells. *FEBS Lett* 1999;**443**(1):75–9.

98. Kacimi R, Gerdes AM. Alterations in G protein and MAP kinase signaling pathways during cardiac remodeling in hypertension and heart failure. *Hypertension* 2003;**41**(4):968–77.

99. Wang X, Phillips MI, Mehta JL. LOX-1 and angiotensin receptors, and their interplay. *Cardiovasc Drugs Ther Spons Int Soc Cardiovasc Pharmacother* 2011;**25**(5):401–17.

100. Chamoux E, Breault L, Lehoux JG, Gallo-Payet N. Involvement of the angiotensin II type 2 receptor in apoptosis during human fetal adrenal gland development. *J Clin Endocrinol Metab* 1999;**84**(12):4722–30.

101. Lehtonen JY, Horiuchi M, Daviet L, Akishita M, Dzau VJ. Activation of the de novo biosynthesis of sphingolipids mediates angiotensin II type 2 receptor-induced apoptosis. *J Biol Chem* 1999;**274**(24):16901–6.

102. Gu X, Olson EC, Spitzer NC. Spontaneous neuronal calcium spikes and waves during early differentiation. *J Neurosci Off J Soc Neurosci* 1994;**14**(11 Pt 1):6325–35.

103. Kaschina E, Grzesiak A, Li J, Foryst-Ludwig A, Timm M, Rompe F, et al. Angiotensin II type 2 receptor stimulation: a novel option of therapeutic interference with the renin-angiotensin system in myocardial infarction? *Circulation* 2008;**118**(24):2523–32.

104. Tan NY, Li J-M, Stocker R, Khachigian LM. Angiotensin II-inducible smooth muscle cell apoptosis involves the angiotensin II type 2 receptor, GATA-6 activation, and FasL-Fas engagement. *Circ Res* 2009;**105**(5):422–30.

105. Sun L, Wang W, Xiao W, Liang H, Yang Y, Yang H. Angiotensin II induces apoptosis in intestinal epithelial cells through the AT2 receptor, GATA-6 and the Bax pathway. *Biochem Biophys Res Commun* 2012;**424**(4):663–8.

106. Ramirez R, Chong T, Curran B, Sadjadi J, Victorino GP. Angiotensin II type 2 receptor decreases ischemia reperfusion induced fluid leak. *J Surg Res* 2007;**138**(2):175–80.

107. Ereso AQ, Ramirez RM, Sadjadi J, Cripps MW, Cureton EL, Curran B, et al. Angiotensin II type 2 receptor provides an endogenous brake during inflammation-induced microvascular fluid leak. *J Am Coll Surg* 2007;**205**(4):527–33.

108. Rompe F, Artuc M, Hallberg A, Alterman M, Ströder K, Thöne-Reineke C, et al. Direct angiotensin II type 2 receptor stimulation acts anti-inflammatory through epoxyeicosatrienoic acid and inhibition of nuclear factor kappaB. *Hypertension* 2010;**55**(4):924–31.

109. Sumners C, Horiuchi M, Widdop RE, McCarthy C, Unger T, Steckelings UM. Protective arms of the renin-angiotensin-system in neurological disease. *Clin Exp Pharmacol Physiol* 2013;**40**(8):580–8.

110. McCarthy CA, Vinh A, Callaway JK, Widdop RE. Angiotensin AT2 receptor stimulation causes neuroprotection in a conscious rat model of stroke. *Stroke J Cereb Circ* 2009;**40**(4):1482–9.

111. Pinzar E, Wang T, Garrido MR, Xu W, Levy P, Bottari SP. Angiotensin II induces tyrosine nitration and activation of ERK1/2 in vascular smooth muscle cells. *FEBS Lett* 2005;**579**(22):5100–4.

112. Kedziora-Kornatowska K. Effect of angiotensin convertase inhibitors and AT1 angiotensin receptor antagonists on the development of oxidative stress in the kidney of diabetic rats. *Clin Chim Acta Int J Clin Chem* 1999;**287**(1-2):19–27.

113. Welch WJ, Wilcox CS. AT1 receptor antagonist combats oxidative stress and restores nitric oxide signaling in the SHR. *Kidney Int* 2001;**59**(4):1257–63.

114. Zafari AM, Ushio-Fukai M, Akers M, Yin Q, Shah A, Harrison DG, et al. Role of NADH/NADPH oxidase-derived H2O2 in angiotensin II-induced vascular hypertrophy. *Hypertension* 1998;**32**(3):488–95.

115. Dandapat A, Hu CP, Chen J, Liu Y, Khan JA, Remeo F, et al. Over-expression of angiotensin II type 2 receptor (agtr2) decreases collagen accumulation in atherosclerotic plaque. *Biochem Biophys Res Commun* 2008;**366**(4):871–7.

116. Ullrich V, Kissner R. Redox signaling: bioinorganic chemistry at its best. *J Inorg Biochem* 2006;**100**(12):2079–86.

117. Csibi A, Communi D, Müller N, Bottari SP. Angiotensin II inhibits insulin-stimulated GLUT4 translocation and Akt activation through tyrosine nitration-dependent mechanisms. *PLoS One* 2010;**5**(4):e10070.

118. Matavelli LC, Huang J, Siragy HM. Angiotensin AT$_2$ receptor stimulation inhibits early renal inflammation in renovascular hypertension. *Hypertension* 2011;**57**(2):308–13.

119. Dhande I, Ali Q, Hussain T. Proximal tubule angiotensin AT2 receptors mediate an anti-inflammatory response via interleukin-10: role in renoprotection in obese rats. *Hypertension* 2013;**61**(6):1218–26.

120. Curato C, Slavic S, Dong J, Skorska A, Altarche-Xifró W, Miteva K, et al. Identification of noncytotoxic and IL-10-producing CD8+AT2R+ T cell population in response to ischemic heart injury. *J Immunol* 2010;**185**(10):6286–93.

121. Valero-Esquitino V, Lucht K, Namsolleck P, Monnet-Tschudic F, Stubbe T, Lucht F, et al. Direct angiotensin AT2-receptor stimulation attenuates T-cell and microglia activation and prevents demyelination in experimental autoimmune encephalomyelitis in mice. *Clin Sci* 2014;.

122. Liu YH, Yang XP, Sharov VG, Nass O, Sabbah HN, Peterson E, et al. Effects of angiotensin-converting enzyme inhibitors and angiotensin II type 1 receptor antagonists in rats with heart failure. Role of kinins and angiotensin II type 2 receptors. *J Clin Invest* 1997;**99**(8):1926–35.

123. Xu J, Carretero OA, Liu Y-H, Shesely EG, Yang F, Kapke A, et al. Role of AT2 receptors in the cardioprotective effect of AT1 antagonists in mice. *Hypertension* 2002;**40**(3):244–50.

124. Nabeshima Y, Tazuma S, Kanno K, Hyogo H, Iwai M, Horiuchi M, et al. Anti-fibrogenic function of angiotensin II type 2 receptor in CCl4-induced liver fibrosis. *Biochem Biophys Res Commun* 2006;**346**(3):658–64.

125. Ulmasov B, Xu Z, Tetri LH, Inagami T, Neuschwander-Tetri BA. Protective role of angiotensin II type 2 receptor signaling in a mouse model of pancreatic fibrosis. *Am J Physiol Gastrointest Liver Physiol* 2009;**296**(2):G284–94.

126. Gelosa P, Pignieri A, Fändriks L, de Gasparo M, Hallberg A, Banfi C, et al. Stimulation of AT2 receptor exerts beneficial effects in stroke-prone rats: focus on renal damage. *J Hypertens* 2009;**27**(12):2444–51.

127. Brassard P, Amiri F, Schiffrin EL. Combined Angiotensin II Type 1 and Type 2 Receptor Blockade on Vascular Remodeling and Matrix Metalloproteinases in Resistance Arteries. *Hypertension* 2005;**46**(3):598–606.

128. Jing T, Wang H, Srivenugopal KS, He G, Liu J, Miao L, et al. Conditional expression of type 2 angiotensin II receptor in rat vascular smooth muscle cells reveals the interplay of the angiotensin system in matrix metalloproteinase 2 expression and vascular remodeling. *Int J Mol Med* 2009;**24**(1):103–10.

129. Lauer D, Slavic S, Sommerfeld M, Thöne-Reineke C, Sharkovska Y, Hallberg A, et al. Angiotensin type 2 receptor stimulation ameliorates left ventricular fibrosis and dysfunction via regulation of tissue inhibitor of matrix metalloproteinase 1/matrix metalloproteinase 9 axis and transforming growth factor β1 in the rat heart. *Hypertension* 2014;**63**(3):e60–7.

130. Kljajic ST, Widdop RE, Vinh A, Welungoda I, Bosnyak S, Jones ES, et al. Direct AT$_2$ receptor stimulation is athero-protective and stabilizes plaque in Apolipoprotein E-deficient mice. *Int J Cardiol* 2013;**169**(4):281–7.

Chapter 5

AT$_1$R–AT$_2$R Cross Talk

Carmine Savoia, Massimo Volpe
Clinical and Molecular Medicine Department, Sapienza University of Rome, Rome, Italy

CROSS REGULATION OF ANGIOTENSIN II TYPE 1 RECEPTOR AND THE ANGIOTENSIN II TYPE 2 RECEPTOR

The renin–angiotensin system (RAS) is an enzymatic and hormonal cascade involved in blood pressure regulation and cardiovascular and renal homeostasis. Angiotensin II (Ang II) is the main effector of the RAS and exerts its biological action by binding with high affinity to two distinct subtype receptors, the Ang II type 1 receptor (AT$_1$R) and the Ang II type 2 receptor (AT$_2$R).[1-5] Both AT$_1$R and AT$_2$R belong to the family of seven-transmembrane G protein-coupled receptors (GPCRs).[1-3]

The functional significance of AT$_1$R and AT$_2$R in cardiovascular and renal pathophysiology has been investigated *in vivo* and *in vitro*.[1-5] It has been recognized that AT$_1$R plays a critical role in the Ang II-mediated actions in the cardiovascular system and the kidney, by inducing vasoconstriction, blood pressure increase, water and sodium retention, growth promotion, fibrosis, and inflammation in several pathophysiological conditions.[3-5] Conversely, the role of AT$_2$R is much less characterized, particularly in humans.[1-3] One of the limitations for the characterization of this subtype receptor is that, in adult life, AT$_2$R is often expressed at the detection limit in few organs and tissues including the coronary arteries, cardiomyocytes, ventricular myocardium,[1,2,6] peripheral vasculature,[7,8] and kidney.[9] Despite its low expression level, the distribution and density of AT$_2$R in adult tissues may be related to the involvement of this subtype receptor in cardiovascular and renal function.

In vitro and *in vivo* studies have shown that AT$_1$R stimulation may regulate the expression of AT$_2$R, suggesting the existence of a complex cross-regulatory mechanism between the two subtype receptors.[2,3,7-11] For instance, Ang II-induced AT$_2$R expression was enhanced in vascular smooth muscle cells (VSMCs) by AT$_1$R blockade, suggesting that AT$_1$R may control AT$_2$R expression.[10] In endothelial cells transfected with the AT$_2$R promoter, Ang II-induced expression of AT$_2$R mRNA was blunted by AT$_1$R stimulation.[11] Low levels of AT$_2$R mRNA and protein have been detected in the vasculature of normotensive and hypertensive rats.[7,10] The expression of the receptor was increased only in hypertensive animals, in which it contributed to vasodilation after *in vivo* chronic AT$_1$R blockade,[7,10] suggesting that the cross-regulatory mechanism between the Ang II subtype receptors works mostly in pathological conditions at transcription level in the vasculature of rodents.

Few studies have investigated the expression and the function of AT$_2$R in humans, particularly in the cardiovascular system. AT$_2$R has been detected in vascular endothelial cells, fibroblasts, and macrophages of the human lung.[12] In the human peripheral arteries, AT$_2$R has been detected in the coronary circulation,[13] particularly in the coronary microcirculation.[14] We have reported[15] that AT$_2$R is upregulated in the vascular wall of peripheral resistance arteries from hypertensive diabetic patients only in the presence of selective AT$_1$R blockade with the angiotensin receptor blocker (ARB) valsartan, suggesting that the cross talk regulatory mechanism between AT$_1$R and AT$_2$R may occur also in humans, particularly in high-risk cardiovascular patients.[15]

In the vasculature, the negative cross talk between AT$_2$R and AT$_1$R may be characterized also by a negative modulation of AT$_1$R expression and AT$_1$R-mediated signaling in the presence of AT$_2$R stimulation and/or overexpression.[16,17] Several lines of evidence support this hypothesis. It was reported that AT$_1$R expression was significantly higher in AT$_2$R knockout mice than in control animals.[16] Overexpression and activation of AT$_2$R downregulated AT1aR expression in rat VSMC in an Ang II-independent manner that involved the activation of the bradykinin (BK)/nitric oxide pathway.[18] Moreover, transfection of the AT$_2$R gene in rat VSMC inhibited AT$_1$R-mediated tyrosine phosphorylation of signal transducers and activators of transcription (STAT) 1a/b, STAT2, and STAT3.[17] The counterregulatory activity of AT$_2$R on AT$_1$R may be relevant also in the renal pathophysiology involving increased activation of RAS. In the kidney, AT$_2$R regulated AT$_1$R expression at transcriptional and functional level via the nitric oxide/cyclic guanosine monophosphate (cGMP)/Sp1 pathway.[19] In particular,

The Protective Arm of the Renin–Angiotensin System (RAS). http://dx.doi.org/10.1016/B978-0-12-801364-9.00005-5

AT_2R attenuated the ability of Ang II to increase Na+/K+/ATPase activity in rat immortalized renal proximal tubule (RPT) cells, effects that were supported also in studies in RPT cells from AT_2R mice.[19] These evidences may further explain the mechanisms of modulation of sodium excretion and blood pressure control.

Apart from the regulation of AT_1R transcription and function via nitric oxide/cGMP pathway, AT_2R may also regulate AT_1R expression via a direct protein–protein interaction. In this regard, it has been shown that GPCRs are able to both homodimerize and heterodimerize.[20–22] For instance, following agonist stimulation, beta-2 adrenergic, muscarinic, dopamine D2, and opioid receptors undergo homodimerization,[21,22] suggesting that it may be possible to modulate receptor function through intermolecular interactions. Homodimers of AT_1R or AT_2R are important for inducing cell signaling. The elevated levels of AT_1R dimers on monocytes promoted atherogenesis in ApoE-deficient mice. In addition, constitutively active homodimerization of AT_2R was localized in the cell membrane without Ang II stimulation and induced apoptosis without changes in receptor conformation.[23] AT_1R may heterodimerize with BK type B2 receptor,[24] dopamine receptor,[25] endothelin type B receptor,[26] Mas receptor,[27] and AT_2R.[28] The functional role of homo- or heterodimerization of Ang II receptors, however, still remains controversial.

There is evidence that heterodimerization may explain the antagonistic functions of AT_1R and AT_2R. The occurrence of a detected AT_1R and AT_2R heterodimerization that leads to AT_1R signal inhibition may be independent of the binding of Ang II to AT_2 receptor. The overexpression of AT_2R induced apoptosis of fibroblasts in the absence of Ang II.[29] Furthermore, the overexpression of the AT_2R after transfection into cultured rat VSMC may induce downregulation of AT1aR expression in a ligand-independent manner that is possibly mediated by the BK/nitric oxide pathway. This, in turn, can reduce the basal DNA synthesis and proliferation of VSMCs and abolish the response of DNA synthesis to Ang II in VSMCs.[18] It is therefore possible that increased production of nitric oxide by overexpression of AT_2R could suppress AT1aR expression by the inhibition of its transcription. These data underline that AT_2R has constitutive activity,[23,29] which might induce heterodimerization independent of Ang II stimulation.

Conversely, it has been shown that AT_2R formed preferably homodimers, rather than heterodimers with AT_1R, on the cell surface, and subsequently induced cell signaling including the antagonism of phospholipase C-b3 phosphorylation.[30] Interestingly, the expression levels of homodimerized AT_1R or AT_2R on the cell surface did not change after treatment with Ang II or with the selective ARBs and PD123319, respectively.[30] Hence, that evidence suggests that Ang II-induced AT_1R signaling may be mainly blocked by AT_2R signaling through their negative cross talk in the cytoplasm rather than by the heterodimerization of both receptors on the cell surface. In this regard, the balance of the expression levels of AT_1R and AT_2R receptors may be critical for the mutual antagonistic actions.

FUNCTIONAL SIGNIFICANCE OF THE AT_1R–AT_2R CROSS TALK

In the vasculature, AT_2R induces vasorelaxation *in vivo* in both resistance and capacitance vessels.[7,10,15,31] *In vitro* experiments in cultured cells and in isolated vessels as well as *in vivo* experiments in mice and rats have shown a link between AT_2R stimulation and the nitric oxide/cGMP-dependent pathway.[31] This occurs either directly[32] or indirectly trough enhanced BK formation[7] or increased eNOS activity/expression.[33]

A large body of evidence suggests that AT_2R may countervail many of the AT_1R-mediated actions in the vasculature under physiological and pathological circumstances. This may occur through mechanisms independent of direct AT_2R stimulation or by a cross-interaction with AT_1R functional expression and signaling,[9] suggesting that the regulatory cross talk between AT_1R and AT_2R is important at a functional level.

Aortic AT_1R expression and vascular response to Ang II are greater in AT_2R knockout than wild-type mice,[16] suggesting that AT_2R may counter the AT_1R-induced effect on blood pressure. Chronic Ang II administration induced vasodilation and hypotension in rats after AT_1R blockade, and this response was abolished by coadministration of the AT_2R antagonist PD123319.[7] Recently, it has been shown that direct AT_2R agonist compound 21 (C21) promoted vasorelaxation *in vitro*, which in turn was associated with vasodepressor responses only in conscious spontaneously hypertensive rats (SHR) previously treated with an ARB.[34] Data from Ang II receptor knockout mice also confirm the existence of a physiological cross talk between AT_1R and AT_2R at the level of the modulation of vascular tone.[9,35] Several reports indicate that the vascular functions of AT_2R are unmasked when AT_1R is inhibited[7,10,33,36] using ARBs. Taken together, those data further underline the occurrence of a functional cross talk between AT_1R and AT_2R in blood vessel function.

AT_2R is highly expressed in the fetal kidney. Its expression gradually decreases after birth, but AT_2R continues to be expressed in the adult kidney, though at low levels.[37] AT_2R is expressed throughout the kidney in both vascular and tubular elements mainly in RPT cells.[34] AT_2R stimulation inhibits the activity of Na+/K+/ATPase in the proximal tubules isolated from adult rats, mice, and rabbits,[38,39] inducing in turn natriuresis, which may also be modulated by the interactions of AT_2R with dopamine D1 receptors.[40]

Selective intrarenal AT$_1$R blockade in rats induced a highly significant natriuretic response that was abolished with intrarenal coinfusion of the AT$_2$R antagonist PD 123319.[39] These results further indicate that, similar to vasodilation, the beneficial natriuretic response to ARB administration is related to AT$_2$R activation. In the 5/6 nephrectomy model, which developed a time-dependent increase in AT$_2$R expression at 7, 15, and 30 days after renal ablation, pretreatment with the ARB losartan showed a further increase in AT$_2$R expression. Treatment with the AT$_2$R antagonist PD123319 was associated with downregulation of AT$_2$R, increased renal damage, and increased blood pressure, suggesting that the AT$_2$R represents a beneficial counterregulatory mechanism to protect the kidney from ischemic injury.[41] Recent evidence showed that the AT$_2$R-mediated renal vasodilator effects after stimulation with C21 are unmasked by ACE inhibition in SHR but not in normotensive rats. Thus, the upregulation of renal vascular AT$_2$R, particularly in hypertension, is associated with a countervailing function opposing the increased AT$_1$R-mediated tonic renal vasoconstriction and underlines the relevance of the occurrence of a functional cross talk between AT$_1$R and AT$_2$R also in the kidney.[42]

A functional counterregulatory mechanism at the expression and signaling level among Ang II subtype receptors and growth-promoting receptors may also occur and contribute to the pathophysiology of the Ang II-induced damage in the cardiovascular system.

It has been well established that Ang II exerts growth-promoting effects via activation of multiple signaling cascades including the activation of ERK and PI 3-K/Akt pathways and increased expression of growth factors including transforming growth factor-β I (TGF-β type I),[43,44] in the cardiovascular system. It has been shown that the Ang II-induced activation of ERK and PI 3-K/Akt cascades required AT$_1$R-operated transactivation of EGFR in VSMC[45] and fibroblasts.[46] These actions were counteracted by AT$_2$R activation in VSMC[47] and in Ang II-stimulated fibroblasts of human hypertrophic scars.[46]

In mouse fibroblasts, both Ang II receptors were able to phosphorylate ERK1/2. However, in the cells expressing only AT$_1$R, the EGF-induced MAPK pathway was enhanced in the presence of Ang II in a synergistic fashion. In contrast, a reduction of EGF-induced MAPK activation was observed in the cells expressing only AT$_2$R. In fibroblasts expressing both Ang II subtype receptors, Ang II promoted an enhancement of EGF-induced MAPK activation. However, in the presence of an ARB, the effect of the EGF receptor was reduced, suggesting the existence of an opposite cross talk of AT$_1$R and AT$_2$R with EGF receptors. This contributes to explain a complex functional interaction between these pathways in the regulation of cellular growth processes.[48] Furthermore, there is evidence that overexpression of AT$_2$R downregulated the AT1aR in VSMCs from WKY but not in cells from SHR. Similarly, overexpression of AT$_2$R abolished DNA synthesis in response to Ang II in VSMCs from WKY, whereas DNA synthesis in response to Ang II was not altered in SHR. These phenomena may be a consequence of the downregulation of the AT1aR only in VSMCs from WKY, and not from SHR. The lack of downregulation of the AT1aR in SHR may contribute, in part, to the exaggerated VSMC growth in SHR.[49]

CONCLUSION

In conclusion, the occurrence of a functional counterregulatory mechanism between AT$_1$R and AT$_2$R, as well as a functional cross talk among Ang II receptors and growth-promoting receptors, may further explain the mechanisms of the pathophysiology of cardiovascular and renal damage linked to RAS activation. These mechanisms are also relevant to understand the pathophysiology and pharmacology of Ang II receptors, which are important to find specific and selective therapies for cardiovascular protection.

REFERENCES

1. de Gasparo M, Catt KJ, Inagami T, Wright JW, Unger T. International Union of Pharmacology XXII. The angiotensin II receptors. *Pharmacol Rev* 2000;**52**:415–72.
2. Volpe M, Musumeci B, De Paolis P, Savoia C, Morganti A. Angiotensin II AT$_2$ receptor subtype: an uprising frontier in cardiovascular disease? *J Hypertens* 2003;**21**:1429–43.
3. Savoia C, Burger D, Nishigaki N, Montezano A, Touyz R. Angiotensin II and the vascular phenotype in hypertension. *Expert Rev Mol Med* 2011;**13**:e11.
4. Callera G, Tostes R, Savoia C, Muscara MN, Touyz R. Vasoactive peptides in cardiovascular (patho)physiology. *Expert Rev Cardiovasc Ther* 2007;**5**(3):531–52.
5. Savoia C, Schiffrin EL. Inhibition of the renin angiotensin system: implications for the endothelium. *Curr Diab Rep* 2006;**6**(4):274–8.
6. Wang ZQ, Moore AF, Ozono R, Siragy HM, Carey RM. Immunolocalization of subtype 2 angiotensin II (AT$_2$) receptor protein in rat heart. *Hypertension* 1998;**32**:78–83.
7. Cosentino F, Savoia C, De Paolis P, Francia P, Russo A, Maffei A, et al. Angiotensin II type 2 receptor contribute to vascular responses in spontaneously hypertensive rats treated with angiotensin II type 1 receptor antagonists. *Am J Hypertens* 2005;**18**:493–9.

8. Viswanathan M, Tsutsumi K, Correa FM, Saavedra JM. Changes in expression of angiotensin receptors subtype in the rat aorta during development. *Biochem Biophys Res Commun* 1991;**179**:1361–7.

9. Padia SH, Carey RM. AT2 receptors: beneficial counter-regulatory role in cardiovascular and renal function. *Pflugers Arch* 2013;**465**(1):99–110.

10. Savoia C, Tabet F, Yao G, Schiffrin EL, Touyz RM. Negative regulation of Rho/Rho kinase by angiotensin II type 2 receptor in vascular smooth muscle cells: role in angiotensin II-induced vasodilation in stroke-prone spontaneously hypertensive rats. *J Hypertens* 2005;**23**:1037–45.

11. De Paolis P, Porcellini A, Gigante B, Giliberti R, Lombardi A, Savoia C, et al. Modulation of the AT_2 subtype receptor gene activation and expression by the AT_1 receptor in endothelial cells. *J Hypertens* 1999;**17**:1873–7.

12. Bullock GR, Steyaert I, Bilbe G, Carey RM, Kips J, De Paepe B, et al. Distribution of type-1 and type-2 angiotensin receptors in the normal human lungs and in lungs from patients with chronic obstructive pulmonary disease. *Histochem Cell Biol* 2001;**115**:117–24.

13. Wharton J, Morgan K, Rutherford RAD, Catravas JD, Chester A, Whitehead BF, et al. Differential distribution of angiotensin AT2 receptors in the normal and failing human heart. *J Pharmacol Exp Ther* 1998;**284**:323–36.

14. Batenburg WW, Garrelds IM, Bernasconi C, Juillerat-Jeanneret L, van Kats JP, Saxena PR, et al. Angiotensin II type 2 receptor-mediated vasodilation in human coronary microarteries. *Circulation* 2004;**109**:2296–301.

15. Savoia C, Touyz RM, Volpe M, Schiffrin EL. Angiotensin type 2 receptor in resistance arteries of type 2 diabetic hypertensive patients. *Hypertension* 2007;**49**(2):341–6.

16. Tanaka M, Tsuchida S, Imai T, Fujii N, Miyazaki H, Ichiki T, et al. Vascular response to angiotensin II is exaggerated through an upregulation of AT1 receptor in AT2 knockout mice. *Biochem Biophys Res Commun* 1999;**258**:194–8.

17. Horiuchi M, Hayashida W, Akishita M, Tamura K, Daviet L, Lehtonen JY, et al. Stimulation of different subtypes of angiotensin II receptors, AT1 and AT2 receptors, regulates STAT activation by negative crosstalk. *Circ Res* 1999;**84**:876–82.

18. Jin XQ, Fukuda N, Su JZ, Lai YM, Suzuki R, Tahira Y, et al. Angiotensin II type 2 receptor gene transfer downregulates angiotensin II type 1a receptor in vascular smooth muscle cells. *Hypertension* 2002;**39**:1021–7.

19. Yang J, Chen C, Ren H, Han Y, He D, Zhou L, et al. Angiotensin II AT2 receptor decreases AT1 receptor expression and function via nitric oxide/cGMP/Sp1 in renal proximal tubule cells from Wistar–Kyoto rats. *J Hypertens* 2012;**30**(6):1176–84.

20. Jordan BA, Devi LA. G-protein-coupled receptor heterodimerization modulates receptor function. *Nature* 1999;**399**:697–700.

21. Hebert TE, Loisel TP, Adam L, Ethier N, Onge SS, Bouvier M. Functional rescue of a constitutively desensitized beta2 AR through receptor dimerization. *Biochem J* 1998;**330**:287–93.

22. Ng GY, OiDowd BF, Lee SP, Chung HT, Brann MR, Seeman P, et al. Dopamine D2 receptor dimers and receptor-blocking peptides. *Biochem Biophys Res Commun* 1996;**277**:200–4.

23. Miura S, Karnik SS, Saku K. Constitutively active homo-oligomeric angiotensin II type 2 receptor induces cell signaling independent of receptor conformation and ligand stimulation. *J Biol Chem* 2005;**280**:18237–44.

24. AbdAlla S, Lother H, el Massiery A, Quitterer U. Increased AT1 receptor heterodimers in preeclampsia mediate enhanced angiotensin II responsiveness. *Nat Med* 2001;**7**:1003–9.

25. Zeng C, Yang Z, Wang Z, Jones J, Wang X, Altea J, et al. Interaction of angiotensin II type 1 and D5 dopamine receptors in renal proximal tubule cells. *Hypertension* 2005;**45**:804–10.

26. Zeng C, Hopfer U, Asico LD, Eisner GM, Felder RA, Jose PA. Altered AT1 receptor regulation of ETB receptors in renal proximal tubule cells of spontaneously hypertensive rats. *Hypertension* 2005;**46**:926–31.

27. Bohlen V. und Halbach O, Walther T, Bader M, Albrecht D, Interaction between Mas and the angiotensin AT1 receptor in the amygdala. *J Neurophysiol* 2000;**83**:2012–21.

28. AbdAlla S, Lother H, Abdel-tawab AM, Quitterer U. The angiotensin II AT2 receptor is an AT1 receptor antagonist. *J Biol Chem* 2001;**276**:39721–6.

29. Miura S, Karnik SS. Ligand-independent signals from angiotensin II type 2 receptor induce apoptosis. *EMBO J* 2000;**19**:4026–35.

30. Miura S, Matsuo Y, Kiya Y, Karnik SS, Saku K. Molecular mechanisms of the antagonistic action between AT1 and AT2 receptors. *Biochem Biophys Res Commun* 2010;**391**:85–90.

31. Gohlke P, Pees C, Unger T. AT_2 receptor stimulation increases aortic cyclic GMP in SHRSP by a kinin-dependent mechanism. *Hypertension* 1998;**31**:349–55.

32. Abadir PM, Carey RM, Siragy HM. Angiotensin AT_2 receptors directly stimulate renal nitric oxide in bradykinin B_2-receptor-null mice. *Hypertension* 2003;**42**:600–4.

33. Savoia C, Ebrahimian T, He Y, Gratton JP, Schiffrin EL, Touyz RM. Angiotensin II/AT_2 receptor-induced vasodilation in stroke-prone spontaneously hypertensive rats involves nitric oxide and cGMP-dependent protein kinase. *J Hypertens* 2006;**24**:2417–22.

34. Bosnyak S, Welungoda IK, Hallberg A, Alterman M, Widdop RE, Jones ES. Stimulation of angiotensin AT2 receptors by the non-peptide agonist, compound 21, evokes vasodepressor effects in conscious spontaneously hypertensive rats. *Br J Pharmacol* 2010;**159**:709–16.

35. Siragy HM, Inagami T, Ichiki T, Carey RM. Sustained hypersensitivity to angiotensin II and its mechanism in mice lacking the subtype-2 (AT2) angiotensin receptor. *Proc Natl Acad Sci U S A* 1999;**96**:6506–10.

36. Widdop RE, Matrougui K, Levy BI, Henrion D. AT2 receptor mediated relaxation is preserved after long-term AT1 receptor blockade. *Hypertension* 2002;**40**:516–20.

37. Miyata N, Park F, Li XF, Cowley Jr AW. Distribution of angiotensin AT1 and AT2 receptor subtypes in the rat kidney. *Am J Physiol* 1999;**277**:F437–46.

38. Hakam AC, Hussain T. Angiotensin II AT2 receptors inhibit proximal tubular Na+−K+−ATPase activity via a NO/cGMP-dependent pathway. *Am J Physiol Renal Physiol* 2006;**290**:F1430–6.

39. Padia SH, Howell NL, Siragy HM, Carey RM. Renal angiotensin type 2 receptors mediate natriuresis via angiotensin III in the angiotensin II type 1 receptor-blocked rat. *Hypertension* 2006;**47**:537–44.

40. Salomone LJ, Howell NL, McGrath HE, Kemp BA, Keller SR, Gildea JJ, et al. Intrarenal dopamine D1-like receptor stimulation induces natriuresis via an angiotensin type-2 receptor mechanism. *Hypertension* 2007;**49**:155–61.

41. Vazquez E, Coronel I, Bautista R, Romo E, Villalon CM, Avila-Casado MC, et al. Angiotensin II-dependent induction of AT(2) receptor expression after renal ablation. *Am J Physiol Renal Physiol* 2005;**288**:F207–13.

42. Brouwers S, Smolders I, Massie A, Dupont AG. Angiotensin II type 2 receptor–mediated and nitric oxide–dependent renal vasodilator response to compound 21 unmasked by angiotensin-converting enzyme inhibition in spontaneously hypertensive rats in vivo. *Hypertension* 2013;**62**:920–6.

43. Satoh C, Fukuda N, Hu W-Y, Kanmatsuse K. Role of endogenous angiotensin II in the increased expression of growth factors in vascular smooth muscle cells from spontaneously hypertensive rats. *J Cardiovasc Pharmacol* 2001;**37**:108–18.

44. Berk BC, Corson MA. Angiotensin II signal transduction in vascular smooth muscle: role of tyrosine kinases. *Circ Res* 1997;**80**(5):607–16.

45. Bokomeyer D, Schmitz U, Kramer HJ. Angiotensin II-induced growth of vascular smooth muscle cells requires an Src-dependent activation of the epidermal growth factor receptor. *Kidney Int* 2000;**58**(2):549–58.

46. Liu HW, Cheng B, Yu WL, Sun RX, Zeng D, Wang J, et al. Angiotensin II regulated phosphoinositide 3 kinase/Akt cascade via a negative crosstalk between AT1 and AT2 receptors in skin fibroblasts of human hypertrophic scars. *Life Sci* 2006;**79**:475–83.

47. Eguchi S, Inagami T. Signal transduction of angiotensin II type 1 receptor through receptor tyrosine kinase. *Regul Pept* 2000;**91**:13–20.

48. De Paolis P, Porcellini A, Savoia C, Lombardi A, Gigante B, Frati G, et al. Functional cross-talk between angiotensin II and epidermal growth factor receptors in NIH3T3 fibroblasts. *J Hypertens* 2002;**20**(4):693–9.

49. Su JZ, Fukuda N, Jin XQ, Lai YM, Suzuki R, Tahira Y, et al. Effect of AT2 receptor on expression of AT1 and TGF-b receptors in VSMCs from SHR. *Hypertension* 2002;**40**:853–8.

Chapter 6

The Angiotensin AT$_2$ Receptor in Myocardial Infarction

Elena Kaschina

Center for Cardiovascular Research (CCR), Institute of Pharmacology, Charité-Universitätsmedizin Berlin, Berlin, Germany

INTRODUCTION

Myocardial infarction (MI) remains the leading cause of morbidity and mortality worldwide.[1] The cardiac renin–angiotensin–aldosterone system (RAAS) is highly activated after MI and in heart failure (HF). It is implicated in all important postischemic pathological processes via AT$_1$ receptor (AT$_1$R) subtype activation, promoting inflammation, proteolysis, apoptosis, fibrosis, and cardiac remodeling (for review, see de Gasparo et al.,[2] Kaschina and Unger[3]). In many cases, the AT$_2$ receptor (AT$_2$R) counteracts—especially due to its anti-inflammatory, antifibrotic, and antigrowth effects—the AT$_1$R in the heart (for review, see Unger[4], Steckelings et al.[5]).

The role of the AT$_2$R in the pathophysiology of MI and HF has been intensively explored by using indirect experimental approaches such as genetically altered animals or indirect AT$_2$R stimulation via treatment with an AT$_1$R blocker (for review, see Carey,[6] Widdop et al.,[7] Kaschina et al.[8]). Based on these studies, the activation of AT$_2$Rs has been considered cardioprotective, although some controversies remain unanswered. During the last 5 years, our knowledge on the AT$_2$R in the heart has been thoroughly expanded, thanks to the synthesis of a nonpeptide direct AT$_2$R agonist compound 21 (C21).[9]

In this short review, the impact of the AT$_2$R on cardiac function and remodeling in the acute and chronic phases after MI will be discussed with a focus on more recent studies in which the nonpeptide AT$_2$R agonist C21 was applied for AT$_2$R stimulation (Figure 1).

AT$_2$ RECEPTOR EXPRESSION IN THE HEART

In the heart, as in other tissues, AT$_2$Rs are developmentally regulated with high expression in the fetal heart, decline after birth, and increase after ischemic injury (for review, see de Gasparo[2]). A study from our laboratory demonstrated *in vivo* that the AT$_2$Rs were expressed in about 10% of rat adult cardiomyocytes before MI as well as 1 day afterward.[10] One week after MI, cardiac expression of the AT$_2$Rs in the rat was enhanced[11,12] and colocalized with inflammatory T cells[13] and progenitor cells[14] in the peri-infarct zone. In the intermediate phase post-MI, relative levels of the AT$_2$Rs have been reported to decrease[15] and left ventricular dysfunction has been associated with AT$_2$R downregulation.[16]

The data on AT$_2$R expression in failing human hearts are inconsistent (for review, see Kaschina et al.[8]). Nevertheless, fibroblasts that are present in the interstitial region are the major cell type responsible for AT$_2$R expression.[17,18] Moreover, regions displaying a relative increase in AT$_2$ binding sites corresponded with areas of fibroblast proliferation and collagen deposition.[19] Collectively, AT$_2$Rs are present in various cardiac cell types and are regulated in the heart in different ways, depending on the time after injury and the phase of remodeling.

ACUTE-PHASE POSTMYOCARDIAL INFARCTION

Cardiac Function

The acute phase after MI is characterized by myocyte cell death, acute inflammation, and scar formation.[20] First studies addressing the role of the AT$_2$R in MI have been performed in genetically altered mice, either AT$_2$R-deficient or overexpressing the AT$_2$R. Although this experimental approach is not without limitations, the vast majority of studies demonstrated cardioprotective effects of the AT$_2$R in the early phase post-MI.[21–24]

The Protective Arm of the Renin–Angiotensin System (RAS). http://dx.doi.org/10.1016/B978-0-12-801364-9.00006-7

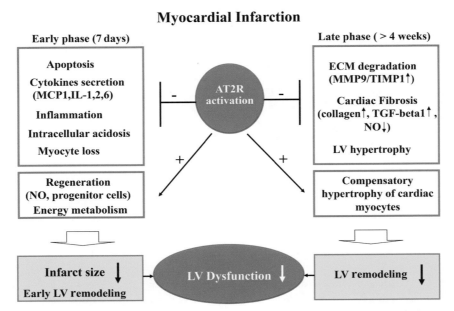

FIGURE 1 Summary of the major effects of the AT₂ receptor activation in the early and late phase after experimental myocardial infarction. Abbreviations: AT_2R, angiotensin type 2 receptor; MCP-1, monocyte chemoattractant protein 1; IL, interleukin; NO, nitric oxide; LV, left ventricular; ECM, extracellular matrix; MMP9, matrix metalloproteinase 9; TIMP1, tissue inhibitor of matrix metalloproteinase 1; TGF-beta1, transforming growth factor beta 1.

The first *in vivo* study using the selective AT_2R agonist C21 looked at the role of the AT_2R on post-MI myocardial function.[25] The effects of C21 were investigated in normotensive rats treated for 1 week post-MI. C21 improved systolic and diastolic cardiac function after the infarct coinciding with a smaller infarct scar measured by MRT. While MI led to an impairment of all parameters measured by echocardiography (LVIDs, EF, and FS for systolic function; LVIDd, E, A, E/A, and EDT for diastolic function) or Millar catheter (contractility and maxd$P$/dt for systolic function; LVEDP and mind$P$/dt for diastolic function), treatment with C21 improved all of these parameters. C21 caused no changes in heart rate or blood pressure. The effect of C21 could be inhibited by cotreatment with the AT_2R antagonist, PD 123319, thus providing evidence that the effects of C21 treatment were AT_2R-specific. In this study, an AT_1R antagonist, candesartan, was used as a reference drug. For a number of parameters, C21 was even more effective than candesartan, which also did not decrease infarct size.[25]

Infarct size reduction and functional improvement could be attributed to several beneficial cellular mechanisms of the AT_2 receptor activation, which will be further discussed.

Bradykinin/Nitric Oxide System

AT_2R stimulation has been suggested to exert cardioprotective properties most likely via bradykinin/nitric oxide (NO)-mediated mechanisms (for review, see Linz et al.,[26] Carey et al.,[6] Rhaleb et al.[27]). An AT_2R-mediated activation of the bradykinin/NO/cGMP system has been reported in endothelial cells,[28] the aorta,[29] and the heart.[30] Jalowy et al.[31] could demonstrate in pigs that indirect AT_2R stimulation reduced infarct size and that this effect was blocked by the B2R antagonist, icatibant. Cardiomyocyte-specific overexpression of the AT_2R activated kininogenase and, consequently, the bradykinin/B2-receptor/NO axis.[32] Several studies in the MI model provided evidence for NO-dependent mechanisms in AT_2R-mediated antifibrosis and antiremodeling[33,30,34,35] as reviewed in detail.[8] Recently, Abadir et al.[36] had shown that AT_2R stimulation with CGP-42112A increased NO production and suppressed respiratory oxygen consumption in isolated mitochondria, which points to the role of intracellular AT_2R.

Intracellular Acidosis

The AT_2 receptor seems to be important also for the regulation of intracellular acidosis, which follows myocardial ischemia. The influence of AT_1 and AT_2 receptor blockade on the expression and activity of proteins that take part in the regulation of the intracellular pH, namely, the Na^+/H^+ exchanger (NHE-1) and the Na^+/HCO_3^- symporter (NBC), has been investigated by Sandmann et al.[37] Using indirect stimulation of the unopposed AT$_2$R with the AT$_1$ antagonist losartan or AT$_2$R blockade with PD 123319 post-MI, the authors demonstrated that the Ang II-induced stimulation of NHE-1 is mediated, at least in part, via activation of the AT$_1$R, whereas AT$_2$R activation induced the stimulation of NBC-1.

Apoptosis

It is widely accepted that inhibition of cardiomyocyte apoptosis may interrupt the pathomechanisms leading to infarct expansion, ventricular remodeling, and HF.[20] Stimulation of AT$_2$R with C21 reduced the MI-induced expression of apoptotic markers, Fas ligand and caspase-3. This effect could be blocked by the selective AT$_2$R antagonist, PD 123319.[25] Furthermore, rescue of p38- and p44/42 MAPK (mitogen-activated protein kinase) by C21 may also be involved in its antiapoptotic effect.[25]

A study by Altarche-Xifró et al.[14] further showed that AT$_2$R stimulation attenuated programmed cell death of cardiomyocytes indirectly by a cardiac c-kit+AT$_2$+ cell population. Stimulation of a subpopulation of cardiac c-kit+AT$_2$+ cells with C21 post-MI *in vivo* resulted in the protection of cardiomyocytes from apoptosis. The antiapoptotic mechanism in this study was attributed to activation of Stat-3 and Akt signaling pathways.[14]

Altogether, recent data from post-MI studies provide evidence that AT$_2$R stimulation limits apoptosis and, therefore, limits LV dysfunction and remodeling after MI. However, since AT$_2$R activation has also been described to act proapoptotic under certain conditions (for review, see Steckelings et al.,[5] Namsolleck et al.[38]), further studies are needed to clarify this controversy.

Inflammation

In the early phase after acute MI, necrosis of cardiac cells sets into motion an inflammatory response that ultimately promotes fibrosis, scar formation, and ultimately HF.[20] Our group has recently reported that 1 week post-MI, C21 treatment decreased cardiac-related marker of inflammation myeloperoxidase and cytokines MCP-1, IL-1ß, IL-2, and IL-6 in plasma and the peri-infarct zone. Importantly, this effect was a direct AT$_2$R effect confirmed by using the AT$_2$R antagonist, PD 123319.[25]

The study by Curato et al.[13] identified 1 week post-MI a CD8$^+$AT$_2$R$^+$ T-cell population, which exhibited downregulated expression of proinflammatory cytokines (IL-2 and INF-γ). AT$_2$R activation by C21 engendered an increase in cardiac CD8$^+$AT$_2$R$^+$ T cells and IL-10 production most likely contributing to reduced heart injury. Moreover, intramyocardial transplantation of CD8$^+$AT$_2$R$^+$ T cells reduced infarct size in recipient rats post-MI.[13] The anti-inflammatory action of direct AT$_2$R stimulation by C21 is attributed to inhibition of JAK/STAT signaling inhibition of NF-kB and inhibition of COX-2 synthesis (for review, see Namsolleck et al.[38]).

Tissue Regeneration

Recent studies also implicate the AT$_2$R in tissue regeneration processes. Altarche-Xifró et al.[14] demonstrated the post-MI expression of AT$_2$ receptors within cardiac c-kit+precursor cells in the peri-infarct zone. This isolated cardiac c-kit+AT$_2$+ cell population was characterized by upregulation of transcription factors responsible for cardiogenic differentiation (Gata-4, Notch-2, and Nkx-2.5) and genes required for self-renewal (Tbx-3, c-Myc, and Akt).[14]

Based on these findings, a novel strategy to improve cardiac repair by preconditioning bone marrow mononuclear cells (BMMNCs) via AT$_2$R stimulation has been proposed.[34] Preconditioning of BMMNCs via AT$_2$R stimulation directly with an AT$_2$R agonist CGP42112A or indirectly with angiotensin II plus the AT$_1$R antagonist valsartan led to ERK activation and increased NO generation via eNOS. Moreover, transplantation of BMMNCs preconditioned via AT$_2$R stimulation in the heart decreased infarct size and improved cardiac function in the late phase post-MI.[34]

LATE-PHASE POSTMYOCARDIAL INFARCTION
Cardiac Function

The effect of long-term, direct AT$_2$R stimulation by C21 on cardiac remodeling has been investigated by our group 6 weeks after experimental MI in the rat.[39] Six-week postinfarct treatment with C21 prevented LV dilatation and dysfunction as evidenced by an improvement in cardiac parameters (LVIDd, LVIDs, EF, FS, dP/dt_{max}, and dP/dt_{min}) and contractility index as well as a reduction in the E/A ratio and LV EDP. Importantly, C21 prevented both systolic and diastolic dysfunctions of the heart.[39] Similarly, AT_2R overexpression via administration of rAAV9-CBA-AT_2R to the peri-infarcted myocardium area attenuated the decrease in fractional shortening and the rate of change of LV pressure, increased LV end-diastolic pressure, and ventricular hypertrophy.[40] In contrast, AT_2R blockade reduced cardiac function after rat MI.[41]

Further studies in genetically altered mice also demonstrated AT_2R-mediated protective effects on cardiac performance in the late phase post-MI.[42,23] However, Tschöpe et al.[43] reported no impact of AT_2R deficiency on post-MI LV function or fibrosis. Jehle et al.[44] also did not observe an improvement of LV remodeling after C21 treatment in a mouse occlusion-reperfusion model. However, the lack of effect in this study might have been due to underdosing.

Cardiac Fibrosis

In the late-phase post-MI, adverse postischemic cardiac remodeling is characterized by fibrotic processes, collagen deposition, and extracellular matrix degradation.[20]

Direct chronic AT_2R stimulation with C21 decreased interstitial fibrosis and collagen accumulation in the remote myocardium 6 weeks post-MI.[39] In addition, C21 reduced the increased augmentation index, which characterizes arterial stiffness and correlates with vascular fibrosis.[39] Treatment with C21[39] as well as AT_2R overexpression[40] attenuated increased collagen I expression post-MI. The antifibrotic effect of the AT_2 agonist was also demonstrated in the aorta of L-NAME-treated hypertensive rats[45] as well as in the vessels and the heart of spontaneously hypertensive rats.[46] Collectively, these findings provide evidence of antifibrotic activity of the AT_2 receptor stimulation in the heart and blood vessels.

In our study, the antifibrotic effect was associated with a massive downregulation of the profibrotic cytokine transforming growth factor beta 1 (TGF-β1).[39] Moreover, C21 attenuated the TGF-β1 mRNA increase in primary cardiac fibroblasts, suggesting its regulation through a direct effect of AT_2R stimulation.[39] A recent study by Xu et al.[34] confirmed the AT_2R-mediated downregulation of TGF-β1 in mice AT_2−/− model of MI and explained this effect via an increased NO production. This group also examined the effect of AT_2R gene overexpression (one, four, or nine copies of the AT_2R transgene) on LV remodeling in mice post-MI. The authors concluded that whether overexpression of AT_2R is beneficial or detrimental to the heart is largely dependent on the expression levels and possibly via regulation of Nox2 and TGF-β1 signaling pathways.[35]

Evidence for the involvement of kinins via B1 and B2 receptors in the AT_2R-mediated antifibrosis and functional improvement has been provided in the rat HF model.[30] More recently, Xu et al.[34] found a relationship between AT_2R, B_1 receptors, B_2 receptors, and cardiac protection via increased NO production and suppressed inflammatory cell infiltration, TGF-$β_1$ protein expression, and ERK1/2 phosphorylation. Although the above studies suggest a functional link between AT_2R and TGF-β1 in remodeling processes, a causal relationship has not been proven until recently.

Cardiac Hypertrophy

Left ventricular remodeling post-MI is associated with hypertrophy in noninfarcted myocytes initiated by wall stress, local stretch, and activation of the local renin–angiotensin system.[20] Activation of the AT_2R was proposed to counteract growth effects of the AT_1R in response to Ang II.[2] In our early studies, the antigrowth effects of the AT_2R have been found in cultured coronary endothelial cells, neonatal cardiomyocytes, and fibroblasts (for review, see Unger[4]). However, until now, the contribution of the AT_2R to cardiac hypertrophy is inconsistent. AT_2R stimulation has been reported to lead to either stimulation of cardiac hypertrophy or its inhibition or may even not affect hypertrophy (for review, see Steckelings et al.[47]). At least part of this inconsistency relates to the experimental approaches used.

AT_2R overexpression by lentiviral gene transfer into the left ventricle of SHR decreased left ventricular wall thickness,[48] and intracardiac gene transfer attenuated cardiac hypertrophy in rats subjected to an Ang II infusion.[49] Consistently, aortic banding in mice with overexpression of AT_2R-TG resulted in decreased cardiomyocyte diameter.[50] Finally, after experimental MI in mice lacking the AT_2R, myocardial cross-sectional areas[32] and heart/body weight ratios were increased.[33,42] The antihypertrophic effects have been attributed to activation of protein tyrosine phosphatase SHP-1 and interaction with an AT_2R binding protein ATBP (reviewed in Funke-Kaiser et al.[51])

However, AT_2R stimulation can also promote growth in the myocardium. D'Amore et al.[52] infected rat neonatal cardiomyocytes with recombinant adenoviruses expressing different ratios of rat AT_1R and AT_2R to study their combined effects on hypertrophy in response to Ang II. Increased AT_2R expression resulted in Ang II-independent hypertrophy of cardiomyocytes.[52] In our study, 6 weeks after MI, the selective AT_2R agonist C21 reinforced post-MI hypertrophy of cardiac myocytes by increasing the cell diameters.[39] A similar trend was observed by Voros et al.[53] in a mouse model of MI. Since hypertrophy after MI is important for compensation of damaged myocardium and preservation of cardiac function, an induction of compensatory hypertrophy post-MI may be considered as a positive mechanism, by which the AT_2R stimulation contributes to functional improvement of the heart.

Collectively, it appears that the AT$_2$R modulates the cardiac hypertrophic process by selectively regulating the expression of growth-promoting and growth-inhibiting factors depending on the pathophysiological conditions.

Extracellular Matrix Degradation

The degradation of matrix components by MMPs is another mechanism that may contribute to LV wall thinning in the remote region after MI.[20] Therefore, it has been suggested that HF can be modified by modulating MMP activity.

The main regulators of extracellular matrix in cardiac tissue are gelatinases MMP2 and MMP9 because they possess substrate affinity for denatured fibrillar collagen and exhibit proteolytic activity against elastin and proteoglycans.[54] In our study, an increased concentration of MMP9 and MMP9/TIMP1 protein expression ratio in cardiac tissue post-MI reflected a proteolytic disbalance, which was ameliorated by treatment with C21.[39] Moreover, experiments performed in single cardiac fibroblasts suggest that C21 primarily activates TIMP1, which in turn inhibits MMP9.[41] Since TIMP1 deficiency plays a negative role in post-MI heart remodeling by promoting LV dilatation and increasing LVEDV,[54] this may be one of the major mediators of the cardioprotective effects of C21.

CONCLUSION

Current studies clearly demonstrate that the AT$_2$R exerts beneficial effects on heart function in the early phase after MI by activating NO production, by suppressing postinfarct inflammatory response, and by antiapoptosis and stimulation of tissue regeneration. Antifibrosis, prevention of extracellular matrix degradation, and attenuation of adverse myocardial remodeling contribute to functional improvement in the late phase after MI. Direct AT$_2$R stimulation could represent a novel therapeutic concept in postischemic left ventricular remodeling. More information is needed on the chronic effects of the ATR stimulation in HF and on possible additional benefits of combined treatment with an AT$_2$R agonist and an AT$_1$R antagonist.

REFERENCES

1. Gajarsa JJ, Kloner RA. Left ventricular remodeling in the post-infarction heart: a review of cellular, molecular mechanisms, and therapeutic modalities. *Heart Fail Rev* 2011;**16**:13–21.
2. De Gasparo M, Catt KJ, Inagami T, Wright JW, Unger T. International union of pharmacology. XXIII. The angiotensin II receptors. *Pharmacol Rev* 2000;**52**(3):415–72.
3. Kaschina E, Unger T. Angiotensin AT1/AT2 receptors: regulation, signalling and function. *Blood Press* 2003;**12**(2):70–88, Review.
4. Unger T. The angiotensin type 2 receptor: variations on an enigmatic theme. *J Hypertens* 1999;**17**(12 Pt 2):1775–86.
5. Steckelings UM, Kaschina E, Unger T. The AT2 receptor—a matter of love and hate. *Peptides* 2005;**26**(8):1401–9.
6. Carey RM. Cardiovascular and renal regulation by the angiotensin type 2 receptor: the AT2 receptor comes of age. *Hypertension* 2005;**45**(5):840–4.
7. Widdop RE, Jones ES, Hannan RE, Gaspari TA. Angiotensin AT2 receptors: cardiovascular hope or hype? *Br J Pharmacol* 2003;**140**:809–24.
8. Kaschina E, Lauer D, Schmerler P, Unger T, Steckelings UM. AT2 receptors targeting cardiac protection post-myocardial infarction. *Curr Hypertens Rep* 2014;**16**(7):441. doi:10.1007/s11906-014-0441-0.
9. Wan Y, Wallinder C, Plouffe B, Beaudry H, Mahalingam AK, Wu X, et al. Design, synthesis, and biological evaluation of the first selective nonpeptide AT2 receptor agonist. *J Med Chem* 2004;**47**(24):5995–6008.
10. Busche S, Gallinat S, Bohle RM, Reinecke A, Seebeck J, Franke F. Expression of angiotensin AT(1) and AT(2) receptors in adult rat cardiomyocytes after myocardial infarction. A single-cell reverse transcriptase-polymerase chain reaction study. *Am J Pathol* 2000;**157**(2):605–11.
11. Nio Y, Matsubara H, Murasawa S, Kanasaki M, Inada M. Regulation of gene transcription of angiotensin II receptor subtypes in myocardial infarction. *J Clin Invest* 1995;**95**(1):46–54.
12. Zhu YZ, Zhu YC, Li J, Schäfer H, Schmidt W, Yao T, et al. Effects of losartan on haemodynamic parameters and angiotensin receptor mRNA levels in rat heart after myocardial infarction. *J Renin Angiotensin Aldosterone Syst* 2000;**1**(3):257–62.
13. Curato C, Slavic S, Dong J, Skorska A, Altarche-Xifró W, Miteva K, et al. Identification of noncytotoxic and IL-10-producing CD8 + AT2R + T cell population in response to ischemic heart injury. *J Immunol* 2010;**185**(10):6286–93.
14. Altarche-Xifró W, Curato C, Kaschina E, Grzesiak A, Slavic S, Dong J, et al. Cardiac c-kit + AT2+ cell population is increased in response to ischemic injury and supports cardiomyocyte performance. *Stem Cells* 2009;**27**:2488–97.
15. Lax CJ, Domenighetti AA, Pavia JM, Di Nicolantonio R, Curl CL, Morris MJ, et al. Transitory reduction in angiotensin AT2 receptor expression levels in postinfarct remodelling in rat myocardium. *Clin Exp Pharmacol Physiol* 2004;**31**(8):512–7.
16. Matsumoto T, Ozono R, Oshima T, Matsuura H, Sueda T, Kajiyama G, et al. Type 2 angiotensin II receptor is downregulated in cardiomyocytes of patients with heart failure. *Cardiovasc Res* 2000;**46**(1):73–81.
17. Tsutsumi Y, Matsubara H, Ohkubo N, Mori Y, Nozawa Y, Murasawa S, et al. Angiotensin II type 2 receptor is upregulated in human heart with interstitial fibrosis, and cardiac fibroblasts are the major cell type for its expression. *Circ Res* 1998;**83**(10):1035–46.

18. Ohkubo N, Matsubara H, Nozawa Y, Mori Y, Murasawa S, Kijima K, et al. Angiotensin type 2 receptors are reexpressed by cardiac fibroblasts from failing myopathic hamster hearts and inhibit cell growth and fibrillar collagen metabolism. *Circulation* 1997;**96**(11):3954–62.

19. Wharton J, Morgan K, Rutherford RA, Catravas JD, Chester A, Whitehead BF, et al. Differential distribution of angiotensin AT2 receptors in the normal and failing human heart. *J Pharmacol Exp Ther* 1998;**284**(1):323–36.

20. Sun Y. Myocardial repair/remodelling following infarction: roles of local factors. *Cardiovasc Res* 2009;**81**:482–90.

21. Adachi Y, Saito Y, Kishimoto I, Harada M, Kuwahara K, Takahashi N, et al. Angiotensin II type 2 receptor deficiency exacerbates heart failure and reduces survival after acute myocardial infarction in mice. *Circulation* 2003;**107**:2406–8.

22. Ichihara S, Senbonmatsu T, Price Jr E, Ichiki T, Gaffney FA, Inagami T. Targeted deletion of angiotensin II type 2 receptor caused cardiac rupture after acute myocardial infarction. *Circulation* 2002;**106**:2244–9.

23. Yang Z, Bove CM, French BA, Epstein FH, Berr SS, DiMaria JM, et al. Angiotensin II type 2 receptor overexpression preserves left ventricular function after myocardial infarction. *Circulation* 2002;**106**:106–11.

24. Bove CM, Gilson WD, Scott CD, Epstein FH, Yang Z, Dimaria JM, et al. The angiotensin II type 2 receptor and improved adjacent region function post-MI. *J Cardiovasc Magn Reson Off J Soc Cardiovasc Magn Reson* 2005;**7**:459–64.

25. Kaschina E, Grzesiak A, Li J, Foryst-Ludwig A, Timm M, Rompe F, et al. Angiotensin II type 2 receptor stimulation: a novel option of therapeutic interference with the renin-angiotensin system in myocardial infarction? *Circulation* 2008;**118**:2523–32.

26. Linz W, Wiemer G, Gohlke P, Unger T, Schölkens BA. Contribution of kinins to the cardiovascular actions of angiotensin-converting enzyme inhibitors. *Pharmacol Rev* 1995;**47**(1):25–49.

27. Rhaleb NE, Yang XP, Carretero OA. The kallikrein-kinin system as a regulator of cardiovascular and renal function. *Compr Physiol* 2011;**1**(2):971–93.

28. Wiemer G, Scholkens BA, Wagner A, et al. The possible role of angiotensin II subtype AT2 receptors in endothelial cells and isolated ischemic rat hearts. *J Hypertens Suppl* 1993;**11**(Suppl 5):S234–5.

29. Gohlke P, Pees C, Unger T. AT2 receptor stimulation increases aortic cyclic GMP in SHRSP by a kinin-dependent mechanism. *Hypertension* 1998;**31**(1 Pt 2):349–55.

30. Liu YH, Yang XP, Sharov VG, Nass O, Sabbah HN, Peterson E, et al. Effects of angiotensin-converting enzyme inhibitors and angiotensin II type 1 receptor antagonists in rats with heart failure. Role of kinins and angiotensin II type 2 receptors. *J Clin Invest* 1997;**99**(8):1926–35.

31. Jalowy A, Schulz R, Dörge H, Behrends M, Heusch G. Infarct size reduction by AT1-receptor blockade through a signal cascade of AT2-receptor activation, bradykinin and prostaglandins in pigs. *J Am Coll Cardiol* 1998;**32**:1787–96.

32. Kurisu S, Ozono R, Oshima T, Kambe M, Ishida T, Sugino H, et al. Cardiac angiotensin II type 2 receptor activates the kinin/NO system and inhibits fibrosis. *Hypertension* 2003;**41**:99–107.

33. Brede M, Roell W, Ritter O, Wiesmann F, Jahns R, Haase A, et al. Cardiac hypertrophy is associated with decreased eNOS expression in angiotensin AT2 receptor-deficient mice. *Hypertension* 2003;**42**:1177–82.

34. Xu Y, Hu X, Wang L, Jiang Z, Liu X, Yu H, et al. Preconditioning via angiotensin type 2 receptor activation improves therapeutic efficacy of bone marrow mononuclear cells for cardiac repair. *PLoS One* 2013;**8**:e82997.

35. Xu J, Sun Y, Carretero OA, Zhu L, Harding P, Shesely EG, et al. Effects of cardiac overexpression of the angiotensin II type 2 receptor on remodeling and dysfunction in mice post-myocardial infarction. *Hypertension* 2014;**63**(6):1251–9.

36. Abadir PM, Foster DB, Crow M, Cooke CA, Rucker JJ, Jain A, et al. Identification and characterization of a functional mitochondrial angiotensin system. *Proc Natl Acad Sci U S A* 2011;**108**(36):14849–54.

37. Sandmann S, Yu M, Kaschina E, Blume A, Bouzinova E, Aalkjaer C, et al. Differential effects of angiotensin AT1 and AT2 receptors on the expression, translation and function of the Na$+$−H$+$ exchanger and Na$+$−HCO3- symporter in the rat heart after myocardial infarction. *J Am Coll Cardiol* 2001;**37**(8):2154–65.

38. Namsolleck P, Recarti C, Foulquier S, Steckelings UM, Unger T. AT(2) receptor and tissue injury: therapeutic implications. *Curr Hypertens Rep* 2014;**16**(2):416.

39. Lauer D, Slavic S, Sommerfeld M, Thöne-Reineke C, Sharkovska Y, Hallberg A, et al. Angiotensin type 2 receptor stimulation ameliorates left ventricular fibrosis and dysfunction via regulation of tissue inhibitor of matrix metalloproteinase 1/matrix metalloproteinase 9 axis and transforming growth factor β1 in the rat heart. *Hypertension* 2014;**63**:e60–7.

40. Qi Y, Li H, Shenoy V, Li Q, Wong F, Zhang L, et al. Moderate cardiac-selective overexpression of angiotensin II type 2 receptor protects cardiac functions from ischaemic injury. *Exp Physiol* 2012;**97**(1):89–101.

41. Kuizinga MC, Smits JF, Arends JW, Daemen MJAP. AT2 receptor blockade reduces cardiac interstitial cell DNA synthesis and cardiac function after rat myocardial infarction. *J Mol Cell Cardiol* 1998;**30**(2):425–34.

42. Oishi Y, Ozono R, Yano Y, Teranishi Y, Akishita M, Horiuchi M, et al. Cardioprotective role of AT2 receptor in postinfarction left ventricular remodeling. *Hypertension* 2003;**41**:814–8.

43. Tschöpe C, Westermann D, Dhayat N, Dhayat S, Altmann C, Steendijk P, et al. Angiotensin AT2 receptor deficiency after myocardial infarction: its effects on cardiac function and fibrosis depend on the stimulus. *Cell Biochem Biophys* 2005;**43**:45–52.

44. Jehle AB, Xu Y, Dimaria JM, French BA, Epstein FH, Berr SS, et al. A nonpeptide angiotensin II type 2 receptor agonist does not attenuate post-myocardial infarction left ventricular remodeling in mice. *J Cardiovasc Pharmacol* 2012;**59**:363–8.

45. Paulis L, Becker ST, Lucht K, Schwengel K, Slavic S, Kaschina E, et al. Direct AT2 receptor stimulation in L-NAME-induced hypertension: the effect on pulse wave velocity and aortic remodeling. *Hypertension* 2012;**59**:485–92.

46. Rehman A, Leibowitz A, Yamamoto N, Rautureau Y, Paradis P, Schiffrin EL. Angiotensin type 2 receptor agonist compound 21 reduces vascular injury and myocardial fibrosis in stroke-prone spontaneously hypertensive rats. *Hypertension* 2012;**59**(2):291–9.

47. Steckelings UM, Widdop RE, Paulis L, Unger T. The angiotensin AT2 receptor in left ventricular hypertrophy. *J Hypertens* 2010 Sep;**28**(Suppl 1): S50–5.

48. Falcon BL, Stewart JM, Bourassa E, Katovich MJ, Walter G, Speth RC, et al. Angiotensin II type 2 receptor gene transfer elicits cardioprotective effects in an angiotensin II infusion rat model of hypertension. *Physiol Genom* 2004;**19**:255–61.

49. Metcalfe BL, Huentelman MJ, Parilak LD, Taylor DG, Katovich MJ, Knot HJ, et al. Prevention of cardiac hypertrophy by angiotensin II type-2 receptor gene transfer. *Hypertension* 2004;**43**:1233–8.

50. Yan X, Schuldt AJ, Price RL, Amende I, Liu FF, Okoshi K, et al. Pressure overload-induced hypertrophy in transgenic mice selectively overexpressing AT2 receptors in ventricular myocytes. *Am J Physiol Heart Circ Physiol* 2008;**294**(3):H1274–81.

51. Funke-Kaiser H, Reinemund J, Steckelings UM, Unger T. Adapter proteins and promoter regulation of the angiotensin AT2 receptor—implications for cardiac pathophysiology. *J Renin-Angiotensin-Aldosterone Syst* 2010;**11**(1):7–17.

52. D'Amore A, Black MJ, Thomas WG. The angiotensin II type 2 receptor causes constitutive growth of cardiomyocytes and does not antagonize angiotensin II type 1 receptor-mediated hypertrophy. *Hypertension* 2005;**46**(6):1347–54.

53. Voros S, Yang Z, Bove CM, Gilson WD, Epstein FH, French BA, et al. Interaction between AT1 and AT2 receptors during postinfarction left ventricular remodeling. *Am J Physiol Heart Circ Physiol* 2006;**290**:H1004–10.

54. Spinale FG. Myocardial matrix remodeling and the matrix metalloproteinases: influence on cardiac form and function. *Physiol Rev* 2007;**87**(4):1285–342.

Chapter 7

AT$_2$R, Vascular Effects, and Blood Pressure

Dhaniel Baraldi*, Emma S. Jones*, Lucinda M. Hilliard†, Mark Del Borgo‡, Tracey A. Gaspari*, Claudia A. McCarthy*, Antony Vinh*, Iresha Welungoda*, Marie-Isabel Aguilar‡, Kate M. Denton†, Robert E. Widdop*

*Department of Pharmacology, Monash University, Clayton Victoria 3800, Australia, †Department of Physiology, Monash University, Clayton Victoria 3800, Australia, ‡Department of Biochemistry and Molecular Biology, Monash University, Clayton Victoria 3800, Australia

INTRODUCTION

The renin–angiotensin system continues to occupy pharmacologists and physicians alike for well over 50 years, as a result of a number of important recent discoveries, including (i) the discovery of subtype 1 and subtype 2 angiotensin receptors (AT$_1$R and AT$_2$R, respectively) that heralded a new era of drug discovery; (ii) the development and study of nonpeptide AT$_1$R antagonists that led to the concept that AT$_2$R stimulation may contribute to the cardiovascular benefit derived from sartan compounds, at least in preclinical studies; and (iii) academic research that championed the view that AT$_2$R agonists may offer therapeutic benefit for a number of cardiovascular diseases, as discussed at length in this book. This chapter will discuss vascular AT$_2$R effects and their impact on blood pressure (BP).

AT$_2$ VASORELAXATION: *IN VITRO* PRECLINICAL STUDIES

Given the well-recognized and ubiquitous expression of AT$_1$Rs within the cardiovascular system and beyond, it took some time for the realization that AT$_2$Rs were indeed expressed in vasculature, albeit at lower levels than AT$_1$Rs.[1] It is generally accepted that AT$_2$Rs are expressed in a variety of blood vessels, including mesenteric, renal, coronary, and uterine arteries, to name a few.[1–3] AT$_2$R signaling pathways are complex, but the most consistently described mechanism occurs in vasculature, whereby AT$_2$R stimulation increases the nitric oxide/cGMP axis, often involving bradykinin as well.[1–3] In most instances, AT$_2$R-mediated relaxation is quite subtle and results in ~30-40% reduction of precontracted vascular tone. In circumstances where direct AT$_2$R-mediated vasorelaxation is not apparent, these effects have been inferred from enhanced Ang II (AT$_1$R)-mediated contraction in the presence of the AT$_2$R antagonist PD 123319.[4,5] In early studies, AT$_2$R actions were elucidated using Ang II in the presence of an AT$_1$R antagonist,[6] but the current best practice uses either the peptide agonist CGP42112[7] or the nonpeptide agonist compound 21 (C21),[8] both of which are highly AT$_2$R-selective on the basis of binding studies.[9]

ACUTE AT$_2$R VASODILATATION/BP EFFECTS: *IN VIVO* PRECLINICAL STUDIES

Surprisingly, acute infusions of AT$_2$R agonists in animals do not generally lower BP. For example, CGP42112 infused intravenously over 4 hours failed to reduce BP in conscious SHR. However, when a low dose of candesartan was administered as a bolus, CGP42112 evoked a depressor effect that was abolished when concomitantly infused with the AT$_2$R antagonist PD 123319.[10] The BP-lowering effects of CGP42112 were caused by peripheral vasodilatation, as was shown in conscious SHR that were chronically instrumented to measure renal, mesenteric, and hindquarter (terminal aorta) blood flows and conductances simultaneously.[11] In those studies, CGP42112 caused regional vasodilatation, again only during low-dose AT$_1$R blockade. Interestingly, PD 123319 itself caused regional vasoconstriction and increased BP, a finding that suggested tonic vasodilator effects of AT$_2$R. Similarly, we first reported that C21 (50–300 ng/kg/min) acutely decreased BP in conscious SHR but only with AT$_1$R blockade, whereas a high dose of C21 (1000 ng/kg/min) increased BP via AT$_1$R stimulation.[12] Of note, C21 relaxed mouse and rat aortic tissue in the absence of AT$_1$R blockade.[12] More recently, Hilliard and colleagues have demonstrated that AT$_2$R blockade enhanced the renal vasoconstrictor responses to angiotensin II in females but not in males[13] and that C21 evoked greater dose-dependent renal vasodilatation in female rats without any AT$_1$R blockade.[14]

In contrast, in a recent study, C21 failed to alter BP or renal flow, with or without AT_1R blockade, when cumulative bolus doses of C21 were given intravenously in anesthetized SHR.[15] These findings are at variance with those of Bosnyak et al.[12] and Hilliard et al.[13,14] and may relate to different methodologies under nonequilibrium conditions. Interestingly, when SHR were given captopril acutely at the commencement of their protocol, Brouwers et al. reported that C21 evoked renal vasodilatation.[15] The lack of effect of C21 seen initially was attributed to high circulating levels of Ang II in response to candesartan, which then competed with C21,[15] although plasma Ang II levels were not actually measured in this study nor is this theory consistent with recent kinetic modeling suggesting that high concentrations of Ang II are needed to activate AT_2Rs.[16]

There are relatively few compounds that have actually been tested in BP assays, apart from CGP42112 and C21. On this point, our group has developed some novel AT_2R ligands, based on β-amino acid substitutions to the native Ang II peptide. From our initial binding screen, we chose to examine the *in vivo* effects of two compounds: β-Tyr⁴-Ang II and β-Ile⁵-Ang II[17] (see Chapter 19). We reported that β-Ile⁵-Ang II decreased BP, in the presence of AT_1R blockade, in an analogous manner to that of CGP42112 and C21, whereas β-Tyr⁴-Ang II failed to lower BP despite exhibiting classical AT_2R/nitric oxide function *in vitro*.[17]

CHRONIC AT_2R EFFECTS ON BP: PRECLINICAL STUDIES

Data from AT_2RKO studies infer that AT_2R has inhibitory effects on BP since increased basal BP and/or potentiated pressor responses to Ang II were seen.[18,19] In mice with targeted overexpression of vascular AT_2R,[20] there was no change in basal BP, although acute Ang II infusion decreased BP. We and others have consistently reported that acute infusions (over several hours) of AT_2R agonists failed to lower BP unless there was a background level of AT_1R blockade, which in itself was subdepressor.[10–12,17] However, Carey and colleagues reported that CGP42112 evoked a reduction in systolic BP in conscious rats over 9 days that was enhanced in the presence of valsartan. Moreover, CGP42112 (100 μg/kg/day), alone, in fact lowered systolic BP over 4–7 days in sodium-replete or sodium-depleted rats.[21] However, due to limited supplies of test compounds, the effect of pharmacological AT_2R stimulation on BP and organ remodeling has only recently been performed in a handful of studies. C21 has been used in a number of experimental models in treatments ranging from 4 days to 6 weeks (see Table 1). In those studies,[23,24,26,28,29] C21 did not reduce BP, either alone or in combination with a sartan, although there was substantial organ protection, including antifibrotic and anti-inflammatory effects, which will be discussed in separate chapters. On the other hand, 2-week treatment with CGP42112 lowered BP in obese hypertensive Zucker rats (anesthetized and conscious states[27,31]), but not in prehypertensive Zucker rats.[30] In other studies, C21 lowered BP when given centrally

TABLE 1 Chronic AT_2R Effects on BP (and Other Indexes) in Preclinical Studies

Experimental Model	Intervention/Protocol	Effects	References
Conscious rats (noninvasive SBP)	CGP42112 (100 μg/kg/min SC) +/−valsartan; sodium-depleted or sodium-replete SD rats	CGP42112 decreased SBP alone (over 4–7 days) and potentiated by valsartan (over 9 days)	[21]
Conscious female SD rats (telemeterized BP)	Ang II (50 and 400 ng/kg/min SC; 2 weeks)	Low-dose Ang II decreased BP in female but not male rats; blocked by PD123319	[22]
Myocardial infarction (MI) (anesthetized BP)	C21 (0.03 and 0.3 mg/kg/day IP; 1 week, post-MI)	C21 did not alter terminal BP but improved cardiac function post-MI	[23]
Salt-fed SP-SHR (noninvasive SBP)	C21 (0.75, 5, and 10 mg/kg/day PO; lifetime)	C21 (10 mg/kg/day) reduced brain damage and preserved kidney structure but SBP unaffected	[24]
Obese Zucker rats (anesthetized BP)	PD123319 (30 μg/kg/min SC; 2 weeks)	PD123319 increased terminal BP	[25]
2K1C rats (noninvasive SBP)	C21 (0.3 mg/kg/day IP; 4 days)	SBP unchanged; C21 reduced renal inflammation	[26]
Obese Zucker rats (anesthetized BP)	CGP42112 (1 μg/kg/min SC; 2 weeks)	CGP42112 reduced terminal BP in obese (hypertensive) but not lean (normotensive) rats; opposite effects on inflammation in 2 models	[27]

TABLE 1 Chronic AT$_2$R Effects on BP (and Other Indexes) in Preclinical Studies—Cont'd

Experimental Model	Intervention/Protocol	Effects	References
Stroke-prone SHR (noninvasive SBP)	C21 (1 mg/kg/day PO; 6 weeks)	C21 initially increased SBP but then waned; no additive effect on BP when combined with ARB; antifibrotic	[28]
L-NAME-treated rats (noninvasive SBP)	C21 (0.3 mg/kg/day IP; 6 weeks)	C21 did not prevent L-NAME-induced hypertension; no additive effect on BP when combined with ARB; caused vascular remodeling	[29]
Prehypertensive obese Zucker rats (noninvasive SBP)	C21 (300 µg/kg/day IP; 2 weeks)	C21 did not alter BP but was anti-inflammatory	[30]
Obese Zucker rats (noninvasive SBP)	CGP42112 (1 µg/kg/min SC; 2 weeks)	CGP42112 decrease SBP and increased sodium excretion	[31]

over 7 days,[32] although the clinical relevance of this effect is not known. Indeed, it is important to note that in virtually all chronic studies thus far reported, there are differences in experimental model, design, treatment duration, and timing, which clearly influence any interpretation. As an example, we have reported that chronic administration of low-dose Ang II lowered BP in only female rats (without AT$_1$R blockade), a seemingly paradoxical effect of Ang II, until one considers that renal AT$_2$Rs are upregulated in females.[22]

HUMAN VASCULAR AT$_2$R FUNCTION: *EX VIVO* AND *IN VIVO* STUDIES

There are limited human data (see Table 2) to compare with preclinical data. Nevertheless, there is evidence that AT$_2$R causes vasorelaxation of isolated vessels from healthy/treated subjects,[33,35] whereas, on the ability of PD 123319 to modify Ang II-induction contraction, AT$_2$R may cause either vasodilatation or vasoconstriction in diseased vessels.[34,36] The majority of acute *in vivo* studies have been performed by one research group and have involved infusing the AT$_2$R antagonist PD

TABLE 2 Vascular AT$_2$R Effects in Humans

Participants/Tissue	Intervention	Effects	References
Ex vivo			
Microcoronaries from healthy donors	Ang II contraction +/− PD123319; AT$_2$R-mediated relaxation	PD123319 potentiated AT$_1$R-mediated contraction; direct AT$_2$R relaxation (Ang II + ARB)	[33]
Internal mammary arteries (diseased)	Ang II contraction +/− PD123319; AT$_2$R-mediated relaxation	PD123319 slightly inhibited Ang II contraction but not significant; CGP42112 did not relax tissue but Ang II + ARB did so	[34]
Subcutaneous arteries from diabetics treated with losartan (1y)	Ang II-mediated relaxation in precontracted vessels	Ang II relaxed vessels from chronic losartan—but not atenolol—treatment; blocked by PD123319; also increased AT$_2$R expression	[35]
Radial arteries (diseased)	Ang II contraction +/− PD123319; AT$_2$R-mediated relaxation	PD123319 caused divergent effects: a predominant marked (~60%) potentiation or small (~20%) inhibition of Ang II contraction; CGP42112 did not relax; AT$_1$R and AT$_2$R upregulated in angiogenic vessels in occluded arteries	[36]

(Continued)

TABLE 2 Vascular AT$_2$R Effects in Humans—Cont'd

Participants/Tissue	Intervention	Effects	References
In vivo			
Healthy male volunteers	Ang II contraction +/− PD123319 (local infusion); venous occlusion plethysmography (VOPG)	Five min infusion of PD123319 (8 μg/min) had no effect on forearm vascular resistance (FVR) or Ang II vasoconstriction	[37]
Elderly women during 3 weeks' candesartan treatment	Ang II contraction +/− PD123319 (local infusion) by VOPG	Tonic AT$_2$R vasodilatation suggested by PD123319 infusion (8 μg/min; 5 min) increasing FVR in candesartan group; Ang II caused vasodilatation in cand group but not blocked by PD123319	[38]
Healthy young volunteers during 1-week placebo or telmisartan	PD123319 (local infusion) by VOPG	PD123319 infusion (10 μg/min) did not alter forearm blood flow (FBF) or FVR in either group but increased systemic BP	[39]
Healthy subjects	IV PD123319 (10 μg/min; 5 min) or placebo on systemic hemodynamics by electrical bioimpedance	No effect of systemic AT$_2$R blockade on BP or hemodynamics; not supporting a role for vascular AT$_2$R in healthy subjects	[40]
Patients with insulin resistance (INSR)	PD123319 infusion (10 μg/min; 3 min) on arterial stiffness in the index finger (digital volume pulse wave)	PD123319 increased arterial stiffness and systemic vascular resistance in INSR patients versus placebo; suggestive of functional AT$_2$R in small vessels	[41]
Patients with insulin resistance (INSR)	IV PD123319 (10 and 20 μg/min; 3 min) +/− Ang II on arterial stiffness in the index finger (digital volume pulse wave)	Increased arterial stiffness with either Ang II or PD123319 but no additive effects; suggestive of increased AT$_1$R vasoconstrictor and AT$_2$R vasodilator effects in small vessels	[42]
Healthy subjects; sequential local infusions	Ang II dose–response +/− telmisartan (40 μg/min); +/− PD123319 (10 μg/min). CGP42112 (2 and 4 μg/min) +/− nitric oxide clamp; +/− PD123319; all by VOPG	PD123319 inhibited Ang II vasoconstriction and increased FBF (small). In contrast, CGP42112 increased FBF dose-dependently, and to a much larger extent (~40%) than PD123319 (~10%)	[43]

123319 locally into the forearm while measuring forearm blood flow and forearm vascular resistance or vascular stiffness (index finger) noninvasively[37–42] in a range of patient populations and healthy subjects. These studies indirectly provide evidence for AT$_2$R-mediated vasodilatation although this finding is not always consistent (see Table 2). Interestingly, Schinzari et al. recently infused CGP42112 for the first time in humans and reported a striking dose-dependent forearm vasodilatation that was blocked by PD 123319.[43] However, in the same study, Ang II-mediated vasoconstriction was attenuated by PD 123319 or converted into a small vasodilator effect in the presence of telmisartan, suggesting that both AT$_2$R-mediated vasoconstriction and vasodilatation could be evoked in response to Ang II.[43]

AT$_2$R VASORELAXATION: WHY IS THERE NO TRANSLATION INTO ANTIHYPERTENSIVE EFFECTS IN CHRONIC STUDIES?

The most obvious explanation is that vascular AT$_1$R mediating vasoconstriction exerts an overriding effect such that AT$_2$R effects on BP are only unmasked when AT$_1$R are blocked. However, often only a subdepressor dose of a sartan is required in order to elicit an AT$_2$R-mediated depressor effect, which may argue against the importance of tonic AT$_1$R vasoconstrictor tone. Based on kinetic modeling, AT$_1$Rs are exquisitely sensitive to low tissue concentrations of Ang II such that AT$_1$R-mediated effects (such as vasoconstriction) would occur at much lower concentrations than required for AT$_2$R activation.[16]

Indeed, it was calculated that high plasma concentrations of Ang II are required to activate AT$_2$R,[16] which fits with the need to inhibit AT$_1$R to unmask vascular AT$_2$R effects. Therefore, it may be appropriate to inhibit AT$_1$Rs to optimize any long-term AT$_2$R effects. Surprisingly, there are relatively few studies that have examined the chronic effects of combined AT$_1$R blockade and AT$_2$R stimulation. Rehman et al. and Paulis et al. administered C21 for 6 weeks in either stroke-prone SHR[28] or L-NAME-treated rats,[29] but neither study reported an additive effect of these treatments on BP, presumably because AT$_1$R blockade exerted maximal effect. This fact, together with the lack of effect of AT$_2$R agonism alone, makes it highly unlikely that AT$_2$R would be considered as a therapeutic target for hypertension per se, although AT$_2$R-mediated organ protection was clearly evident in all studies that have given C21 for any extended period (Table 1). In this context, it is conceivable that cardioprotective effects (e.g., anti-inflammatory/antifibrotic/antiremodeling/natriuretic) of AT$_2$R (discussed in other chapters; see also Table 1) may contribute more subtle secondary reductions in BP, although current study protocols have not been designed to address this issue.

Another factor that may limit any antihypertensive effect of a novel AT$_2$R agonist clinically is the heterogeneous nature of vascular AT$_2$R function per se in humans, which may also relate to the diverse cardiovascular status of both volunteer and patient groups. While an AT$_2$R contractile phenotype has been noted under certain conditions using vessels obtained from experimental models, such as aging and hypertension,[44–46] AT$_2$R-mediated vasodilatation is more consistently reported. However, even with limited human data, it is apparent that AT$_2$R can exert opposing vascular effects (Table 2). Therefore, while highly speculative, it is conceivable that AT$_2$R-mediated vasoconstriction may offset AT$_2$R-mediated vasodilatation.

In addition, the lack of in vivo effects with some AT$_2$R ligands may relate to metabolic stability. An interesting example in this context is the action of β-Tyr^4-Ang II, which we reported to be highly selective for AT$_2$R over AT$_1$R (~1000-fold) and cause AT$_2$R-mediated vasorelaxation.[17] However, when given by infusion, it failed to lower BP in conscious SHR (even against a background of AT$_1$R blockade[17]), presumably because it was rapidly broken down. This was evident from in vitro stability studies where β-Tyr^4-Ang II exhibited only marginally greater stability than the native peptide, Ang II. On the other hand, another AT$_2$R-selective compound, β-Ile^5-Ang II, was ten-fold more stable than Ang II, and it also caused a marked BP-lowering effect during low-dose AT$_1$R blockade.[17] However, this factor is unlikely to explain the reason why the nonpeptide C21 did not alter BP when given acutely.[12]

In summary, the paucity of AT$_2$R ligands has meant that relatively few compounds have actually been tested in BP assays, except for CGP42112 and C21. Therefore, subtle variations in BP effects of AT$_2$R agonists as a class may yet arise with the development of more compounds, as has occurred when detailed analysis of many sartan compounds has revealed additional mechanisms of action for particular compounds.

In any case, even with unresolved differences between acute and chronic effects of AT$_2$R agonists on BP, functional bioassays, such as acute relaxation in vitro and acute BP reductions in vivo, still represent some of the most reliable methods by which to detect AT$_2$R function for future drug candidates!

ACKNOWLEDGMENTS

Studies from the authors' laboratories were supported in part by grants from the National Health and Medical Research Council of Australia and the National Heart Foundation of Australia.

REFERENCES

1. Widdop RE, Jones ES, Hannan RE, Gaspari TA. Angiotensin AT2 receptors: cardiovascular hope or hype? *Br J Pharmacol* 2003;**140**:809–24.
2. Henrion D, Kubis N, Levy BI. Physiological and pathophysiological functions of the AT(2) subtype receptor of angiotensin II: from large arteries to the microcirculation. *Hypertension* 2001;**38**(5):1150–7.
3. Jones ES, Vinh A, McCarthy CA, Gaspari TA, Widdop RE. AT2 receptors: functional relevance in cardiovascular disease. *Pharmacol Ther* 2008;**120**(3):292–316.
4. Zwart AS, Davis EA, Widdop RE. Modulation of AT1 receptor-mediated contraction of rat uterine artery by AT2 receptors. *Br J Pharmacol* 1998;**125**(7):1429–36.
5. Hannan RE, Davis EA, Widdop RE. Functional role of angiotensin II AT2 receptor in modulation of AT1 receptor-mediated contraction in rat uterine artery: involvement of bradykinin and nitric oxide. *Br J Pharmacol* 2003;**140**(5):987–95.
6. de Gasparo M, Catt KJ, Inagami T, Wright JW, Unger T. International union of pharmacology. XXIII. The angiotensin II receptors. *Pharmacol Rev* 2000;**52**(3):415–72.
7. Whitebread S, Mele M, Kamber B, de Gasparo M. Preliminary biochemical characterization of two angiotensin II receptor subtypes. *Biochem Biophys Res Commun* 1989;**163**(1):284–91.
8. Wan Y, Wallinder C, Plouffe B, Mahalingam AK, Wu X, Johansson B, et al. Design, synthesis, and biological evaluation of the first selective nonpeptide AT2 receptor agonist. *J Med Chem* 2004;**47**(24):5995–6008.

9. Bosnyak S, Jones ES, Christopoulos A, Aguilar MI, Thomas WG, Widdop RE. Relative affinity of angiotensin peptides and novel ligands at AT1 and AT2 receptors. *Clin Sci* 2011;**121**(7):297–303.

10. Barber MN, Sampey DB, Widdop RE. AT(2) receptor stimulation enhances antihypertensive effect of AT(1) receptor antagonist in hypertensive rats. *Hypertension* 1999;**34**(5):1112–16.

11. Li XC, Widdop RE. AT2 receptor-mediated vasodilatation is unmasked by AT1 receptor blockade in conscious SHR. *Br J Pharmacol* 2004;**142**(5):821–30.

12. Bosnyak S, Welungoda IK, Hallberg A, Alterman M, Widdop RE, Jones ES. Stimulation of angiotensin AT2 receptors by the non-peptide agonist, Compound 21, evokes vasodepressor effects in conscious spontaneously hypertensive rats. *Br J Pharmacol* 2010;**159**(3):709–16.

13. Hilliard LM, Nematbakhsh M, Kett MM, Teichman E, Sampson AK, Widdop RE, Evans RG, Denton KM. Gender differences in pressure-natriuresis and renal autoregulation: role of the angiotensin type 2 receptor. *Hypertension* 2011;**57**(2):275–82.

14. Hilliard LM, Jones ES, Steckelings UM, Unger T, Widdop RE, Denton KM. Sex-specific influence of angiotensin type 2 receptor stimulation on renal function. *Hypertension* 2012;**59**(2):409–14.

15. Brouwers S, Smolders I, Massie A, Dupont AG. Angiotensin II type 2 receptor-mediated and nitric oxide-dependent renal vasodilator response to compound 21 unmasked by angiotensin-converting enzyme inhibition in spontaneously hypertensive rats in vivo. *Hypertension* 2013;**62**(5):920–6.

16. Schalekamp MA, Danser AH. How does the angiotensin II type 1 receptor 'trump' the type 2 receptor in blood pressure control? *J Hypertens* 2013;**31**(4):705–12.

17. Jones ES, Del Borgo MP, Kirsch JF, Clayton D, Bosnyak S, Welungoda I, et al. A single beta-amino acid substitution to angiotensin II confers AT(2) receptor selectivity and vascular function. *Hypertension* 2011;**57**(3):570–6.

18. Hein L, Barsh GS, Pratt RE, Dzau VJ, Kobilka BK. Behavioural and cardiovascular effects of disrupting the angiotensin II type-2 receptor in mice [published erratum appears in Nature 1996;380(6572):366]. *Nature* 1995;**377**(6551):744–7.

19. Ichiki T, Labosky PA, Shiota C, Okuyama S, Imagawa Y, Fogo A, et al. Effects on blood pressure and exploratory behaviour of mice lacking angiotensin II type-2 receptor. *Nature* 1995;**377**(6551):748–50.

20. Tsutsumi Y, Matsubara H, Masaki H, Kurihara H, Murasawa S, Takai S, et al. Angiotensin II type 2 receptor overexpression activates the vascular kinin system and causes vasodilation. *J Clin Invest* 1999;**104**(7):925–35.

21. Carey RM, Howell NL, Jin XH, Siragy HM. Angiotensin type 2 receptor-mediated hypotension in angiotensin type-1 receptor-blocked rats. *Hypertension* 2001;**38**(6):1272–7.

22. Sampson AK, Moritz KM, Jones ES, Flower RL, Widdop RE, Denton KM. Enhanced angiotensin II type 2 receptor mechanisms mediate decreases in arterial pressure attributable to chronic low-dose angiotensin II in female rats. *Hypertension* 2008;**52**(4):666–71.

23. Kaschina E, Grzesiak A, Li J, Foryst-Ludwig A, Timm M, Rompe F, et al. Angiotensin II type 2 receptor stimulation: a novel option of therapeutic interference with the renin-angiotensin system in myocardial infarction? *Circulation* 2008;**118**(24):2523–32.

24. Gelosa P, Pignieri A, Fändriks L, de Gasparo M, Hallberg A, Banfi C, et al. Stimulation of AT2 receptor exerts beneficial effects in stroke-prone rats: focus on renal damage. *J Hypertens* 2009;**27**(12):2444–51.

25. Siddiqui AH, Ali Q, Hussain T. Protective role of angiotensin II subtype 2 receptor in blood pressure increase in obese Zucker rats. *Hypertension* 2009;**53**(2):256–61.

26. Matavelli LC, Huang J, Siragy HM. Angiotensin AT(2) receptor stimulation inhibits early renal inflammation in renovascular hypertension. *Hypertension* 2011;**57**(2):308–13.

27. Sabuhi R, Ali Q, Asghar M, Al-Zamily NR, Hussain T. Role of the angiotensin II AT2 receptor in inflammation and oxidative stress: opposing effects in lean and obese Zucker rats. *Am J Physiol Renal Physiol* 2011;**300**(3):F700–6.

28. Rehman A, Leibowitz A, Yamamoto N, Rautureau Y, Paradis P, Schiffrin EL. Angiotensin type 2 receptor agonist compound 21 reduces vascular injury and myocardial fibrosis in stroke-prone spontaneously hypertensive rats. *Hypertension* 2012;**59**(2):291–9.

29. Paulis L, Becker STR, Lucht K, Schwengel K, Slavic S, Kaschina E, et al. Direct angiotensin II type 2 receptor stimulation in N-nitro-l-arginine-methyl ester- induced hypertension. *Hypertension* 2012;**59**(2):485–92.

30. Dhande I, Ali Q, Hussain T. Proximal tubule angiotensin AT2 receptors mediate an anti-inflammatory response via interleukin-10: role in renoprotection in obese rats. *Hypertension* 2013;**61**(6):1218–26.

31. Ali Q, Wu Y, Hussain T. Chronic AT2 receptor activation increases renal ACE2 activity, attenuates AT1 receptor function and blood pressure in obese Zucker rats. *Kidney Int* 2013;**84**(5):931–9.

32. Gao J, Zhang H, Le KD, Chao J, Gao L. Activation of central angiotensin type 2 receptors suppresses norepinephrine excretion and blood pressure in conscious rats. *Am J Hypertens* 2011;**24**(6):724–30.

33. Batenburg WW, Garrelds IM, Bernasconi CC, Juillerat-Jeanneret L, van Kats JP, Saxena PR, et al. Angiotensin II type 2 receptor-mediated vasodilation in human coronary microarteries. *Circulation* 2004;**109**(19):2296–301.

34. van de Wal RM, van der Harst P, Wagenaar LJ, Wassmann S, Morshuis WJ, Nickenig G, et al. Angiotensin II type 2 receptor vasoactivity in internal mammary arteries of patients with coronary artery disease. *J Cardiovasc Pharmacol* 2007;**50**(4):372–9.

35. Savoia C, Touyz RM, Volpe M, Schiffrin EL. Angiotensin type 2 receptor in resistance arteries of type 2 diabetic hypertensive patients. *Hypertension* 2007;**49**(2):341–6.

36. Zulli A, Hare DL, Buxton BF, Widdop RE. Vasoactive role for angiotensin II type 2 receptors in human radial artery. *Int J Immunopathol Pharmacol* 2014;**27**(1):79–85.

37. Phoon S, Howes LG. Role of angiotensin type 2 receptors in human forearm vascular responses of normal volunteers. *Clin Exp Pharmacol Physiol* 2001;**28**(9):734–6.

38. Phoon S, Howes LG. Forearm vasodilator response to angiotensin II in elderly women receiving candesartan: role of AT(2)- receptors. *J Renin-Angiotensin-Aldosterone Syst* 2002;**3**(1):36–9.

39. Gilles R, Vingerhoedt N, Howes J, Griffin M, Howes LG. Increase in systemic blood pressure during intra-arterial PD123319 infusion: evidence for functional expression of angiotensin type 2 receptors in normal volunteers. *Blood Press* 2004;**13**(2):110–14.

40. Brillante DG, Johnstone MT, Howes LG. Effects of intravenous PD 123319 on haemodynamic and arterial stiffness indices in healthy volunteers. *J Renin-Angiotensin-Aldosterone Syst* 2005;**6**(2):102–6.

41. Brillante DG, O'Sullivan AJ, Johnstone MT, Howes LG. Evidence for functional expression of vascular angiotensin II type 2 receptors in patients with insulin resistance. *Diabetes Obesity Metabol* 2008;**10**(2):143–50.

42. Brillante DG, O'Sullivan AJ, Johnstone MT, Howes LG. Arterial stiffness and haemodynamic response to vasoactive medication in subjects with insulin-resistance syndrome. *Clin Sci* 2008;**114**(2):139–47.

43. Schinzari F, Tesauro M, Rovella V, Adamo A, Mores N, Cardillo C. Coexistence of functional angiotensin II type 2 receptors mediating both vasoconstriction and vasodilation in humans. *J Hypertens* 2011;**29**(9):1743–8.

44. Touyz RM, Endemann D, He G, Li JS, Schiffrin EL. Role of AT2 receptors in angiotensin II-stimulated contraction of small mesenteric arteries in young SHR. *Hypertension* 1999;**33**(1 Pt 2):366–72.

45. You D, Loufrani L, Baron C, Levy BI, Widdop RE, Henrion D. High blood pressure reduction reverses angiotensin II type 2 receptor-mediated vasoconstriction into vasodilation in spontaneously hypertensive rats. *Circulation* 2005;**111**(8):1006–11.

46. Pinaud F, Bocquet A, Dumont O, Retailleau K, Baufreton C, Andriantsitohaina R, et al. Paradoxical role of angiotensin II type 2 receptors in resistance arteries of old rats. *Hypertension* 2007;**50**(1):96–102.

Chapter 8

AT$_2$R in Nervous System

Pawel Namsolleck*, Juraj Culman†, Thomas Unger*

*CARIM School for Cardiovascular Diseases, Maastricht University, Maastricht, The Netherlands, †Institute of Experimental and Clinical Pharmacology, University Hospitals of Schleswig-Holstein, Campus Kiel, Kiel, Germany

INTRODUCTION

This chapter focuses on two processes, neuroprotection and neuroregeneration, that are mediated by the angiotensin AT$_2$ receptor (AT$_2$R). Neuroprotection can be defined as a process that directly prevents necrotic or apoptotic neuronal cell death (primary neuroprotection) or affords protection of myelin, axons, and neurons by, for example, anti-inflammation (secondary neuroprotection). Neuroregeneration can be defined as a complex process restoring the interrupted neuronal connectivity and resulting in functional recovery.

AT$_2$R EXPRESSION IN CNS

Distribution of the AT$_2$R in the central nervous system is species- and age-dependent (summarized by[1,2]). The expression of the AT$_2$R in the brain has a very distinct pattern and only a limited overlap with the angiotensin AT$_1$ receptor (AT$_1$R).[3] In the brains of adult rats, AT$_2$R mRNA is predominantly expressed in the thalamic nuclei, the medial geniculate nucleus, the nucleus of the optic tract, the subthalamic nucleus, the interposed nucleus of the cerebellum, and the inferior olive.[4,5] A similar distribution of the AT$_2$R was found in the mouse brain.[6] In contrast to rodents, in the human brain, AT$_2$R expression was reported to be mainly limited to the molecular layer of the cerebellar cortex, where it is coexpressed with the AT$_1$R.[7]

Differences in AT$_2$R expression during ontogeny have been studied in 2- and 8-week-old rats.[8] In all investigated brain regions, the AT$_2$R was much higher expressed in young than in adult animals.[8] When comparing embryonic life with postnatal life, the AT$_2$R was transiently expressed in differentiating the lateral hypothalamic area, the superior olivary complex, and the red nucleus.[9] In contrast, persistent expression of the AT$_2$R during late fetal life and neonatal life was found in the thalamus, cerebellum, and motor facial nucleus.[9]

The cell-specific expression of the AT$_2$R in the CNS was studied with quantitative PCR and Western blot. The AT$_2$R was expressed in all of the cerebral cell types investigated including neurons,[10] astrocytes,[10] microglia,[10,11] endothelial cells,[12,13] and smooth muscle cells.[13]

AT$_2$R EXPRESSION IN NEURONAL INJURY

The AT$_2$R expression in peripheral neuronal injury has been investigated in a rat model of sciatic nerve crush. The spontaneous regeneration of the injured sciatic nerves correlated with the spatiotemporal regulation of the AT$_2$R expression, reaching the maximum at 5 days in the proximal segment and at 20 days in the distal segment, as measured from a lesion center.[14] The expression of AT$_2$R in CNS injury has been studied in a model of global ischemia in rats.[15] In animals subjected to ischemic injury, the AT$_2$R mRNA levels increased strongly in the cortex and hippocampus and moderately in the striatum, amygdala, and cerebellum.[15] After experimental middle cerebral artery occlusion (MCAO) in rats, the AT$_2$R was upregulated 1 and 3 days after injury, with maximum expression at 24 h in the infarcted area of the cerebral cortex.[16] Similarly, Li et al. had shown that mRNA and protein expression of the AT$_2$R were higher in the peri-infarct area 48 h after MCAO as compared to sham animals.[17] AT$_2$R immunoreactivity was present in the NeuN-positive neurons but not in GFAP-positive astrocytes. Interestingly, the elevated AT$_2$R expression was associated with an enhanced neurite outgrowth in neurons of the ischemic striatum, suggesting a neuroregenerative potential of the AT$_2$R.[17] Similarly, the expression of the AT$_2$R was elevated in the peri-infarct area of MCAO mice with a maximum at

The Protective Arm of the Renin–Angiotensin System (RAS). http://dx.doi.org/10.1016/B978-0-12-801364-9.00008-0

72 h, but unchanged in the ischemic core.[18] An elevated expression of the AT_2R in CNS injury has also been observed in nonneuronal cells. Following MCAO in rats, AT_2R immunopositive microglial cells were present in infarcted and penumbral areas of the ischemic cortex and were characterized by the hypertrophic and amoeboid morphology indicating their activated state.[19] In contrast, no immunopositive microglia cells were found in sham-operated animals, suggesting that the elevated AT_2 expression in microglia counterbalances inflammatory processes related to the ischemic injury.[19] In addition to the anti-inflammatory activity of the microglial AT_2R, the elevated receptor expression may lead to an enhanced synthesis of neurotrophic factors (e.g., BDNF) and thus provide a neuroprotective environment for the neuronal cells.[20] In pertussis toxin-treated MBP/CCL2 transgenic mice, an animal model of blood–brain barrier (BBB) breakdown, the AT_2R colocalized with the astrocytic marker GFAP in the glia limitans of the perivascular space supporting the idea that that the increased AT_2R expression at the BBB may prevent leakage of blood components into the brain parenchyma.[21]

AT_2R-MEDIATED EFFECTS IN CNS CELLS

AT_2R stimulation elicits multiple beneficial effects in CNS-derived cells, including regulation of ion currents/channels,[22] anti-inflammation,[23] neuroprotection,[24] and neuroregeneration.[25]

The impact of the AT_2R stimulation on ion currents was studied with whole-cell voltage clamp in neonate rat neurons.[26] Selective activation of AT_2R by angiotensin II (Ang II) in the presence of AT_1R blocker caused potentiation of both transient K^+ current I_A and voltage-dependent delayed rectifier K+ current $I_{K(v)}$.[26] This is in agreement with the study, in which the selective AT_2R stimulation with CGP42112A also elicited stimulatory effect on I_A and $I_{K(v)}$.[27] On the other hand, AT_1R stimulation elicited an opposing effect on the I_A and $I_{K(v)}$.[28] The abovementioned studies suggest that the AT_2R stimulation leads to a hyperpolarization or decrease in neuronal excitability.

Whether or not AT_2R stimulation elicits pro- or antiapoptotic processes in CNS-derived cells depends inter alia on the presence or absence of the growth factors. For instance, we have shown early on that in PC12W cells, grown in the presence of NGF, AT_2R stimulation induced neuronal apoptosis via increased ceramide levels.[29] In contrast, in N-methyl-D-aspartate (NMDA)-mediated cell death, AT_2R stimulation inhibited apoptosis in differentiated neuronal cell lines NG108-15 and N1E115.[30] In primary cortical neurons subjected to glucose deprivation, AT_2R stimulation with CGP42112 significantly reduced cell death.[31] In chemically induced hypoxia, cortical neurons exposed to sodium azide showed increased apoptosis that was inhibited by AT_2R stimulation.[32] The underlying neuroprotective molecular mechanisms involved the delayed rectifier K^+ channel, Na^+/Ca^{2+} exchanger, and Na^+/K^+ ATPase.[33]

The anti-inflammatory potential of the AT_2R has been shown in primary astrocytes and microglial cells. In primary rat astrocytes, selective AT_2R stimulation with compound 21 (C21) inhibited LPS-induced IL-6 and TNFα upregulation.[23] Similarly, C21 inhibited microglial activation, as shown by reduced NO release and abolished cell migration,[11,34] and inhibited the synthesis of proinflammatory cytokines.[11] These mechanisms seem to be involved in AT_2R-mediated myelin protection. In aggregating brain cell cultures, LPS treatment and INFγ treatment induce massive demyelination as shown by the reduced immunoreactivity for myelin basic protein (MBP) and myelin oligodendrocyte glycoprotein (MOG).[11] C21 treatment prevented demyelination and promoted remyelination[11] pointing to a therapeutic potential of the direct AT_2R stimulation in demyelinating diseases.

Neurite outgrowth is a hallmark of regenerative potential that might be pharmacologically induced in neuronal cell culture. In 1996, AT_2R-mediated neurite elongation has been demonstrated for the first time independently by two research groups.[35,36] In the exclusively AT_2R-harboring NG108-15 cell line, stimulation with Ang II or CGP42112 induced morphological differentiation, as shown by the enhanced outgrowth of neurites.[35] Similarly, Ang II induced neurite outgrowth in exclusively AT_2R-harboring PC12W cell line, which was accompanied by reduced proliferation.[36] This might be explained by a reduced expression of the middle-sized neurofilament subunit (NF-M) as observed in PC12W cells upon AT_2R stimulation.[37] Other components of the neuronal cytoskeleton are also triggered by the AT_2R. In quiescent PC12W cells, stimulation of the AT_2R upregulated polymerized beta-tubulin and microtubule-associated protein 2 (MAP2) but downregulated MAP1B.[38] The underlying molecular mechanisms involve four distinct signaling cascades that may occur sequentially or simultaneously during AT_2R-mediated neurite outgrowth.[25] The first pathway involves an AT_2R-interacting protein ATIP (ATBP)[39,40] and Src homology 2 domain-containing protein-tyrosine phosphatase 1 (SHP-1), leading to transactivation of methyl methanesulfonate sensitive 2 (MMS2) enzyme.[41,42] In the second pathway, AT_2R stimulation leads to a decreased activity of PKCα followed by a decreased p21^{RAS} activity.[43] The third pathway requires phosphorylation of p42/p44^{mapk} mediated by Rap1/B-Raf.[44] This may be initiated via phosphorylation of neutrophin receptors[45] since inhibition of tyrosine kinases with K252a abolished AT$_2$R-mediated neurite outgrowth.[46] The fourth pathway, finally, involves activation of nNOS and cGMP.[47]

AT$_2$R IN PERIPHERAL NEURONAL INJURY

The first study documenting a neuroregenerative potential of AT$_2$R stimulation *in vivo* was published by our group in 1998.[48] In Wistar rats, the optic nerve was crushed behind its bulbar exit and treated locally for 14 days either with Ang II alone or in combination with an AT$_1$R antagonist, losartan, or AT$_2$R antagonist, PD123177, using a collagen foam placed in the vitreous body. Stimulation with Ang II alone or in combination with losartan induced regrowth of lesioned axons to the same extent, as assessed by the Gap43 immunostaining. The effect of Ang II was completely abolished by cotreatment with PD123177, indicating that the axonal regrowth was mediated specifically by the AT$_2$R.[48]

Following this work, Reineke et al. demonstrated a functional recovery from severe neuronal injury mediated by AT$_2$R stimulation *in vivo*.[49] In the Sprague-Dawley rats, the sciatic nerve crush was performed unilaterally at the level of the hip. Animals were treated locally using osmotic minipumps with either vehicle, Ang II alone, or in combination with losartan or with PD123319 for 2 weeks. Treatment with Ang II alone or in combination with losartan improved locomotor function, as measured by toe spread distance during walking, and improved sensorimotor function assessed with a reflex withdrawal test. The functional recovery was abolished in animals cotreated with Ang II and PD123319. A morphometric analysis of the sciatic nerves distal to the lesion showed increased axonal diameter and elevated degree of myelination in animals treated either with Ang II or with a combination of AngII and losartan, as compared to vehicle treatment. The effect of Ang II on neuroregeneration was abolished by cotreatment with PD123319, indicating involvement of the AT$_2$R in the regenerative processes.[49]

AT$_2$R STIMULATION IN STROKE

A number of studies have reported on the neuroprotective and neuroregenerative potential of the renin–angiotensin system in stroke by the so-called protective arm of the RAS, which involves AT$_2$ and Mas receptors counteracting the deleterious effects of the AT$_1$R and other systems (for review, see McCarthy et al.[50]).

In 1999 and in 2000, we first demonstrated that the AT$_2$R may contribute to the beneficial effects of chronic central AT$_1$R blockade in stroke.[51,52] Blockade of brain AT$_1$R with antagonists, losartan or irbesartan, reduced neurological deficits after MCAO in rats. Cotreatment with the AT$_2$R antagonist, PD123177, abolished the advantageous effects of losartan or irbesartan on neurological outcome, stroke volume, and the expression of c-Fos and c-Jun after MCAO.[51,52] Both transcription factors are elevated in ischemic neurons, and increased expression of c-Jun, and in some cases also c-Fos, plays a crucial role in the induction of neuronal apoptosis.[53] Indeed, sustained blockade of brain AT$_1$Rs by irbesartan reduced inflammatory responses, neuronal injury, and apoptosis in rats exposed to focal cerebral ischemia.[54] Ischemic lesions of the nervous system increase the expression of AT$_2$R in the brain.[16]

Similar to the abovementioned study, systemic pretreatment with a nonhypotensive dose of the ARB, valsartan, reduced ischemic area and neurological deficits 24 h after MCAO in wild-type C57BL/6J mice.[55] This was accompanied by an improved blood flow in the peripheral region of the middle cerebral artery territory and a reduced superoxide anion production and NADPH oxidase activity in the ischemic area, indicating reduced oxidative stress. Interestingly, almost all of the beneficial effects mediated by valsartan were abolished in AT$_2$R knockout animals suggesting again a crucial role of the AT$_2$R in ARB-mediated neuroprotection.[55]

In another study, the impact of AT$_2$R stimulation on the infarct volume and neurological deficits was analyzed in Wistar rats.[17] Animals received an intracerebroventricular infusion of vehicle, the ARB irbesartan, alone or in combination with an AT$_2$R antagonist, PD123317, for 5 consecutive days before and 24 h after MCAO. Irbesartan significantly reduced infarct volume and neurological deficits 24 h after MCAO, and this effect was attenuated in animals cotreated with the AT$_2$R antagonist, PD123317.[17]

In a study performed by Lu et al., the impact of RAS inhibition on infarct volume and neurological outcome was evaluated in spontaneous hypertensive rats subjected to MCAO.[56] Animals were treated p.o. with various equipotent doses of candesartan or ramipril for 4 weeks prior to MCAO. At medium equipotent doses (SBP reduction of 40 mmHg), only candesartan caused a significant reduction in infarct volume and neurological improvement as evaluated 24 h after MCAO. The neuroprotective effect of candesartan was almost completely abolished by an AT$_2$R antagonist underlining once again the central role of the AT$_2$R in AT$_1$R blocker-mediated cerebroprotection.[56] Interestingly, the cortical expression of AT$_1$R mRNA was significantly downregulated in animals treated with candesartan, whereas AT$_2$R mRNA was significantly upregulated, as compared to the vehicle-treated group. This is in agreement with a previously published study, in which the expression of AT$_2$R was upregulated in Wistar rats treated p.o. with candesartan for 4 weeks prior to MCAO.[57]

In another study, 10 days of s.c. pretreatment with candesartan at low, nonhypotensive dose increased AT$_2$R expression 3 h after MCAO and reduced stroke volume and neurological deficits 24 h after MCAO in male C57BL/6J mice.[58] Interestingly, no differences in AT$_2$R expression, stroke volume, or neurological outcome were observed in animals treated with a low, nonhypotensive dose of the ACE inhibitor enalapril. Additionally, 24 h after MCAO, the elevated superoxide anion production in the cerebral cortex was reduced in animals treated with candesartan, but not with enalapril. Discrepancies between neuroprotective effects of ARBs and ACE inhibitors might be partially explained by the involvement of the AT$_2$R in ARB-mediated neuroprotection and/or by the Ang II production by chymase after long-term ACE inhibitor administration.[58]

A similar question was addressed in the study performed by Faure et al.[59] Embolic stroke was induced in Sprague-Dawleys rats by injection of polystyrene microspheres in the right internal carotid artery. Animals were treated i.p. either with the vehicle candesartan or with ACEi lisinopril for 5 days prior to stroke induction. As expected, candesartan treatment reduced mortality, neurological deficits, and infarct volume 24 h after embolic stroke. This was AT$_2$R-dependent since cotreatment with PD123319 abolished the protective effect of AT$_1$R blockade. Interestingly, lisinopril treatment had a deleterious effect as shown by the elevated mortality, increased infarct volume, and reduced neurological performance, as compared to the vehicle-treated animals. The detrimental effect of lisinopril was blunted in animals pretreated with candesartan. The authors concluded that the neuroprotective effect of AT$_1$R blockade was AT$_2$R-dependent: AT$_1$R blockade resulted in a progressive Ang II production stimulating the unopposed AT$_2$R. In contrast, inhibition of the Ang II synthesis by ACE inhibition would reduce stimulation not only of the detrimental AT$_1$R but also of cerebroprotective AT$_2$R.[59]

In 2006, Mogi et al. described for the first time that MMS2 was actively involved in AT$_2$R-mediated neuroprotection.[60] Wild-type and AT$_2$R-KO mice were treated i.p. with valsartan for 10 days prior to MCAO. In wild-type animals, an elevated MMS2 mRNA expression was observed in the ischemic cortex as compared to the contralateral nonischemic hemisphere. In AT$_2$R-KO animals, no differences in MMS2 expression between ipsilateral and contralateral hemispheres were observed. In wild-type animals treated with valsartan, an enhanced expression of MMS2 was present in both ischemic and nonischemic regions, when compared to the vehicle-treated animals. As compared to the wild-type animals, the AT$_2$R-KO animals exhibited elevated impairment of cognitive function after stroke, as determined by the passive avoidance test. In wild-type animals, valsartan treatment significantly improved cognitive dysfunction, whereas MMS2 *in vivo* knockdown enhanced the decline in cognitive function after stroke.[60] Experiments with primary neurons prepared from the rat cerebral cortex showed that MMS2 knockdown reduced neurite outgrowth and synapse formation indicating its role in neuroregenerative processes.[42]

The impact of direct and selective AT$_2$R stimulation on neuroprotection was studied in a conscious rat model of stroke.[61] Spontaneously hypertensive rats were treated intracerebroventricularly either with the peptidic AT$_2$R agonist, CGP42112, alone or in combination with the AT$_2$R antagonist, PD123319, beginning 5 days before stroke induction. Cerebral ischemia was induced by administration of the vasoconstrictor endothelin-1. As compared to the vehicle treatment, CGP42112 reduced locomotor deficits and infarct area and improved neuronal survival in the cortical region of the ipsilateral hemisphere 72 h after the stroke. The neuroprotective effect of AT$_2$R stimulation was abolished by cotreatment with PD123319.[61]

The impact of anti-inflammation mediated by the AT$_2$R on neuroprotection in stroke was studied by Iwanami et al.[62] Here, the authors focused on the protective action of bone marrow stromal cells (MSCs), previously demonstrated as an effective therapeutic option in a number of experimental stroke studies. Cerebral ischemia was induced in wild-type mice by MCAO. Immediately after reperfusion, the MSCs obtained either from AT$_2$R wild-type (*Agtr2$^+$*) or from AT$_2$R-KO (*Agtr2$^-$*) animals were injected through the tail vein. Treatment with *Agtr2$^+$* MSCs reduced mortality as compared to the vehicle-treated animals, whereas *Agtr2$^-$* MSCs reduced the survival rate 6 days after MCAO. An impaired neurological performance was observed in animals injected with saline and was improved by the treatment with *Agtr2$^+$* MSCs but not with *Agtr2$^-$* MSCs. Surprisingly, stroke volume was not affected by any of the treatment regimens. The elevated mRNA expression of proinflammatory cytokines, TNFα and MCP-1, after MCAO was reduced in *Agtr2$^+$* MSCs-treated animals but remained unchanged upon *Agtr2$^-$* MSCs treatment.[62] This study suggests that the neuroprotective action of the MSCs in stroke was mediated by the anti-inflammatory actions of the AT$_2$R.

The anti-inflammatory action of the AT$_2$R stimulation in stroke was also studied with nonpeptide AT$_2$R agonist C21.[63] Following MCAO, animals were treated systemically either with vehicle or C21 for 5 days. Without affecting blood pressure, C21 reduced stroke volume, improved neurological outcome and cerebral blood flow, and decreased BBB permeability and cerebral edema, as compared to the vehicle. This was accompanied by the reduced superoxide anion production and expression of proinflammatory cytokines, TNFα and MCP-1.[63] The effectiveness of the nonpeptide AT$_2$R agonist C21 suggested a therapeutic potential of AT$_2$R stimulation in the prevention of ischemic brain damage after acute stroke.

In a further study, Schwengel et al. studied mechanisms involved in AT$_2$R-mediated neuroprotection in C57/BL-6 mice using C21.[64] Starting 45 min after MCAO, animals were treated systemically either with vehicle or with C21 for 4 days. Pharmacological stimulation of the AT$_2$R did not alter infarct size; however, it improved neurological performance and dramatically reduced mortality rate, as compared to the vehicle-treated animals. In the ipsilateral hemisphere, treatment with C21 significantly reduced the elevated expression of the proinflammatory cytokine IL-6. AT$_2$R stimulation increased expression of the neurotrophin BDNF and the neurite outgrowth marker Gap43 and downregulated the expression of the inhibitor of axonal growth NogoA and its receptor NogoR, suggesting an induction of neuroprotective and neuroregenerative processes by AT$_2$R stimulation in the ischemic brain.[64]

The impact of delayed AT$_2$R stimulation on neuroprotection was further evaluated in hypertensive and normotensive animals. For the induction of stroke, McCarthy et al.[65] used an ET-1 infusion in conscious, spontaneously hypertensive rats, as described above.[61] Delayed intracerebroventricular (i.c.v.) treatment with C21 did not alter locomotor deficits and the number of activated microglial cells; however, it reduced the cortical and striatal infarct areas and improved neuronal survival, as compared to vehicle treatment. These effects were abolished by the cotreatment with PD123319.[65] In the study by Joseph et al., the impact of AT$_2$R stimulation on neuroprotection was tested in normotensive rats.[66] Animals were treated systemically either with C21 or with vehicle 4, 24, and 48 hours after the ET-1-induced stroke. C21 reduced infarct size and neurological deficits, as compared to the vehicle treatment. This was accompanied by the reduction of monocyte/microglial recruitment markers (CCL2 and its receptor CCR2).[66] These studies indicate that delayed central or systemic C21 administration provides robust and significant neuroprotection in normotensive and hypertensive animal models of stroke.

AT$_2$R IN SPINAL CORD INJURY

The impact of direct AT$_2$R stimulation on neurological performance was studied in an animal model of spinal cord injury (SCI).[46] SCI was induced in Balb/C mice by precise spinal cord compression followed by corticospinal tract tracing performed by injection of biotinylated dextran amine (BDA) into the motor cortex. Four weeks after SCI, the C21-treated animals featured an improved locomotor performance compared to vehicle-treated mice. Moreover, in animals treated with C21, a significantly higher number of BDA-positive axons were found caudally to the site of injury. The number of BDA-positive fibers in individual animals positively correlated with the BMS score suggesting functionality of these fibers. Treatment with C21 also led to a significantly increased immunoreactivity of the neutrophin receptor, TrkB, and improved neuronal survival within the injured and in perilesional tissue.[46] Thus, selective AT$_2$R stimulation with C21 engendered neuroprotection and neuroregeneration in the injured spinal cord tissue.

AT$_2$R IN MULTIPLE SCLEROSIS

The effectiveness of a direct AT$_2$R stimulation with C21 on myelination and inflammatory processes was evaluated in experimental autoimmune encephalomyelitis (EAE), an animal model of multiple sclerosis.[11] C57BL-6 mice immunized with myelin oligodendrocyte protein (MOG) displayed progressive neurological impairment accompanied by the immune cell infiltration and demyelination of the spinal cord tissue. Treatment with C21 for 4 weeks led to decreased microglia activation, ameliorated T-cell infiltration, and reduction of demyelinated areas as compared to the vehicle-treated animals. These processes may have a crucial impact on the neurological improvement in animals treated with C21 starting from the second week after MOG immunization. This study together with the abovementioned publication of Reinecke et al.[49] demonstrated that AT$_2$R stimulation can protect against demyelination and may be actively involved in remyelination processes.

AT$_2$R IN TRAUMATIC BRAIN INJURY

In heat-acclimated (HA) animals subjected to traumatic brain injury (TBI), the neurotrophin pathway has been suggested as a key neuroprotective mechanism of AT$_2$R stimulation.[67] Previous studies had shown that long-term exposure to mild heat induced the expression of hypoxia-inducible factor 1α (HIF-1α), BDNF, and Akt phosphorylation. This was accompanied by an improved locomotor function and cognitive outcome and reduced lesion volume and edema in an animal model of TBI.[68,69] Interestingly, HA increased hypothalamic AT$_2$R expression accompanied by the enhanced neurogenesis.[67] Inhibition of the AT$_2$R with PD123319 diminished the improvements in motor and cognitive recovery and abolished the reduction in lesion volume and neurogenesis after TBI. In HA animals, the elevated expression of the neurotrophin receptors TrkA and TrkB of the neurotrophins BDNF and NGF and of HIF-1α was abolished by AT$_2$R antagonism.[67] Taken together, an endogenous AT$_2$R stimulation seems to be involved in HA-mediated neuroprotection and neuroregeneration after TBI.

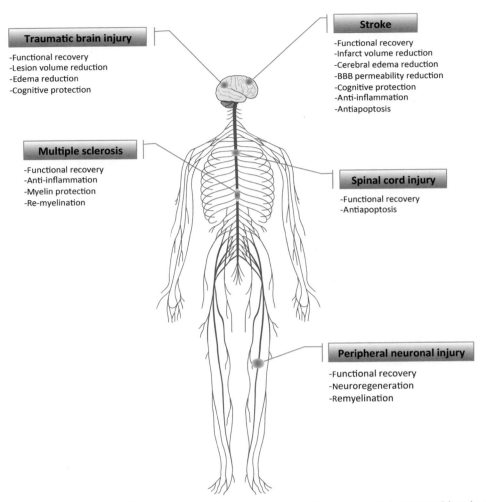

FIGURE 1 Beneficial effects of AT_2R stimulation on neuroprotective and neuroregenerative processes as demonstrated in animal models of human neurological diseases.

CONCLUSION

This chapter demonstrates that selective AT_2R stimulation provokes anti-inflammatory, neuroprotective, and neuroregenerative effects in the central nervous system and peripheral nervous system (Figure 1). The underlying molecular mechanisms are very complex by nature and require further investigation. Future work should address the molecular cross talk between AT_2R and other receptors of the renin–angiotensin system, including the Mas receptor, which could provide further information regarding signaling and clarify the complexity of the beneficial effects mediated by the "protective RAS" in the nervous system.

REFERENCES

1. Steckelings UM, Bottari SP, Unger T. Angiotensin receptor subtypes in the brain. *Trends Pharmacol Sci* 1992;**13**:365–8.
2. De Gasparo M, Catt KJ, Inagami T, Wright JW, Unger T. International union of pharmacology. XXIII. The angiotensin II receptors. *Pharmacol Rev* 2000;**52**(3):415–72.
3. Lenkei Z, Palkovits M, Corvol P, Llorens-Cortès C. Expression of angiotensin type-1 (AT1) and type-2 (AT2) receptor mRNAs in the adult rat brain: a functional neuroanatomical review. *Front Neuroendocrinol* 1997;**18**(4):383–439.
4. Jöhren O, Inagami T, Saavedra JM. AT1A, AT1B, and AT2 angiotensin II receptor subtype gene expression in rat brain. *Neuroreport* 1995;**6**(18):2549–52.
5. Jöhren O, Inagami T, Saavedra JM. Localization of AT2 angiotensin II receptor gene expression in rat brain by in situ hybridization histochemistry. *Brain Res Mol Brain Res* 1996;**37**(1-2):192–200.

6. Häuser W, Jöhren O, Saavedra JM. Characterization and distribution of angiotensin II receptor subtypes in the mouse brain. *Eur J Pharmacol* 1998;**348**(1):101–14.

7. MacGregor DP, Murone C, Song K, Allen AM, Paxinos G, Mendelsohn FA. Angiotensin II receptor subtypes in the human central nervous system. *Brain Res* 1995;**675**(1-2):231–40.

8. Tsutsumi K, Saavedra JM. Characterization and development of angiotensin II receptor subtypes (AT1 and AT2) in rat brain. *Am J Physiol* 1991;**261**(1 Pt 2):R209–16.

9. Nuyt AM, Lenkei Z, Palkovits M, Corvol P, Llorens-Cortés C. Ontogeny of angiotensin II type 2 receptor mRNA expression in fetal and neonatal rat brain. *J Comp Neurol* 1999;**407**(2):193–206.

10. Garrido-Gil P, Valenzuela R, Villar-Cheda B, Lanciego JL, Labandeira-Garcia JL. Expression of angiotensinogen and receptors for angiotensin and prorenin in the monkey and human substantia nigra: an intracellular renin-angiotensin system in the nigra. *Brain Struct Funct* 2013;**218**(2):373–88.

11. Valero-Esquitino V, Lucht K, Namsolleck P, Monnet-Tschudi F, Stubbe T, Lucht F, et al. Direct angiotensin AT2-receptor stimulation attenuates T-cell and microglia activation and prevents demyelination in experimental autoimmune encephalomyelitis in mice. *Clin Sci (Lond)* 2015;**128**(2):95–109.

12. Foulquier S, Dupuis F, Perrin-Sarrado C, Maguin Gaté K, Merhi-Soussi F, Liminana P, et al. High salt intake abolishes AT(2)-mediated vasodilation of pial arterioles in rats. *J Hypertens* 2011;**29**(7):1392–9.

13. Wackenfors A, Vikman P, Nilsson E, Edvinsson L, Malmsjö M. Angiotensin II-induced vasodilatation in cerebral arteries is mediated by endothelium-derived hyperpolarising factor. *Eur J Pharmacol* 2006;**531**(1-3):259–63.

14. Gallinat S, Yu M, Dorst A, Unger T, Herdegen T. Sciatic nerve transection evokes lasting up-regulation of angiotensin AT2 and AT1 receptor mRNA in adult rat dorsal root ganglia and sciatic nerves. *Brain Res Mol Brain Res* 1998;**57**(1):111–22.

15. Makino I, Shibata K, Ohgami Y, Fujiwara M, Furukawa T. Transient upregulation of the AT2 receptor mRNA level after global ischemia in the rat brain. *Neuropeptides* 1996;**30**(6):596–601.

16. Zhu YZ, Chimon GN, Zhu YC, Lu Q, Li B, Hu HZ, et al. Expression of angiotensin II AT2 receptor in the acute phase of stroke in rats. *Neuroreport* 2000;**11**(6):1191–4.

17. Li J, Culman J, Hörtnagl H, Zhao Y, Gerova N, Timm M, et al. Angiotensin AT2 receptor protects against cerebral ischemia-induced neuronal injury. *FASEB J* 2005;**19**(6):617–19.

18. Miyamoto N, Zhang N, Tanaka R, Liu M, Hattori N, Urabe T. Neuroprotective role of angiotensin II type 2 receptor after transient focal ischemia in mice brain. *Neurosci Res* 2008;**61**(3):249–56.

19. Wu C-Y, Zha H, Xia Q-Q, Yuan Y, Liang X-Y, Li J-H, et al. Expression of angiotensin II and its receptors in activated microglia in experimentally induced cerebral ischemia in the adult rats. *Mol Cell Biochem* 2013;**382**(1-2):47–58.

20. McCarthy CA, Widdop RE, Deliyanti D, Wilkinson-Berka JL. Brain and retinal microglia in health and disease: an unrecognized target of the renin-angiotensin system. *Clin Exp Pharmacol Physiol* 2013;**40**(8):571–9.

21. Füchtbauer L, Toft-Hansen H, Khorooshi R, Owens T. Expression of astrocytic type 2 angiotensin receptor in central nervous system inflammation correlates with blood-brain barrier breakdown. *J Mol Neurosci* 2010;**42**(1):89–98.

22. Sumners C, Gelband CH. Neuronal ion channel signalling pathways: modulation by angiotensin II. *Cell Signal* 1998;**10**(5):303–11.

23. Steckelings UM, Larhed M, Hallberg A, Widdop RE, Jones ES, Wallinder C, et al. Non-peptide AT2-receptor agonists. *Curr Opin Pharmacol* 2011;**11**(2):187–92.

24. Sumners C, Horiuchi M, Widdop RE, McCarthy C, Unger T, Steckelings UM. Protective arms of the renin-angiotensin-system in neurological disease. *Clin Exp Pharmacol Physiol* 2013;**40**(8):580–8.

25. Namsolleck P, Recarti C, Foulquier S, Steckelings UM, Unger T. AT(2) receptor and tissue injury: therapeutic implications. *Curr Hypertens Rep* 2014;**16**(2):416.

26. Kang J, Sumners C, Posner P. Angiotensin II type 2 receptor-modulated changes in potassium currents in cultured neurons. *Am J Physiol Cell Physiol* 1993;**265**(3):C607–16.

27. Kang J, Posner P, Sumners C. Angiotensin II type 2 receptor stimulation of neuronal K+ currents involves an inhibitory GTP binding protein. *Am J Physiol Cell Physiol* 1994;**267**(5):C1389–97.

28. Sumners C, Zhu M, Gelband CH, Posner P. Angiotensin II type 1 receptor modulation of neuronal K+ and Ca2+ currents: intracellular mechanisms. *Am J Physiol* 1996;**271**(1 Pt 1):C154–63.

29. Gallinat S, Busche S, Schütze S, Krönke M, Unger T. AT2 receptor stimulation induces generation of ceramides in PC12W cells. *FEBS Lett* 1999;**443**(1):75–9.

30. Schelman WR, Andres R, Ferguson P, Orr B, Kang E, Weyhenmeyer JA. Angiotensin II attenuates NMDA receptor-mediated neuronal cell death and prevents the associated reduction in Bcl-2 expression. *Brain Res Mol Brain Res* 2004;**128**(1):20–9.

31. Lee S, Brait VH, Arumugam TV, Evans MA, Kim HA, Widdop RE, et al. Neuroprotective effect of an angiotensin receptor type 2 agonist following cerebral ischemia in vitro and in vivo. *Exp Transl Stroke Med* 2012;**4**(1):16.

32. Grammatopoulos T, Morris K, Ferguson P, Weyhenmeyer J. Angiotensin protects cortical neurons from hypoxic-induced apoptosis via the angiotensin type 2 receptor. *Brain Res Mol Brain Res* 2002;**99**(2):114–24.

33. Grammatopoulos TN, Johnson V, Moore SA, Andres R, Weyhenmeyer JA. Angiotensin type 2 receptor neuroprotection against chemical hypoxia is dependent on the delayed rectifier K+ channel, Na+/Ca2+ exchanger and Na+/K+ ATPase in primary cortical cultures. *Neurosci Res* 2004;**50**(3):299–306.

34. Namsolleck P, Valero-Esquitino V, Lucht K, Unger T, Steckelings U. Inhibition of microglia activation and migration by direct AT2-receptor stimulation. *J Hypertens* 2012;e269.

35. Laflamme L, Gasparo M, Gallo JM, Payet MD, Gallo-Payet N. Angiotensin II induction of neurite outgrowth by AT2 receptors in NG108-15 cells. Effect counteracted by the AT1 receptors. *J Biol Chem* 1996;**271**(37):22729–35.

36. Meffert S, Stoll M, Steckelings UM, Bottari SP, Unger T. The angiotensin II AT2 receptor inhibits proliferation and promotes differentiation in PC12W cells. *Mol Cell Endocrinol* 1996;**122**(1):59–67.

37. Gallinat S, Csikos T, Meffert S, Herdegen T, Stoll M, Unger T. The angiotensin AT2 receptor down-regulates neurofilament M in PC12W cells. *Neurosci Lett* 1997;**227**(1):29–32.

38. Stroth U, Meffert S, Gallinat S, Unger T. Angiotensin II and NGF differentially influence microtubule proteins in PC12W cells: role of the AT2 receptor. *Brain Res Mol Brain Res* 1998;**53**(1-2):187–95.

39. Nouet S, Amzallag N, Li J-M, Louis S, Seitz I, Cui T-X, et al. Trans-inactivation of receptor tyrosine kinases by novel angiotensin II AT2 receptor-interacting protein, ATIP. *J Biol Chem* 2004;**279**(28):28989–97.

40. Wruck CJ, Funke-Kaiser H, Pufe T, Kusserow H, Menk M, Schefe JH, et al. Regulation of transport of the angiotensin AT2 receptor by a novel membrane-associated Golgi protein. *Arterioscler Thromb Vasc Biol* 2005;**25**(1):57–64.

41. Horiuchi M, Iwanami J, Mogi M. Regulation of angiotensin II receptors beyond the classical pathway. *Clin Sci* 2012;**123**(4):193–203.

42. Li J-M, Mogi M, Tsukuda K, Tomochika H, Iwanami J, Min L-J, et al. Angiotensin II-induced neural differentiation via angiotensin II type 2 (AT2) receptor-MMS2 cascade involving interaction between AT2 receptor-interacting protein and Src homology 2 domain-containing protein-tyrosine phosphatase 1. *Mol Endocrinol* 2007;**21**(2):499–511.

43. Beaudry H, Gendron L, Guimond M-O, Payet MD, Gallo-Payet N. Involvement of protein kinase C alpha (PKC alpha) in the early action of angiotensin II type 2 (AT2) effects on neurite outgrowth in NG108-15 cells: AT2-receptor inhibits PKC alpha and p21ras activity. *Endocrinology* 2006;**147**(9):4263–72.

44. Gendron L, Laflamme L, Rivard N, Asselin C, Payet MD, Gallo-Payet N. Signals from the AT2 (angiotensin type 2) receptor of angiotensin II inhibit p21ras and activate MAPK (mitogen-activated protein kinase) to induce morphological neuronal differentiation in NG108-15 cells. *Mol Endocrinol* 1999;**13**(9):1615–26.

45. Plouffe B, Guimond M-O, Beaudry H, Gallo-Payet N. Role of tyrosine kinase receptors in angiotensin II AT2 receptor signaling: involvement in neurite outgrowth and in p42/p44mapk activation in NG108-15 cells. *Endocrinology* 2006;**147**(10):4646–54.

46. Namsolleck P, Boato F, Schwengel K, Paulis L, Matho KS, Geurts N, et al. AT2-receptor stimulation enhances axonal plasticity after spinal cord injury by upregulating BDNF expression. *Neurobiol Dis* 2013;**51**:177–91.

47. Gendron L, Côté F, Payet MD, Gallo-Payet N. Nitric oxide and cyclic GMP are involved in angiotensin II AT(2) receptor effects on neurite outgrowth in NG108-15 cells. *Neuroendocrinology* 2002;**75**(1):70–81.

48. Lucius R, Gallinat S, Rosenstiel P, Herdegen T, Sievers J, Unger T. The angiotensin II type 2 (AT2) receptor promotes axonal regeneration in the optic nerve of adult rats. *J Exp Med* 1998;**188**(4):661–70.

49. Reinecke K, Lucius R, Reinecke A, Rickert U, Herdegen T, Unger T. Angiotensin II accelerates functional recovery in the rat sciatic nerve in vivo: role of the AT2 receptor and the transcription factor NF-kappaB. *FASEB J* 2003;**17**(14):2094–6.

50. McCarthy CA, Facey LJ, Widdop RE. The protective arms of the Renin-angiotensin system in stroke. *Curr Hypertens Rep* 2014;**16**(7):440.

51. Dai WJ, Funk A, Herdegen T, Unger T, Culman J. Blockade of central angiotensin AT(1) receptors improves neurological outcome and reduces expression of AP-1 transcription factors after focal brain ischemia in rats. *Stroke* 1999;**30**(11):2391–8, Discussion, pp. 2398–9.

52. Blume A, Funk A, Gohlke P, Unger T, Culman J. AT2 receptor inhibition in the rat brain reverses the beneficial effects of AT1 receptor blockade on neurological outcome after focal brain ischemia. *Hypertension* 2000;**36**:656.

53. Love S. Apoptosis and brain ischaemia. *Prog Neuropsychopharmacol Biol Psychiatry* 2003;**27**(2):267–82.

54. Lou M, Blume A, Zhao Y, Gohlke P, Deuschl G, Herdegen T, et al. Sustained blockade of brain AT1 receptors before and after focal cerebral ischemia alleviates neurologic deficits and reduces neuronal injury, apoptosis, and inflammatory responses in the rat. *J Cereb Blood Flow Metab* 2004;**24**(5):536–47.

55. Iwai M, Liu H-W, Chen R, Ide A, Okamoto S, Hata R, et al. Possible inhibition of focal cerebral ischemia by angiotensin II type 2 receptor stimulation. *Circulation* 2004;**110**(7):843–8.

56. Lu Q, Zhu Y-Z, Wong P. Neuroprotective effects of candesartan against cerebral ischemia in spontaneously hypertensive rats. *Neuroreport* 2005;**16**(17):1963–7.

57. Lu Q, Zhu Y-Z, Wong PT-H. Angiotensin receptor gene expression in candesartan mediated neuroprotection. *Neuroreport* 2004;**15**(17):2643–6.

58. Hamai M, Iwai M, Ide A, Tomochika H, Tomono Y, Mogi M, et al. Comparison of inhibitory action of candesartan and enalapril on brain ischemia through inhibition of oxidative stress. *Neuropharmacology* 2006;**51**(4):822–8.

59. Faure S, Bureau A, Oudart N, Javellaud J, Fournier A, Achard J-M. Protective effect of candesartan in experimental ischemic stroke in the rat mediated by AT2 and AT4 receptors. *J Hypertens* 2008;**26**(10):2008–15.

60. Mogi M, Li J-M, Iwanami J, Min L-J, Tsukuda K, Iwai M, et al. Angiotensin II type-2 receptor stimulation prevents neural damage by transcriptional activation of methyl methanesulfonate sensitive 2. *Hypertension* 2006;**48**(1):141–8.

61. McCarthy CA, Vinh A, Callaway JK, Widdop RE. Angiotensin AT2 receptor stimulation causes neuroprotection in a conscious rat model of stroke. *Stroke* 2009;**40**(4):1482–9.

62. Iwanami J, Mogi M, Li J-M, Tsukuda K, Min L-J, Sakata A, et al. Deletion of angiotensin II type 2 receptor attenuates protective effects of bone marrow stromal cell treatment on ischemia–reperfusion brain injury in mice. *Stroke* 2008;**39**(9):2554–9.

63. Min L-J, Mogi M, Tsukuda K, Jing F, Ohshima K, Nakaoka H, et al. Direct Stimulation of angiotensin ii type 2 receptor initiated after stroke ameliorates ischemic brain damage. *Am J Hypertens* 2014;**27**(8):1036–44.

64. Schwengel K, Thoene-Reineke C, Lucht K, Namsolleck P, Müller S, Horiuchi M, et al. Direct angiotensin AT2-receptor stimulation improves survival and neurological outcome in experimental stroke (MCAO) in mice. *J Hypertens* 2011;**29**:e50–1.

65. McCarthy CA, Vinh A, Miller AA, Hallberg A, Alterman M, Callaway JK, et al. Direct angiotensin AT2 receptor stimulation using a novel AT2 receptor agonist, Compound 21, evokes neuroprotection in conscious hypertensive rats. *PLoS One* 2014;**9**(4):e95762.

66. Joseph JP, Mecca AP, Regenhardt RW, Bennion DM, Rodríguez V, Desland F, et al. The angiotensin type 2 receptor agonist Compound 21 elicits cerebroprotection in endothelin-1 induced ischemic stroke. *Neuropharmacology* 2014;**81**:134–41.

67. Umschweif G, Shabashov D, Alexandrovich AG, Trembovler V, Horowitz M, Shohami E. Neuroprotection after traumatic brain injury in heat-acclimated mice involves induced neurogenesis and activation of angiotensin receptor type 2 signaling. *J Cereb Blood Flow Metab* 2014;**34**(8):1381–90.

68. Horowitz M. Heat acclimation, epigenetics, and cytoprotection memory. *Compr Physiol* 2014;**4**(1):199–230.

69. Umschwief G, Shein NA, Alexandrovich AG, Trembovler V, Horowitz M, Shohami E. Heat acclimation provides sustained improvement in functional recovery and attenuates apoptosis after traumatic brain injury. *J Cereb Blood Flow Metab* 2010;**30**(3):616–27.

Chapter 9

Roles of AT$_2$R in Cognitive Function

Masatsugu Horiuchi, Masaki Mogi

Molecular Cardiovascular Biology and Pharmacology, Ehime University Graduate School of Medicine, Tohon, Shitsukawa, Ehime, Japan

INTRODUCTION

Although hypertension is well known to be a cause of vascular dementia, recent findings highlight the role of hypertension in the pathogenesis of Alzheimer's disease (AD) and mild cognitive impairment and that disruption of diurnal blood pressure variation is closely associated with cognitive impairment via injury of small cerebral arteries.[1,2] Antihypertensive treatment could be beneficial to cognitive function by lowering blood pressure and by a specific neuroprotective effect. Angiotensin II type 2 (AT$_2$) receptor stimulation by unbound angiotensin II could be expected during treatment with angiotensin II type 1 (AT$_1$) receptor blockers (ARBs), and there is accumulating evidence suggesting that the AT$_2$ receptor (AT$_2$R) not only opposes the AT$_1$ receptor (AT$_1$R) but also has unique effects beyond interaction with AT$_1$R signaling in the brain.[3] Accordingly, we reviewed the effects of activation of the AT$_2$R on cognitive function.

PREVENTIVE EFFECTS OF ARB ON COGNITIVE DECLINE AND DEMENTIA

Stroke is one of the major causes of cognitive decline. Fuentes et al. reported that patients taking antihypertensive drugs at stroke onset had a lower rate of poor outcome than those not on antihypertensive treatment and those taking an ARB had better outcomes than those without an ARB.[4] Levi Marpillat et al. reported that antihypertensive treatment had beneficial effects on cognitive decline and in the prevention of dementia and indicated that these effects may differ between drug classes, with ARBs possibly being the most effective.[5] Li et al. reported that ARBs were associated with a significant reduction in the incidence and progression of AD and dementia compared with angiotensin-converting enzyme (ACE) inhibitors and other cardiovascular drugs.[6] Hajjar et al. reported that participants with or without AD who were treated with ARBs showed lower levels of amyloid deposition markers compared with those treated with other antihypertensive medications, including ACE inhibitors.[7] The ambiguous effect of ACE inhibitors on cognitive decline and in dementia prevention may be explained by the fact that brain ACE is not specific for angiotensin I and that brain ACE also catabolizes cognition-enhancing brain peptides and converts toxic amyloid beta (Aβ) 42 into less toxic Aβ40, suggesting that ACE inhibitors may have short-term cognition-enhancing properties but may increase the long-term Aβ42 brain burden and cognitive decline.[8] These results suggest that specific AT$_1$R blockade by ARBs and AT$_2$R stimulation is more important for preventing cognitive decline, rather than blockade of angiotensin II synthesis.

PROTECTION AGAINST ISCHEMIC BRAIN DAMAGE BY AT$_2$R STIMULATION

AT$_2$R stimulation supported neuronal survival and neurite outgrowth in response to ischemia-induced neuronal injury.[9] We observed that the ischemic brain area was significantly larger in AT$_2$R-deficient mice after middle cerebral artery (MCA) occlusion, with a decrease in cerebral blood flow (CBF) and an increase in superoxide anion production compared with wild-type mice.[10] We also reported that treatment with valsartan prevented cognitive decline after MCA occlusion in wild-type mice, but this effect was weaker in AT$_2$R-null mice, suggesting that AT$_2$R stimulation during ARB treatment is important to prevent cognitive decline after stroke.[11] AT$_2$R agonists such as compound 21 (C21) have been newly developed and are expected to be useful agents for improving pathological disorders.[12] Gelosa et al. reported that administration of C21 delayed the occurrence of brain damage and prolonged survival without affecting blood pressure in spontaneously

The Protective Arm of the Renin–Angiotensin System (RAS). http://dx.doi.org/10.1016/B978-0-12-801364-9.00009-2

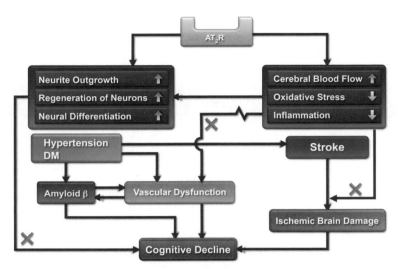

FIGURE 1 Protective effects of AT$_2$R stimulation against vascular dysfunction and neural damage. AT$_2$R activation increases CBF and inhibits oxidative stress and inflammation, thereby protecting against vascular dysfunction and neural damage. AT$_2$R activation also enhances neurite outgrowth, regeneration of neurons, and neural differentiation.

hypertensive stroke-prone rats (SHRSP) fed a high-salt diet.[13] Consistent with this result, we also observed that treatment with C21 initiated even after MCA occlusion significantly reduced the ischemic area, with improvement of neurological deficit.[14] The protective effects of AT$_2$R stimulation on vascular dysfunction and neural damage are summarized in Figure 1. More detailed mechanisms of AT$_2$R-mediated vascular and neural protective effects are described in other chapters.

ROLE OF AT$_2$R STIMULATION IN COGNITIVE FUNCTION

The AT$_2$R is reported to be expressed in areas related to learning and control of motor activity in the brain, in addition to the vasculature. In addition to the protective effects of AT$_2$R stimulation against ischemic brain damage, AT$_2$R activation is known to be involved in axonal regeneration, memory and behavior, promotion of neural differentiation, neurite outgrowth, and regeneration of neuronal tissue.[15–20] Vervoort et al. reported that expression of the AT$_2$R gene was absent in a female patient with mental retardation and that male patients with mental retardation were found to have sequence changes in the AT$_2$R gene, indicating a role of the AT$_2$R in brain development and cognitive function in humans.[21] We demonstrated that the passive avoidance rate after MCA occlusion was significantly impaired in AT$_2$R-deficient mice compared with wild-type mice and AT$_2$R mRNA expression was significantly increased in the ischemic side of the brain. In ischemic brain regions, methyl methanesulfonate sensitive 2 (MMS2), which belongs to a family of ubiquitin-conjugating enzyme variants, was increased in wild-type mice but not in AT$_2$R-deficient mice, and intracerebroventricular administration of MMS2 siRNA further impaired the avoidance rate after MCA occlusion compared with that in control siRNA-transfected mice.[11] Maul et al. also demonstrated that performance in a spatial memory task and in a one-way active avoidance task was significantly impaired in AT$_2$R-deficient mice.[22] Administration of candesartan prevented scopolamine-induced amnesia, restored CBF and acetylcholine (ACh) level, and decreased acetylcholinesterase, whereas memory, CBF, ACh level, and oxidative stress were blunted by concomitant blockade of the AT$_2$R.[23] AbdAlla et al. demonstrated that Aβ induced formation of cross-linked AT$_2$R oligomers that contributed to dysfunction of Galphaq/11 in an animal AD model and stereotactic inhibition of AT$_2$ oligomers by RNA interference prevented the impairment of Galphaq/11 and delayed Tau phosphorylation.[24]

We observed that intraperitoneal injection of C21 significantly increased spatial learning memory in wild-type mice, with increases in CBF and hippocampal field-excitatory postsynaptic potential (f-EPSP), and that treatment with C21 promoted neurite outgrowth of cultured mouse hippocampal neurons.[25] C21-induced cognitive enhancement was attenuated by coadministration of icatibant, a bradykinin B2R antagonist. We also investigated the pathological relevance of C21 using an AD mouse model with intracerebroventricular injection of Aβ(1-40) and observed that treatment with C21 prevented cognitive decline in this model. Recently, Steckeling's and Unger's groups demonstrated that treatment with C21 induced axonal regeneration, with increases in antiapoptotic Bcl-2, brain-derived neurotrophic factor, the neurotrophinRs TrkA and TrkB, and a marker of neurite growth, GAP43, but not TrkC.[26]

It is reported that AT$_2$R stimulation induces PPAR-γ activation in PC12W cells,[27] suggesting that possible cross talk of AT$_2$R activation and PPAR-γ stimulation in the brain could contribute to more exaggerated protective effects against ischemic brain damage and cognitive impairment. PPAR-γ activation in the brain is known to prevent brain damage via anti-inflammatory effects, antioxidative actions and improvement of endothelial function, increased Aβ clearance, and enhanced neural stem cell proliferation. We demonstrated that direct AT$_2$R stimulation by C21 accompanied by PPAR-γ activation ameliorated insulin resistance in type 2 diabetic mice.[28] However, the possible beneficial roles of AT$_2$R activation with PPAR-γ stimulation in improving cognitive function need to be investigated and clarified in more detail.[29]

ROLE OF AT$_2$R-ASSOCIATED PROTEINS IN COGNITIVE FUNCTION

The AT$_2$R does not couple in a typical manner to classical heterotrimeric G proteins and elicits unusual signaling cascades involving activation of protein phosphatases, inhibition of protein kinases and RhoA GTPase, and/or production of nitric oxide. Proteins interacting with the AT$_2$R have been highlighted as factors regulating this unique receptor, suggesting the possibility that AT$_2$R-interacting proteins play key roles in these diverse mechanisms of AT$_2$R signaling.[30] We have cloned ATIP (AT$_2$R-interacting protein) as a protein interacting with the C-terminal tail of the AT$_2$R, using a yeast two-hybrid system.[31] It has been shown to cooperate with the AT$_2$R to transinactivate receptor tyrosine kinases independent of G proteins.

We examined the possible signaling mechanism by which AT$_2$R stimulation could exert neuroprotective effects and enhance neural differentiation. We reported that after AT$_2$R stimulation, ATIP, also known as AT$_2$R-binding protein (ATBP),[32] and SHP-1 (Src homology 2 domain-containing protein-tyrosine phosphatase 1) were translocated into the nucleus following formation of their complex and transactivated MMS2.[33] Furthermore, increased MMS2 expression mediates an inhibitor of DNA binding 1 proteolysis and promotes DNA repair, resulting in neural protection and also neural differentiation. We also demonstrated that ATIP1 could inhibit vascular smooth muscle cell senescence, with an increase in MMS2 expression and SHP-1 activity.[34] Moreover, we reported that ATIP1 could exert anti-inflammatory effects in adipose tissue via macrophage polarization, associated with improvement of insulin resistance in type 2 diabetic mice.[35] All ATIP members are expressed in brain tissues and carry a conserved domain able to interact with the AT$_2$R intracellular tail, suggesting that dysfunction in the AT$_2$R/ATIP axis may be involved in mental retardation.[36] In addition, we demonstrated the possibility that ATIP1 overexpression decreased neointimal formation independent of AT$_2$R activation using ATIP1-Tg mice.[37] It is necessary to elucidate the detailed mechanisms of the involvement of ATIP/ATBP in AT$_2$R-mediated improvement of cognitive function. Possible roles of ATIP/ATBP and PPAR-γ activation associated with the angiotensin II/AT$_2$R axis in cognitive function are shown in Figure 2.

CONCLUSION

In conclusion, the angiotensin II/AT$_2$R axis could play a role in protection against cognitive decline beyond blood pressure control, and its more detailed mechanisms and pathophysiological relevance should be investigated. Moreover, the roles of

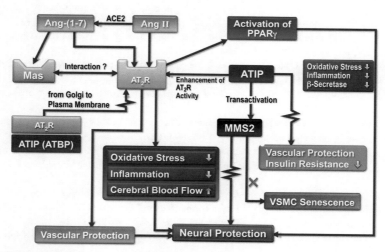

FIGURE 2   Angiotensin II/AT$_2$R and ACE2/angiotensin-(1-7)/Mas axis could play a role in protection against cognitive decline. Roles of ATIP/ATBP including possible direct effects, AT$_2$R-mediated PPAR-γ activation, and interaction of the AT$_2$R and Mas in terms of neurovascular coupling are interesting issues to be addressed.

ATIP/ATBP including its possible direct effects, AT_2R-mediated PPAR-γ activation, and the interaction of the AT_2R and angiotensin-(1-7) (which are described in detail in other chapters) in terms of neurovascular protection are interesting issues to be addressed. Further elucidation of these mechanisms of the functions of angiotensin II receptors beyond the classical ACE/angiotensin II/AT_1R axis could provide possibilities for the development of novel drugs that regulate the renin–angiotensin system in a more sophisticated manner, thereby preventing cognitive decline and treating patients with impaired cognitive function.

ACKNOWLEDGMENTS

This study was partially supported by grants from the Ministry of Education, Culture, Sports, Science and Technology of Japan to M.H., M.M., and J.I. and research grants from pharmaceutical companies: Ajinomoto Pharmaceuticals Co., Ltd.; Astellas Pharma Inc.; Bayer Yakuhin, Ltd.; Daiichi-Sankyo Pharmaceutical Co., Ltd.; Nippon Boehringer Ingelheim Co., Ltd.; Novartis Pharma K. K.; Shionogi & Co., Ltd.; and Takeda Pharmaceutical Co., Ltd.

REFERENCES

1. Novak V, Hajjar I. The relationship between blood pressure and cognitive function. *Nat Rev Cardiol* 2010;**7**:686–98.
2. Nagai M, Hoshide S, Kario K. Hypertension and dementia. *Am J Hypertens* 2010;**23**:116–24.
3. Horiuchi M, Mogi M. Role of angiotensin II receptor subtype activation in cognitive function and ischaemic brain damage. *Br J Pharmacol* 2011;**163**:1122–30.
4. Fuentes B, Fernández-Domínguez J, Ortega-Casarrubios MA, SanJosé B, Martínez-Sánchez P, Díez-Tejedor E. Treatment with angiotensin receptor blockers before stroke could exert a favourable effect in acute cerebral infarction. *J Hypertens* 2010;**28**:575–81.
5. Levi Marpillat N, Macquin-Mavier I, Tropeano AI, Bachoud-Levi AC, Maison P. Antihypertensive classes, cognitive decline and incidence of dementia: a network meta-analysis. *J Hypertens* 2013;**31**:1073–82.
6. Li NC, Lee A, Whitmer RA, Kivipelto M, Lawler E, Kazis LE, et al. Use of angiotensin receptor blockers and risk of dementia in a predominantly male population: prospective cohort analysis. *BMJ* 2010;**340**:b5465.
7. Hajjar I, Brown L, Mack WJ, Chui H. Impact of angiotensin receptor blockers on Alzheimer disease neuropathology in a large brain autopsy series. *Arch Neurol* 2012;**10**:1–7.
8. Kehoe PG, Wilcock GK. Is inhibition of the renin-angiotensin system a new treatment option for Alzheimer's disease? *Lancet Neurol* 2007;**6**:373–8.
9. Li J, Culman J, Hortnagl H, Zhao Y, Gerova N, Timm M, et al. Angiotensin AT_2R protects against cerebral ischemia-induced neuronal injury. *FASEB J* 2005;**19**:617–19.
10. Iwai M, Liu HW, Chen R, Ide A, Okamoto S, Hata R, et al. Possible inhibition of focal cerebral ischemia by angiotensin II type 2 receptor stimulation. *Circulation* 2004;**110**:843–8.
11. Mogi M, Li JM, Iwanami J, Min LJ, Tsukuda K, Iwai M, et al. Angiotensin II type-2 receptor stimulation prevents neural damage by transcriptional activation of methyl methanesulfonate sensitive 2. *Hypertension* 2006;**48**:141–8.
12. Wan Y, Wallinder C, Plouffe B, Beaudry H, Mahalingam AK, Wu X, et al. Design, synthesis, and biological evaluation of the first selective nonpeptide AT_2R agonist. *J Med Chem* 2004;**47**:5995–6008.
13. Gelosa P, Pignieri A, Fändriks L, de Gasparo M, Hallberg A, Banfi C, et al. Stimulation of AT_2R exerts beneficial effects in stroke-prone rats: focus on renal damage. *J Hypertens* 2009;**27**:2444–51.
14. Min LJ, Mogi M, Tsukuda K, Jing F, Ohshima K, Nakaoka H, et al. Direct stimulation of angiotensin II type 2 receptor initiated after stroke ameliorates ischemic brain damage. *Am J Hypertens* 2014;**27**:1036–44.
15. Lucius R, Gallinat S, Rosenstiel P, Herdegan T, Sievers J, Unger T. The angiotensin II type 2 (AT_2) receptor promotes axonal regeneration in the optic nerve of adult rats. *J Exp Med* 1998;**188**:661–70.
16. Cote F, Laflamme L, Payet MD, Gallo-Payet N. Nitric oxide, a new second messenger involved in the action of angiotensin II on neuronal differentiation of NG108-15 cells. *Endocr Res* 1998;**24**:403–7.
17. Cote F, Do TH, Laflamme L, Gallo JM, Gallo-Payet N. Activation of the AT(2) receptor of angiotensin II induces neurite outgrowth and cell migration in microexplant cultures of the cerebellum. *J Biol Chem* 1999;**274**:31686–92.
18. Gendron L, Laflamme L, Rivard N, Asselin C, Payet MD, Gallo-Payet N. Signals from the AT_2 (angiotensin type 2) receptor of angiotensin II inhibit p21ras and activate MAPK (mitogen-activated protein kinase) to induce morphological neuronal differentiation in NG108-15 cells. *Mol Endocrinol* 1999;**13**:1615–26.
19. Wright JW, Reichert JR, Davis CJ, Hatding JW. Neural plasticity and the brain renin-angiotensin system. *Neurosci Biobehav Rev* 2002;**26**:529–52.
20. Reinecke K, Lucius R, Reinecke A, Rickert U, Herdegen T, Unger T. Angiotensin II accelerates functional recovery in the rat sciatic nerve in vivo: role of the AT_2R and the transcription factor NF-kappaB. *FASEB J* 2003;**17**:2094–6.
21. Vervoort VS, Beachem MA, Edwards PS, Ladd S, Miller KE, de Mollerat X, et al. AGTR2 mutations in X-linked mental retardation. *Science* 2002;**296**:2401–3.
22. Maul B, von Bohlen und Halbach O, Becker A, Sterner-Kock A, Voigt JP, Siems WE, et al. Impaired spatial memory and altered dendritic spine morphology in angiotensin II type 2 receptor-deficient mice. *J Mol Med* 2008;**86**:563–71.

23. Tota S, Hanif K, Kamat PK, Najmi AK, Nath C. Role of central angiotensin receptors in scopolamine-induced impairment in memory, cerebral blood flow, and cholinergic function. *Psychopharmacology* 2012;**222**:185–202.

24. AbdAlla S, Lother H, el Missiry A, Langer A, Sergeev P, el Faramawy Y, et al. Angiotensin II AT$_2$R oligomers mediate G-protein dysfunction in an animal model of Alzheimer disease. *J Biol Chem* 2009;**284**:6554–65.

25. Jing F, Mogi M, Sakata A, Iwanami J, Tsukuda K, Ohshima K, et al. Direct stimulation of angiotensin II type 2 receptor enhances spatial memory. *J Cereb Blood Flow Metab* 2012;**32**:248–55.

26. Namsolleck P, Boato F, Schwengel K, Paulis L, Matho KS, Geurts N, et al. AT$_2$-receptor stimulation enhances axonal plasticity after spinal cord injury by upregulating BDNF expression. *Neurobiol Dis* 2013;**51**:177–91.

27. Zhao Y, Foryst-Ludwig A, Bruemmer D, Culman J, Bader M, Unger T, et al. Angiotensin II induces peroxisome proliferator-activated receptor gamma in PC12W cells via angiotensin type 2 receptor activation. *J Neurochem* 2005;**94**:1395–401.

28. Ohshima K, Mogi M, Jing F, Iwanami J, Tsukuda K, Min LJ, et al. Direct angiotensin II type 2 receptor stimulation ameliorates insulin resistance in type 2 diabetes mice with PPARγ activation. *PLoS One* 2012;**7**:e4837.

29. Sumners C, Horiuchi M, Widdop RE, McCarthy C, Unger T, Steckelings UM. Protective arms of the renin-angiotensin-system in neurological disease. *Clin Exp Pharmacol Physiol* 2013;**40**:580–8.

30. Horiuchi M, Iwanami J, Mogi M. Regulation of angiotensin II receptors beyond the classical pathway. *Clin Sci (Lond)* 2012;**123**:193–203.

31. Nouet S, Amzallag N, Li JM, Louis S, Seitz I, Cui TX, et al. Trans-inactivation of receptor tyrosine kinases by novel angiotensin II AT$_2$R-interacting protein, ATIP. *J Biol Chem* 2004;**279**:28989–97.

32. Wruck CJ, Funke-Kaiser H, Pufe T, Kusserow H, Menk M, Schefe JH, et al. Regulation of transport of the angiotensin AT$_2$R by a novel membrane-associated Golgi protein. *Arterioscler Thromb Vasc Biol* 2005;**25**:57–64.

33. Li JM, Mogi M, Tsukuda K, Tomochika H, Iwanami J, Min LJ, et al. Angiotensin II-induced neural differentiation via angiotensin II type 2 (AT$_2$) receptor-MMS2 cascade involving interaction between AT$_2$R-interacting protein and Src homology 2 domain-containing protein-tyrosine phosphatase 1. *Mol Endocrinol* 2007;**21**:499–511.

34. Min LJ, Mogi M, Iwanami J, Jing F, Tsukuda K, Ohshima K, et al. Angiotensin II type 2 receptor-interacting protein prevents vascular senescence. *J Am Soc Hypertens* 2012;**6**:179–84.

35. Jing F, Mogi M, Min LJ, Ohshima K, Nakaoka H, Tsukuda K, et al. Effect of angiotensin II type 2 receptor-interacting protein on adipose tissue function via modulation of macrophage polarization. *PLoS One* 2013;**8**:e60067.

36. Rodrigues-Ferreira S, le Rouzic E, Pawlowski T, Srivastava A, Margottin-Goguet F, Nahmias C. AT$_2$R-interacting proteins ATIPs in the brain. *Int J Hypertens* 2013;**2013**:513047.

37. Fujita T, Mogi M, Min LJ, Iwanami J, Tsukuda K, Sakata A, et al. Attenuation of cuff-induced neointimal formation by overexpression of angiotensin II type 2 receptor-interacting protein 1. *Hypertension* 2009;**53**:688–93.

Chapter 10

The Protective Role of Angiotensin II (AT$_2$) Receptors in Renal Disease: Molecular Mechanisms and Indications for AT$_2$ Agonist Therapy

Isha S. Dhande, Tahir Hussain
University of Hosuton, Houston, Texas, USA

INTRODUCTION

The key role of angiotensin II (Ang II) in mediating renal injury via the AT$_1$ receptor (AT$_1$R) is well characterized. Ang II is a potent inducer of systemic and glomerular hypertension. In addition to its effects on blood pressure and regulation of fluid volume, Ang II is known to exert numerous nonhemodynamic effects that participate in the progression of renal disease via AT$_1$R. These include the induction of proteinuria, oxidative stress, renal inflammation evidenced by increased cytokine and chemokine expression, expansion of mesangial matrix, and fibrosis via increased transforming growth factor-β (TGF-β), which leads to glomerulosclerosis and tubulointerstitial fibrosis and ultimately nephron loss. Thus, blockade of Ang II effector functions, by either ACE inhibitors or AT$_1$R blockers (ARBs), is the therapy of choice in patients with progressive renal disease.

While a majority of the AT$_1$R-mediated pathways exert detrimental effects on renal structure and function, emerging data suggest that downstream signaling in response to Ang II may not necessarily be deleterious. In addition to AT$_1$R, Ang II is also capable of activating the AT$_2$ receptor (AT$_2$R), which is considered to be a functional antagonist of AT$_1$R. Stimulation of AT$_2$R can counteract the effects of AT$_1$R by promoting vasodilatation, antigrowth, anti-inflammatory, and antifibrotic effects. This provides an additional mechanism that contributes to the therapeutic efficacy of ARBs in kidney disease. Though some controversy exists regarding its beneficial role in the context of renal injury, accumulating evidence suggests a protective role for AT$_2$R. The involvement of AT$_2$Rs in renal disease in light of these data will be discussed in the subsequent sections.

RENAL AT$_2$R EXPRESSION

AT$_2$R is believed to play a role in fetal development and is ubiquitously expressed in fetal mesenchymal tissues but declines rapidly after birth.[1,2] Thus, in the adult kidney, AT$_1$R is the dominant Ang II receptor in the absence of disease. However, AT$_2$R may still be detected in the proximal and distal tubules.[3–6] The expression of AT$_2$Rs has been reported in the adventitia of the preglomerular arcuate and interlobular arteries.[7] AT$_2$R expression in the glomerulus has been reported by Carey and colleagues[8,9] and has been found to be expressed in mesangial cells.[10] On the other hand, Miyata et al.[11] reported a lack of glomerular AT$_2$R expression. The findings of Ruiz-Ortega et al.[4] suggest that normally, glomerular AT$_2$R expression may be lacking or is below detection limits; however, it is markedly upregulated in response to renal injury.

The involvement of AT$_2$R in renal pathophysiology is suggested by the fact that its expression is found to be upregulated in the kidneys in renal injury, irrespective of the causative factors.[4] Ang II infusion in AT$_1$R knockout mice results in the upregulation of AT$_2$R.[4] A number of stimuli have been reported to influence AT$_2$R expression. For example, in R3T3 fibroblasts, IL-1β, insulin, and Ang II are capable of inducing AT$_2$R expression, while serum (10%), fibroblast growth factors, and phorbol ester decrease it.[12] Persistent proteinuria was also reported to increase AT$_2$R expression in the proximal

The Protective Arm of the Renin–Angiotensin System (RAS). http://dx.doi.org/10.1016/B978-0-12-801364-9.00010-9

tubule.[13] Interferon regulatory factor 1 (IRF-1) is a transcription factor that has also been shown to promote AT_2R expression to induce apoptosis, while IRF-2 negatively regulates this effect.[14] In response to high glucose, IRF-1 is activated and results in increased AT_2R expression in proximal tubule epithelial cells.[15] Vazquez et al.[16] reported an Ang II-dependent induction of renal AT_2R in the renal ablation model, though the physiological consequence is not clear. Based on the expression data alone, it may be speculated that AT_2R itself participates in the progression of renal injury. However, a large body of work using AT_2R agonists supports the notion that the upregulation of AT_2R in renal injury may be a compensatory mechanism to counterregulate the deleterious effects of enhanced AT_1R function.

FINDINGS FROM AT_2R KNOCKOUT (AT_2KO) MICE

AT_2KO mice have been used to clarify the role of AT_2R in progressive renal disease. Renal injury, evidenced by albuminuria, increased renal inflammatory cytokine and chemokine production, impaired renal function, mesangial matrix accumulation, and glomerulosclerosis, was found to be exacerbated in AT_2KO mice after undergoing subtotal nephrectomy.[17] Also, AT_2R has been reported to exert antifibrotic effects, which was demonstrated using AT_2KO mice subjected to unilateral ureteral obstruction.[18] In line with these data, overexpression of AT_2R in the vasculature was found to ameliorate renal injury in the remnant kidney model.[19] Ongoing work in our laboratory points towards a protective role for AT_2R in attenuating inflammation and renal injury. AT_2KO mice fed with a high-fat diet had higher levels of renal TNF-α and lower IL-10 compared with wild-type mice on the same diet. This was associated with increased urinary albumin excretion and glomerulosclerosis.[20] These findings suggest a protective role for AT_2R in the pathogenesis of renal disease and provide a strong rationale for the activation of AT_2R as a therapeutic strategy for attenuating progressive renal damage.

AT_2R-MEDIATED SIGNALING PATHWAYS

Our knowledge of the downstream events in response to AT_2R activation continues to grow as novel mechanisms of AT_2R-mediated renoprotection are elucidated. G protein-dependent and G protein-independent pathways have been reported.[21–23] The major pathways that mediate renoprotection linked to AT_2R stimulation involve the activation of NO/cGMP,[24] phosphatases,[25,26] and phospholipase A2[27] signaling. AT_2R stimulation mediates diverse, and even opposing, physiological effects, depending upon the cell type and experimental conditions. For example, AT_2R has been shown to promote cell differentiation, while conversely, it has also been linked to the inhibition of growth and apoptosis. This is because AT_2R-linked various mediators, such as cGMP, ERK1/2, and serine/threonine or tyrosine phosphatases, could individually exert opposing effects.

RENOPROTECTIVE MECHANISMS OF AT_2R STIMULATION

Blockade of AT_1R is a powerful therapeutic strategy to slow the progression of renal injury. A potential mechanism involves the unopposed activation of AT_2R by ARB-induced increase in Ang II levels. More recently, stimulation of AT_2R by a selective AT_2R agonist compound 21 (C21) has also been shown to be protective in a number of models of renal injury. Multiple cellular mechanisms exist by which AT_2R stimulation may preserve renal function and structural integrity. These include the attenuation of inflammation and oxidative stress, matrix accumulation, fibrosis, and proteinuria. AT_2R stimulation may also improve the Ang II-mediated alteration of renal hemodynamics and sodium handling, which also contribute to declining renal function. Some of the key renoprotective pathways activated by AT_2R are discussed below.

Inflammation and Oxidative Stress

Inflammation is a key initiating factor that is involved in renal disease. Infiltration of inflammatory cells, especially macrophages, in the renal tissue is associated with increased renal damage.[28] Recently, the anti-inflammatory effect of AT_2R stimulation has gained recognition. Two principal pathways involving transcription factors have been identified that mediate the anti-inflammatory response to AT_2R stimulation. The first is the inhibition of NF-κB via increased epoxyeicosatrienoic acid synthesis and activation of protein phosphatases[29] in dermal fibroblasts. The second involves the inhibition of STAT3 signaling and decreased TNF-α production in PC12W cells.[30] Further, Curato et al.[31] had postulated that an increase in IL-10 may be a possible mechanism by which AT_2R agonist C21 could attenuate proinflammatory cytokine production.

In the context of renal injury, however, the anti-inflammatory role of AT_2R is somewhat controversial. Earlier studies using AT_2R blockade have demonstrated a pathogenic role of AT_2R in renal injury. AT_2R antagonist PD123319 was shown to attenuate macrophage infiltration and NF-κB activation,[32] possibly via the inhibition of the chemokine RANTES in glomerular endothelial cells.[33] Esteban and colleagues[34] also reported an increase in renal NF-κB activation and inflammation in the unilateral ureteral obstruction model of renal injury. While interpreting these findings, it is important to consider the suitability of PD123319 as an AT$_2$R antagonist, since at higher doses, PD123319 has been shown to lose specificity for AT$_2$R and could potentially exert agonistic actions via AT$_1$R, which could complicate the *in vivo* results.[35] Furthermore, it has been suggested that PD123319 may be a partial agonist rather than an antagonist of AT$_2$R.[36]

On the other hand, recent studies using C21 have revealed a distinct anti-inflammatory role for AT$_2$R in various models of renal injury. Treatment with C21 for 4 days significantly attenuated renal proinflammatory cytokines in a model of renovascular hypertension independent of changes in BP. This was associated with decreased inflammatory cell infiltrates.[37] Similar effects on renal cytokines were obtained in hypertensive obese Zucker rats treated for 2 weeks with AT$_2$R agonist CGP42112A.[38] Gelosa et al.[39] demonstrated that C21 treatment reduced renal inflammatory cell infiltration and was renoprotective in stroke-prone spontaneously hypertensive rats (SHRSP). C21 treatment was found to be as effective as the ARB losartan in this model.[40] Moreover, in addition to the decline in renal proinflammatory markers, AT$_2$R agonist C21 promoted the production of IL-10 by proximal tubule epithelial cells in a NO-dependent manner.[6] Emerging data from our laboratory suggest that pretreatment with C21 may attenuate systemic and renal inflammation in response to LPS administration in mice. Furthermore, proximal tubule AT$_2$Rs via increased IL-10 may potentially inhibit macrophage activation in response to LPS suggesting that renal AT$_2$Rs may play an important part in determining the inflammatory environment in the kidney (unpublished data). There are also indications that AT$_2$R stimulation may directly attenuate inflammatory cytokine production by macrophages and thus also lower systemic inflammation that accompanies chronic kidney disease.[41]

Reactive oxygen species (ROS) are known to promote renal injury in glomerular and tubulointerstitial nephritis,[42] ischemia-reperfusion injury,[43] chronic kidney disease,[44,45] and endotoxin-induced acute renal failure.[46] Ang II via AT$_1$R has been shown to induce ROS production via the activation of NADPH oxidase.[47] This increase in oxidative stress can further activate MAPKs to promote renal tubular hypertrophy that culminates in tubular atrophy and fibrosis.[48] Thus, the inhibition of oxidative stress is another mechanism that may be responsible for the renoprotective effect of the AT$_2$R agonist. This was demonstrated by Dandapat and colleagues[49] where the overexpression of AT$_2$R attenuated ROS generation by the inhibition of NADPH oxidase leading to a decrease in phosphorylation of ERK1/2 and p38 MAPK and lower inflammation. In agreement with these findings, Sabuhi et al.[38] reported a decrease in renal NADPH oxidase activation and lower TNF-α and IL-6 in obese Zucker rats treated chronically with CGP42112A.

These data collectively provide compelling evidence that selective activation of AT$_2$R may exert renoprotective effects in progressive renal injury by way of anti-inflammatory and antioxidant mechanisms.

Fibrosis and Extracellular Matrix Accumulation

Inflammation ultimately leads to renal fibrosis and nephron loss. Ang II via AT$_1$R promotes renal fibrosis in rodent models of progressive renal disease.[50,51] Murine mesangial cells express increased fibronectin and TGF-β levels upon incubation with Ang II and this was found to be mediated by AT$_1$R.[52] Investigations of the role of AT$_2$Rs in renal fibrosis suggest an antifibrotic response. Overexpression of AT$_2$R on rat mesangial cells prevented macrophage-induced production of fibronectin and TGF-β.[53] Morrissey and Klahr[54] demonstrated that blockade of AT$_2$R promoted renal fibrosis and proliferation of tubular cells via the inhibition of apoptosis in a model of unilateral ureteral obstruction. Stimulation of AT$_2$R on mesangial cells from WKY, but not SHRSP rats, exerted potent antiproliferative effects, which may partially explain the susceptibility of the latter strain to hypertensive renal injury.[55] More recently, Matavelli et al.[37] have shown that treatment with AT$_2$R agonist C21 markedly reduced renal TGF-β expression in a model of renovascular hypertension. Similarly, chronic C21 treatment attenuated while AT$_2$R antagonist PD123319 worsened mesangial matrix expansion in obese Zucker rats.[6] Preliminary data from our laboratory also suggest a decrease in LPS-induced TGF-β levels in proximal tubule epithelial cells in response to C21 treatment (unpublished data).

The antifibrotic effect of AT$_2$R has been attributed to an increased activation of tissue inhibitors of metalloproteinase (TIMP), specifically TIMP1 and TIMP2, which leads to an inhibition of matrix metalloproteinases (MMPs), specifically MMP9 and MMP2. This has been demonstrated in the vasculature[56] and in models of infarct-induced heart failure[57] and atherosclerosis.[58] On the other hand, Rehman et al.[40] reported a decrease in fibrosis without a change in MMP expression in the heart and aorta from SHRSP. However, while much is known about the pathways involved in the antifibrotic mechanisms of AT$_2$R stimulation in the cardiovascular system, the precise cellular mechanisms that attenuate renal fibrosis are not clear. A very recent study demonstrated that the attenuation of renal TGF-β expression and fibrosis in a model of unilateral ureteral obstruction was a result of heterodimers formed between AT$_2$R and the receptor for the hormone relaxin.[59]

AT$_2$R-Mediated Modulation of Other RAS Components

While the ACE-Ang II-AT$_1$R axis of the RAS is widely acknowledged to promote renal injury and dysregulated renal function, recent efforts are aimed at unraveling the protective effects of the beneficial arm of the RAS, the ACE2-Ang-(1-7)-MasR axis. ACE2 catalyzes the conversion of Ang II to Ang-(1-7); the latter then activates the Mas receptor. In the absence of disease, renal ACE2 expression, particularly on the proximal tubules, is high but is found to be downregulated under pathogenic conditions.[60,61] Chronic inhibition of ACE2 worsens glomerular injury.[62] These effects on renal structural integrity were found to be AT$_1$R-mediated. Ang-(1-7) via the MasR exerts physiological effects that are similar to those elicited by AT$_2$R activation. Infusion of Ang-(1-7) antagonizes Ang II-induced glomerulosclerosis in experimental glomerulonephritis.[63] Su et al.[64] reported an attenuation of Ang II-mediated ERK1/2 activation in proximal tubule epithelial cells, which may prevent glomerular inflammation and injury. Further, genetic deletion of the MasR results in a predisposition to glomerular hyperfiltration, renal fibrosis, and microalbuminuria.[65] These findings indicate an important role for ACE2-Ang-(1-7)-MasR signaling in preserving renal structure and function.

Recent reports suggest a functional and, potentially, a structural interaction between the MasR and AT$_2$R in the heart and vasculature.[66,67] In the kidney, Ali et al.[68] demonstrated a marked upregulation of renal ACE2 activity, Ang-(1-7) levels, and MasR expression in obese Zucker rats treated with AT$_2$R agonist CGP42112A for 2 weeks. There is also a significant decrease in AT$_1$R expression and renin activity in proximal tubule epithelial cells treated with CGP42112A. These alterations in RAS components in response to AT$_2$R stimulation were reversed with concomitant AT$_2$R antagonist administration. These findings suggest that a "fine-tuning" of the RAS, which is often dysregulated in renal disease, may be an important mechanism by which AT$_2$R signaling amplifies its downstream signaling to elicit a physiological response.

Additional Mechanisms Involved in AT$_2$R-Mediated Renoprotection

Stimulation of AT$_2$R is postulated to exert renoprotection via multiple mechanisms. The most widely studied of these mechanisms are the anti-inflammatory and antifibrotic pathways. Additional mechanisms include vasodilation and counteracting the hemodynamic effects of increased AT$_1$R function, such as sodium reabsorption.[5,68] AT$_2$R agonist treatment has been linked to upregulation of nephrin to maintain podocyte integrity and prevent slit diaphragm dysfunction and proteinuria.[69] In addition, AT$_2$R stimulation by Ang II has been found to promote albumin endocytosis in proximal tubule epithelial cells via PKB activation.[70] Albumin is often used as a marker of renal injury; however, on its own, albumin can promote tubulointerstitial inflammation and fibrosis. While it is not clear how regulation of tubular albumin endocytosis by AT$_2$R may impact renal injury, it has been recently demonstrated by Schießl and Castrop[71] that AT$_2$R partially attenuates AT$_1$R-mediated increase in the glomerular filtration of albumin. Further, AT$_2$R stimulation has been shown to improve metabolic parameters, such as insulin sensitivity and lipid metabolism, which can also contribute to renal injury in the metabolic syndrome.[72,73]

AT$_2$R IN RENAL DISEASE ASSOCIATED WITH OBESITY/DIABETES

Recent investigations have been directed to understanding the role of AT$_2$R in renal injury associated with the metabolic syndrome. AT$_2$R expression has been reported to increase in various tissues, including obese and diabetic kidneys.[5,6,74] The primary metabolic abnormalities of obesity/diabetes, including hyperinsulinemia, hyperglycemia, and dyslipidemia, are also major independent risk factors for chronic kidney disease and end-stage renal failure. Recently, AT$_2$R activation in animal models of obesity and insulin resistance has been shown to lower plasma insulin and improve blood glucose and lipid profiles[72,73] while protecting against renal injury.[6,20] Accumulating evidence also indicates that AT$_2$R activation may be beneficial in obesity-related hypertension, another major risk factor for kidney injury.[75] However, the molecular mechanisms involved and the long-term effects of AT$_2$R agonist on these outcomes in the metabolic syndrome are an open area of investigation.

CONCLUSION

The availability of the selective AT$_2$R agonist C21 has facilitated a greater understanding of the protective role of the AT$_2$R in renal pathophysiology. The molecular mechanisms involved and potential indications for AT$_2$R agonists as a viable therapeutic strategy for renal injury are also being clarified. In this regard, AT$_2$R agonist treatment may be especially beneficial in renal damage associated with obesity/diabetes owing to its ability to improve metabolic parameters in addition to protecting renal structural and functional integrity. Furthermore, it is reasonable to speculate that the addition of AT$_2$R agonists to existing therapeutic regimens using ARBs may further enhance the efficacy of these drugs and in turn can keep more patients from progressing from chronic kidney disease to end-stage renal failure.

REFERENCES

1. Grady EF, Sechi LA, Griffin CA, Schambelan M, Kalinyak JE. Expression of AT2 receptors in the developing rat fetus. *J Clin Invest* 1991;**88**:921–33.
2. Nahmias C, Strosberg AD. The angiotensin AT2 receptor: searching for signal-transduction pathways and physiological function. *Trends Pharmacol Sci* 1995;**16**:223–5.
3. Cao Z, Bonnet F, Candido R, Nesteroff SP, Burns WC, Kawachi H, et al. Angiotensin type 2 receptor antagonism confers renal protection in a rat model of progressive renal injury. *J Am Soc Nephrol* 2002;**13**:1773–87.
4. Ruiz-Ortega M, Esteban V, Suzuki Y, Ruperez M, Mezzano S, Ardiles L, et al. Renal expression of angiotensin type 2 (AT2) receptors during kidney damage. *Kidney Int Suppl* 2003;**86**:S21–6.
5. Hakam AC, Hussain T. Renal angiotensin II type-2 receptors are upregulated and mediate the candesartan-induced natriuresis/diuresis in obese Zucker rats. *Hypertension* 2005;**45**:270–5.
6. Dhande I, Ali Q, Hussain T. Proximal tubule angiotensin AT2 receptors mediate an anti-inflammatory response via interleukin-10: role in renoprotection in obese rats. *Hypertension* 2013;**61**:1218–26.
7. Zhuo J, Dean R, MacGregor D, Alcorn D, Mendelsohn FA. Presence of angiotensin II AT2 receptor binding sites in the adventitia of human kidney vasculature. *Clin Exp Pharmacol Physiol Suppl* 1996;**3**:S147–54.
8. Ozono R, Wang ZQ, Moore AF, Inagami T, Siragy HM, Carey RM. Expression of the subtype 2 angiotensin (AT2) receptor protein in rat kidney. *Hypertension* 1997;**30**:1238–46.
9. Wang ZQ, Millatt LJ, Heiderstadt NT, Siragy HM, Johns RA, Carey RM. Differential regulation of renal angiotensin subtype AT1A and AT2 receptor protein in rats with angiotensin-dependent hypertension. *Hypertension* 1999;**33**:96–101.
10. Goto M, Mukoyama M, Suga S, Matsumoto T, Nakagawa M, Ishibashi R, et al. Growth-dependent induction of angiotensin II type 2 receptor in rat mesangial cells. *Hypertension* 1997;**30**(3 Pt 1):358–62.
11. Miyata N, Park F, Li XF, Cowley Jr. AW. Distribution of angiotensin AT1 and AT2 receptor subtypes in the rat kidney. *Am J Physiol* 1999;**277**(3 Pt 2):F437–46.
12. Ichiki T, Kambayashi Y, Inagami T. Multiple growth factors modulate mRNA expression of angiotensin II type-2 receptor in R3T3 cells. *Circ Res* 1995;**77**:1070–6.
13. Tejera N, Gómez-Garre D, Lázaro A, Gallego-Delgado J, Alonso C, Blanco J, et al. Persistent proteinuria up-regulates angiotensin II type 2 receptor and induces apoptosis in proximal tubular cells. *Am J Pathol* 2004;**164**:1817–26.
14. Horiuchi M, Koike G, Yamada T, Mukoyama M, Nakajima M, Dzau VJ. The growth-dependent expression of angiotensin II type 2 receptor is regulated by transcription factors interferon regulatory factor-1 and −2. *J Biol Chem* 1995;**270**:20225–2030.
15. Ali Q, Sabuhi R, Hussain T. High glucose up-regulates angiotensin II subtype 2 receptors via interferon regulatory factor-1 in proximal tubule epithelial cells. *Mol Cell Biochem* 2010;**344**:65–71.
16. Vázquez E, Coronel I, Bautista R, Romo E, Villalón CM, Avila-Casado MC, et al. Angiotensin II-dependent induction of AT(2) receptor expression after renal ablation. *Am J Physiol Renal Physiol* 2005;**288**:F207–13.
17. Benndorf RA, Krebs C, Hirsch-Hoffmann B, Schwedhelm E, Cieslar G, Schmidt-Haupt R, et al. Angiotensin II type 2 receptor deficiency aggravates renal injury and reduces survival in chronic kidney disease in mice. *Kidney Int* 2009;**75**:1039–49.
18. Ma J, Nishimura H, Fogo A, Kon V, Inagami T, Ichikawa I. Accelerated fibrosis and collagen deposition develop in the renal interstitium of angiotensin type 2 receptor null mutant mice during ureteral obstruction. *Kidney Int* 1998;**53**:937–44.
19. Hashimoto N, Maeshima Y, Satoh M, Odawara M, Sugiyama H, Kashihara N, et al. Overexpression of angiotensin type 2 receptor ameliorates glomerular injury in a mouse remnant kidney model. *Am J Physiol Renal Physiol* 2004;**286**:F516–25.
20. Dhande IS, Khan MA, Nag S, Hussain T. Role of the Angiotensin AT2 receptor in obesity-linked inflammation and renal injury: effect of gender. *FASEB J* 2013;**27**:1114.2.
21. Hansen JL, Servant G, Baranski TJ, Fujita T, Iiri T, Sheikh SP. Functional reconstitution of the angiotensin II type 2 receptor and G(i) activation. *Circ Res* 2000;**87**:753–9.
22. Kambayashi Y, Bardhan S, Takahashi K, Tsuzuki S, Inui H, Hamakubo T, et al. Molecular cloning of a novel angiotensin II receptor isoform involved in phosphotyrosine phosphatase inhibition. *J Biol Chem* 1993;**268**:24543–6.
23. Berry C, Touyz R, Dominiczak AF, Webb RC, Johns DG. Angiotensin receptors: signaling, vascular pathophysiology, and interactions with ceramide. *Am J Physiol Heart Circ Physiol* 2001;**281**:H2337–65.
24. Siragy HM, Carey RM. The subtype-2 (AT2) angiotensin receptor regulates renal cyclic guanosine 3', 5'-monophosphate and AT1 receptor-mediated prostaglandin E2 production in conscious rats. *J Clin Invest* 1996;**97**:1978–82.
25. Hayashida W, Horiuchi M, Dzau VJ. Intracellular third loop domain of angiotensin II type-2 receptor. Role in mediating signal transduction and cellular function. *J Biol Chem* 1996;**271**:21985–92.
26. Bedecs K, Elbaz N, Sutren M, Masson M, Susini C, Strosberg AD, et al. Angiotensin II type 2 receptors mediate inhibition of mitogen-activated protein kinase cascade and functional activation of SHP-1 tyrosine phosphatase. *Biochem J* 1997;**325**:449–54.
27. Jacobs LS, Douglas JG. Angiotensin II type 2 receptor subtype mediates phospholipase A2-dependent signaling in rabbit proximal tubular epithelial cells. *Hypertension* 1996;**28**:663–8.
28. Kluth DC, Erwig LP, Rees AJ. Multiple facets of macrophages in renal injury. *Kidney Int* 2004;**66**:542–57.
29. Rompe F, Artuc M, Hallberg A, Alterman M, Stroder K, Thone-Reineke C, et al. Direct angiotensin II type 2 receptor stimulation acts as anti-inflammatory through epoxyeicosatrienoic acid and inhibition of nuclear factor κB. *Hypertension* 2010;**55**:924–31.

30. Abadir PM, Walton JD, Carey RM, Siragy HM. Angiotensin II type 2 receptors modulate inflammation through signal transducer and activator transcription proteins 3 phosphorylation and TNF-α production. *J Interferon Cytokine Res* 2011;**31**:471–4.

31. Curato C, Slavic S, Dong J, Skorska A, Altarche-Xifró W, Miteva K, et al. Identification of noncytotoxic and IL-10-producing CD8 + AT2R+ T cell population in response to ischemic heart injury. *J Immunol* 2010;**185**:6286–93.

32. Ruiz-Ortega M, Lorenzo O, Rupérez M, Blanco J, Egido J. Systemic infusion of angiotensin II into normal rats activates nuclear factor-kappaB and AP-1 in the kidney: role of AT(1) and AT(2) receptors. *Am J Pathol* 2001;**158**:1743–56.

33. Wolf G, Ziyadeh FN, Thaiss F, Tomaszewski J, Caron RJ, Wenzel U, et al. Angiotensin II stimulates expression of the chemokine RANTES in rat glomerular endothelial cells. Role of the angiotensin type 2 receptor. *J Clin Invest* 1997;**100**:1047–58.

34. Esteban V, Lorenzo O, Rupérez M, Suzuki Y, Mezzano S, Blanco J, et al. Angiotensin II, via AT1 and AT2 receptors and NF-kappaB pathway, regulates the inflammatory response in unilateral ureteral obstruction. *J Am Soc Nephrol* 2004;**15**:1514–29.

35. Wenzel UO, Krebs C, Benndorf R. The angiotensin II type 2 receptor in renal disease. *J Renin Angiotensin Aldosterone Syst* 2010;**11**:37–41.

36. Zhou J, Ernsberger P, Douglas JG. A novel angiotensin receptor subtype in rat mesangium. Coupling to adenylyl cyclase. *Hypertension* 1993;**21**(6 Pt 2):1035–8.

37. Matavelli LC, Jiang H, Siragy HM. Angiotensin AT2 receptor stimulation inhibits early renal inflammation in renovascular hypertension. *Hypertension* 2011;**57**:308–13.

38. Sabuhi R, Ali Q, Asghar M, Al-Zamily NRH, Hussain T. Role of angiotensin II AT2 receptor in inflammation and oxidative stress: Opposing effects in lean and obese rats. *Am J Renal Physiol* 2011;**300**:F700–6.

39. Gelosa P, Pignieri A, Fändriks L, de Gasparo M, Hallberg A, Banfi C, et al. Stimulation of AT2 receptor exerts beneficial effects in stroke-prone rats: focus on renal damage. *J Hypertens* 2009;**27**:2444–51.

40. Rehman A, Leibowitz A, Yamamoto N, Rautureau Y, Paradis P, Schiffrin EL. Angiotensin type 2 receptor agonist compound 21 reduces vascular injury and myocardial fibrosis in stroke-prone spontaneously hypertensive rats. *Hypertension* 2012;**59**:291–9.

41. Dhande IS, Ma W, Hussain T. Angiotensin AT2 receptor stimulation is anti-inflammatory in lipopolysaccharide-activated THP-1 macrophages via increased interleukin-10 production. *Hypertens Res* 2015;**38**:21–9.

42. Klahr S. Oxygen radicals and renal diseases. *Miner Electrolyte Metab* 1997;**23**:140–3.

43. Dobashi K, Ghosh B, Orak JK, Singh I, Singh AK. Kidney ischemia-reperfusion: modulation of antioxidant defenses. *Mol Cell Biochem* 2000;**205**:1–11.

44. Ceballos-Picot I, Witko-Sarsat V, Merad-Boudia M, Nguyen AT, Thevenin M, Jaudon MC, et al. Glutathione antioxidant system as a marker of oxidative stress in chronic renal failure. *Free Radic Biol Med* 1996;**21**:845–53.

45. Mimic-Oka J, Simic T, Djukanovic L, Reljic Z, Davicevic Z. Alteration in plasma antioxidant capacity in various degrees of chronic renal failure. *Clin Nephrol* 1999;**51**:233–41.

46. Wiesel P, Patel AP, DiFonzo N, Marria PB, Sim CU, Pellacani A, et al. Endotoxin-induced mortality is related to increased oxidative stress and end-organ dysfunction, not refractory hypotension, in heme oxygenase-1-deficient mice. *Circulation* 2000;**102**:3015–22.

47. Pendergrass KD, Gwathmey TM, Michalek RD, Grayson JM, Chappell MC. The angiotensin II-AT1 receptor stimulates reactive oxygen species within the cell nucleus. *Biochem Biophys Res Commun* 2009;**384**:149–54.

48. Hannken T, Schroeder R, Zahner G, Stahl RA, Wolf G. Reactive oxygen species stimulate p44/42 mitogen-activated protein kinase and induce p27(Kip1): role in angiotensin II-mediated hypertrophy of proximal tubular cells. *J Am Soc Nephrol* 2000;**11**:1387–97.

49. Dandapat A, Hu CP, Chen J, Liu Y, Khan JA, Remeo F, et al. Over-expression of angiotensin II type 2 receptor (agtr2) decreases collagen accumulation in atherosclerotic plaque. *Biochem Biophys Res Commun* 2008;**366**:871–7.

50. Ishidoya S, Morrissey J, McCracken R, Reyes A, Klahr S. Angiotensin II receptor antagonist ameliorates renal tubulointerstitial fibrosis caused by unilateral ureteral obstruction. *Kidney Int* 1995;**47**:1285–94.

51. Satoh M, Kashihara N, Yamasaki Y, Maruyama K, Okamoto K, Maeshima Y, et al. Renal interstitial fibrosis is reduced in angiotensin II type 1 a receptor-deficient mice. *J Am Soc Nephrol* 2001;**12**:317–25.

52. Perlman A, Lawsin LM, Kolachana P, Saji M, Moore Jr J, Ringel MD. Angiotensin II regulation of TGF-beta in murine mesangial cells involves both PI3 kinase and MAP kinase. *Ann Clin Lab Sci* 2004;**34**:277–86.

53. Pawluczyk IZ, Harris KP. Effect of angiotensin type 2 receptor over-expression on the rat mesangial cell fibrotic phenotype: effect of gender. *J Renin Angiotensin Aldosterone Syst* 2012;**13**:221–31.

54. Morrissey JJ, Klahr S. Effect of AT2 receptor blockade on the pathogenesis of renal fibrosis. *Am J Physiol* 1999;**276**(1 Pt 2):F39–45.

55. Goto M, Mukoyama M, Sugawara A, Suganami T, Kasahara M, Yahata K, et al. Expression and role of angiotensin II type 2 receptor in the kidney and mesangial cells of spontaneously hypertensive rats. *Hypertens Res* 2002;**25**:125–33.

56. Brassard P, Amiri F, Schiffrin EL. Combined angiotensin II type 1 and type 2 receptor blockade on vascular remodeling and matrix metalloproteinases in resistance arteries. *Hypertension* 2005;**46**:598–606.

57. Lauer D, Slavic S, Sommerfeld M, Thöne-Reineke C, Sharkovska Y, Hallberg A, et al. Angiotensin type 2 receptor stimulation ameliorates left ventricular fibrosis and dysfunction via regulation of tissue inhibitor of matrix metalloproteinase 1/matrix metalloproteinase 9 axis and transforming growth factor β1 in the rat heart. *Hypertension* 2014;**63**:e60–7.

58. Kljajic ST, Widdop RE, Vinh A, Welungoda I, Bosnyak S, Jones ES, et al. Direct AT2 receptor stimulation is athero-protective and stabilizes plaque in Apolipoprotein E-deficient mice. *Int J Cardiol* 2013;**169**:281–7.

59. Chow BS, Kocan M, Bosnyak S, Sarwar M, Wigg B, Jones ES, et al. Relaxin requires the angiotensin II type 2 receptor to abrogate renal interstitial fibrosis. *Kidney Int* 2014;**86**:75–85.

60. Dilauro M, Zimpelmann J, Robertson SJ, Genest D, Burns KD. Effect of ACE2 and angiotensin-(1–7) in a mouse model of early chronic kidney disease. *Am J Physiol Renal Physiol* 2010;**298**:F1523–32.

61. da Silveira KD, Pompermayer Bosco KS, Diniz LR, Carmona AK, Cassali GD, et al. ACE2-angiotensin-(1–7)-Mas axis in renal ischaemia/reperfusion injury in rats. *Clin Sci (Lond)* 2010;**119**:385–94.

62. Soler MJ, Wysocki J, Ye M, Lloveras J, Kanwar Y, Batlle D. ACE2 inhibition worsens glomerular injury in association with increased ACE expression in streptozotocin-induced diabetic mice. *Kidney Int* 2007;**72**:614–23.

63. Zhang J, Noble NA, Border WA, Huang Y. Infusion of angiotensin-(1–7) reduces glomerulosclerosis through counteracting angiotensin II in experimental glomerulonephritis. *Am J Physiol Renal Physiol* 2010;**298**:F579–88.

64. Su JZ, Fukuda N, Jin XQ, Lai YM, Suzuki R, et al. Effect of AT2 receptor on expression of AT1 and TGF-beta receptors in VSMCs from SHR. *Hypertension* 2002;**40**:853–8.

65. Pinheiro SV, Ferreira AJ, Kitten GT, da Silveira KD, da Silva DA, Santos SH, et al. Genetic deletion of the angiotensin-(1–7) receptor Mas leads to glomerular hyperfiltration and microalbuminuria. *Kidney Int* 2009;**75**:1184–93.

66. Walters PE, Gaspari TA, Widdop RE. Angiotensin-(1–7) acts as a vasodepressor agent via angiotensin II type 2 receptors in conscious rats. *Hypertension* 2005;**45**:960–6.

67. Castro CH, Santos RA, Ferreira AJ, Bader M, Alenina N, Almeida AP. Evidence for a functional interaction of the angiotensin-(1–7) receptor Mas with AT1 and AT2 receptors in the mouse heart. *Hypertension* 2005;**46**:937–42.

68. Ali Q, Wu Y, Hussain T. Chronic AT2 receptor activation increases renal ACE2 activity, attenuates AT1 receptor function and blood pressure in obese Zucker rats. *Kidney Int* 2013;**84**:93193–9.

69. Suzuki K, Han GD, Miyauchi N, Hashimoto T, Nakatsue T, Fujioka Y, et al. Angiotensin II type 1 and type 2 receptors play opposite roles in regulating the barrier function of kidney glomerular capillary wall. *Am J Pathol* 2007;**170**:1841–53.

70. Caruso-Neves C, Kwon SH, Guggino WB. Albumin endocytosis in proximal tubule cells is modulated by angiotensin II through an AT2 receptor-mediated protein kinase B activation. *Proc Natl Acad Sci U S A* 2005;**102**:17513–18.

71. Schießl IM, Castrop H. Angiotensin II AT2 receptor activation attenuates AT1 receptor-induced increases in the glomerular filtration of albumin: a multiphoton microscopy study. *Am J Physiol Renal Physiol* 2013;**305**:F1189–200.

72. Ohshima K, Mogi M, Jing F, Iwanami J, Tsukuda K, Min LJ, et al. Direct angiotensin II type 2 receptor stimulation ameliorates insulin resistance in type 2 diabetes mice with PPARγ activation. *PLoS One* 2012;**7**:e48387.

73. Samuel P, Khan MA, Nag S, Inagami T, Hussain T. Angiotensin AT(2) receptor contributes towards gender bias in weight gain. *PLoS One* 2013;**8**:e48425.

74. Hakam AC, Siddiqui AH, Hussain T. Renal angiotensin II AT2 receptors promote natriuresis in streptozotocin-induced diabetic rats. *Am J Physiol Renal Physiol* 2006;**290**:F503–8.

75. Siddiqui AH, Ali Q, Hussain T. Protective role of angiotensin II subtype 2 receptor in blood pressure increase in obese Zucker rats. *Hypertension* 2009;**53**:256–61.

Chapter 11

AT$_2$ Receptors and Natriuresis

Robert M. Carey

Department of Medicine, University of Virginia, Charlottesville, Virginia, USA

INTRODUCTION

Renal sodium (Na$^+$) excretion is critical for the maintenance of body fluid volume and the control of blood pressure (BP).[1,2] Under normal circumstances, Na$^+$ retention raises BP by expanding extracellular fluid and plasma volumes. The resulting increase in renal perfusion pressure, through incompletely understood mechanisms, induces natriuresis, termed pressure natriuresis, which returns BP towards normal.[3] In normal individuals, the pressure-natriuresis mechanism completely restores BP to normal. In hypertension, the pressure-natriuresis mechanism is defective such that an increase in renal perfusion pressure is not offset completely by the increase in Na$^+$ excretion.[4,5] Thus, in hypertension, Na$^+$ excretion normalizes at the expense of increased BP.

According to the hypothesis originally developed by Guyton, a chronic increase in BP is achievable only if there is a defect in the renal handling of Na$^+$.[4,5] Over many years, this hypothesis has been validated by kidney transplantation studies in genetically hypertensive rats showing that hypertension follows the abnormal transplanted kidney.[6–8] Na$^+$ excretion is regulated by both renal hemodynamic and tubular mechanisms, but under physiological conditions, tubular mechanisms predominate and provide the fine-tuning of renal Na$^+$ transport necessary to sustain normal extracellular fluid volume and BP.

A number of hormonal systems participate in the control of renal Na$^+$ transporters and excretion, including norepinephrine, the renin–angiotensin system (RAS), and aldosterone. Hall et al., through an elegant series of servo-control studies, defined pressure natriuresis as the mechanism of "escape" from the Na$^+$-retaining actions of these hormonal influences.[9–11] Thus, pressure natriuresis is important for the overall regulation of total body sodium and fluid balance as well as BP.[11]

The RAS, especially within the kidney, is an important hormonal system in the control of renal Na$^+$ excretion.[12] Studies have demonstrated that Ang II via AT$_1$ receptors (AT$_1$Rs) induces Na$^+$ reabsorption at virtually all segments of the nephron.[12] Recent work by Coffman et al. employing cross transplantation studies has demonstrated the essential role of renal AT$_1$Rs in the development of hypertension induced by Ang II.[13–15] Renal AT$_1$Rs were required to sustain hypertensive responses to continuous Ang II infusion for 2 weeks in mice. Mice with AT$_1$Rs specifically deleted from renal proximal tubule cells using Cre-lox technology under the control of a proximal tubule-specific *Pepck* promoter had approximately a 10 mm Hg reduction in BP associated with a 36% reduction in proximal tubule Na$^+$ reabsorption, together with reduced expression of the most important proximal tubule apical Na$^+$ transporter, Na+/H$^+$ exchanger isoform 3 (NHE-3).[15] These mice were protected from the antinatriuretic and pressor actions of infused Ang II.[15] Sigmund et al. also demonstrated that AT$_1$R overexpression in the proximal tubule increased BP by approximately 15 mm Hg.[16,17] These studies underline the importance of renal proximal tubule AT$_1$Rs in the control of BP by means of their effects on Na$^+$ reabsorption, especially in the proximal nephron.

While AT$_1$R activation via Ang II induces renal tubule Na$^+$ reabsorption and increases BP, many studies have demonstrated that AT$_2$ receptors (AT$_2$Rs) may oppose this action.[18–28] This chapter will review evidence that AT$_2$Rs exert a counterregulatory force on Na$^+$ reabsorption, predominantly in the proximal tubule, opposing the action of Ang II to induce Na$^+$ retention.

RENAL EXPRESSION OF AT$_2$RS

In fetal life, AT$_2$Rs are highly expressed throughout the kidney in the glomeruli, the tubules, and the renal interstitial space.[29–31] However, their expression levels decrease markedly during the postnatal period.[29–31] Nevertheless, during

The Protective Arm of the Renin–Angiotensin System (RAS). http://dx.doi.org/10.1016/B978-0-12-801364-9.00011-0

adulthood, AT_2Rs continue to be expressed in the glomeruli and in the vascular and tubular cells, although in smaller quantities.[30,31] During adulthood, the highest renal expression level of AT_2Rs is in the renal proximal tubule cells.[30,31] The expression of AT_2R mRNA in the kidney is relatively small compared with that of AT_1Rs.[31] Although AT_2Rs were found to have low mRNA, the AT_2R protein was consistently expressed, similar to that of renal dopamine receptors.

Little is currently known about the regulation of renal AT_2R expression. Early studies demonstrated that renal AT_2Rs are quantitatively upregulated by dietary Na^+ restriction, which activates the RAS and increases the circulating level of Ang II.[30] AT_2R upregulation has also been observed in AT_1R-null mice in which Ang II production is increased with availability to bind to and activate unblocked AT_2Rs.[32] Studies also have suggested that estrogen may upregulate both AT_1Rs and AT_2Rs.[33]

AT_2R INDUCTION OF NATRIURESIS: INITIAL EVIDENCE

Unambiguous evidence now exists that renal AT_2Rs mediate natriuresis.[34–39] This possibility was initially suggested by AT_2R knockout studies in mice, wherein the receptor-null mice demonstrated antinatriuretic hypersensitivity to chronically infused Ang II.[40] In addition, these mice had pressor hypersensitivity to Ang II and their pressure–natriuresis relationship was markedly impaired. Earlier studies had strongly suggested that AT_2R activation induces an extracellular signaling cascade including bradykinin (BK), nitric oxide (NO), and cyclic GMP (cGMP).[22–27] The mechanism of natriuresis in the aforementioned AT_2R-null mice was identified as a deficiency of intrarenal generation of BK, NO, and cGMP in response to Ang II.[40] Later studies confirmed that the pressure–natriuresis curves were shifted to the right (less sensitive) in AT_2R-null mice compared with wild-type mice with intact AT_2Rs.[41]

Continuing investigation of the role of endogenous renal AT_2Rs in natriuresis employed direct renal interstitial micro-infusion of pharmacological and molecular agents that enabled the direct evaluation of Na^+ excretion and renal function without interference by systemic hemodynamic or hormonal influences. Using these techniques, selective intrarenal AT_1R blockade (with candesartan) in rats resulted in a highly significant natriuresis that was blocked completely by concurrent intrarenal administration (low dose) of specific AT_2R antagonist PD-123319 .[34] These results strongly suggested that the natriuretic response to intrarenal AT_1R blockade is dependent upon intrarenal Ang II-induced activation of unblocked AT_2Rs.

ROLE OF ANGIOTENSIN III IN AT_2R-INDUCED NATRIURESIS

On the basis of these results, we expected that intrarenal administration of Ang II would engender a robust natriuresis in rats that had been concurrently blocked systemically with AT_1R inhibitor candesartan. However, we were unable to obtain a natriuretic response to infused Ang II, at any dose level, and even at high infusion rates.[34] Because the angiotensin heptapeptide [des-aspartyl]¹-Ang II (angiotensin III (Ang III)), an Ang II metabolite, had been shown to be active at AT_2Rs in the brain, we studied whether Ang III might be an AT_2R agonist peptide in the kidney. With systemic AT_1Rs blocked, direct intrarenal interstitial infusion of Ang III engendered a robust natriuretic response that was abolished by intrarenal coinfusion of PD.[34] These results suggested that Ang III might be an endogenous agonist for AT_2Rs in the kidney. However, there was still a requirement for systemic AT_1R blockade to be able to observe AT_2R-mediated natriuresis.[34]

The effectiveness of Ang III but not of Ang II required an explanation as both were considered equally potent in activating AT_2Rs *in vitro*. We hypothesized that Ang II must be converted to Ang III in order to interact with AT_2Rs within the kidney.[35] The heptapeptide Ang III is formed from Ang II by means of the action of aminopeptidase A (APA) and is degraded to the hexapeptide angiotensin IV by aminopeptidase N (APN). In the presence of systemic AT_1R blockade, direct intrarenal infusion of Ang III again induced a robust natriuretic response that was augmented by intrarenal coinfusion of APN blocker 2-amino-4-methylsulfonyl-butane-thiol, methane-thiol (PC-18).[35] The higher level of natriuresis in response to PC-18 and Ang III was also abolished by intrarenal PD, indicating the specific AT_2R activation of the combination.[35] Intrarenal infusion of the canonical angiotensin receptor ligand Ang II only induced natriuresis in the presence of APN inhibition with PC-18, and this response was completely abrogated by intrarenal coinfusion of APA inhibitor 3-amino-4-thio-butyl sulfonate (EC-33).[35] These results strongly suggested that Ang III, and not Ang II, is an important ligand for AT_2R activation in the kidney and that Ang II must be converted to Ang III to induce an AT_2R-mediated natriuretic response (Figure 1).

The importance of Ang III as a selective AT_2R ligand was recently confirmed.[42] The mechanism that allows Ang III but not Ang II to induce natriuresis is not currently understood, but likely involves the role of the N-terminal arginine in stabilizing the AT_2R binding pocket and allowing optimal binding of the remaining C-terminal amino acids of the peptide to the receptor. However, the role of Ang III in inducing an AT_2R-mediated response has not been confined to the kidney. In the

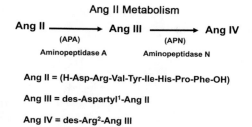

Ang II Metabolism

Ang II ——(APA)——→ Ang III ——(APN)——→ Ang IV

(APA) Aminopeptidase A

(APN) Aminopeptidase N

Ang II = (H-Asp-Arg-Val-Tyr-Ile-His-Pro-Phe-OH)

Ang III = des-Aspartyl^1-Ang II

Ang IV = des-Arg^2-Ang III

FIGURE 1 Schematic representation of the metabolism of angiotensin II (Ang II) to angiotensin III (Ang III) by aminopeptidase A (APA) and of Ang III to angiotensin IV (Ang IV) by aminopeptidase N (APN).

coronary microcirculation, Ang III rather than Ang II is the preferred ligand to induce AT$_2$R-mediated vasodilation; in the heart, to activate AT$_2$Rs to induce ANP secretion; and in the adrenal cortex, for AT$_2$R-mediated aldosterone secretion.[43–45]

NEPHRON SITES AND MECHANISMS OF AT$_2$R-INDUCED NATRIURESIS VIA ANG III

AT$_2$Rs are expressed maximally in the proximal tubule, and on the basis of endogenous Na$^+$/lithium clearance studies, renal AT$_2$Rs induce natriuresis mainly by reducing Na$^+$ reabsorption in this nephron segment.[46] These studies failed to demonstrate any change in renal blood flow or glomerular filtration rate. However, AT$_2$Rs may also induce natriuresis via an action in the thick ascending loop of Henle,[47,48] although AT$_2$R expression levels are relatively low in this nephron segment.[29–31]

The signaling mechanisms of AT$_2$Rs in natriuresis have recently been documented. *In vivo*, Ang III-induced natriuresis was associated with a significant increase in renal interstitial cGMP levels and was completely abrogated by soluble guanylyl cyclase inhibitor 1H-[1,2,4]oxadiazolo[4,3-α]quinoxalin-1-one (ODQ), suggesting that AT$_2$R-induced natriuresis via endogenous Ang III is mediated by increased renal cGMP production.[46] Additional *in vitro* evidence supporting the involvement of the common NO/cGMP pathway includes studies in Wistar-Kyoto rats wherein the ubiquitous transcription factor Sp1 was also involved.[49] In addition, AT$_2$R activation results in heterodimerization with AT$_1$Rs reducing their expression by direct protein–protein interaction at the cell surface.[12] Thus, renal AT$_2$Rs may oppose the action of AT$_1$Rs to reabsorb Na$^+$ by several signaling mechanisms.

AT$_2$R-DOPAMINE D$_1$-LIKE RECEPTOR INTERACTIONS

The renal dopaminergic system is another important regulator of renal Na$^+$ excretion and couples with AT$_2$Rs to oppose AT$_1$R-induced antinatriuresis. There are two families of dopamine receptors, D$_1$-like (D$_{1-like}$; D$_1$ and D$_5$) and D$_2$-like (D$_2$, D$_3$, and D$_4$) receptors.[50] Renal dopaminergic tone has been shown to account for approximately 50–60% of basal Na$^+$ excretion: Intrarenal administration of the highly selective dopamine D$_{1-like}$ receptor antagonist SCH-23390 induces antinatriuresis of this magnitude dose-dependently under conditions of moderate Na$^+$ balance, whereas intrarenal administration of highly selective D$_{1-like}$ receptor agonist fenoldopam induces natriuresis that is completely blocked by SCH-23390 in the Na$^+$-loaded rat.[51–53] Interestingly, the natriuretic response to fenoldopam is abolished by concomitant intrarenal AT$_2$R blockade with PD, indicating that D$_{1-like}$ receptor-induced natriuresis is dependent upon AT$_2$R activation.[52,53]

In the previous study, fenoldopam administration was associated with trafficking of AT$_2$Rs from intracellular sites to the apical plasma membranes of renal proximal tubule cells without any alteration in total cell or basolateral membrane expression of AT$_2$Rs.[53] This observation suggested that dopaminergic stimulation of its receptor also involves activation of AT$_2$Rs. Fenoldopam-induced natriuresis is also accompanied by translocation of D$_1$Rs to the plasma membrane and both D$_1$R translocation and AT$_2$R translocation are dependent upon an intact microtubule network.[53,54]

Dopamine D$_1$Rs signal through cAMP/protein kinase A (PKA) to mediate natriuresis.[50] Thus, it was possible that the dependence of D$_{1-like}$ receptor-induced natriuresis on AT$_2$R activation involved the generation of cAMP and the stimulation of the PKA pathway, which in turn would translocate and activate AT$_2$Rs. This was indeed demonstrated in experimental animals by direct activation of adenylyl cyclase with forskolin in the presence of 3-isobutyl-1-methylxanthine (IBMX) to inhibit cAMP degradation, which resulted in a significant and sustained AT$_2$R-mediated natriuretic response.[53] Direct dopamine agonist stimulation of D$_{1-like}$ receptors was not necessary for AT$_2$R-induced natriuresis because forskolin+IBMX-mediated natriuresis persisted in the presence of D$_{1-like}$ receptor inhibition with SCH-23390.[53] These results were confirmed in human renal proximal tubule cells by showing that D$_1$ receptor stimulation of AT$_2$R recruitment is cAMP- and protein phosphatase 2A-dependent and the recruitment of AT$_2$Rs to the apical plasma membrane is required for them to function.[55]

Thus, the mechanism by which AT_2R and $D_{1\text{-like}}$ receptors interact appears to be $D_{1\text{-like}}$ receptor/cAMP signaling, which furnishes the stimulus for AT_2R translocation and natriuresis.

PHYSIOLOGICAL SIGNIFICANCE OF RENAL AT_2RS IN RENAL NA^+ EXCRETION

In the majority of available studies, the demonstration of Ang III-induced natriuresis has required the presence of concurrent systemic AT_1R blockade.[34–36,46] The only exception so far is that Ang III-induced natriuresis can be shown in the absence of systemic AT_1R blockade when intrarenal Ang III levels are increased by concurrently inhibiting APN.[46] Thus, in a normal physiological sense, intrarenal Ang III and AT_2Rs probably do not play a role when AT_1Rs are completely intact because the antinatriuresis generated by Ang II-induced renal AT_1R activation would be expected to overwhelm AT_2R-induced natriuresis under these circumstances. However, AT_2R activation would be expected to induce natriuresis in the presence of AT_1R blockade or when APN is inhibited in the absence of AT_1R blockade. Importantly, AT_2Rs would also be expected to induce natriuresis in the presence of an activated RAS, as demonstrated in certain disease states discussed below.

Incidentally, there is little or no evidence supporting the induction of natriuresis by alternative RAS pathways such as the angiotensin-converting enzyme-2 (ACE-2), angiotensin-(1-7), and *Mas* receptor pathway.[46]

AT_2R-INDUCED NATRIURESIS IN DISEASE STATES

AT_2R-induced natriuresis has been studied in the spontaneously hypertensive rat (SHR).[56,57] As opposed to the Wistar-Kyoto (WKY) control rat with robust natriuretic responses to intrarenal Ang III administration, hypertensive SHR has absent natriuretic responses to Ang III.[56] While WKY demonstrates Ang III-induced recruitment of AT_2Rs to the apical plasma membranes of renal proximal tubule cells, hypertensive SHR fails to demonstrate this response.[56] While normotensive WKY responds to intrarenal AT_1R blockade with a significant natriuresis, hypertensive SHR fails to mount a significant natriuretic response to intrarenal AT_1R blockade.[56] However, a natriuretic response to intrarenal Ang III administration was enabled in SHR by inhibiting the metabolism of Ang III with PC-18.[57] The defect in AT_2R-induced Na^+ excretion in SHR was present in young SHR prior to the onset of hypertension and was restored to normal WKY responses with intrarenal PC-18 administration in young normotensive SHR.[57] In SHR, the Na^+ retention was localized to the renal proximal tubule.[57] Thus, increased renal proximal tubule Na^+ retention is observed before the onset of hypertension in SHRs and inhibition of Ang III metabolism ameliorates the pathophysiological defect in Na^+ excretion. This is consistent with the repeated observations from several laboratories that SHRs have activation of the intrarenal RAS, as manifested by increased renal proximal tubule AT_1R expression, elevated renal Ang II content, and increased Ang II-AT_1R activation of NHE-3.[58–67] Further studies are required to elucidate the specific role and mechanisms of defective AT_2R-mediated natriuresis in SHR, because similar to humans, SHR develops hypertension as it ages and is widely employed as a model to study the development and maintenance of human primary hypertension.

Another disease state in which AT_2R-induced natriuresis has been extensively studied is the model of obesity hypertension provided by the obese Zucker rat.[37–39,68] Zucker rats, a genetic model of obesity and insulin resistance, exhibit a hyperactive RAS that is manifested in only mildly elevated BP. Systemic AT_1R blockade in obese Zucker rats induces a higher level of natriuresis than in lean control animals, likely due to the upregulation of AT_2Rs in the renal proximal tubules of obese animals. This same phenomenon can be observed in streptozotocin-induced type 1 diabetic rats. In obese Zucker rats, AT_2R upregulation in the renal proximal tubule was mediated via the common NO/cGMP/PKG pathway that directly inhibited tubule Na^+/K^+-ATPase (NKA) activity. Thus, the results suggest that in obesity and hypertension, renal proximal tubule AT_2Rs may be upregulated and may protect the animal from severe hypertension by inducing natriuresis and opposing the action of proximal tubule AT_1Rs to retain Na^+ and increase BP.

AT_2R AGONIST ADMINISTRATION AND NATRIURESIS

Because renal AT_2Rs are natriuretic receptors, opposing the predominant antinatriuretic actions of Ang II via AT_1Rs, there has been a great deal of interest in whether the activation of AT_2Rs with an exogenous agonist can induce and sustain a natriuretic response that might be therapeutically useful. Such a potential compound is compound 21 (C21), the first available highly selective nonpeptide AT_2R agonist.[69,70] Systemic administration of C21 increases urinary Na^+ excretion by 12-fold in normal male uninephrectomized Sprague-Dawley rats.[70] The natriuretic response to systemic C21 administration is abolished with concomitant intrarenal PD, indicating that C21 induces natriuresis specifically by stimulating renal AT_2Rs.[70] C21 was also reported to increase Na^+ excretion in male and female Sprague-Dawley and also in obese Zucker rats that were neutralized by systemic coadministration of PD.[71,72]

Recently, both systemic administration and intrarenal administration of C21 were demonstrated to be natriuretic acutely in Sprague-Dawley rats and chronically in mice.[71] These natriuretic responses were reversed by intrarenal AT$_2$R inhibition and by the deletion of the AT$_2$R gene.[71] One of the surprising findings of these studies was that the AT$_2$R-mediated natriuresis occurred in the absence of AT$_1$R blockade.[71] Another important finding was that the natriuretic response was localized predominantly to the renal proximal tubule, which reabsorbs a large fraction (at least 60%) of filtered Na$^+$ and for which there is currently no available efficacious diuretic/natriuretic agent.[71] As shown in the past studies, cardiovascular and/or renal responses to AT$_2$R activation have been demonstrable only when the RAS is activated and/or AT$_1$Rs are concurrently blocked.

Systemically administered C21 also prevented Na$^+$ retention and lowered BP in the commonly employed Ang II infusion model of experimental hypertension.[71] Intrarenal C21 not only abrogated the initial Ang II-induced antinatriuresis but also augmented Na$^+$ excretion chronically in this model.[71] Therefore, intrarenal AT$_2$R activation improved the pressure–natriuresis relationship in this model.

C21-induced natriuresis occurs principally through AT$_2$R activation in the renal proximal tubule through the bradykinin-NO-cGMP-Src-ERK signaling cascade, which has been found to apply to virtually all existing cellular AT$_2$R actions.[71] In contrast to other AT$_2$R responses, however, C21-induced natriuresis is similar in male and female animals.[71] C21-induced AT$_2$R activation translocates AT$_2$Rs to the apical plasma membranes of renal proximal tubule cells and internalizes/inactivates NHE-3 and NKA Na$^+$ transport systems in these cells.[71] Because no clinically effective diuretic/natriuretic agents acting in the renal proximal tubule are currently available, systematic AT$_2$R activation potentially represents a unique opportunity for treatment of volume-overload/edema-forming states and hypertension in humans.

CONCLUSION

The major points of this chapter are summarized in Figure 2. AT$_2$Rs are present within the kidney, particularly in the proximal tubule cells, and are upregulated by increased activity of the RAS. Virtually all studies demonstrate that AT$_2$Rs engender natriuresis. The natriuretic response to AT$_2$R activation does not involve Ang-(1-7) or the *Mas* receptor. Renal AT$_2$R activation enhances the sensitivity of the pressure–natriuresis mechanism; in the absence of AT$_2$Rs, marked Na$^+$ retention occurs in response to RAS activation. Ang III, not Ang II, is the preferred endogenous AT$_2$R agonist in the production of natriuresis. Endogenous renal Ang III activates AT$_2$Rs to induce natriuresis in a proximal tubule-specific manner via a renal BK, NO, and cGMP signaling pathway. Interestingly, dopaminergic stimulation of natriuresis is completely dependent upon AT$_2$R translocation to the apical plasma membrane and activation by dopamine D$_1$ receptors. This mechanism may

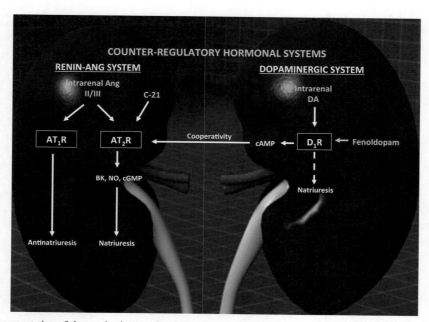

FIGURE 2   Schematic representation of the mechanisms and actions of renal AT$_2$Rs in the control of Na$^+$ excretion. AT$_1$R, AT$_1$ receptor; AT$_2$R, AT$_2$ receptor; DA, dopamine; D$_1$R, D$_1$ receptor; intrarenal Ang II/Ang III, angiotensin II/angiotensin III; cAMP, cyclic AMP; C-21, compound 21; fenoldopam, D$_1$R-selective agonist.

be important in the tonic regulation of basal Na^+ excretion since at least 50% of which is regulated by renal dopaminergic mechanisms. Physiologically, AT_2R stimulation induces natriuresis predominantly when AT_1Rs concurrently are blocked and/or during RAS activation.

AT_2Rs fail to induce natriuresis in SHR but do so in their WKY control animals. The mechanisms of absent AT_2R-induced natriuresis in SHR are not currently understood, but one potential mechanism is increased intrarenal APN activity leading to enhanced Ang III degradation in SHR. On the other hand, AT_2R-induced natriuresis appears to be exaggerated in the obese Zucker rat as opposed to its lean control. This may be a compensatory mechanism that allows increased natriuresis in the face of metabolic syndrome and/or diabetes mellitus.

Pharmacologically, AT_2R stimulation induces natriuresis in normal animals in the absence of concurrent AT_1R blockade or RAS activation. A nonpeptide AT_2R agonist is effective systemically but is dependent on renal AT_2R activation predominantly (or exclusively) in the proximal tubule. Proximal tubule cellular action of nonpeptide AT_2R agonist administration includes AT_2R recruitment from intracellular sites to the apical plasma membrane and is dependent on a BK-NO-cGMP signaling pathway, which at its terminus may involve the activation of Src family kinase and ERK signaling mechanisms. Internalization of NHE-3 and inactivation of NKA transporters in the proximal tubule have been demonstrated. Pharmacological AT_2R activation effectively lowers BP and prevents Na^+ retention in the Ang II infusion model of experimental hypertension. Because no clinically effective diuretic/natriuretic agents acting at proximal tubule sites are currently available, systemic AT_2R agonist therapy represents a meaningful potential therapeutic target for human disease involving fluid retention and/or hypertension.

REFERENCES

1. Roman RJ. Pressure diuresis mechanism in the control of renal function and arterial pressure. *Fed Proc* 1986;**45**:2878–84.
2. Lieb DC, Kemp BA, Howell NL, Gildea JJ, Carey RM. Reinforcing feedback loop in renal cyclic guanosine 3',5'-monophosphate and interstitial hydrostatic pressure in pressure-natriuresis. *Hypertension* 2009;**54**:1278–83.
3. Evans RG, Majid DS, Eppel GA. Mechanisms mediating pressure-natriuresis: what we know and what we need to find out. *Clin Exp Pharmacol Physiol* 2005;**32**:400–9.
4. Guyton AC, Coleman TG, Cowley Jr AW, Scheel KW, Manning Jr RD, Norman Jr RA. Arterial pressure regulation. Overriding dominance of the kidneys in long-term regulation and in hypertension. *Am J Med* 1972;**52**:584–94.
5. Guyton AC. The surprising kidney-fluid mechanism for pressure control: its infinite gain. *Hypertension* 1990;**16**:725–30.
6. Graf C, Maser-Gluth C, de Muinck Keizer W, Rettig R. Sodium retention and hypertension after kidney transplantation in rats. *Hypertension* 1993;**21**:724–30.
7. Kopf D, Waldherr R, Rettig R. Source of kidney determines blood pressure in young renal transplanted rats. *Am J Physiol Renal Physiol* 1993;**265**:F104–11.
8. Rettig R, Folberth CG, Stauss H, Kopf D, Waldherr R, Baldhauf G, et al. Hypertension in rats induced by renal grafts from renovascular hypertensive donors. *Hypertension* 1990;**15**:429–35.
9. Hall JE, Granger JP, Smith Jr MJ, Premen AJ. Role of renal hemodynamics and arterial pressure in aldosterone "escape". *Hypertension* 1984;**6**:1183–92.
10. Hall JE, Granger JP, Hester RL, Coleman TG, Smith Jr MJ, Cross RB. Mechanisms of escape from sodium retention during angiotensin II hypertension. *Am J Physiol Renal Physiol* 1984;**246**:F627–34.
11. Mizelle HL, Montani JP, Hester RL, Didlake RH, Hall JE. Role of pressure natriuresis in long-term control of renal electrolyte excretion. *Hypertension* 1993;**22**:102–10.
12. Carey RM, Siragy HM. Newly recognized components of the renin-angiotensin system: potential roles in cardiovascular and renal disease. *Endocr Rev* 2003;**24**:261–71.
13. Crowley SD, Gurley SB, Oliverio MI, Pazmino AK, Griffiths R, Flannery PJ, et al. Distinct roles for the kidney and systemic tissues in blood pressure regulation by the renin-angiotensin system. *J Clin Invest* 2005;**115**:1092–9.
14. Crowley SD, Gurley SB, Hererra MJ, Ruiz P, Griffiths R, Kumar AP, et al. Angiotensin II causes hypertension and cardiac hypertrophy through its receptors in the kidney. *Proc Natl Acad Sci U S A* 2006;**103**:17985–90.
15. Gurley SB, Riquier-Brison AD, Schnermann J, Sparks MA, Allen AM, Haase VH, et al. AT1A angiotensin receptors in the renal proximal tubule regulate blood pressure. *Cell Metab* 2011;**13**:469–75.
16. Li H, Weatherford ET, Davis DR, Keen HL, Grobe JL, Daugherty A, et al. Renal proximal tubule angiotensin AT1A receptors regulate blood pressure. *Am J Physiol Regul Integr Comp Physiol* 2011;**301**:R1067–77.
17. Ding Y, Davisson RL, Hardy DO, Zhu LJ, Merrill DC, Catterall Jr JF, et al. The kidney androgen-regulated protein promoter confers renal proximal tubule cell-specific and highly androgen-responsive expression on the human angiotensinogen gene in transgenic mice. *J Biol Chem* 1977;**272**:28142–8.
18. Carey RM. Cardiovascular and renal regulation by the angiotensin type 2 receptor: the AT2 receptor comes of age. *Hypertension* 2005;**45**:840–4.
19. Jones ES, Vinh A, McCarthy CA, Gaspari TA, Widdop RE. AT2 receptors: functional relevance in cardiovascular disease. *Pharmacol Ther* 2008;**120**:292–316.

20. Siragy HM, Xue C, Abadir P, Carey RM. Angiotensin subtype-2 receptors inhibit renin biosynthesis and angiotensin II formation. *Hypertension* 2005;**45**:133–7.

21. Berry C, Touyz R, Dominiczak AF, Webb RC, Johns DG. Angiotensin receptors: signaling, vascular pathophysiology, and interactions with ceramide. *Am J Physiol Heart Circ Physiol* 2001;**281**:H2337–65.

22. Siragy HM, Carey RM. The subtype-2 (AT2) angiotensin receptor regulates renal cyclic guanosine 3', 5'-monophosphate and AT1 receptor-mediated prostaglandin E2 production in conscious rats. *J Clin Invest* 1996;**97**:1978–82.

23. Siragy HM, Jaffa AA, Margolius HS, Carey RM. Renin-angiotensin system modulates renal bradykinin production. *Am J Physiol Reg Int Comp Physiol* 1996;**271**(Pt 2):R1090–5.

24. Siragy HM, Carey RM. The subtype 2 (AT2) angiotensin receptor mediates renal production of nitric oxide in conscious rats. *J Clin Invest* 1997;**100**:264–9.

25. Siragy HM, Jaffa AA, Margolius HS. Bradykinin B2 receptor modulates renal prostaglandin E2 and nitric oxide. *Hypertension* 1997;**29**:757–62.

26. Tsutsumi Y, Matsubara H, Masaki H, Kurihara H, Murasawa S, Takai S, et al. Angiotensin II type 2 receptor overexpression activates the vascular kinin system and causes vasodilation. *J Clin Invest* 1999;**104**:925–35.

27. Abadir PM, Carey RM, Siragy HM. Angiotensin AT2 receptors directly stimulate renal nitric oxide in bradykinin B2-receptor-null mice. *Hypertension* 2003;**42**:600–4.

28. Widdop RE, Jones ES, Hannan RE, Gaspari TA. Angiotensin AT2 receptors: cardiovascular hope or hype? *Br J Pharmacol* 2003;**140**:809–24.

29. Zhuo JAA, Alcorn D, Aldred GP, MacGregor DP, Mendelsohn FA. The distribution of angiotensin II receptors. *Hypertension* 1995;**35**:155–63.

30. Ozono R, Wang Z-Q, Moore AF, Inagami T, Siragy HM, Carey RM. Expression of the subtype 2 angiotensin (AT2) receptor protein in rat kidney. *Hypertension* 1997;**30**:1238–46.

31. Miyata N, Park F, Li XF, Cowley Jr AW. Distribution of angiotensin AT1 and AT2 receptor subtypes in the rat kidney. *Am J Physiol Renal Physiol* 1999;**277**:F437–46.

32. Tanaka M, Tsachida S, Imai T, Fujii N, Miyazaki H, Ichiki T, et al. Vascular response to angiotensin II is exaggerated through an upregulation of AT1 receptor in AT2 knockout mice. *Biochem Biophys Res Commun* 1999;**258**:194–8.

33. Balardi G, Macova M, Armando I, Ando H, Tyurmin D, Saavedra JM. Estrogen upregulates renal angiotensin II AT1 and AT2 receptors in the rat. *Regul Pept* 2005;**124**:7–17.

34. Padia SH, Howell NL, Siragy HM, Carey RM. Renal angiotensin type 2 receptors mediate natriuresis via angiotensin III in the angiotensin II type 1 receptor-blocked rat. *Hypertension* 2006;**47**:537–44.

35. Padia SH, Kemp BA, Howell NL, Siragy HM, Fournie-Zaluski MC, Roques BP, et al. Intrarenal aminopeptidase N inhibition augments natriuretic responses to angiotensin III in angiotensin type 1 receptor-blocked rats. *Hypertension* 2007;**49**:625–30.

36. Padia SH, Kemp BA, Howell NL, Fournie-Zaluski M-C, Roques BP, Carey RM. Conversion of renal angiotensin II to angiotensin III is critical for AT2 receptor-induced natriuresis in rats. *Hypertension* 2008;**51**:460–5.

37. Hakam AC, Hussain T. Renal angiotensin II type-2 receptors are upregulated and mediate the candesartan-induced natriuresis/diuresis in obese Zucker rats. *Hypertension* 2005;**45**:270–5.

38. Hakam AC, Siddiqui AH, Hussain T. Renal angiotensin II AT2 receptors promote natriuresis in streptozotocin-induced diabetic rats. *Am J Physiol Renal Physiol* 2006;**290**:F503–8.

39. Hakam AC, Hussain T. Angiotensin II AT2 receptors inhibit proximal tubular Na$+$$-K+$$-$ATPase activity via a NO/cGMP-dependent pathway. *Am J Physiol Renal Physiol* 2006;**290**:F1430–6.

40. Siragy HM, Inagami T, Ichiki T, Carey RM. Sustained hypersensitivity to angiotensin II and its mechanism in mice lacking the subtype-2 (AT2) angiotensin receptor. *Proc Natl Acad Sci U S A* 1999;**96**:6506–10.

41. Gross V, Schunck W-H, Honeck H, Milia AF, Kargel E, Walther T, et al. Inhibition of pressure natriuresis in mice lacking the AT2 receptor. *Kidney Int* 2000;**57**:191–202.

42. Bosnyak S, Jones ES, Christopoulos A, Aguilar MI, Thomas WG, Widdop RE. Relative affinity of angiotensin peptides and novel ligands at AT1 and AT2 receptors. *Clin Sci (Lond)* 2011;**121**:297–303.

43. Batenburg WW, Garrelds IM, Bunasconi CL, Juillerat-Jeanneret L, van Kats JP, Saxena PR, et al. Angiotensin II type 2 receptor-mediated vasodilation in human coronary microarteries. *Circulation* 2004;**109**:2296–301.

44. Yatabe J, Yoneda M, Yatabe MS, Watanabe T, Felder RA, Jose PA, et al. Angiotensin III stimulates aldosterone secretion from adrenal gland partially via angiotensin II type 2 receptor but not angiotensin II type 1 receptor. *Endocrinology* 2011;**152**:1582–8.

45. Park BM, Oh YB, Gao S, Cha SA, Kang KP, Kim SH. Angiotensin III stimulates high stretch-induced ANP secretion via angiotensin type 2 receptor. *Peptides* 2013;**42**:131–7.

46. Kemp BA, Bell JF, Rottkamp DM, Howell NL, Shao W, Navar LG, et al. Intrarenal angiotensin II is the predominant agonist for proximal tubule angiotensin type 2 receptors. *Hypertension* 2012;**60**:387–95.

47. Herrera M, Garvin JL. Angiotensin II stimulates thick ascending limb NO production via AT(2) receptors and Akt1-dependent nitric oxide synthase 3 (NOS3) activation. *J Biol Chem* 2010;**285**:14932–40.

48. Hong NJ, Garvin JL. Angiotensin II type 2 receptor-mediated inhibition of NaCl absorption is blunted in thick ascending limbs from Dahl salt-sensitive rats. *Hypertension* 2012;**60**:765–9.

49. Yang J, Chen C, Ren H, Han Y, He D, Zhou L, et al. Angiotensin II AT2 receptor decreases AT1 receptor expression and function via nitric oxide/cGMP/Sp1 in renal proximal tubule cells from Wistar-Kyoto rats. *J Hypertens* 2012;**30**:1176–84.

50. Carey RM. Theodore Cooper Lecture: renal dopamine system: paracrine regulator of sodium homeostasis and blood pressure. *Hypertension* 2001;**38**:297–302.

51. Siragy HM, Felder RA, Howell NL, Chevalier RL, Peach MJ, Carey RM. Evidence that intrarenal dopamine acts as a paracrine substance at the renal tubule. *Am J Physiol Renal Physiol* 1989;**257**:F469–77.
52. Salomone LJ, Howell NL, McGrath HE, Kemp BA, Keller SR, Gildea JJ, et al. Intrarenal dopamine D1-like receptor stimulation induces natriuresis via an angiotensin type-2 receptor mechanism. *Hypertension* 2007;**49**:155–61.
53. Padia SH, Kemp BA, Howell NL, Keller SR, Gildea JJ, Carey RM. Mechanisms of dopamine D(1) and angiotensin type 2 receptor interaction in natriuresis. *Hypertension* 2012;**59**:437–45.
54. Brismar H, Asghar M, Carey RM, Greengard P, Aperia A. Dopamine-induced recruitment of dopamine D1 receptors to the plasma membrane. *Proc Natl Acad Sci U S A* 1998;**95**:5573–8.
55. Gildea JJ, Wang X, Shah N, Tran H, Spinosa M, Van Sciver R, et al. Dopamine and angiotensin type 2 receptors cooperatively inhibit sodium transport in human renal proximal tubule cells. *Hypertension* 2012;**60**:396–403.
56. Padia SH, Kemp BA, Howell NL, Gildea JJ, Keller SR, Carey RM. Intrarenal angiotensin III infusion induces natriuresis and angiotensin type 2 receptor translocation in Wistar-Kyoto but not spontaneously hypertensive rats. *Hypertension* 2009;**53**:338–43.
57. Padia SH, Howell NL, Kemp BA, Fournie-Zaluski MC, Roques BP, Carey RM. Intrarenal aminopeptidase N inhibition restores defective angiotensin II type 2-mediated natriuresis in spontaneously hypertensive rats. *Hypertension* 2010;**55**:474–80.
58. Meng QC, Durand J, Chen Y-F, Oparil S. Effects of dietary salt on angiotensin peptides in the kidney. *J Am Soc Nephrol* 1995;**6**:1209–15.
59. Cheng H-F, Wang J-L, Vinson GP, Harris RC. Young SHR express increased type 1 angiotensin II receptors in renal proximal tubule. *Am J Physiol Renal Physiol* 1998;**274**:F10–17.
60. Yoneda M, Sanada H, Yatabe J, Midorikawa S, Hashimoto S, Sasaki M, et al. Differential effects of angiotensin II type-1 receptor antisense oligonucleotides on renal function in spontaneously hypertensive rats. *Hypertension* 2005;**46**:58–65.
61. Kobori H, Ozawa Y, Suzuke Y, Nishiyama A. Enhanced intrarenal angiotensinogen contributes to early renal injury in spontaneously hypertensive rats. *J Am Soc Nephrol* 2005;**16**:2073–80.
62. Varagic J, Frohlich ED, Susic D, Ahn J, Matavelli L, Lopez B, et al. AT1 receptor antagonism attenuates target organ effects of salt excess in SHRs without affecting pressure. *Am J Physiol Heart Circ Physiol* 2008;**2994**:H853–8.
63. Landgraf SS, Wengert M, Silva JS, Zapata-Sudo G, Sudo RT, Takiya CM, et al. Changes in angiotensin receptors expression play a pivotal role in the renal damage observed in spontaneously hypertensive rats. *Am J Physiol Renal Physiol* 2011;**300**:F499–510.
64. Susic D, Zhou X, Frohlich ED. Angiotensin blockade prevents salt-induced injury of the renal circulation in spontaneously hypertensive rats. *Am J Nephrol* 2009;**20**:639–45.
65. Susic D, Frohlich ED, Kobori H, Shao W, Seth D, Navar LG. Salt-induced renal injury in SHRs is mediated by AT1 receptor activation. *J Hypertens* 2011;**29**:716–23.
66. Lee H-A, Cho H-M, Lee D-Y, Kim K-C, Han HS, Kim IK. Tissue-specific upregulation of angiotensin-converting enzyme 1 in spontaneously hypertensive rats through histone code modifications. *Hypertension* 2012;**59**:621–6.
67. Hakam AC, Hussein T. Angiotensin type 2 receptor agonist directly inhibits proximal tubule sodium pump activity in obese but not lean Zucker rats. *Hypertension* 2006;**47**:1117–24.
68. Wan Y, Wallinder C, Plouffe B, Beaudry H, Mahalingam AK, Wu X, et al. Design, synthesis and biological evaluation of the first selective nonpeptide AT2 receptor agonist. *J Med Chem* 2004;**47**:5995–6008.
69. Steckelings UM, Ludovit P, Namsolleck P, Unger T. AT2 receptor agonists: hypertension and beyond. *Curr Opin Nephrol Hypertens* 2012;**21**:142–6.
70. Kemp BA, Howell NL, Gildea JJ, Keller SR, Padia SH, Carey RM. AT2 receptor activation induces natriuresis and lowers blood pressure. *Circ Res* 2014;**115**(3):388–99.
71. Hilliard LM, Jones ES, Steckelings UM, Unger TM, Widdop RE, Denton KM. Sex-specific influence of angiotensin type 2 receptor stimulation on renal function: a novel therapeutic target for hypertension. *Hypertension* 2012;**59**:409–14.
72. Ali Q, Hussein T. AT2 receptor non-peptide agonist C21 promotes natriuresis in obese Zucker rats. *Hypertens Res* 2012;**35**:654–60.

Chapter 12

The Role of the AT₂R in Vascular Remodeling

Sébastien Foulquier,* Ludovit Paulis†

*CARIM, School for Cardiovascular Diseases, Maastricht University, Maastricht, The Netherlands, †Institute of Pathophysiology, Faculty of Medicine, Comenius University, Bratislava, Slovakia

INTRODUCTION

Structural and functional remodeling of the vasculature is the hallmark of many cardiovascular diseases such as hypertension and atherosclerosis. Vascular remodeling represents the rearrangement of the different components of the vascular wall. Although primarily adaptive, vascular remodeling may cause vessel narrowing, increased vascular resistance, augmented wall stiffness, or—on the contrary—augmented wall distensibility with aneurysm development or atherosclerotic plaque development and rupture. The sequelae of these events are adverse cardiovascular conditions and outcomes such as hypertension, myocardial infarction, cerebral ischemia, and vascular death. Therefore, the prevention and reversion of vascular remodeling represent a pertinent theoretical target for cardiovascular therapy.

THE THEORETICAL CONCEPT OF THE ROLE OF AT₂R IN VASCULAR REMODELING

The exact type and mechanisms of vascular remodeling depend on the particular pathophysiological setting. In hypertension, remodeling affects both small resistance arteries and larger recoil arteries.[1] Small artery remodeling involves structural rearrangement with increased media/lumen ratio, mostly without altered cross-sectional area (inward eutrophic remodeling) leading to reduced vascular reserve and target-organ damage.[2] In large arteries, the remodeling presents with increased wall stiffness, increased left ventricle afterload, impaired organ perfusion, and enhanced systolic blood pressure that promotes further small artery remodeling. It was hypothesized that the main trigger for small artery remodeling was the vascular tone, while the remodeling of larger arteries was more blood-pressure-dependent.[2] In addition to blood-pressure-induced changes and endothelial damage, additional risk factors, such as dyslipidemia, increased oxidative load, and proinflammatory state, promote the development of atherosclerotic remodeling in large- and medium-sized arteries. The process of remodeling within the atherosclerotic plaque will then determine its stability against possible rupture.[3] Under certain conditions, such as Marfan's syndrome, remodeling of the vascular wall may predispose it to the development of aneurysms and determine their stability.[4]

The mechanisms of vascular remodeling involve smooth muscle cell migration, apoptosis, and altered collagen turnover (expression, synthesis, and/or degradation). Collagen synthesis is promoted by mechanical stretch and by neurohumoral activation and inflammatory phenotype. One of the most important inducers of vascular fibrosis is the transforming growth factor beta (TGF-β), *via* Smad and mitogen-activated protein kinase (MAPK) activation. These pathways and TGF-β production are stimulated by angiotensin II (Ang II) *via* the AT₁ receptor (AT₁R).[3] TGF-β release with profibrotic effects is stimulated by proinflammatory stimuli such as interferon gamma (IFN-γ) and promoted by T-helper 1 (T$_{h1}$) cells producing interleukin-6 (IL-6), tumor necrosis factor-α (TNF-α), and monocyte chemoattractant protein-1 (MCP-1). On the other hand, T regulatory (T$_{reg}$) cells expressing inhibitory IL-10 may oppose these effects by attenuating the inflammatory response.[5] IL-10 production induced by TGF-β released from T$_{reg}$ cells may override the profibrotic effects of TGF-β per se, resulting in a net anti-inflammatory and antifibrotic action.[6] The produced collagen is cleaved by collagenases (matrix metalloproteinases (MMPs) 1, 8, 13, and 18) and gelatinases (MMPs 2 and 9). MMPs are upregulated by the activated MAPKs, NADPH oxidase or nuclear factor (NF)-κB. MMPs 2 and 9 also trigger the release of TGF-β, which in turn not only upregulates their expression (and attenuates the expression of MMPs 1 and 3) but also stimulates the production

The Protective Arm of the Renin–Angiotensin System (RAS). http://dx.doi.org/10.1016/B978-0-12-801364-9.00012-2

of tissue inhibitors of metalloproteinases (TIMPs).[3] Besides hypertensive remodeling, the interplay between MMPs and TGF-β determines plaque stability in atherosclerotic lesions,[3] and the enhanced TGF-β signaling due to impaired sequestration by fibrillin-1 is responsible for aortic dilation and aneurysm development in Marfan's syndrome.[4]

AT$_2$ receptor (AT$_2$R) is associated with several pathways that provide the theoretical background for its involvement in vascular remodeling:

1. The NO/cyclic guanosine monophosphate (cGMP) pathway[7] with its antiproliferative and antifibrotic effects and anti-inflammatory action
2. The activation of protein phosphatases (MAPK phosphatase-1, Src homology region 2 domain-containing phosphatase, and protein phosphatase 2A) that inactivate the profibrotic MAPKs or antiapoptotic Bcl-2[8]
3. Downregulation of MAPKs with NADPH oxidase inhibition and subsequent attenuation of oxidative load[9]
4. Inhibition of NF-κB activity by epoxidation of 11,12-epoxyeicosatrienoic acid[10]
5. Direct and indirect anti-inflammatory action with augmented IL-10 production[11,12] and T-cell differentiation to the T$_{reg}$ phenotype[13]
6. Heterodimerization of the AT$_2$R with AT$_1$R that abrogates the AT$_1$R-dependent profibrotic effects[14] (Figure 1).

The relevance of these pathways depends on vascular AT$_2$R expression, which is low in the adult vasculature but which is upregulated in hypertension and vascular injury.[15–17] Moreover, in arterial[18] and pulmonary hypertension models,[19] a functional cross talk between the AT$_2$R and Mas receptor axis was suggested.

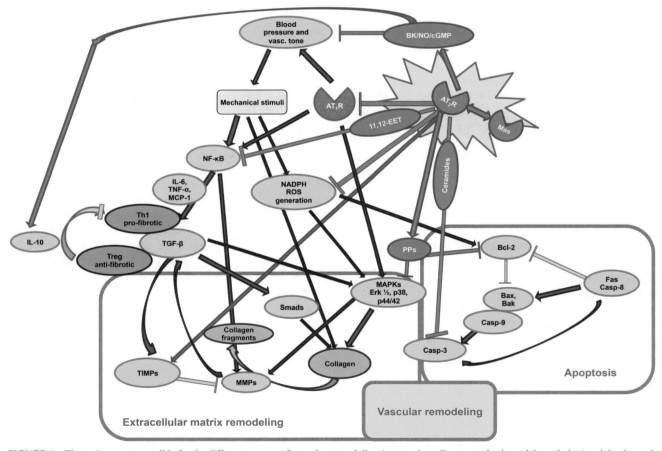

FIGURE 1 The pathways responsible for the different aspects of vascular remodeling (apoptosis, collagen synthesis, and degradation) and the theoretical concept of the interference of the AT$_2$R with these mechanisms. NF-κB, nuclear factor-κB; IL-6, interleukin-6; IL-10, interleukin-10; TNF-α, tumor necrosis factor-α; MCP-1, monocyte chemoattractant protein-1; T$_{h1}$ cells, T-helper 1 cells; T$_{reg}$ cells, T regulatory cells; TGF-β, transforming growth factor-β; NADPH oxidase, nicotinamide adenine dinucleotide phosphate oxidase; ROS, reactive oxygen species; MAPKs, mitogen-activated protein kinases; TIMPs, tissue inhibitors of metalloproteinases; MMPs, matrix metalloproteinases; Bcl-2, B-cell lymphoma 2; Bax, Bak, Bcl-2 associated proteins; Casp-3, caspase 3; Casp-9, caspase 9; PK, protein kinase; PPs, protein phosphatases; 11,12-EET, 11,12-epoxyeicosatrienoic acid; BK, bradykinin; cGMP, cyclic guanosine monophosphate; AT$_1$R, angiotensin II type 1 receptor; AT$_2$R, angiotensin II type 2 receptor.

Here, we review the evidence for the role of AT$_2$R in vascular remodeling *in vivo* in hypertension, atherosclerosis, or aneurysm development in Marfan's syndrome.

AT$_2$R AND HYPERTENSION-INDUCED VASCULAR REMODELING

Vascular remodeling is closely associated with hypertension and its adverse outcomes. Increased media/lumen ratio predicted cardiovascular events[20–22] with increased cross-sectional area being an additional risk predictor.[23] Increased pulse wave velocity (indicator of arterial stiffness) represents an independent cardiovascular risk factor.[24] Although there is not much evidence that remodeling could be used as a surrogate to guide therapy,[25] on-treatment structure of small arteries might predict cardiovascular risk in hypertensive patients,[26] thereby rendering the prevention of vascular remodeling an additional therapeutic target in hypertension.

The early evidence for the involvement of AT$_2$R in vascular remodeling was indirect, using the AT$_2$R antagonist PD 123319 and AT$_2$R knockout (KO) mice. Surprisingly, chronic AT$_2$R blockade by PD 123319 (30 mg/kg/day) antagonized the effect of Ang II on arterial hypertrophy and fibrosis suggesting vasotropic effect of the AT$_2$R.[27] However, the selectivity of PD 123319 for AT$_2$R is low, and at higher concentrations, it might antagonize the AT$_1$R as well. A 3x lower dose of PD 123319 abolished the protective effect of candesartan against media/lumen augmentation in the aorta (but not in intramyocardial arteries) of aged rats[28] and aged SHR (here with some effect on intramyocardial arteries as well).[29] Accordingly, a nonselective blockade of AT$_1$R and AT$_2$R in Sprague-Dawley rats after Ang II infusion increased vascular stiffness, fibronectin, and MMP2 activity compared with selective AT$_1$R blockade.[30]

In AT$_2$R-KO mice without blood pressure challenge, hypertrophy of vascular smooth muscle cells without any effect on vascular smooth muscle count or media thickness was observed.[31] When chronic pressure overload was applied by aortic banding for 6 or 12 weeks, the increase in wall/lumen ratio in small and medium-sized coronary arteries and the perivascular fibrosis were attenuated by the presence of AT$_2$R in the wild-type mice compared with the AT$_2$R-KO mice without any significant blood pressure difference between the strains.[32] In this model, valsartan (in a dose without blood pressure effect) completely prevented the remodeling of coronary arteries and perivascular fibrosis in wild-type mice. In the AT$_2$R-KO mice, however, the effect of valsartan was less expressed and comparable to unselective angiotensin II receptor blockade.[33] Thus, the AT$_2$R seems to attenuate vascular remodeling and to be responsible for at least some of the effects of AT$_1$R blockade.

There are two preclinical studies investigating the direct effects of AT$_2$R stimulation with compound 21 on hypertension-induced remodeling. In L-NAME-induced hypertension, compound 21 (0.3 mg/kg/day) attenuated the increase in wall/lumen ratio, the augmentation of pulse wave velocity, and hydroxyproline accumulation in the aorta. Although compound 21 did not alter blood pressure, these effects were comparable to olmesartan that completely prevented hypertension. In addition, there were a numerically larger attenuation of wall/lumen ratio and pulse wave velocity and even a significantly lower hydroxyproline concentration in the compound 21+olmesartan combination arm, compared with olmesartan alone.[34] In stroke-prone SHR, compound 21 (1 mg/kg/day) did not alter media/lumen ratio or cross-sectional area in small mesenteric arteries. On the other hand, it reduced mesenteric artery stiffness, numerically reduced media collagen fraction in the coronary artery, and significantly reduced media collagen fraction and fibronectin expression in the aorta. Again, despite no significant blood pressure effect by compound 21, the decrease in aortic collagen was comparable to the losartan arm, and there was an additional effect on the reduction of the media collagen fraction in the aorta when compound 21 was added to losartan, compared with losartan alone.[35]

These studies support the concept that AT$_2$R stimulation prevents/reverses vascular remodeling in hypertension. The effect might be more pronounced in larger arteries and is very likely associated with the modulation of collagen turnover. AT$_2$R-induced vascular remodeling does not seem to be mediated by altered vascular tone or NO production (besides SHR, the effects were observed in a NO-deficient model). The effects might have been mediated either *via* direct effects of AT$_2$R on collagen turnover or, more likely, *via* the already reported anti-inflammatory effects of AT$_2$R that involve TGF-β modulation. The role of MMPs, however, remains controversial. According to the theoretical expectations and previous studies observing reduced arterial stiffness by compound 21 in a model of infarction-induced heart failure[36] or the effects of CGP42112 in atherosclerotic plaque,[37] MMP 2 and 9 activities should be attenuated. However, in L-NAME hypertension, there were numerically higher MMP 2 and 9 expressions (our unpublished results), while in the SHR model, there was no change in the expression of MMPs.[35] The altered MMP mRNA expression, however, does not necessarily need to translate into proportional change of MMP protein expression or MMP activity.[36]

The results of the studies in arterial hypertension are consistent with the reports of the prevention of vascular remodeling in monocrotaline-induced pulmonary hypertension.[19] Finally, and from the clinical perspective very importantly, AT$_2$R stimulation exerted some additive effects to the AT$_1$R blockade.

The clinical relevance of this combined renin–angiotensin system (RAS) modulation is supported by enhanced vascular AT_2R expression following AT_1R blockade (in contrast to β-receptor blockade) in hypertensive diabetic patients. Moreover, in these patients, the AT_2Rs were not only upregulated but also functional: they mediated modest vasodilator effect and were associated with attenuated remodeling of resistance arteries.[38]

AT₂R AND ATHEROSCLEROSIS

Atherosclerosis is a chronic inflammatory disease that affects the vascular wall, *via* the formation and progression of plaques with a subsequent blood flow reduction. It implies the oxidation of low-density lipoproteins (LDLs) and the subsequent promotion of a proinflammatory environment. The RAS plays a major role in the pathogenesis of atherosclerosis *via* the action of Ang II on the AT_1R. The activation of the AT_1R promotes atherogenesis by the induction of inflammation, endothelial dysfunction, and LDL oxidation.[39,40] This is confirmed by the atheroprotection (attenuated atherosclerosis, plaque disruption, lipid deposition, and macrophage accumulation) induced by AT_1R deficiency[41] or AT_1R blockade.[42,43] A part of AT_1R blockade-induced atheroprotection might be mediated by an indirect stimulation of the AT_2R by unbound Ang II (Figure 2).

Iwai et al.[44] and Sales et al.[45] in 2005 observed an increased AT_2R expression in atherosclerotic lesions, supporting a role of the AT_2R in this pathology.[44–47] In the aorta of AT_2R/ApoE-KO mice after a high cholesterol diet for 10 weeks, the atherosclerotic changes were exaggerated compared to ApoE-KO mice, independently of any change in plasma cholesterol or blood pressure levels but in association with an increased superoxide production, NADPH oxidase activity, and expression. In addition, the atheroprotective effects of valsartan were lower in the AT_2R-KO mice. Thus, a part of the atheroprotection mediated by AT_1R blockade may result from AT_2R stimulation.[44] In aortic arch lesions, the absence of the AT_2R did not alter the size of the lesion but increased the macrophage, smooth muscle, and collagen content, suggesting AT_2R-mediated modulation of the proliferation/apoptosis ratio.[45]

In LDL receptor (LDLR)-KO mice fed with a high-cholesterol diet, AT_2R overexpression reduced collagen deposition in the aorta by 50% as well as procollagen I, osteopontin, and fibronectin expressions.[9] In addition, the increased MMP 2 and 9 expressions and activities observed in the LDLR-KO mice were normalized by the AT_2R overexpression as was the reduction of the superoxide dismutase activity.[9] The expression of NADPH oxidase, nitrotyrosine, and NF-κB was increased in LDLR-KO but not in LDLR-KO mice with AT_2R upregulation.[48] The expression of endothelial nitric oxide synthase (eNOS) and heme oxygenase-1 (HO-1), an enzyme protecting against oxidative injury and involved in

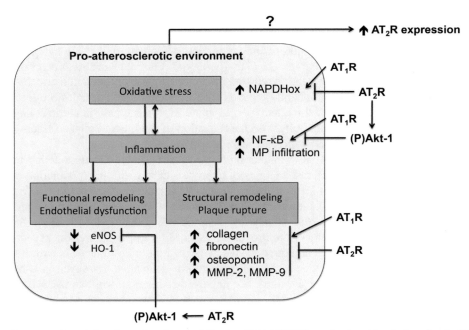

FIGURE 2 The putative attenuation of the atherosclerotic remodeling by AT_2R. NADPHox, nicotinamide adenine dinucleotide phosphate oxidase; NF-κB, nuclear factor-kappa B; MP, macrophage; MMP, matrix metalloproteases; (P)Akt-1, phosphorylated Akt-1; eNOS, endothelial nitric oxide synthase; HO-1, heme oxygenase-1.

the production of the strong vasodilator carbon monoxide, was also altered in LDLR-KO mice but not in mice with AT$_2$R overexpression.[48] These data support the role of AT$_2$R in modulating oxidative stress, inflammation, and endothelial function limiting the progression of atherosclerotic lesions. Further, Akt-1 phosphorylation, a key signaling element involved in NF-κB and eNOS signaling, was reduced in the LDLR-KO mice but not in mice overexpressing the AT$_2$R, suggesting that Akt-1 activation could be involved in the AT$_2$R-mediated atheroprotection. Finally, the increased expression of LOX-1 in LDLR-KO was normalized by AT$_2$R overexpression. LOX-1 (receptor for oxidized LDL (ox-LDL), responsible for the internalization and degradation of ox-LDL in endothelial cells) expression is upregulated by Ang II via AT$_1$R activation.[49]

Recent studies confirmed the antiatherogenic role of the AT$_2$R. In ApoE-KO mice overexpressing the AT$_2$R, the formation of the atherosclerotic lesions was reduced compared to control ApoE-KO mice.[50] The expression of vascular cell adhesion molecule-I (VCAM-I), monocyte/macrophage accumulation, and a superoxide production after Ang II stimulation were also reduced.[50] These antiatherogenic actions were abolished by the inhibition of NOS or the bradykinin receptor B2, suggesting that these are involved in AT$_2$R atheroprotection.[50] In another recent study, ApoE-KO mice fed with a high-fat diet for 16 weeks were treated with CGP42112, a direct peptide AT$_2$R agonist, during the last 4 weeks. CGP42112 decreased the progression of the atherosclerotic lesions, decreased the lipid content, and increased the collagen content in plaques, with a beneficial effect on plaque stability. In addition, CGP42112 reduced ICAM-1 expression in the vessel wall, as well as the aortic expression of MMPs 2 and 9.[37] The vascular endothelial function (enhanced acetylcholine-induced vasodilation) was improved in CGP42112-treated mice, possibly due to an increased eNOS expression and decreased superoxide production in the aorta. These effects were observed without any change of blood pressure and were reversed by the AT$_2$R antagonist PD123319[37] (Figure 2).

Altogether, the AT$_2$R may constitute an important therapeutic target to limit atherosclerotic progression. However, there are some studies that do not support this conclusion. In ApoE-KO mice infused with Ang II, the AT$_2$R expression in the plaques was increased by six- to ninefold but AT$_2$R blockade with PD123319 did not affect the plaque area in the aortic root.[51] In streptozotocin-induced diabetic ApoE-KO mice, the expression of the AT$_2$R was significantly increased in the aorta, but its genetic deletion or its blockade with PD123319 surprisingly attenuated the plaque size and reduced proinflammatory and profibrotic markers, independently of any change on glycemia or blood pressure.[51] These conflicting results might be due to differences in animal models, diets, and study design. Especially, the indirect strategies to investigate the role of AT$_2$R are limited by the low selectivity of PD123319 for the AT$_2$R. Further studies using direct AT$_2$R stimulation are required to clarify its impact on atherosclerotic damage before considering the clinical use of AT$_2$R agonists to limit atherosclerosis progression.

AT$_2$R AND AORTIC ANEURYSM

Although Ang II is implied in the progression of aortic aneurysm, the particular role of the AT$_2$R has been only sparsely studied in this context.[52] Marfan's syndrome is a disease that predisposes for aortic root aneurysm and rupture.[53] In a crossed model of AT$_2$R-KO mice with a genetic mouse model of Marfan's syndrome, a larger aortic root diameter at an early age and at 1 year of age was observed. In addition, these mice had an increased death rate compared with Marfan's syndrome mice without AT$_2$R-KO.[53] The ACE inhibitor, enalapril, was less effective against aneurysm progression compared to losartan that afforded protective effects dependent on the AT$_2$R expression and the inhibition of TGF-β-dependent Erk activation.[53] On the other hand, AT$_2$R deficiency had no effect on Ang II-induced abdominal aortic aneurysms in a recent study.[54] Further studies implying different animal models and exploring different pathways are required to better decipher the role of AT$_2$R in aortic aneurysm and related diseases such as Marfan's syndrome.

CONCLUSION

Vascular remodeling is triggered and regulated by mechanical and neurohumoral stimuli. One of the key neurohumoral systems involved is definitely the RAS. The AT$_2$R represents a part of a relatively recently described "protective" RAS. The available evidence presented in this chapter supports the theoretical concept of AT$_2$R-mediated protection in arterial and pulmonary hypertension, atherosclerosis, or aneurysm development. These data not only provide additional rationale for the explanation of the beneficial effects of AT$_1$R blockade in a clinical setting but also open new horizons for the putative exploitation of the "protective" RAS stimulation as add-on therapy to the blockade of the "harmful" RAS. Further experimental data are required to tackle some remaining controversies regarding the AT$_2$R. Their outcomes and further clinical data will determine to which extent the concept of AT$_2$R stimulation will be established in the ongoing fight with cardiovascular diseases.

REFERENCES

1. Briet M, Schiffrin EL. Treatment of arterial remodeling in essential hypertension. *Curr Hypertens Rep* 2013;**15**(1):3–9.
2. Mulvany MJ. Small artery remodelling in hypertension. *Basic Clin Pharmacol Toxicol* 2012;**110**(1):49–55.
3. Lan T-H, Huang X-Q, Tan H-M. Vascular fibrosis in atherosclerosis. *Cardiovasc Pathol* 2013;**22**(5):401–7.
4. Jones JA, Spinale FG, Ikonomidis JS. Transforming growth factor-beta signaling in thoracic aortic aneurysm development: a paradox in pathogenesis. *J Vasc Res* 2009;**46**(2):119–37.
5. Schiffrin EL. Immune mechanisms in hypertension and vascular injury. *Clin Sci* 2014;**126**(4):267–74.
6. Wynn TA, Ramalingam TR. Mechanisms of fibrosis: therapeutic translation for fibrotic disease. *Nat Med* 2012;**18**(7):1028–40.
7. Gohlke P, Pees C, Unger T. AT2 receptor stimulation increases aortic cyclic GMP in SHRSP by a kinin-dependent mechanism. *Hypertension* 1998;**31**(1 Pt 2):349–55.
8. Fischer TA, Singh K, O'Hara DS, Kaye DM, Kelly RA. Role of AT1 and AT2 receptors in regulation of MAPKs and MKP-1 by ANG II in adult cardiac myocytes. *Am J Physiol* 1998;**275**(3 Pt 2):H906–16.
9. Dandapat A, Hu CP, Chen J, Liu Y, Khan JA, Remeo F, et al. Over-expression of angiotensin II type 2 receptor (agtr2) decreases collagen accumulation in atherosclerotic plaque. *Biochem Biophys Res Commun* 2008;**366**(4):871–7.
10. Rompe F, Artuc M, Hallberg A, Alterman M, Ströder K, Thöne-Reineke C, et al. Direct angiotensin II type 2 receptor stimulation acts anti-inflammatory through epoxyeicosatrienoic acid and inhibition of nuclear factor kappaB. *Hypertension* 2010;**55**(4):924–31.
11. Curato C, Slavic S, Dong J, Skorska A, Altarche-Xifró W, Miteva K, et al. Identification of noncytotoxic and IL-10-producing CD8+AT2R+ T cell population in response to ischemic heart injury. *J Immunol* 2010;**185**(10):6286–93.
12. Dhande I, Ali Q, Hussain T. Proximal tubule angiotensin AT2 receptors mediate an anti-inflammatory response via interleukin-10: role in renoprotection in obese rats. *Hypertension* 2013;**61**(6):1218–26.
13. Valero-Esquitino V, Lucht K, Namsolleck P, Monnet-Tschudic F, Stubbed T, Lucht F, et al. Direct angiotensin AT2-receptor stimulation attenuates T-cell and microglia activation and prevents demyelination in experimental autoimmune encephalomyelitis in mice. *Clinical Science* 2014 (accepted).
14. AbdAlla S, Lother H, Abdel-tawab AM, Quitterer U. The angiotensin II AT2 receptor is an AT1 receptor antagonist. *J Biol Chem* 2001;**276**(43):39721–6.
15. Nakajima M, Hutchinson HG, Fujinaga M, Hayashida W, Morishita R, Zhang L, et al. The angiotensin II type 2 (AT2) receptor antagonizes the growth effects of the AT1 receptor: gain-of-function study using gene transfer. *Proc Natl Acad Sci U S A* 1995;**92**(23):10663–7.
16. Widdop RE, Vinh A, Henrion D, Jones ES. Vascular angiotensin AT2 receptors in hypertension and ageing. *Clin Exp Pharmacol Physiol* 2008;**35**(4):386–90.
17. Yayama K, Hiyoshi H, Imazu D, Okamoto H. Angiotensin II stimulates endothelial NO synthase phosphorylation in thoracic aorta of mice with abdominal aortic banding via type 2 receptor. *Hypertension* 2006;**48**(5):958–64.
18. Ohshima K, Mogi M, Nakaoka H, Iwanami J, Min L-J, Kanno H, et al. Possible role of angiotensin-converting enzyme 2 and activation of angiotensin II type 2 receptor by angiotensin-(1–7) in improvement of vascular remodeling by angiotensin II type 1 receptor blockade. *Hypertension* 2014;**63**(3):e53–9.
19. Bruce E, Shenoy V, Francis J, Steckelings UM, Unger T, Sumners C, et al. Stimulation of angiotensin type 2 receptor as a potential therapy for pulmonary hypertension. *Abstract Circulation* 2012;**126**, A18903.
20. Mathiassen ON, Buus NH, Sihm I, Thybo NK, Mørn B, Schroeder AP, et al. Small artery structure is an independent predictor of cardiovascular events in essential hypertension. *J Hypertens* 2007;**25**(5):1021–6.
21. Mathiassen ON, Buus NH, Larsen ML, Mulvany MJ, Christensen KL. Small artery structure adapts to vasodilatation rather than to blood pressure during antihypertensive treatment. *J Hypertens* 2007;**25**(5):1027–34.
22. De Ciuceis C, Porteri E, Rizzoni D, Rizzardi N, Paiardi S, Boari GEM, et al. Structural alterations of subcutaneous small-resistance arteries may predict major cardiovascular events in patients with hypertension. *Am J Hypertens* 2007;**20**(8):846–52.
23. Izzard AS, Rizzoni D, Agabiti-Rosei E, Heagerty AM. Small artery structure and hypertension: adaptive changes and target organ damage. *J Hypertens* 2005;**23**(2):247–50.
24. Vlachopoulos C, Aznaouridis K, Stefanadis C. Prediction of cardiovascular events and all-cause mortality with arterial stiffness: a systematic review and meta-analysis. *J Am Coll Cardiol* 2010;**55**(13):1318–27.
25. Laurent S, Briet M, Boutouyrie P. Arterial stiffness as surrogate end point: needed clinical trials. *Hypertension* 2012;**60**(2):518–22.
26. Buus NH, Mathiassen ON, Fenger-Grøn M, Præstholm MN, Sihm I, Thybo NK, et al. Small artery structure during antihypertensive therapy is an independent predictor of cardiovascular events in essential hypertension. *J Hypertens* 2013;**31**(4):791–7.
27. Levy BI, Benessiano J, Henrion D, Caputo L, Heymes C, Duriez M, et al. Chronic blockade of AT2-subtype receptors prevents the effect of angiotensin II on the rat vascular structure. *J Clin Invest* 1996;**98**(2):418–25.
28. Jones ES, Black MJ, Widdop RE. Angiotensin AT2 receptor contributes to cardiovascular remodelling of aged rats during chronic AT1 receptor blockade. *J Mol Cell Cardiol* 2004;**37**(5):1023–30.
29. Jones ES, Black MJ, Widdop RE. Influence of Angiotensin II Subtype 2 Receptor (AT(2)R) Antagonist, PD123319, on Cardiovascular Remodelling of Aged Spontaneously Hypertensive Rats during Chronic Angiotensin II Subtype 1 Receptor (AT(1)R) Blockade. *Int J Hypertens* 2012;**2012**:543062.
30. Brassard P, Amiri F, Schiffrin EL. Combined angiotensin II type 1 and type 2 receptor blockade on vascular remodeling and matrix metalloproteinases in resistance arteries. *Hypertension* 2005;**46**(3):598–606.
31. Brede M, Hadamek K, Meinel L, Wiesmann F, Peters J, Engelhardt S, et al. Vascular hypertrophy and increased P70S6 kinase in mice lacking the angiotensin II AT(2) receptor. *Circulation* 2001;**104**(21):2602–7.

32. Akishita M, Iwai M, Wu L, Zhang L, Ouchi Y, Dzau VJ, et al. Inhibitory effect of angiotensin II type 2 receptor on coronary arterial remodeling after aortic banding in mice. *Circulation* 2000;**102**(14):1684–9.

33. Wu L, Iwai M, Nakagami H, Chen R, Suzuki J, Akishita M, et al. Effect of angiotensin II type 1 receptor blockade on cardiac remodeling in angiotensin II type 2 receptor null mice. *Arterioscler Thromb Vasc Biol* 2002;**22**(1):49–54.

34. Paulis L, Becker STR, Lucht K, Schwengel K, Slavic S, Kaschina E, et al. Direct angiotensin II type 2 receptor stimulation in Nω-nitro-L-arginine-methyl ester-induced hypertension: the effect on pulse wave velocity and aortic remodeling. *Hypertension* 2012;**59**(2):485–92.

35. Rehman A, Leibowitz A, Yamamoto N, Rautureau Y, Paradis P, Schiffrin EL. Angiotensin type 2 receptor agonist compound 21 reduces vascular injury and myocardial fibrosis in stroke-prone spontaneously hypertensive rats. *Hypertension* 2012;**59**(2):291–9.

36. Lauer D, Slavic S, Sommerfeld M, Thöne-Reineke C, Sharkovska Y, Hallberg A, et al. Angiotensin type 2 receptor stimulation ameliorates left ventricular fibrosis and dysfunction via regulation of tissue inhibitor of matrix metalloproteinase 1/matrix metalloproteinase 9 axis and transforming growth factor β1 in the rat heart. *Hypertension* 2014;**63**(3):e60–7.

37. Kljajic ST, Widdop RE, Vinh A, Welungoda I, Bosnyak S, Jones ES, et al. Direct AT2 receptor stimulation is athero-protective and stabilizes plaque in Apolipoprotein E-deficient mice. *Int J Cardiol* 2013;**169**(4):281–7.

38. Savoia C, Touyz RM, Volpe M, Schiffrin EL. Angiotensin type 2 receptor in resistance arteries of type 2 diabetic hypertensive patients. *Hypertension* 2007;**49**(2):341–6.

39. Chen J, Li D, Schaefer R, Mehta JL. Cross-talk between dyslipidemia and renin-angiotensin system and the role of LOX-1 and MAPK in atherogenesis studies with the combined use of rosuvastatin and candesartan. *Atherosclerosis* 2006;**184**(2):295–301.

40. Singh BM, Mehta JL. Interactions between the renin-angiotensin system and dyslipidemia: relevance in the therapy of hypertension and coronary heart disease. *Arch Intern Med* 2003;**163**(11):1296–304.

41. Tiyerili V, Becher UM, Aksoy A, Lütjohann D, Wassmann S, Nickenig G, et al. AT1-receptor-deficiency induced atheroprotection in diabetic mice is partially mediated via PPARγ. *Cardiovasc Diabetol* 2013;**12**:30.

42. Fukuda D, Enomoto S, Hirata Y, Nagai R, Sata M. The angiotensin receptor blocker, telmisartan, reduces and stabilizes atherosclerosis in ApoE and AT1aR double deficient mice. *Biomed Pharmacother* 2010;**64**(10):712–7.

43. Johnstone MT, Perez AS, Nasser I, Stewart R, Vaidya A, Al Ammary F, et al. Angiotensin receptor blockade with candesartan attenuates atherosclerosis, plaque disruption, and macrophage accumulation within the plaque in a rabbit model. *Circulation* 2004;**110**(14):2060–5.

44. Iwai M, Chen R, Li Z, Shiuchi T, Suzuki J, Ide A, et al. Deletion of angiotensin II type 2 receptor exaggerated atherosclerosis in apolipoprotein E-null mice. *Circulation* 2005;**112**(11):1636–43.

45. Sales VL, Sukhova GK, Lopez-Ilasaca MA, Libby P, Dzau VJ, Pratt RE. Angiotensin type 2 receptor is expressed in murine atherosclerotic lesions and modulates lesion evolution. *Circulation* 2005;**112**(21):3328–36.

46. Zulli A, Burrell LM, Widdop RE, Black MJ, Buxton BF, Hare DL. Immunolocalization of ACE2 and AT2 receptors in rabbit atherosclerotic plaques. *J Histochem Cytochem* 2006;**54**(2):147–50.

47. Johansson ME, Fagerberg B, Bergström G. Angiotensin type 2 receptor is expressed in human atherosclerotic lesions. *J Renin Angiotensin Aldosterone Syst* 2008;**9**(1):17–21.

48. Hu C, Dandapat A, Chen J, Liu Y, Hermonat PL, Carey RM, et al. Over-expression of angiotensin II type 2 receptor (agtr2) reduces atherogenesis and modulates LOX-1, endothelial nitric oxide synthase and heme-oxygenase-1 expression. *Atherosclerosis* 2008;**199**(2):288–94.

49. Li DY, Zhang YC, Philips MI, Sawamura T, Mehta JL. Upregulation of endothelial receptor for oxidized low-density lipoprotein (LOX-1) in cultured human coronary artery endothelial cells by angiotensin II type 1 receptor activation. *Circ Res* 1999;**84**(9):1043–9.

50. Takata H, Yamada H, Kawahito H, Kishida S, Irie D, Kato T, et al. Vascular angiotensin II type 2 receptor attenuates atherosclerosis via a kinin/NO-dependent mechanism. *J Renin Angiotensin Aldosterone Syst* 2013. Epub ahead of print.

51. Johansson ME, Wickman A, Fitzgerald SM, Gan L, Bergström G. Angiotensin II, type 2 receptor is not involved in the angiotensin II-mediated pro-atherogenic process in ApoE−/− mice. *J Hypertens* 2005;**23**(8):1541–9.

52. Daugherty A, Manning MW, Cassis LA. Angiotensin II promotes atherosclerotic lesions and aneurysms in apolipoprotein E-deficient mice. *J Clin Investig* 2000;**105**(11):1605–12.

53. Habashi JP, Doyle JJ, Holm TM, Aziz H, Schoenhoff F, Bedja D, et al. Angiotensin II type 2 receptor signaling attenuates aortic aneurysm in mice through ERK antagonism. *Science* 2011;**332**(6027):361–5.

54. Daugherty A, Rateri DL, Howatt DA, Charnigo R, Cassis LA. PD123319 augments angiotensin II-induced abdominal aortic aneurysms through an AT2 receptor-independent mechanism. *PLoS One* 2013;**8**(4):e61849.

Chapter 13

The AT$_2$ Receptor and Inflammation

Veronica Valero Esquitino,* Leon Alexander Danyel*, Ulrike M. Steckelings†

*Center for Cardiovascular Research (CCRI), Institute of Pharmacology, Charité-Universitätsmedizin Berlin, Germany, †IMM—Department of Cardiovascular and Renal Research, University of Southern Denmark, Odense, Denmark

INTRODUCTION

Inflammation is one of the first reactions of the immune system to tissue/cell damage produced by pathogens, chemicals, or physical injury.[1] While acute inflammation is considered as a healthy process indicative of protection, chronic inflammation has been directly associated with a wide range of disorders, such as cardiovascular, renal, and autoimmune diseases. The renin–angiotensin system, which has been traditionally linked to blood pressure control and volume/electrolyte balance, also plays an important role in inflammatory processes and the immune response.[2,3] As in most (patho)physiological settings, it is also in the context of inflammation that the two main receptor subtypes for angiotensin II (Ang II), the angiotensin type 1 (AT$_1$) receptor (AT$_1$R) and angiotensin type 2 (AT$_2$) receptor (AT$_2$R), mediate opposing actions: while AT$_1$R activation is linked to proinflammatory effects, the AT$_2$R exerts anti-inflammatory effects.[3] Key mechanisms involved in the AT$_2$R-mediated anti-inflammatory effects seem to be the reduction of proinflammatory cytokine release[3] or of oxidative stress.[4]

AT$_2$R SIGNALING IN INFLAMMATION

Nuclear factor kappa-B (NF-κB) is a protein complex, which controls the transcription of different proinflammatory mediators such as cytokines, chemokines, and adhesion molecules.[5] Activation of NF-κB is triggered by proinflammatory cytokines and microbial products and involves the activation of inhibitors of kappa-B kinases (IKKs), which in turn phosphorylate inhibitors of NF-κB proteins (IκB). This phosphorylation leads to the release of RelA/p50 subunits, which can then translocate into the nucleus and initiate transcription of proinflammatory mediators.[5]

In 2004, Wu and colleagues demonstrated that AT$_2$R stimulation by Ang II or the AT$_2$R agonist CGP42112A led to dephosphorylation of the I-κB molecule by Src homology protein tyrosine phosphatase-1 (SHP-1), thus preventing NF-κB activation.[6] In accordance with these results, our group found that AT$_2$R stimulation by the nonpeptide AT$_2$R agonist compound 21 (C21) reduced TNF-α-induced IL-6 expression by inhibiting nuclear translocation of the NF-κB p50 subunit.[7] This anti-inflammatory effect was absent in the presence of an inhibitor of tyrosine phosphatases, okadaic acid, or an inhibitor of serine/threonine phosphatases, sodium orthovanadate, pointing again to the involvement of protein phosphatases in the inhibitory effect of AT$_2$R stimulation on NF-κB activity (Figure 1).

Apart from protein phosphatases, epoxyeicosatrienoic acid (EET), an anti-inflammatory mediator known to generally inhibit NF-κB,[8] seems to play a role in AT$_2$R-mediated NF-κB inhibition, too, because AT$_2$R-mediated NF-κB inhibition was prevented by the blockade of EET synthesis.[7]

In contrast to these observations, there are also data showing an AT$_2$R-mediated increase in NF-κB activity.[9,10] These contradicting results may be due to different experimental conditions, depending on whether NF-κB mediates proinflammatory effects or—as shown by Reinecke et al.—promotes neuroregeneration.[9]

The JAK-STAT signaling pathway represents another molecular mechanism involved in cytokine synthesis, inflammation, and immune response.[11] Horiuchi et al. showed that AT$_2$R stimulation inactivated STAT transcription factors by tyrosine and serine dephosphorylation.[12] Interestingly, AT$_2$R-mediated inactivation of STAT seemed not dependent on the original STAT-activating stimulus, but worked for STAT activation by ANG II, IFN-γ, PDGF, or EGF.[12]

Oxidative stress, that is, the production and accumulation of reactive oxygen species (ROS), is another inducer of inflammation, for example, through increasing activity of transcription factors such as NF-κB or AP-1.[13] Continued oxidative stress can promote chronic inflammation and thus be an underlying contributor to chronic cardiovascular, neurological, and renal diseases or diabetes. Chabrashvili et al. conducted a study in order to generally look at the impact of AT$_1$R and AT$_2$R

The Protective Arm of the Renin–Angiotensin System (RAS). http://dx.doi.org/10.1016/B978-0-12-801364-9.00013-4

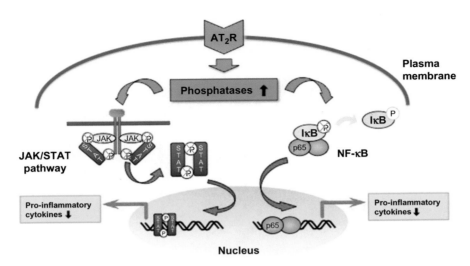

FIGURE 1 Activation of phosphatases leads to dephosphorylation of tyrosine and serine residues of components, which are essential in the signaling pathways leading to JAK-STAT and NF-κB activation. Decreased transcriptional activity of STAT and NF-κB results in reduced synthesis of proinflammatory cytokines.

on the generation of oxidative stress.[4] They found that the expression of certain factors involved in the Ang II-induced, AT_1R-mediated generation of ROS, namely, the NADPH subunits p22phox, Nox-1, and p67phox, is even more increased when AT_2R is blocked, pointing again to an inhibitory impact of AT_2R on ROS production.

Specific examples of anti-inflammatory or antioxidant effects of AT_2R stimulation are reviewed in the following.

INFLAMMATION IN CARDIOVASCULAR DISEASE AND THE ROLE OF THE AT_2 RECEPTOR

Atherosclerosis

AT_2Rs are expressed in atherosclerotic plaques and modify the inflammatory component of atherosclerosis.[14] This anti-inflammatory effect was shown, for example, in ApoE−/− mice on a cholesterol-rich diet with a targeted overexpression of AT_2Rs in the vascular smooth muscle cells, in which atherosclerotic lesion development was significantly attenuated through a kinin/NO-dependent mechanism.[15] In contrast, AT_2R deficiency in ApoE−/− mice aggravated atherosclerotic lesions by increasing superoxide production and macrophage infiltration.[14,16] AT_2R stimulation by Ang II in ApoE−/− mice deficient of AT_1R reduced atherogenesis, but had no effect on macrophage/T-cell infiltration and markers of oxidative stress.[17] However, direct AT_2R stimulation with the peptide AT_2R agonist CGP42112 in ApoE−/− mice in a study by Kljajic et al. not only significantly reduced atherosclerotic lesion progression but also resulted in improved plaque stability, reduced oxidative stress, reduced vascular cell adhesion, and improved endothelial function.[18]

Myocardial Infarction: Heart Failure

Cardiac tissue injury after myocardial infarction (MI) occurs in several phases: initial acute ischemia and related cell death and local proteolysis, an inflammatory response with massive invasion of inflammatory cells and eventually a fibrotic response, which involves not only the infarcted but also the peri-infarct area.[19] The location and size of the scar and the course of remodeling determine whether chronic heart failure will develop as a result of MI.

A beneficial effect of AT_2R stimulation (directly by AT_2R agonists or indirectly by ARBs) on cardiac function post-MI has been demonstrated in a number of studies.[20] However, while most studies focused on cardiac function, hypertrophy, and fibrosis, only few studies looked at the effect on MI-related inflammation. In a study by our group, treatment of rats with C21 for 1 week after MI resulted in significantly improved cardiac function, and this coincided with a strong and significant attenuation in elevated levels of proinflammatory cytokines IL-1β, IL-2, and IL-6 in the peri-infarct zone.[21] Moreover, oxidative stress was reduced as shown by lowered plasma levels of myeloperoxidase. These effects were blocked by the AT_2R antagonist PD 123319 pointing to AT_2R specificity of the C21 effects.[21]

In addition to this classical anti-inflammatory effect, AT_2R stimulation may further act by modifying the cellular immune response in ischemic heart injury. Curato et al. isolated CD8+ T cells from the peri-infarct area of rat hearts and found that a subpopulation of these cells (CD8$^+$AT$_2$R$^+$) expressed AT$_2$Rs and at the same time changed phenotype to being noncytotoxic.[22] These cells were further characterized by increased levels of IL-10 and decreased levels of IL-2 and INF-γ when compared with CD8$^+$AT$_2$R$^-$ T cells. The CD8$^+$AT$_2$R$^+$ subpopulation was enhanced upon AT$_2$R stimulation *in vivo*. Furthermore, intramyocardial transplantation of CD8$^+$AT$_2$R$^+$ T cells after MI led to a decrease in infarct size. In addition to this CD8$^+$AT$_2$R$^+$ cell population, a population of c-Kit$^+$AT$_2$R$^+$ cells was also found to be increased in hearts of infarcted rats treated with C21 and to act antiapoptotic on cardiomyocytes.[23]

Stroke

Therapeutic measures in relation to stroke can be either the prevention of occurrence of stroke, for example, through control of hypertension or antithrombotic therapy; acute recanalization by thrombolysis, which however is only possible in 5–7% of patients; or reduction of ischemic tissue injury and attenuation of neurological deficits. Stroke prevention is successfully performed in current cardiovascular medicine, whereas an effective neuroprotective treatment is still a huge, unmet therapeutic need.

In stroke, ischemic injury leads to excitotoxicity and oxidative damage followed by disintegration of the blood–brain barrier and a pronounced inflammatory response.[24] This chain of events determines the actual amount of permanent damage, that is, therapeutic measures to interrupt this fatal cascade may diminish neuronal loss and long-term neurological deficits.

Several studies with various experimental approaches have shown that AT$_2$R stimulation is able to reduce infarct size and neurological deficits after stroke and that the attenuation of inflammation and oxidative stress seem to be major protective mechanisms in this context.[25] For example, in AT$_2$R-knockout mice undergoing permanent middle cerebral artery occlusion (MCAO), a greater ischemic area and more severe neurological deficits (compared with wild-type mice) were associated with increased superoxide anion production and NADPH oxidase activity.[26] The reverse experimental approach in the same model, which was AT$_2$R stimulation by systemic application of C21 initiated after stroke, resulted in a marked attenuation of superoxide anion production and of proinflammatory cytokine expression including monocyte chemoattractant protein 1 and tumor necrosis factor-α.[27] Similar results were obtained in another stroke model and another species, which was transient, endothelin-induced MCAO in rats.[28] In this study, peripheral and central administration of C21 reduced infarct size and neurological deficits, which coincided with an anti-inflammatory effect as shown by reduced levels of chemokine CCL2 and its receptor CCR2 and of inducible nitric oxide synthase.

Microglia, the macrophages of the brain, are rapidly activated after stroke. They not only are essentially involved in stroke-induced inflammation but also can have neuroprotective actions by the release of BDNF. Microglia express AT$_2$Rs. However, data are controversial currently about the effect of AT$_2$R stimulation on microglia activation in the context of stroke, since both a decrease and an increase in microglia markers have been observed.[28–30] Interestingly, enhanced AT$_2$R-mediated microglia activation coinciding with a diminished infarct size mainly seems to involve the protective phenotype of microglia.[30]

Renal Injury

Renal diseases of various causes such as hypertension, diabetes, and autoimmunity are associated with and driven by an inflammatory response. Several studies have found that AT$_2$R stimulation attenuates the inflammatory response in the kidney. The first evidence pointing to an anti-inflammatory effect of the AT$_2$R in kidney disease came from a model of ureteral obstruction, which in AT$_2$R-deficient mice resulted in a more pronounced increase in macrophage infiltration than in wild-type mice.[31] The same observation was made in a model of renal ablation in AT$_2$R-knockout mice.[32] The effect of direct AT$_2$R stimulation was first studied in hypertension-related nephropathy in stroke-prone spontaneously hypertensive rats (SHRSP) fed a high-sodium diet.[33] Six weeks after the initiation of this diet, massive inflammatory infiltration was detected in the kidneys by immunohistochemistry, and a certain fraction of proteins indicative for renal inflammation was significantly increased in urine samples of hypertensive compared with control rats. Oral treatment with the AT$_2$R agonist C21 attenuated both the infiltration of inflammatory cells into the renal tissue and the increase in the urine markers of inflammation. While the study in SHRSP looked on a rather long-term effect of AT$_2$R stimulation in hypertension-related nephropathy, Matavelli et al. studied the effect of an only 4-day treatment with C21 in a two-kidney, one-clip hypertension model in rats and found a significant reduction of elevated TNF-α and interleukin-6 (IL-6) levels in the renal interstitial fluid.[34] The group of Tahir Hussain has performed several studies showing a renoprotective effect of AT$_2$R stimulation in obese Zucker rats, a model of metabolic syndrome.[35–37] These protective effects included AT$_2$R-mediated anti-inflammation. For example, a 2-week subcutaneous infusion of the peptide AT$_2$R agonist CGP42112A significantly reduced elevated TNF-α and IL-6 protein levels in plasma and the renal cortex.[36] These results could be confirmed in a similar study using the nonpeptide agonist C21.[35] Based on *in vivo* and *in vitro* data, the latter study further suggested that the AT_2R-mediated repression of proinflammatory cytokines is based on an increased synthesis of the anti-inflammatory cytokine interleukin-10.[35]

INFLAMMATION IN AUTOIMMUNE DISEASE AND THE ROLE OF THE AT_2 RECEPTOR

In addition to the inhibitory effect on inflammation, AT_2R also seems to have an immunomodulatory role, for example, in autoimmune diseases. AT_2Rs were found to be expressed in cells controlling innate and specific immune responses such as monocytes,[38] macrophages,[39] microglia,[40] and T cells.[41] Moreover, recent studies using animal models of autoimmunity have provided evidence of AT_2R-mediated immunomodulatory actions. For instance, in a model of immune-mediated glomerulonephritis, indirect AT_2R stimulation during AT_1R blockade resulted in reduced MCP-1 expression and inhibition of ERK phosphorylation.[42] These effects were absent in AT_2R-deficient mice. Recent data about the effect of direct AT_2R stimulation in two further autoimmune models, experimental autoimmune encephalomyelitis (EAE; a mouse model of multiple sclerosis) and collagen-induced rheumatoid arthritis, also indicate an anti-inflammatory immunomodulatory effect of the AT_2R, which involves attenuated T-cell infiltration and promotion of anti-inflammatory Foxp3 regulatory T cells.[43–45]

CONCLUSIONS

Current data overwhelmingly support that the AT_2R exerts anti-inflammatory effects in a broad range of diseases, in which inflammation is a main contributor to the pathology. Such diseases comprise atherosclerosis, myocardial infarction (MI) and post-MI heart failure, stroke, renal disease, and autoimmune diseases. In autoimmune disease, AT_2R stimulation seems to act additionally through modulation of T-cell differentiation and the immune response. Inhibition of NF-κB activity seems a major mechanism of action of AT_2R in inflammation.

All in all, AT_2R stimulation may provide a new promising treatment approach for diseases in which pharmacological targeting of the inflammatory component is therapeutically effective.

REFERENCES

1. Medzhitov R. Origin and physiological roles of inflammation. *Nature* 2008;**24**(454):428–35.
2. Capettini LSA, Montecucco F, Mach F, Stergiopulos N, Santos RAS, da Silva RF. Role of renin-angiotensin system in inflammation, immunity and aging. *Curr Pharm Des* 2012;**18**:963–70.
3. Rompe F, Unger T, Steckelings UM. The angiotensin AT2 receptor in inflammation. *Drug News Perspect* 2010;**23**:104–11.
4. Chabrashvili T, Kitiyakara C, Blau J, Karber A, Aslam S, Welch WJ, et al. Effects of ANG II type 1 and 2 receptors on oxidative stress, renal NADPH oxidase, and SOD expression. *Am J Physiol Regul Integr Comp Physiol* 2003;**285**:R117–24.
5. Lawrence T. The nuclear factor NF-κB pathway in inflammation. *Cold Spring Harb Perspect Biol* 2009;**1**:a001651.
6. Wu L, Iwai M, Li Z, Shiuchi T, Min L-J, Cui T-X, et al. Regulation of inhibitory protein-kappaB and monocyte chemoattractant protein-1 by angiotensin II type 2 receptor-activated Src homology protein tyrosine phosphatase-1 in fetal vascular smooth muscle cells. *Mol Endocrinol Baltim MD* 2004;**18**:666–78.
7. Rompe F, Artuc M, Hallberg A, Alterman M, Ströder K, Thöne-Reineke C, et al. Direct angiotensin II type 2 receptor stimulation acts anti-inflammatory through epoxyeicosatrienoic acid and inhibition of nuclear factor kappaB. *Hypertension* 2010;**55**:924–31.
8. Node K, Huo Y, Ruan X, Yang B, Spiecker M, Ley K, et al. Anti-inflammatory properties of cytochrome P450 epoxygenase-derived eicosanoids. *Science* 1999;**285**:1276–9.
9. Reinecke K, Lucius R, Reinecke A, Rickert U, Herdegen T, Unger T. Angiotensin II accelerates functional recovery in the rat sciatic nerve in vivo: role of the AT2 receptor and the transcription factor NF-kappaB. *FASEB J* 2003;**17**:2094–6.
10. Esteban V, Lorenzo O, Rupérez M, Suzuki Y, Mezzano S, Blanco J, et al. Angiotensin II, via AT1 and AT2 receptors and NF-kappaB pathway, regulates the inflammatory response in unilateral ureteral obstruction. *J Am Soc Nephrol* 2004;**15**:1514–29.
11. Schindler C, Levy DE, Decker T. JAK-STAT signaling: from interferons to cytokines. *J Biol Chem* 2007;**282**:20059–63.
12. Horiuchi M, Hayashida W, Akishita M, Tamura K, Daviet L, Lehtonen JYA, et al. Stimulation of Different Subtypes of Angiotensin II Receptors, AT1 and AT2 Receptors, Regulates STAT Activation by Negative Crosstalk. *Circ Res* 1999;**84**:876–82.
13. Reuter S, Gupta SC, Chaturvedi MM, Aggarwal BB. Oxidative stress, inflammation, and cancer: how are they linked? *Free Radic Biol Med* 2010;**49**:1603–16.
14. Sales VL, Sukhova GK, Lopez-Ilasaca MA, Libby P, Dzau VJ, Pratt RE. Angiotensin type 2 receptor is expressed in murine atherosclerotic lesions and modulates lesion evolution. *Circulation* 2005;**112**:3328–36.
15. Takata H, Yamada H, Kawahito H, Kishida S, Irie D, Kato T, et al. Vascular angiotensin II type 2 receptor attenuates atherosclerosis via a kinin/NO-dependent mechanism. *J Renin-Angiotensin-Aldosterone Syst* 2013;June 4 [Epub ahead of print].
16. Iwai M, Chen R, Li Z, Shiuchi T, Suzuki J, Ide A, et al. Deletion of angiotensin II type 2 receptor exaggerated atherosclerosis in apolipoprotein E-null mice. *Circulation* 2005;**112**:1636–43.

17. Tiyerili V, Mueller CFH, Becher UM, Czech T, van Eickels M, Daiber A, et al. Stimulation of the AT2 receptor reduced atherogenesis in ApoE(−/−)/AT1A(−/−) double knock out mice. *J Mol Cell Cardiol* 2012;**52**:630–7.
18. Kljajic ST, Widdop RE, Vinh A, Welungoda I, Bosnyak S, Jones ES, et al. Direct AT$_2$ receptor stimulation is athero-protective and stabilizes plaque in Apolipoprotein E-deficient mice. *Int J Cardiol* 2013;**169**:281–7.
19. Sun Y. Myocardial repair/remodelling following infarction: roles of local factors. *Cardiovasc Res* 2009;**81**:482–90.
20. Kaschina E, Lauer D, Schmerler P, Unger T, Steckelings UM. AT2 receptors targeting cardiac protection post-myocardial infarction. *Curr Hypertens Rep* 2014;**16**:441.
21. Kaschina E, Grzesiak A, Li J, Foryst-Ludwig A, Timm M, Rompe F, et al. Angiotensin II type 2 receptor stimulation: a novel option of therapeutic interference with the renin-angiotensin system in myocardial infarction? *Circulation* 2008;**118**:2523–32.
22. Curato C, Slavic S, Dong J, Skorska A, Altarche-Xifró W, Miteva K, et al. Identification of noncytotoxic and IL-10-producing CD8+AT2R+ T cell population in response to ischemic heart injury. *J Immunol Baltim MD 1950* 2010;**185**:6286–93.
23. Altarche-Xifró W, Curato C, Kaschina E, Grzesiak A, Slavic S, Dong J, et al. Cardiac c-kit+AT2+ cell population is increased in response to ischemic injury and supports cardiomyocyte performance. *Stem Cells Dayt OH* 2009;**27**:2488–97.
24. Lakhan SE, Kirchgessner A, Hofer M. Inflammatory mechanisms in ischemic stroke: therapeutic approaches. *J Transl Med* 2009;**7**:97.
25. Sumners C, Horiuchi M, Widdop RE, McCarthy C, Unger T, Steckelings UM. Protective arms of the renin-angiotensin-system in neurological disease. *Clin Exp Pharmacol Physiol* 2013;**40**:580–8.
26. Iwai M, Liu H-W, Chen R, Ide A, Okamoto S, Hata R, et al. Possible inhibition of focal cerebral ischemia by angiotensin II type 2 receptor stimulation. *Circulation* 2004;**110**:843–8.
27. Min L-J, Mogi M, Tsukuda K, Jing F, Ohshima K, Nakaoka H, et al. Direct stimulation of angiotensin II type 2 receptor initiated after stroke ameliorates ischemic brain damage. *Am J Hypertens* 2014;**27**(8):1036–44.
28. Joseph JP, Mecca AP, Regenhardt RW, Bennion DM, Rodríguez V, Desland F, et al. The angiotensin type 2 receptor agonist Compound 21 elicits cerebroprotection in endothelin-1 induced ischemic stroke. *Neuropharmacology* 2014;**81**:134–41.
29. McCarthy CA, Vinh A, Broughton BRS, Sobey CG, Callaway JK, Widdop RE. Angiotensin II type 2 receptor stimulation initiated after stroke causes neuroprotection in conscious rats. *Hypertension* 2012;**60**:1531–7.
30. McCarthy CA, Vinh A, Miller AA, Hallberg A, Alterman M, Callaway JK, et al. Direct angiotensin AT2 receptor stimulation using a novel AT2 receptor agonist, compound 21 evokes neuroprotection in conscious hypertensive rats. *PLoS One* 2014;**9**:e95762.
31. Ma J, Nishimura H, Fogo A, Kon V, Inagami T, Ichikawa I. Accelerated fibrosis and collagen deposition develop in the renal interstitium of angiotensin type 2 receptor null mutant mice during ureteral obstruction. *Kidney Int* 1998;**53**:937–44.
32. Benndorf RA, Krebs C, Hirsch-Hoffmann B, Schwedhelm E, Cieslar G, Schmidt-Haupt R, et al. Angiotensin II type 2 receptor deficiency aggravates renal injury and reduces survival in chronic kidney disease in mice. *Kidney Int* 2009;**75**:1039–49.
33. Gelosa P, Pignieri A, Fändriks L, de Gasparo M, Hallberg A, Banfi C, et al. Stimulation of AT2 receptor exerts beneficial effects in stroke-prone rats: focus on renal damage. *J Hypertens* 2009;**27**:2444–51.
34. Matavelli LC, Huang J, Siragy HM. Angiotensin AT$_2$ receptor stimulation inhibits early renal inflammation in renovascular hypertension. *Hypertension* 2011;**57**:308–13.
35. Dhande I, Ali Q, Hussain T. Proximal tubule angiotensin AT2 receptors mediate an anti-inflammatory response via interleukin-10: role in renoprotection in obese rats. *Hypertension* 2013;**61**:1218–26.
36. Sabuhi R, Ali Q, Asghar M, Al-Zamily NRH, Hussain T. Role of the angiotensin II AT2 receptor in inflammation and oxidative stress: opposing effects in lean and obese Zucker rats. *Am J Physiol Renal Physiol* 2011;**300**:F700–6.
37. Siddiqui AH, Ali Q, Hussain T. Protective role of angiotensin II subtype 2 receptor in blood pressure increase in obese Zucker rats. *Hypertension* 2009;**53**:256–61.
38. Marino F, Maresca AM, Cosentino M, Castiglioni L, Rasini E, Mongiardi C, et al. Angiotensin II type 1 and type 2 receptor expression in circulating monocytes of diabetic and hypercholesterolemic patients over 3-month rosuvastatin treatment. *Cardiovasc Diabetol* 2012;**11**:153.
39. Jing F, Mogi M, Min L-J, Ohshima K, Nakaoka H, Tsukuda K, et al. Effect of angiotensin II type 2 receptor-interacting protein on adipose tissue function via modulation of macrophage polarization. *PLoS One* 2013;**8**:e60067.
40. McCarthy CA, Widdop RE, Deliyanti D, Wilkinson-Berka JL. Brain and retinal microglia in health and disease: an unrecognized target of the renin-angiotensin system. *Clin Exp Pharmacol Physiol* 2013;**40**:571–9.
41. Jurewicz M, McDermott DH, Sechler JM, Tinckam K, Takakura A, Carpenter CB, et al. Human T and natural killer cells possess a functional renin-angiotensin system: further mechanisms of angiotensin II-induced inflammation. *J Am Soc Nephrol* 2007;**18**:1093–102.
42. Okada H, Inoue T, Kikuta T, Watanabe Y, Kanno Y, Ban S, et al. A possible anti-inflammatory role of angiotensin II type 2 receptor in immune-mediated glomerulonephritis during type 1 receptor blockade. *Am J Pathol* 2006;**169**:1577–89.
43. Valero-Esquitino Veronica, Cirera-Salinas D, Lucht Kristin, Pascual A, Curato Caterina, Arkink Joeri, et al. Direct angiotensin type 2 receptor stimulation inhibits T-lymphocyte recruitment to the central nervous system and modulates T-cell differentiation. *J Hypertens Suppl* 2013;**31**(e-Suppl A):e110.
44. Valero-Esquitino V, Lucht K, Namsolleck P, Monnet-Tschudi F, Stubbe T, Lucht F, et al. Direct angiotensin AT2-receptor stimulation attenuates T-cell and microglia activation and prevents demyelination in experimental autoimmune encephalomyelitis in mice. *Clin Sci* 2014;**128**(2):95–109.
45. Sehnert B, Schett G, Voll RE, Steckelings UM. A non-peptide AT2 receptor agonist attenuates clinical and histological signs in murine collagen-induced arthritis. *Ann Rheum Dis* 2012;**71**:A88.

Chapter 14

The AT$_2$ Receptor and Interacting Proteins (ATIPs) in Cancer

Sylvie Rodrigues-Ferreira, Angie Molina, Anne Nehlig, Clara Nahmias
Department of Molecular Medicine, Institut Gustave Roussy, INSERM U981, Université Paris Sud, 94800 Villejuif, France

THE RENIN–ANGIOTENSIN SYSTEM AND CANCER

The renin–angiotensin system (RAS), a major regulator of blood pressure and cardiovascular homeostasis, is a complex enzymatic cascade that ultimately generates bioactive angiotensin peptides in the systemic circulation as well as locally in various tissues.[1] The "classical RAS" generates the angiotensin II (AngII) octapeptide through the cleavage of inactive AngI by angiotensin-converting enzyme (ACE) (Figure 1), whereas the "alternative RAS axis"[2] produces Ang-(1-7) by subsequent cleavage of AngII by the ACE homologue ACE2.

AngII binds to two receptor subtypes (AT$_1$ and AT$_2$) and Ang-(1-7) binds to the Mas receptor (Figure 1). AT$_1$, AT$_2$, and Mas all belong to the superfamily of seven-transmembrane domain receptors, but they display different and often opposite effects inside the cells. It is generally accepted that the classical ACE/AngII/AT$_1$ axis promotes the majority of RAS actions on cardiovascular, renal, and cerebral functions and that these effects are opposed by the activation of the AT$_2$ receptor[3,4] and of the "protective" ACE2/Ang-(1-7)/Mas axis.[2]

The complexity of the RAS underlies its essential functions in a number of physiopathologic situations. Not surprisingly, over the past 10 years, the picture has emerged that dysregulation of RAS components may also play a role in various aspects of cancer.[5–7] Overexpression or activation of the ACE/AngII/AT$_1$ axis has been shown to facilitate tumor growth, vascularization, inflammation, and metastatic progression, by acting on both cancer cells and the tumor microenvironment.[7,8] In animal models, targeting the AT$_1$ receptor with selective receptor blockers has proved valuable to reduce tumor growth, macrophage recruitment, and metastatic dissemination. It has also been suggested that treatment with AT$_1$ receptor blockers may enhance chemotherapy efficacy by decompressing tumor vessels, thereby increasing drug delivery.[9] On the other hand, a number of studies have pointed to a beneficial role of the ACE2/Ang-(1-7)/Mas axis against cancer, with pleiotropic effects on both tumor and stroma (review in Ref.[10]).

Given the complex interplay between different components of the classical and new arms of the RAS and the potential benefit of using AT$_1$ inhibitors and Mas agonists in cancer patients, it is of utmost importance to address the question of the role of the AT$_2$ receptor in cancer progression. Protective effects of AT$_2$ stimulation in cancer are expected due to the antiproliferative, antimigratory, and AT$_1$-opposing effects often attributed to this receptor in various *in vivo* and cellular models. However, despite extensive studies, the consequences of AT$_2$ receptor overexpression or stimulation in cancer remain controversial. In the following section, we will summarize major studies that have been undertaken to evaluate whether the AT$_2$ receptor may be detrimental or protective against cancer.

THE AT$_2$ RECEPTOR AND CANCER

The AT$_2$ receptor is abundantly expressed during fetal development and its expression levels fall abruptly at birth. In the adult, AT$_2$ expression is upregulated in pathophysiological situations involving tissue remodeling, inflammation, and/or repair. In line with these observations, AT$_2$ expression has been reported to be upregulated in cancer, a disease associated with extensive tissue remodeling and recruitment of inflammatory cells and molecules. Increased levels of AT$_2$ receptors have been detected by immunohistochemistry in a variety of human tumors (Table 1), including ductal and invasive breast carcinoma,[11] gastric cancer and lymph node metastases,[12] high-grade astrocytoma,[13] and high-grade renal cell carcinoma.[14] In some cases, increased AT$_2$ expression correlated with decreased survival in patients,[13,14] suggesting that this receptor may

The Protective Arm of the Renin–Angiotensin System (RAS). http://dx.doi.org/10.1016/B978-0-12-801364-9.00014-6

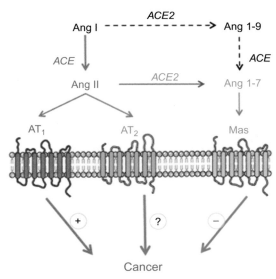

FIGURE 1 The renin–angiotensin system and cancer.

TABLE 1 AT$_2$ Receptor Expression in Human Cancer Samples

Cancer Type	IHC	ISH	PCR	References
Breast cancer	Increased	Increased	-	[11]
Gastric cancer	Increased	-	-	[12]
Astrocytoma	Increased	-	Increased	[13]
Kidney (RCCC)	Increased	-	No change	[14]
Prostate cancer	Decreased	-	-	[15]

IHC, immunohistochemistry; ISH, *in situ* hybridization; PCR, polymerase chain reaction.

be considered as a biomarker for tumor progression. To date, prostate cancer is the only malignancy in which AT$_2$ expression has been found to be downregulated rather than upregulated in high-grade tumors.[15]

It should be noted however that the specificity of commercially available anti-AT$_2$ antibodies has recently been challenged,[16] and caution should thus be taken in interpreting the results of IHC studies. Molecular analysis of AT$_2$ mRNA levels in human tumors has in some, but not all, cases confirmed the IHC results (Table 1). Further development of highly specific anti-AT$_2$R antibodies is recommended before these studies can be translated into the clinic.

At the functional level, AT$_2$ receptor activation or overexpression in tumor cell lines of various origins has been shown to inhibit cell proliferation and induce apoptosis (Table 2), suggesting protective effects of this receptor on cancer cell growth.[17–19] Only one study, using AT$_2$ receptor deficiency or pharmacological blockade, suggested that AT$_2$ may be protumorigenic *in vitro* and *in vivo*.[20]

Animal studies led to highly contradictory results, possibly due to complex interactions between the tumor and stromal microenvironment and to different experimental settings. Indeed, AT$_2$ receptor-knockout mice studies[20–23] address the role of host stromal AT$_2$ receptors in cancer, whereas AT$_2$ receptor overexpression—by means of adenoviral transduction or nanoparticle injection—may target both cancer cells and the stromal microenvironment.[24,25] In different models of chemically induced mouse tumors, AT$_2$ deletion delayed tumor growth[21,22] through an indirect effect on metabolism of the carcinogen rather than by directly affecting cancer progression.

Thus, using different cancer cell types and mouse models, AT$_2$ was reported to be either protective or detrimental against cancer (Table 2), and to date, no clear picture can be drawn from these studies. It may be interesting to consider AT$_2$ effects in the context of the expression and/or activation of other RAS components in different cancer models, since activation of endogenous AT$_1$ or Mas receptors may compensate for AT$_2$ deficiency or be masked by AT$_2$ receptor overexpression.

TABLE 2 AT$_2$ Receptor Functional Effects in Cancer

Cancer Type	Cell Line	Experimental Model	*In Vitro* Effects	*In Vivo* Effects	References
Prostate	PC3, LnCap	AT$_2$ Ligands	Protective	nd	17
Prostate	DU145, PC3, LnCap	Ad-AT$_2$	Protective	nd	18
Gastric	AGS, MKN28 NK87, MKN45	AT$_2$ antagonist	Protective Proinvasive	nd	12
Lung	A549, H358	Ad-AT$_2$, nanoparticle	Protective	nd	19
Colon	Carcinogen	AT$_2$-null mice	nd	Detrimental	21
Lung	Carcinogen	AT$_2$-null mice	nd	Detrimental	22
Fibrosarcoma	Carcinogen	AT$_2$-null mice	Detrimental	Detrimental	20
Lung	LL/2	Antagonist PD		Detrimental	
Lung	LCC	Nanoparticle	Protective	Protective	24
Pancreatic	PAN02	AT$_2$-null mice	Protective	Protective	23
Liver (HCC)	SMMC7221	Ad-AT$_2$	Protective	Detrimental	25

Ad-AT$_2$, adenoviral vector expressing AT$_2$; nd, not determined.

AT$_2$ RECEPTOR-INTERACTING PROTEINS (ATIPs) AND CANCER

Since its molecular cloning in 1995, AT$_2$ has been recognized as an atypical seven-transmembrane domain receptor coupled to unconventional signaling pathways. The search for AT$_2$ transducing molecules using yeast two-hybrid system has led to the identification of several AT$_2$ receptor-interacting partners that may serve as scaffolds to regulate its biological functions. AT$_2$ partners include the epidermal growth factor receptor ERBB3,[26] the promyelocytic leukemia zinc finger (PLZF) protein,[27] the scaffold protein CNK1,[28] the tissue inhibitor of metalloproteinase 3 (TIMP3),[29] and a family of AT$_2$ receptor-interacting proteins designated ATIPs[30,31] or ATBP.[32] All these proteins have been shown to interact with the C-terminal intracellular portion of the AT$_2$ receptor (Figure 2), raising the interesting possibility that AT$_2$ partners may form large macromolecular complexes with the receptor or may compete with each other for AT$_2$ interaction. However, to date, only a few of them, namely, PLZF and ATIPs/ATBP, have been functionally linked to AT$_2$ receptor functions.

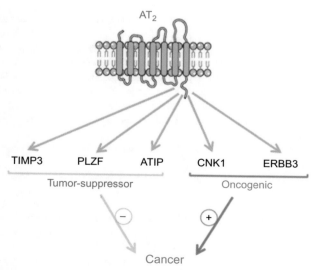

FIGURE 2   AT$_2$ receptor-interacting partners and cancer.

Of note, all identified AT$_2$ partners are related to cancer: ERBB3 (also designated HER3) and CNK1 are oncogenic, whereas PLZF, TIMP3, and ATIPs are associated with tumor suppression (Figure 2). Potential interactions of the AT$_2$ receptor with proteins showing either oncogenic or tumor-suppressor effects make it difficult to hypothesize on the functional consequences of AT$_2$ receptor expression or stimulation in cancer. It also remains unknown whether these interactions may result in the activation or inhibition of AT$_2$ signaling. Based on these observations, it may be possible that contradictory results regarding AT$_2$ effects in various cancer models may be due to relative expression levels and/or competition of different AT$_2$ receptor-interacting proteins for binding to the receptor.

ATIPs constitute to date the most extensively studied family of AT$_2$ receptor-interacting partners. Three major ATIP members (ATIP1, ATIP3, and ATIP4) are encoded by alternative promoter usage and exon splicing of the microtubule-associated tumor suppressor (*MTUS1*) gene.[30,31,33] All ATIPs are identical in their C-terminal portion containing the AT$_2$ receptor-interacting domain; however, only ATIP1 has been firmly demonstrated to interact with AT$_2$ in living cells. Notably, ATIP1 constitutively associates with the AT$_2$ receptor at the cell membrane and dissociates from the receptor upon AngII stimulation.[30,31] This isoform is involved in the transport of AT$_2$ receptor from the Golgi to the cell membrane[32] and contributes to AT$_2$ effects on cell growth inhibition, neuronal differentiation, neointima formation, vascular senescence, and adipose function.[31,34,35] A role for ATIP1 as a tumor suppressor in pancreatic cancer has also been reported independently of the AT$_2$ receptor.[36]

Downregulation of *MTUS1* gene expression in a variety of human cancers, including breast,[37,38] colon,[39] head-and-neck,[40] digestive system,[41] and bladder[42,43] tumors, supports its role as a major tumor suppressor in solid cancers, with the exception of prostate cancer[15] in which upregulation of *MTUS1* has been associated with tumor progression. Most studies however have not examined whether alterations of *MTUS1* expression affect ATIP1 or ATIP3 levels, or both.

Our group has identified the ATIP3 isoform as being the major *MTUS1* product downregulated in high-grade and metastatic breast tumors[37] and a novel prognostic biomarker of patient survival.[38] Its reexpression in ATIP3-negative breast cancer cells has profound inhibitory effects on breast cancer cell proliferation and migration, tumor growth, and metastasis,[38] making it an interesting therapeutic target for new breast cancer treatments. Growth-inhibitory effects of ATIP3 are AT$_2$-independent, but it may be interesting to investigate whether a subpopulation of tumors coexpressing AT$_2$ and ATIP3 may benefit from AT$_2$ receptor stimulation.

ACKNOWLEDGMENTS

We thank the Inserm, the CNRS, the Institut Gustave Roussy, and the association Odyssea and Prolific.

REFERENCES

1. Crowley SD, Coffman TM. Recent advances involving the renin-angiotensin system. *Exp Cell Res* 2012;**318**:1049–56.
2. Santos RA, Ferreira AJ, Verano-Braga T, Bader M. Angiotensin-converting enzyme 2, angiotensin-(1–7) and Mas: new players of the renin-angiotensin system. *J Endocrinol* 2013;**216**:R1–17.
3. Porrello ER, Delbridge LM, Thomas WG. The angiotensin II type 2 (AT2) receptor: an enigmatic seven transmembrane receptor. *Front Biosci (Landmark Ed)* 2009;**14**:958–72.
4. McCarthy CA, Widdop RE, Denton KM, Jones ES. Update on the angiotensin AT(2) receptor. *Curr Hypertens Rep* 2013;**15**:25–30.
5. Deshayes F, Nahmias C. Angiotensin receptors: a new role in cancer? *Trends Endocrinol Metab* 2005;**16**:293–9.
6. George AJ, Thomas WG, Hannan RD. The renin-angiotensin system and cancer: old dog, new tricks. *Nat Rev Cancer* 2010;**10**:745–59.
7. Wegman-Ostrosky T, Soto-Reyes E, Vidal-Millán S, Sánchez-Corona J. The renin-angiotensin system meets the hallmarks of cancer. *J Renin Angiotensin Aldosterone Syst* 2013, Epub ahead of print (doi:10.1177/1470320313496858).
8. Rodrigues-Ferreira S, Abdelkarim M, Dillenburg-Pilla P, Luissint AC, di-Tommaso A, Deshayes F, et al. Angiotensin II facilitates breast cancer cell migration and metastasis. *PLoS One* 2012;**7**:e35667.
9. Chauhan VP, Martin JD, Liu H, Lacorre DA, Jain SR, Kozin SV, et al. Angiotensin inhibition enhances drug delivery and potentiates chemotherapy by decompressing tumour blood vessels. *Nat Commun* 2013;**4**:2516.
10. Gallagher PE, Arter AL, Deng G, Ann Tallant E. Angiotensin-(1–7): A peptide hormone with anti-cancer activity. *Curr Med Chem* 2014. [Epub ahead of print].
11. De Paepe B, Verstraeten VM, De Potter CR, Bullock GR. Increased angiotensin II type-2 receptor density in hyperplasia, DCIS and invasive carcinoma of the breast is paralleled with increased iNOS expression. *Histochem Cell Biol* 2002;**117**:13–9.
12. Carl-McGrath S, Ebert MP, Lendeckel U, Röcken C. Expression of the local angiotensin II system in gastric cancer may facilitate lymphatic invasion and nodal spread. *Cancer Biol Ther* 2007;**6**:1218–26.
13. Arrieta O, Pineda-Olvera B, Guevara-Salazar P, Hernández-Pedro N, Morales-Espinosa D, Cerón-Lizarraga TL, et al. Expression of AT1 and AT2 angiotensin receptors in astrocytomas is associated with poor prognosis. *Br J Cancer* 2008;**99**:160–6.

14. Dolley-Hitze T, Jouan F, Martin B, Mottier S, Edeline J, Moranne O, et al. Angiotensin-2 receptors (AT1-R and AT2-R), new prognostic factors for renal clear-cell carcinoma? *Br J Cancer* 2010;**103**:1698–705.

15. Guimond MO, Battista MC, Nikjouitavabi F, Carmel M, Barres V, Doueik AA, et al. Expression and role of the angiotensin II AT2 receptor in human prostate tissue: in search of a new therapeutic option for prostate cancer. *Prostate* 2013;**73**:1057–68.

16. Hafko R, Villapol S, Nostramo R, Symes A, Sabban EL, Inagami T, et al. Commercially available angiotensin II At$_2$ receptor antibodies are nonspecific. *PLoS One* 2013;**8**:e69234.

17. Chow L, Rezmann L, Imamura K, Wang L, Catt K, Tikellis C, et al. Functional angiotensin II type 2 receptors inhibit growth factor signaling in LNCaP and PC3 prostate cancer cell lines. *Prostate* 2008;**68**:651–60.

18. Li H, Qi Y, Li C, Braseth LN, Gao Y, Shabashvili AE, et al. Angiotensin type 2 receptor-mediated apoptosis of human prostate cancer cells. *Mol Cancer Ther* 2009;**8**:3255–65.

19. Pickel L, Matsuzuka T, Doi C, Ayuzawa R, Maurya DK, Xie SX, et al. Overexpression of angiotensin II type 2 receptor gene induces cell death in lung adenocarcinoma cells. *Cancer Biol Ther* 2010;**9**(4):277–85.

20. Clere N, Corre I, Faure S, Guihot AL, Vessières E, Chalopin M, et al. Deficiency or blockade of angiotensin II type 2 receptor delays tumorigenesis by inhibiting malignant cell proliferation and angiogenesis. *Int J Cancer* 2010;**127**:2279–91.

21. Takagi T, Nakano Y, Takekoshi S, Inagami T, Tamura M. Hemizygous mice for the angiotensin II type 2 receptor gene have attenuated susceptibility to azoxymethane-induced colon tumorigenesis. *Carcinogenesis* 2002;**23**:1235–41.

22. Kanehira T, Tani T, Takagi T, Nakano Y, Howard EF, Tamura M. Angiotensin II type 2 receptor gene deficiency attenuates susceptibility to tobacco-specific nitrosamine-induced lung tumorigenesis: involvement of transforming growth factor-beta-dependent cell growth attenuation. *Cancer Res* 2005;**65**:7660–5.

23. Doi C, Egashira N, Kawabata A, Maurya DK, Ohta N, Uppalapati D, et al. Angiotensin II type 2 receptor signaling significantly attenuates growth of murine pancreatic carcinoma grafts in syngeneic mice. *BMC Cancer* 2010;**10**:67.

24. Kawabata A, Baoum A, Ohta N, Jacquez S, Seo GM, Berkland C, et al. Intratracheal administration of a nanoparticle-based therapy with the angiotensin II type 2 receptor gene attenuates lung cancer growth. *Cancer Res* 2012;**72**:2057–67.

25. Du H, Liang Z, Zhang Y, Jie F, Li J, Fei Y, et al. Effects of angiotensin II type 2 receptor overexpression on the growth of hepatocellular carcinoma cells in vitro and in vivo. *PLoS One* 2013;**8**:e83754.

26. Knowle D, Ahmed S, Pulakat L. Identification of an interaction between the angiotensin II receptor sub-type AT2 and the ErbB3 receptor, a member of the epidermal growth factor receptor family. *Regul Pept* 2000;**87**:73–82.

27. Senbonmatsu T, Saito T, Landon EJ, Watanabe O, Price Jr E, Roberts RL, et al. A novel angiotensin II type 2 receptor signaling pathway: possible role in cardiac hypertrophy. *EMBO J* 2003;**22**:6471–82.

28. Fritz RD, Radziwill G. The scaffold protein CNK1 interacts with the angiotensin II type 2 receptor. *Biochem Biophys Res Commun* 2005;**338**:1906–12.

29. Kang KH, et al. Tissue inhibitor of metalloproteinases-3 interacts with angiotensin II type 2 receptor and additively inhibits angiogenesis. *Cardiovasc Res* 2008;**79**:150–60.

30. Nouet S, Amzallag N, Li JM, Louis S, Seitz I, Cui TX, et al. Trans-inactivation of receptor tyrosine kinases by novel angiotensin II AT2 receptor-interacting protein. ATIP. *J Biol Chem* 2004;**279**:28989–97.

31. Rodrigues-Ferreira S, Nahmias C. An ATIPical family of angiotensin II AT2 receptor-interacting proteins. *Trends Endocrinol Metab* 2010;**21**:684–90.

32. Wruck CJ, Funke-Kaiser H, Pufe T, Kusserow H, Menk M, Schefe JH, et al. Regulation of transport of the angiotensin AT2 receptor by a novel membrane-associated Golgi protein. *Arterioscler Thromb Vasc Biol* 2005;**25**:57–64.

33. Di Benedetto M, Bièche I, Deshayes F, Vacher S, Nouet S, Collura V, et al. Structural organization and expression of human MTUS1, a candidate 8p22 tumor suppressor gene encoding a family of angiotensin II AT2 receptor-interacting proteins. ATIP. *Gene* 2006;**380**:127–36.

34. Min LJ, Mogi M, Iwanami J, Jing F, Tsukuda K, Ohshima K, et al. Angiotensin II type 2 receptor-interacting protein prevents vascular senescence. *J Am Soc Hypertens* 2012;**6**:179–84.

35. Jing F, Mogi M, Min LJ, Ohshima K, Nakaoka H, Tsukuda K, et al. Effect of angiotensin II type 2 receptor-interacting protein on adipose tissue function via modulation of macrophage polarization. *PLoS One* 2013;**8**:e60067.

36. Seibold S, Rudroff C, Weber M, Galle J, Wanner C, Marx M. Identification of a new tumor suppressor gene located at chromosome 8p21.3-22. *FASEB J* 2003;**17**:1180–2.

37. Rodrigues-Ferreira S, Di Tommaso A, Dimitrov A, Cazaubon S, Gruel N, Colasson H, et al. 8p22 MTUS1 gene product ATIP3 is a novel anti-mitotic protein underexpressed in invasive breast carcinoma of poor prognosis. *PLoS One* 2009;**4**:e7239.

38. Molina A, Velot L, Ghouinem L, Abdelkarim M, Bouchet BP, Luissint AC, et al. ATIP3, a novel prognostic marker of breast cancer patient survival, limits cancer cell migration and slows metastatic progression by regulating microtubule dynamics. *Cancer Res* 2013;**73**:2905–15.

39. Zuern C, Heimrich J, Kaufmann R, Richter KK, Settmacher U, Wanner C, et al. Down-regulation of MTUS1 in human colon tumors. *Oncol Rep* 2010;**23**:183–9.

40. Ding X, Zhang N, Cai Y, Li S, Zheng C, Jin Y, et al. Down-regulation of tumor suppressor MTUS1/ATIP is associated with enhanced proliferation, poor differentiation and poor prognosis in oral tongue squamous cell carcinoma. *Mol Oncol* 2012;**6**:73–80.

41. Li X, Liu H, Yu T, Dong Z, Tang L, Sun X. Loss of MTUS1 in gastric cancer promotes tumor growth and metastasis. *Neoplasma* 2014;**61**:128–35.

42. Xiao J, Chen JX, Zhu YP, Zhou LY, Shu QA, Chen LW. Reduced expression of MTUS1 mRNA is correlated with poor prognosis in bladder cancer. *Oncol Lett* 2012;**4**:113–8.

43. Rogler A, Hoja S, Giedl J, Ekici AB, Wach S, Taubert H, et al. Loss of MTUS1/ATIP expression is associated with adverse outcome in advanced bladder carcinomas: data from a retrospective study. *BMC Cancer* 2014;**14**:214.

Chapter 15

AT$_2$R and Sympathetic Outflow

Lie Gao, Irving H. Zucker

Department of Cellular and Integrative Physiology, University of Nebraska Medical Center, Omaha, Nebraska, USA

INTRODUCTION

The control of sympathetic outflow is a complex and multifactorial process. In addition to central sympathetic rhythms,[1] a variety of modulating factors participate in the physiological adjustments of the sympathetic nervous system to physical and emotional stress and in multiple disease states.[2,3] Ultimately, the level of sympathetic outflow is determined by three components. These are (1) the magnitude of neurotransmitter release from cells projecting to the presympathetic neurons in the rostral ventrolateral medulla (RVLM) and spinal cord, (2) circulating factors that gain access to presympathetic neurons through the cerebrospinal fluid or blood in areas of the brain that have low blood–brain barriers, and (3) modulation of sympathetic discharge by virtue of input from homeostatic cardiovascular receptors and reflexes and nociceptive input from the periphery.

The renin–angiotensin system (RAS) has been implicated in the sympathoexcitatory process for many years.[4] Angiotensin II (Ang II) modulates sympathetic function in the central nervous system, at peripheral sympathetic ganglia and at the neuroeffector junction in peripheral target tissues. Most components of the RAS have been localized to the central nervous system.[5] Overexpression and knockdown of the brain prorenin receptor have been associated with modulation of sympathetic tone and hypertension.[6,7] On the other hand, overexpression of angiotensin-converting enzyme 2 has been shown to blunt the sympathoexcitatory and hypertension in response to Ang II infusion,[8] ostensibly through the generation of angiotensin 1-7 (Ang 1-7).

The major angiotensin peptides, Ang II and Ang 1-7, signal through three primary membrane receptors: the G protein-coupled angiotensin type 1 (AT$_1$) receptor, the AT$_2$ receptor, and the Mas receptor. As far as sympathoexcitation is concerned, the bulk of research has focused on the role of the AT$_1$R in mediating increases in sympathetic outflow. Unfortunately, much less work has been done on the role of the angiotensin type 2 receptor (AT$_2$R). This is due primarily to the notion that the AT$_2$R is expressed at significantly lower levels in the adult animal compared to the fetus or neonate.[9] As will be shown below, this notion has recently been challenged[10] and it appears that signaling through the AT$_2$R does indeed provide a major sympathoinhibitory effect, especially in the face of AT$_1$R blockade.[11,12] In the setting of chronic heart failure (CHF) and hypertension, there appears to be an increase in the ratio of AT$_1$ to AT$_2$R in the medulla contributing to an increase in the discharge sensitivity of presympathetic neurons.[11]

AT$_2$R ONTOGENY IN RODENT BRAIN

The AT$_2$R has long been believed to be a vestigial receptor and to be of limited physiological significance in mature individuals.[13] This well-accepted concept implies (i) that the AT$_2$R is highly expressed in the fetus and dramatically declines within 24h of birth and almost disappears in all adult tissues and (ii) that the AT$_2$R is the predominant Ang II receptor subtype in the fetus. This concept originally stemmed from a widely cited autoradiographic study by Grady et al.[9] who showed a markedly higher density of AT$_2$R-radioactive signal in the rat embryo with rapid disappearance after birth. This dogma has been reinforced by the fact that the net effect of AT$_2$R activation is difficult to observe in intact adult animals due to the fact that the AT$_2$R has no exclusive natural ligand and competes with the AT$_1$R for Ang II binding. Most of the biological effects of AT$_1$R stimulation sharply counteract those of AT$_2$R stimulation. Because the AT$_1$R is still the predominant receptor subtype, its activation masks the function of AT$_2$R activation.

Unfortunately, the idea of a reduction in AT$_2$R expression after birth has been accepted for two decades. However, biochemical and molecular evidence for this observation is lacking. A series of experiments recently performed in our laboratory evaluated the developmental change of AT$_2$R protein expression and found clear contrary evidence for this notion.

The Protective Arm of the Renin–Angiotensin System (RAS). http://dx.doi.org/10.1016/B978-0-12-801364-9.00015-8

Employing Western blotting analysis, we demonstrated a significantly higher AT_2R protein expression in the brains and other major organs of adult rats and mice as compared to the fetus.[10,14] This expression pattern was further confirmed in membrane and cytosolic extracts of rodent brain.[14] Interestingly, when we examined AT_2R expression in the skin, we found a higher AT_2R protein level in the fetus as compared with the adult,[15] similar to the conventional concept.[13] Indeed, after closely inspecting the autoradiographic data of Grady et al.,[9] it is clear that the majority of the radioactive signal was concentrated in the outline of the fetus, suggesting that the traditional concept of AT_2R ontogeny may be valid only in the skin. Our data suggest that during development, the AT_2R is upregulated in the brain and other major organs, whereas it is downregulated in the skin. Nevertheless, the AT_1R still dominates over the AT_2R. These data have two important implications: (1) The AT_1R is the predominant angiotensin receptor subtype in the fetal brain, and (2) a functional AT_2R exists in adult animals. Indeed, in addition to the biochemical data, functional experiments from our laboratory and others also support this concept, especially in the nervous system. For example, in the fetal sheep, Zhu et al. demonstrated that i.c.v. Ang II evoked a significant increase in arterial blood pressure,[16,17] which was completely abolished by the AT_1R antagonist losartan, but not by the AT_2R antagonist, PD123319.[17] These data clearly demonstrated that, in the fetus, Ang II exerts its central action through the AT_1R and not through the AT_2R. Thus, the AT_1R is the predominant functional receptor subtype. On the other hand, in adult rats, we found that i.c.v. infusion of the AT_2 agonist compound 21 (C21)[18] or microinjection of the agonist CGP42114 into the RVLM[11] evoked a hypotensive and sympathoinhibitory response, suggesting the presence of functional AT_2Rs in the brains of mature animals. In adult rats, Smith et al. recently demonstrated AT_2R expression in the lumbar dorsal root ganglia.[19] They further found that EMA300, a new AT_2R antagonist, produced dose-dependent pain relief in the chronic constriction injury rat model of neuropathic pain[19,20] and prostate cancer-induced bone pain.[21] Taken together, the above data suggest functional AT_2Rs in the mature animal. The next question is which organ(s) and what physiological function(s) does the AT_2R exert? After characterizing the relative expression level of AT_2R in various tissues and organs, we recently demonstrated that the central nervous system is one of the organs in adult rats that express the highest levels of AT_2R protein (the other two are the pancreas and testicle),[22] suggesting an important role for this receptor in neural control.

EFFECT OF AT_2R ACTIVATION ON NEURONAL ELECTROPHYSIOLOGICAL CHARACTERISTICS

Current data suggest a profound influence of AT_2R stimulation on neuronal electrophysiological characteristics. Employing whole-cell patch-clamp techniques and primarily cultured neurons from the hypothalamus and brain stem of newborn rats, Kang et al.[23] found that Ang II stimulated both transient K^+ current (I_A) and delayed-rectifier K^+ current (I_k). This effect was completely blocked by PD123177, an AT_2R antagonist, but not by losartan, an AT_1R antagonist, suggesting that AT_2R facilitates neuronal potassium channels. Our recent data also confirmed that AT_2R activation induced an increase in whole-cell K^+ current following treatment with C21 (a nonpeptide AT_2R agonist) in a catecholaminergic (CATH.a) neuronal cell line.[18] Using the cell-attached and inside-out patch configurations of the patch-clamp recording technique, Martens et al.[24] further demonstrated that the effects of AT_2R stimulation on whole-cell K^+ current were mediated by an increased open probability of single 56-pS K^+ channels. It should be emphasized here that opposing influences between AT_2R and AT_1R activations on the regulation of neuronal K^+ channel function have also been demonstrated.[25] Taken together, these data suggest that AT_2R activation increases K^+ current and thereby, we presume, decreases neuronal excitability and activity. Indeed, in a brain slice preparation, Xiong and Marshall[26,27] demonstrated that Ang II, through the AT_2R, depressed glutamate-evoked depolarizations and excitatory postsynaptic potentials of locus coeruleus (LC) neurons. In bulbospinal RVLM neurons of AT_1R knockout mice, Matsuura et al.[28] found hyperpolarization and slower basal spontaneous firing rate as compared with the wild-type mice. They further demonstrated that these AT_1R-null neurons displayed a hyperpolarized membrane potential and decreased discharge frequency in response to Ang II treatment.[28]

The AT_2R belongs to the seven-transmembrane G protein-coupled receptor family, which employs specific intracellular signaling pathways to exert effects. Several key signaling molecules have been identified that mediate the influence of the AT_2R on neuronal electrophysiological characteristics. Even though the AT_2R has been shown to exert G protein-independent effects,[29] the majority of the neuronal AT_2R downstream signaling pathways are mediated by inhibitory G proteins (G_i).[30] In cultured neurons, intracellular application of anti-$G_{i\alpha}$ antibody abolished AT_2R-mediated stimulation of neuronal I_k, whereas the antibodies against other G proteins, such as the $G_{o\alpha}$ and $G_{q/11\alpha}$, displayed no effect.[31] Further studies suggested that the third intracellular loop of AT_2R protein is the key component connecting with $G_{i\alpha}$. Kang et al.[32] reported that intracellular injection of a 22-amino-acid peptide corresponding to this segment elicited a similar increase in I_k as did overexpression of AT_2R. These authors concluded that the third intracellular loop of the AT_2R is held in a position inaccessible to the G$_i$ protein until binding of Ang II causes a conformational change in the receptor, which then permits receptor-G$_i$ coupling. The alteration of this segment would be necessary to allow the AT$_2$R to be continually present in the membrane and to detect changes in the extracellular level of Ang II without continually stimulating G$_i$.[32] Following activation of G$_i$, phospholipase A$_2$ (PLA$_2$), protein phosphatase 2A (PP2A), and arachidonic acid (AA) act as the downstream signaling transduction molecules to mediate the facilitation of AT$_2$R on neuronal K$^+$ channel. Zhu et al.[33] demonstrated that AT$_2$R activation promoted AA release from cultured neurons, which was blocked by inhibition of PLA$_2$. They further indicated that the AT$_2$R-elicited increase in neuronal I_k was attenuated by selective inhibitors of PLA$_2$ and was mimicked by application of AA.[33] Moreover, inhibition of lipooxygenase (LO) significantly reduced both AT$_2$R-induced I_k and AA-induced I_k, whereas the 12-LO metabolite of AA 12S-hydroxyeicosatetraenoic acid (12S-HETE) stimulated I_k.[33] These data implicate the involvement of a PLA2, AA, and LO metabolite as the intracellular pathway in the AT$_2$R-mediated stimulation of neuronal I_k. Finally, they found that inhibition of PP2A abolished the stimulatory effects of AT$_2$R, AA, and 12S-HETE on neuronal I_K but did not alter AT$_2$R-stimulated AA release, suggesting that PP2A is a distal event in this pathway.[33]

AT$_2$R AND SYMPATHETIC REGULATION IN NORMAL STATES

Because of its marked effects on neuronal electrophysiological characteristics, the AT$_2$R likely plays an important role in the regulation of sympathetic outflow and cardiovascular function. It has been reported that mice with the AT$_2$R gene deletion exhibit increased arterial blood pressure as compared with wild-type mice.[34,35] While the precise mechanism for this slight hypertension is not clear, a central mechanism was suggested because these mice exhibited behavioral abnormalities, including reduced exploration, increased fear, and hypoalgesia.[34] Indeed, i.c.v. injection of Ang II evoked a significantly smaller pressor response in wild-type mice as compared with AT$_2$R knockout mice, suggesting a negative regulation of central AT$_2$R on blood pressure.[36] In addition, i.c.v. Ang II-evoked cardiovascular responses were significantly enhanced by AT$_2$R blockade.[37,38] Moreover, in normal mice, i.c.v. injection of Ang II plus PD123319 initiated an exaggerated pressor response than that induced by Ang II alone,[36] again suggesting an antihypertensive effect of central AT$_2$R stimulation, which we assume was due to sympathoinhibition. In a recent study by the authors, we provided direct evidence linking the central AT$_2$R to sympathetic modulation in conscious intact rats. I.c.v. infusion of C21, a nonpeptide AT$_2$R agonist, significantly decreased norepinephrine (NE) excretion.[18] These C21-treated rats also displayed a slight but significant decreased arterial blood pressure with an upregulated nNOS protein expression in PVN and RVLM.[18] In a recent study, we found that gene transfer-induced overexpression of AT$_2$R in the RVLM of normal rats significantly decreased arterial blood pressure and reduced NE excretion.[39] The inhibitory effect of AT$_2$R overexpression in the RVLM on sympathetic nerve activity was further confirmed in anesthetized normal rats in which microinjection of CGP42112 into the RVLM evoked a profound drop of renal sympathetic nerve activity (RSNA) and blood pressure.[11] This effect was completely abolished by PD123319 and partially attenuated by 5,8,11,14-eicosatetraynoic acid (a general inhibitor of AA metabolism).[11] On the other hand, microinjection of PD123319 into the RVLM of normal rats evoked a significant elevation of blood pressure and enhanced Ang II-induced hypertension, suggesting a tonic inhibitory effect of this receptor on sympathetic tone.[40] In addition to the RVLM, the AT$_2$R also mediates sympathoinhibition in the intermediolateral cell column (IML) of the spinal cord. The IML expresses higher AT$_2$R protein than other regions of the spinal cord in normal rats.[41] Consistent with this finding microinjection of CGP42112 into the IML induced a dramatic decrease in RSNA and blood pressure.[41] This effect was abolished by PD123319 and attenuated by L-NAME, suggesting the involvement of AT$_2$R and the NO/nNOS signaling pathway.[41] More importantly, PD123319 itself significantly increased RSNA and blood pressure as well as exaggerated the response to Ang II, implying a tonic sympathoinhibitory effect of AT$_2$R in the IML.[41] The influence of central AT$_2$Rs on sympathetic outflow in intact rats is mediated, in part, by the NO-dependent enhancement of arterial baroreflex function. Abdulla and Johns recently reported that i.c.v. infusion of CGP42112 significantly increased baroreflex sensitivities for RSNA and HR, which was completely abolished by L-NAME.[42]

The involvement of the AT$_2$R in sympathetic regulation was further confirmed in AT$_2$R-deficient mice by direct recording of sympathetic nerve activity. In anesthetized AT$_2$R knockout mice (generously supplied by Dr. Thomas Walther, University College Cork, Ireland; University of Leipzig, Germany), we found a significant increase in basal RSNA and blood pressure as compared with wild-type mice (Figure 1, Panel a). Interestingly, AT$_2$R-deficient mice displayed a complete failure of baroreflex control of RSNA and significantly impaired baroreflex control of heart rate (Figure 1, Panel b). The precise mechanism of sympathetic dysregulation in AT$_2$R-deficient mice is unclear due, in part, to the global nature of this deletion. However, a central mechanism is most likely involved in this phenomenon in that AT$_2$R-deficient mice display exaggerated pressor and sympathoexcitatory responses to i.c.v. Ang II and an elevated ROS production in the NTS (Figure 2).

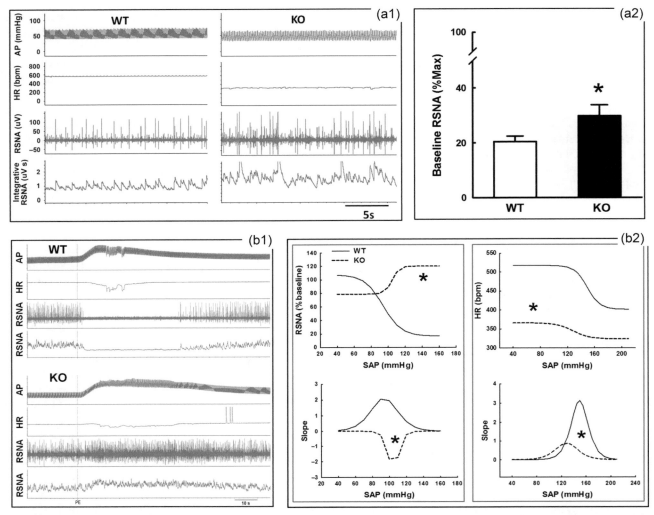

FIGURE 1 Basal renal sympathetic nerve activity (RSNA) and baroreflex function in wild-type and AT₂R knockout mice. (a1 and b1) Representative recordings. (a2) Mean data of basal RSNA. (b2) Mean data of baroreflex sensitivity for RSNA control (left panel) and heart rate control (right panel). *$P < 0.05$ vs. wild-type mice; $n = 7$ per group.

EFFECTS OF THE AT₂R ON SYMPATHETIC OUTFLOW IN PATHOLOGICAL CONDITIONS AND THERAPEUTIC POTENTIAL

Although the pathological significance of central AT₁R activation in cardiovascular disorders has been intensively investigated,[43] the effects of AT₂R stimulation are largely unknown. Biochemical and pharmacological evidence has shown alterations of AT₂R expression in the sympathetic regions of the brain in hypertension and heart failure animal models. For instance, in the cold-induced hypertensive rat model, Peng et al.[44] found that AT₂R mRNA expression in the brain stem was significantly downregulated as compared with normotensive rats. Employing autoradiography, they further demonstrated that AT₂R binding in the LC and inferior olive of these hypertensive rats was significantly lower.[44] In the electric foot-shock hypertensive rat model, Du et al.[40] reported a significant downregulation of AT₂R protein in the RVLM. Downregulated AT₂R protein was also found in the RVLM in rats with coronary ligation-induced heart failure in recent experiments from our laboratory.[11]

Consistent with the change in AT₂R expression, rats with cardiovascular disorders display an impaired AT₂R function in sympathetic nuclei. In hypertensive rats, microinjection of PD123319 into the RVLM evoked a smaller increase in blood pressure as compared with control rats, whereas losartan induced an increase in hypotension in these rats.[40] These data suggest attenuated AT₂R signaling in the RVLM of hypertension rats that may contribute to the pathogenesis of this disease by sympathetic hyperactivity. Direct evidence implicating the AT₂R in the RVLM on sympathetic outflow stemmed

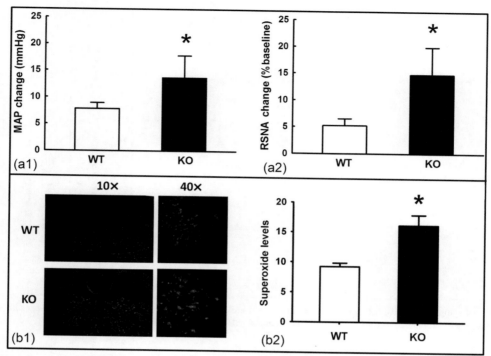

FIGURE 2   I.c.v. Ang II-evoked pressor and sympathoexcitatory responses (panel a) and NTS superoxide production by DHE staining (panel b) in wild-type and AT₂R knockout mice. *$P < 0.05$ vs. wild-type mice; $n = 5$ per group.

from studies in heart failure rats. We found that in normal rats, administration of Ang II plus PD123319 into the RVLM evoked a larger increase in RSNA as compared with Ang II alone, suggesting a tonic inhibition of AT₂R on sympathetic tone.[11] In heart failure rats, however, we could not detect a significant difference in the sympathetic response to Ang II plus PD123319 and Ang II alone, suggesting a loss of sympathetic inhibition by the AT₂R in the RVLM.[11] Indeed, microinjection of CGP42112 into the RVLM of heart failure rats did not evoke a significant change in RSNA, whereas this application in normal rats markedly inhibited sympathetic outflow.[11] Alterations in AT₂R expression and function in the RVLM of heart failure rats, we believe, contribute to sympathoexcitation. Pharmacological activation of central AT₂Rs could therefore be a therapeutic option for diseases characterized by sympathoexcitation, such as neurogenic hypertension and heart failure. Indeed, a recent study from this laboratory demonstrated that, in rats with heart failure i.c.v. infusion of C21 significantly suppressed RSNA, decreased urinary NE excretion and improved spontaneously baroreflex gain in the conscious state and enhanced induced baroreflex control of sympathetic nerve activity and heart rate under anesthesia.[12] C21-treated heart failure rats exhibited a downregulated AT₁R expression and upregulated nNOS expression in the PVN and RVLM, suggesting a contribution of suppressed AT₁R signaling and activation of the nNOS-NO pathway in response to central AT₂R activation.[12] Restoration of AT₂R expression in the RVLM of heart failure rats by gene transfer also ameliorated sympathoexcitation. Microinjection of an AT₂R viral vector (Ad5-SYN-AT₂R-IRES-EGFP, generously supplied by Dr. Colin Sumners, University of Florida) into the RVLM of rats upregulated AT₂R expression in this nucleus.[39] As can be seen in Figure 3, AT₂R overexpression in the RVLM of heart failure rats significantly suppressed basal RSNA, whereas this effect was not observed in normal rats (sham). Interestingly, in normal rats, microinjection of Ang II into the RVLM following gene transfer-induced AT₂R overexpression evoked a sympathoinhibition, rather than sympathoexcitation as usually observed. In heart failure rats following AT₂R overexpression, Ang II-evoked sympathoexcitation was significantly attenuated, but not reversed (Figure 4). These data suggest that, in the normal state, gene transfer-induced AT₂R overexpression may reverse the ratio of AT₁R to AT₂R expression in the RVLM and the latter becomes the predominant receptor subtype. In this case, Ang II evoked an AT₂R-mediated sympathoinhibition. In heart failure rats, gene transfer-induced AT₂R overexpression only reduced the ratio of AT₁R to AT₂R due to the high endogenous AT₁R expression, and accordingly, the Ang II-evoked sympathoexcitation was attenuated. Taken together, these data suggest that activation or upregulation of AT₂Rs in sympathetic nuclei has therapeutic potential for cardiovascular diseases that are characterized by sympathoexcitation. Figure 5[45] provides a schematic overview of a potential therapeutic role of overexpression of the AT₂R in the modulation of sympathetic outflow in disease states.

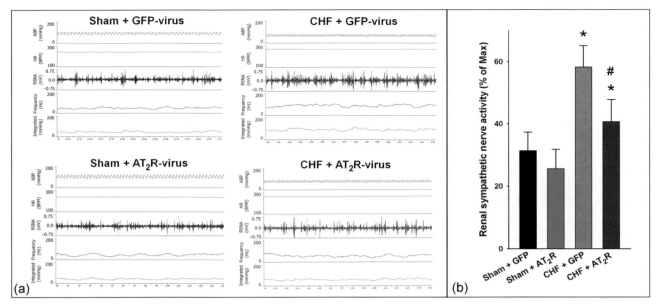

FIGURE 3 Basal renal sympathetic nerve activity (RSNA) of rats with overexpression of AT_2R in the rostral ventrolateral medulla (RVLM) by gene transfer. (a) Representative recording; (b) mean data. $*P < 0.05$ vs. sham; $^\#P < 0.05$ vs. CHF + GFP virus. $n = 5$ per group.

CONCLUSION AND PERSPECTIVES

The discussion above suggests that the AT_2R plays an important role in the regulation of sympathetic outflow by modulating neuronal activity in areas of the medulla and hypothalamus that control sympathetic circuitry. This is especially true in disease states such as CHF and hypertension. As indicated in "Introduction," membrane potential and discharge sensitivity of presympathetic neurons are determined by multiple factors including the magnitude and types of neurotransmitter release from presynaptic terminals and expression of various excitatory and inhibitory membrane receptors and their downstream signaling pathways. Ultimately, these factors must translate into changes in ion conductance at the membrane level. Because activation of the AT_1R in neurons has been clearly shown to reduce K^+ currents and thereby increase discharge sensitivity,[46–49] it is notable that activation of the AT_2R does the opposite (i.e., increases K^+ currents) or has no effect.[23,31,50] Therefore, all things being equal, the balance between the AT_1 and the AT_2 receptors is a critical factor in determining the influence of central Ang II on sympathetic outflow. The data summarized in this review clearly point to abnormal Ang II receptor expression playing a key role in the modulation of K^+ currents and neuronal activity. AT_1R is increased, while AT_2R is decreased in the heart failure state[11] tipping the balance between sympathoexcitation and sympathoinhibition. Interestingly, this observation has also been made in peripheral tissues from animals with CHF.[51] The regulation of AT_1R expression is rather well understood.[52–55] However, the mechanism by which the AT_2R is controlled is less clear. In order to maximize signaling through this "good" receptor, we must completely understand its regulation. Furthermore, it appears that the downstream signaling pathway is, in part, dependent on activation of nNOS and NO.[42,56,57] Given that NO is a sympathoinhibitory substance and can attenuate glutamate signaling[58–60] in the central nervous system, the sympathoinhibition of AT_2R activation may operate through this pathway.

It should be kept in mind that the central RAS is not just about AT_1 and AT_2R signaling, but other components also play a role. The relative activities of ACE and ACE2 also influence sympathetic outflow in disease states through the modulation of the balance between Ang II and Ang 1-7,[61–65] the latter also operating through an NO-dependent pathway via the Mas receptor.[66–68] Finally, new data implicate prorenin and the prorenin receptor in the modulation of sympathetic outflow in disease states.[69,70]

The use of novel small molecule activators of the AT_2R such as C21 has the potential to provide protection against the deleterious effects of Ang II on the AT_1R. Given the fact that AT_1R protein seems to be more abundant than AT_2R protein under any condition, it is important to amplify the AT_2R signal using these new pharmacological entities. Future work needs to be targeted at enhancing the specificity and delivery of AT_2R agonists to the sites where they are needed in cardiovascular disease.

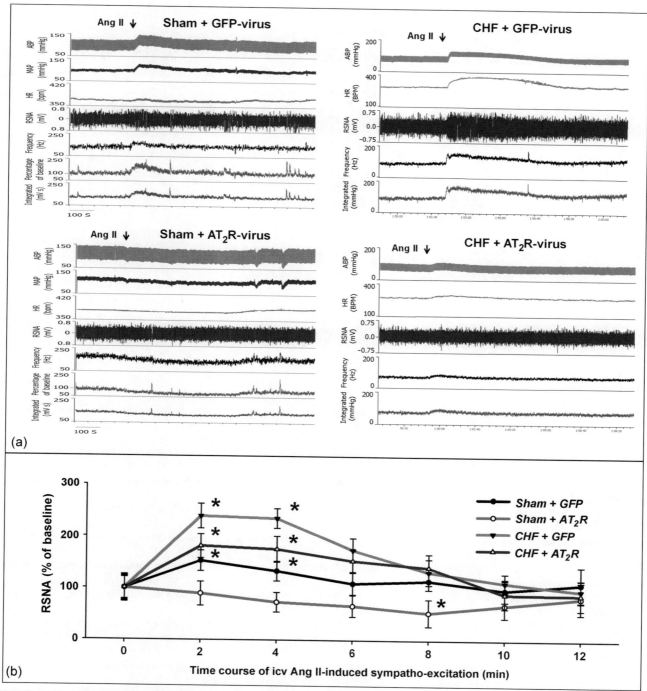

FIGURE 4   I.c.v. Ang II-evoked blood pressure and sympathetic responses in the rats with overexpression of AT$_2$R in the rostral ventrolateral medulla (RVLM) by gene transfer. (a) Representative recording; (b) mean data. *$P < 0.05$ vs. baseline nerve activity. $n = 5$ per group.

ACKNOWLEDGMENTS

Some of the experiments described in this review were supported by grants PO-1 HL62222 (Dr. Irving H. Zucker) and RO-1 HL093028 (Dr. Lie Gao) from the National Institutes of Health. The authors thank Dr. Colin Sumners (University of Florida, the United States) for his generous donation of AT$_2$R adenoviral vector. The AT$_2$R knockout mice were kindly supplied by Dr. Thomas Walther (University College Cork, Ireland; University of Leipzig, Germany). The authors appreciate the critical support from Dr. Walther.

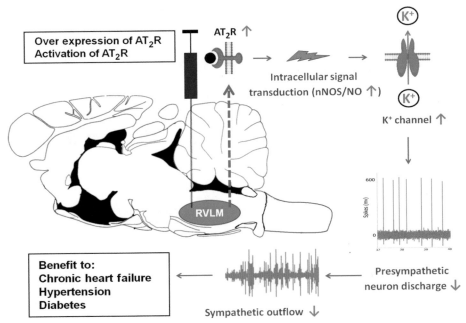

FIGURE 5 Outline of therapeutic strategy by which activation of overexpressed central AT_2Rs reduces sympathoexcitation in cardiovascular diseases such as heart failure and hypertension. RVLM: rostral ventrolateral medulla. *From Ref. [45], with permission.*

REFERENCES

1. Barman SM, Gebber GL. Role of ventrolateral medulla in generating the 10-Hz rhythm in sympathetic nerve discharge. *Am J Physiol Regul Integr Comp Physiol* 2007;**293**:R223–33.
2. Zucker IH. Novel mechanisms of sympathetic regulation in chronic heart failure. *Hypertension* 2006;**48**:1005–11.
3. Zucker IH, Schultz HD, Patel KP, Wang W, Gao L. Regulation of central angiotensin type 1 receptors and sympathetic outflow in heart failure. *Am J Physiol Heart Circ Physiol* 2009;**297**:H1557–66.
4. Reid IA. Interactions between ANG II, sympathetic nervous system, and baroreceptor reflexes in regulation of blood pressure. *Am J Physiol* 1992;**262**:E763–78.
5. Grobe JL, Xu D, Sigmund CD. An intracellular renin-angiotensin system in neurons: fact, hypothesis, or fantasy. *Physiology (Bethesda)* 2008;**23**:187–93.
6. Li W, Peng H, Mehaffey EP, Kimball CD, Grobe JL, van Gool JM, et al. Neuron-specific (pro)renin receptor knockout prevents the development of salt-sensitive hypertension. *Hypertension* 2014;**63**:316–23.
7. Li W, Peng H, Cao T, Sato R, McDaniels SJ, Kobori H, et al. Brain-targeted (pro)renin receptor knockdown attenuates angiotensin II-dependent hypertension. *Hypertension* 2012;**59**:1188–94.
8. Feng Y, Xia H, Cai Y, Halabi CM, Becker LK, Santos RA, et al. Brain-selective overexpression of human angiotensin-converting enzyme type 2 attenuates neurogenic hypertension. *Circ Res* 2010;**106**:373–82.
9. Grady EF, Sechi LA, Griffin CA, Schambelan M, Kalinyak JE. Expression of AT2 receptors in the developing rat fetus. *J Clin Invest* 1991;**88**:921–33.
10. Yu L, Zheng M, Wang W, Rozanski GJ, Zucker IH, Gao L. Developmental changes in AT1 and AT2 receptor-protein expression in rats. *J Renin Angiotensin Aldosterone Syst* 2010;**11**:214–21.
11. Gao L, Wang WZ, Wang W, Zucker IH. Imbalance of angiotensin type 1 receptor and angiotensin II type 2 receptor in the rostral ventrolateral medulla: potential mechanism for sympathetic overactivity in heart failure. *Hypertension* 2008;**52**:708–14.
12. Gao J, Zucker IH, Gao L. Activation of central angiotensin type 2 receptors by compound 21 improves arterial baroreflex sensitivity in rats with heart failure. *Am J Hypertens* 2014;**27**(10):1248–56, PMID: 24687998.
13. de GM, Catt KJ, Inagami T, Wright JW, Unger T. International union of pharmacology. XXIII. The angiotensin II receptors. *Pharmacol Rev* 2000;**52**:415–72.
14. Gao J, Chao J, Parbhu KJ, Yu L, Xiao L, Gao F, et al. Ontogeny of angiotensin type 2 and type 1 receptor expression in mice. *J Renin Angiotensin Aldosterone Syst* 2012;**13**:341–52.
15. Yu L, Shao C, Gao L. Developmental expression patterns for angiotensin receptors in mouse skin and brain. *J Renin Angiotensin Aldosterone Syst* 2012;**15**(2):139–49, (in press, PMID:23204186).
16. Xu Z, Shi L, Hu F, White R, Stewart L, Yao J. In utero development of central ANG-stimulated pressor response and hypothalamic fos expression. *Brain Res Dev Brain Res* 2003;**145**:169–76.

17. Shi L, Mao C, Thornton SN, Sun W, Wu J, Yao J, et al. Effects of intracerebroventricular losartan on angiotensin II-mediated pressor responses and c-fos expression in near-term ovine fetus. *J Comp Neurol* 2005;**493**:571–9.

18. Gao J, Zhang H, Le KD, Chao J, Gao L. Activation of central angiotensin type 2 receptors suppresses norepinephrine excretion and blood pressure in conscious rats. *Am J Hypertens* 2011;**24**:724–30.

19. Smith MT, Woodruff TM, Wyse BD, Muralidharan A, Walther T. A small molecule angiotensin II type 2 receptor (AT R) antagonist produces analgesia in a rat model of neuropathic pain by inhibition of p38 mitogen-activated protein kinase (MAPK) and p44/p42 MAPK activation in the dorsal root ganglia. *Pain Med* 2013;**14**:1557–68.

20. Smith MT, Wyse BD, Edwards SR. Small molecule angiotensin II type 2 receptor (AT(2)R) antagonists as novel analgesics for neuropathic pain: comparative pharmacokinetics, radioligand binding, and efficacy in rats. *Pain Med* 2013;**14**:692–705.

21. Muralidharan A, Wyse BD, Smith MT. Analgesic efficacy and mode of action of a selective small molecule angiotensin II type 2 receptor antagonist in a rat model of prostate cancer-induced bone pain. *Pain Med* 2014;**15**:93–110.

22. Shao C, Zucker IH, Gao L. Angiotensin type 2 receptor in pancreatic islets of adult rats: a novel insulinotropic mediator. *Am J Physiol Endocrinol Metab* 2013;**305**:E1281–91.

23. Kang J, Sumners C, Posner P. Angiotensin II type 2 receptor-modulated changes in potassium currents in cultured neurons. *Am J Physiol* 1993;**265**:C607–16.

24. Martens JR, Wang D, Sumners C, Posner P, Gelband CH. Angiotensin II type 2 receptor-mediated regulation of rat neuronal K+ channels. *Circ Res* 1996;**79**:302–9.

25. Gelband CH, Zhu M, Lu D, Reagan LP, Fluharty SJ, Posner P, et al. Functional interactions between neuronal AT1 and AT2 receptors. *Endocrinology* 1997;**138**:2195–8.

26. Xiong H, Marshall KC. Angiotensin II depresses glutamate depolarizations and excitatory postsynaptic potentials in locus coeruleus through angiotensin II subtype 2 receptors. *Neuroscience* 1994;**62**:163–75.

27. Xiong HG, Marshall KC. Angiotensin II modulation of glutamate excitation of locus coeruleus neurons. *Neurosci Lett* 1990;**118**:261–4.

28. Matsuura T, Kumagai H, Onimaru H, Kawai A, Iigaya K, Onami T, et al. Electrophysiological properties of rostral ventrolateral medulla neurons in angiotensin II 1a receptor knockout mice. *Hypertension* 2005;**46**:349–54.

29. Gallinat S, Busche S, Raizada MK, Sumners C. The angiotensin II type 2 receptor: an enigma with multiple variations. *Am J Physiol Endocrinol Metab* 2000;**278**:E357–74.

30. Gelband CH, Sumners C, Lu D, Raizada MK. Angiotensin receptors and norepinephrine neuromodulation: implications of functional coupling. *Regul Pept* 1998;**73**:141–7.

31. Kang J, Posner P, Sumners C. Angiotensin II type 2 receptor stimulation of neuronal K+ currents involves an inhibitory GTP binding protein. *Am J Physiol* 1994;**267**:C1389–97.

32. Kang J, Richards EM, Posner P, Sumners C. Modulation of the delayed rectifier K+ current in neurons by an angiotensin II type 2 receptor fragment. *Am J Physiol* 1995;**268**:C278–82.

33. Zhu M, Gelband CH, Moore JM, Posner P, Sumners C. Angiotensin II type 2 receptor stimulation of neuronal delayed-rectifier potassium current involves phospholipase A2 and arachidonic acid. *J Neurosci* 1998;**18**:679–86.

34. Ichiki T, Labosky PA, Shiota C, Okuyama S, Imagawa Y, Fogo A, et al. Effects on blood pressure and exploratory behaviour of mice lacking angiotensin II type-2 receptor. *Nature* 1995;**377**:748–50.

35. Siragy HM, Inagami T, Ichiki T, Carey RM. Sustained hypersensitivity to angiotensin II and its mechanism in mice lacking the subtype-2 (AT2) angiotensin receptor. *Proc Natl Acad Sci U S A* 1999;**96**:6506–10.

36. Li Z, Iwai M, Wu L, Shiuchi T, Jinno T, Cui TX, et al. Role of AT2 receptor in the brain in regulation of blood pressure and water intake. *Am J Physiol Heart Circ Physiol* 2003;**284**:H116–21.

37. Widdop RE, Gardiner SM, Bennett T. Effects of angiotensin II AT1- or AT2-receptor antagonists on drinking evoked by angiotensin II or water deprivation in rats. *Brain Res* 1994;**648**:46–52.

38. Toney GM, Porter JP. Functional roles of brain AT1 and AT2 receptors in the central angiotensin II pressor response in conscious young spontaneously hypertensive rats. *Brain Res Dev Brain Res* 1993;**71**:193–9.

39. Gao L, Wang W, Wang W, Li H, Sumners C, Zucker IH. Effects of angiotensin type 2 receptor overexpression in the rostral ventrolateral medulla on blood pressure and urine excretion in normal rats. *Hypertension* 2008;**51**:521–7.

40. Du D, Chen J, Liu M, Zhu M, Jing H, Fang J, et al. The effects of angiotensin II and angiotensin-(1–7) in the rostral ventrolateral medulla of rats on stress-induced hypertension. *PLoS One* 2013;**8**:e70976.

41. Chao J, Gao J, Parbhu KJ, Gao L. Angiotensin type 2 receptors in the intermediolateral cell column of the spinal cord: negative regulation of sympathetic nerve activity and blood pressure. *Int J Cardiol* 2013;**168**:4046–55.

42. Abdulla MH, Johns EJ. Nitric oxide impacts on angiotensin AT2 receptor modulation of high-pressure baroreflex control of renal sympathetic nerve activity in anaesthetized rats. *Acta Physiol (Oxf)* 2014;**210**:832–44.

43. Veerasingham SJ, Raizada MK. Brain renin-angiotensin system dysfunction in hypertension: recent advances and perspectives. *Br J Pharmacol* 2003;**139**:191–202.

44. Peng JF, Phillips MI. Opposite regulation of brain angiotensin type 1 and type 2 receptors in cold-induced hypertension. *Regul Pept* 2001;**97**:91–102.

45. Gao L, Zucker IH. AT2 receptor signaling and sympathetic regulation. *Curr Opin Pharmacol* 2011;**11**:124–30.

46. Gao L, Li Y, Schultz HD, Wang WZ, Wang W, Finch M, et al. Downregulated Kv4.3 expression in the RVLM as a potential mechanism for sympathoexcitation in rats with chronic heart failure. *Am J Physiol Heart Circ Physiol* 2010;**298**:H945–55.

47. Su F, Shi M, Yan Z, Ou D, Li J, Lu Z, et al. Simvastatin modulates remodeling of Kv4.3 expression in rat hypertrophied cardiomyocytes. *Int J Biol Sci* 2012;**8**:236–48.

48. Jiang N, Shi P, Desland F, Kitchen-Pareja MC, Sumners C. Interleukin-10 inhibits angiotensin II-induced decrease in neuronal potassium current. *Am J Physiol Cell Physiol* 2013;**304**:C801–7.

49. Moreira TH, Cruz JS, Weinreich D. Angiotensin II increases excitability and inhibits a transient potassium current in vagal primary sensory neurons. *Neuropeptides* 2009;**43**:193–9.

50. Sonoyama K, Ninomiya H, Igawa O, Kaetsu Y, Furuse Y, Hamada T, et al. Inhibition of inward rectifier K+ currents by angiotensin II in rat atrial myocytes: lack of effects in cells from spontaneously hypertensive rats. *Hypertens Res* 2006;**29**:923–34.

51. Clayton SC, Haack KK, Zucker IH. Renal denervation modulates angiotensin receptor expression in the renal cortex of rabbits with chronic heart failure. *Am J Physiol Renal Physiol* 2011;**300**:F31–9.

52. Mitra AK, Gao L, Zucker IH. Angiotensin II-induced upregulation of AT(1) receptor expression: sequential activation of NF-kappaB and Elk-1 in neurons. *Am J Physiol Cell Physiol* 2010;**299**:C561–9.

53. Zhang H, Sun GY. Expression and regulation of AT1 receptor in rat lung microvascular endothelial cell. *J Surg Res* 2006;**134**:190–7.

54. Haack KK, Mitra AK, Zucker IH. NF-kappaB and CREB are required for angiotensin II type 1 receptor upregulation in neurons. *PLoS One* 2013;**8**:e78695.

55. Haack KK, Engler CW, Papoutsi E, Pipinos II, Patel KP, Zucker IH. Parallel changes in neuronal AT1R and GRK5 expression following exercise training in heart failure. *Hypertension* 2012;**60**:354–61.

56. Wainford RD. Angiotensin AT2 receptors and the baroreflex control of renal sympathetic nerve activity. *Acta Physiol (Oxf)* 2014;**210**:714–6.

57. Ratliff BB, Sekulic M, Rodebaugh J, Solhaug MJ. Angiotensin II regulates NOS expression in afferent arterioles of the developing porcine kidney. *Pediatr Res* 2010;**68**:29–34.

58. Jendzjowsky NG, DeLorey DS. Role of neuronal nitric oxide in the inhibition of sympathetic vasoconstriction in resting and contracting skeletal muscle of healthy rats. *J Appl Physiol (1985)* 2013;**115**:97–106.

59. Patel KP, Schultz HD. Angiotensin peptides and nitric oxide in cardiovascular disease. *Antioxid Redox Signal* 2013;**19**:1121–32.

60. Zheng H, Liu X, Li Y, Sharma NM, Patel KP. Gene transfer of neuronal nitric oxide synthase to the paraventricular nucleus reduces the enhanced glutamatergic tone in rats with chronic heart failure. *Hypertension* 2011;**58**:966–73.

61. Zucker IH, Xiao L, Haack KK. The central renin-angiotensin system and sympathetic nerve activity in chronic heart failure. *Clin Sci (Lond)* 2014;**126**:695–706.

62. Sun HJ, Li P, Chen WW, Xiong XQ, Han Y. Angiotensin II and angiotensin-(1–7) in paraventricular nucleus modulate cardiac sympathetic afferent reflex in renovascular hypertensive rats. *PLoS One* 2012;**7**:e52557.

63. Han Y, Sun HJ, Li P, Gao Q, Zhou YB, Zhang F, et al. Angiotensin-(1–7) in paraventricular nucleus modulates sympathetic activity and cardiac sympathetic afferent reflex in renovascular hypertensive rats. *PLoS One* 2012;**7**:e48966.

64. Guimaraes PS, Santiago NM, Xavier CH, Velloso EP, Fontes MA, Santos RA, et al. Chronic infusion of angiotensin-(1–7) into the lateral ventricle of the brain attenuates hypertension in DOCA-salt rats. *Am J Physiol Heart Circ Physiol* 2012;**303**:H393–400.

65. Xiao L, Gao L, Lazartigues E, Zucker IH. Brain-selective overexpression of angiotensin-converting enzyme 2 attenuates sympathetic nerve activity and enhances baroreflex function in chronic heart failure. *Hypertension* 2011;**58**:1057–65.

66. Chen J, Zhao Y, Chen S, Wang J, Xiao X, Ma X, et al. Neuronal over-expression of ACE2 protects brain from ischemia-induced damage. *Neuropharmacology* 2014;**79**:550–8.

67. Costa A, Galdino G, Romero T, Silva G, Cortes S, Santos R, et al. Ang-(1–7) activates the NO/cGMP and ATP-sensitive K(+) channels pathway to induce peripheral antinociception in rats. *Nitric Oxide* 2014;**37**:11–6.

68. Li P, Chappell MC, Ferrario CM, Brosnihan KB. Angiotensin-(1–7) augments bradykinin-induced vasodilation by competing with ACE and releasing nitric oxide. *Hypertension* 1997;**29**:394–400.

69. Peng H, Li W, Seth DM, Nair AR, Francis J, Feng Y. (Pro)renin receptor mediates both angiotensin II-dependent and -independent oxidative stress in neuronal cells. *PLoS One* 2013;**8**:e58339.

70. Li W, Peng H, Seth DM, Feng Y. The prorenin and (pro)renin receptor: new players in the brain renin-angiotensin system? *Int J Hypertens* 2012;**2012**:290635.

Chapter 16

Metabolic Effects of AT$_2$R Stimulation in Adipose Tissue

Anna Foryst-Ludwig, Ulrich Kintscher

Center for Cardiovascular Research (CCR), Institute of Pharmacology, Charité-Universitätsmedizin Berlin, Berlin, Germany

INTRODUCTION

The obesity and obesity-related metabolic disorders, such as type 2 diabetes mellitus, dyslipidemia, and hypertension, collectively known as metabolic syndrome, are regarded as a global health dilemma.[1] Recently, the World Health Organization pointed out that overweight and obesity are among the fifth principal risk factors for worldwide morbidity and mortality (https://apps.who.int/infobase). The continuous imbalance between energy intake and energy expenditure, considered as the main etiologic factor of the obesity,[2] results in the uptake of fatty acids (FAs) and their accumulation in the form of triglycerides (TG) in adipose tissue, leading to the hypertrophic response of adipocytes. FAs could be released from the fat tissue upon demand in the process of lipolysis and subsequently utilized by beta-oxidation.[3,4] Attenuated potential of adipose tissue to store TG in adipocytes leads to the ectopic accumulation of lipids in other metabolically important organs and tissues, such as the liver, skeletal muscle, and heart, known as lipotoxicity.[5] The main functional role of the adipose tissue relies not only on its preferential uptake of FFA (free fatty acid), stored in form of TGs. Adipose tissue is also considered as an endocrine organ controlling the metabolic status of the whole organism mostly due to its ability to release in a paracrine and systemic manner sets of cytokines (MCP1, IL-6, and TNF-alpha) and adipocytokines (adiponectin, leptin, resistin, RBP4, and many others). As adipose tissue consists not only of adipocytes but also of preadipocytes, endothelial cells, fibroblasts, and immune cells, the obesity-related metabolic disorders not only affect adipocyte hypertrophy, as discussed above, but also are linked with chronic adipose tissue inflammation.[6,7]

The renin–angiotensin–aldosterone system (RAAS) has been shown to mediate distinct pathophysiological effects related to obesity and adipose tissue. All components of the RAAS, including both angiotensin receptors AT$_1$R and AT$_2$R, are expressed in fat tissue.[8–10] Moreover, angiotensin II (Ang II) plasma level positively correlates with the progression of obesity (reviewed by Goossens et al.[11]). Therapeutic modulation of the RAAS activity in adipose tissue could potentially improve the metabolic outcome of obese subjects.

METABOLIC FUNCTION OF AT$_2$R: THE LESSONS FROM ARBs

AT$_1$R mediates most of the deleterious effects of Ang II in adipose tissue, such as vasoconstriction, endothelial damage, and inhibition of adipocyte differentiation and adipogenesis.[12] The decreased rate of the adipogenesis in adipose tissue was recently linked with the reduced storage capacity of adipocytes leading to systemic lipotoxicity. Importantly, pharmacological blockade of the AT$_1$R with AT$_1$R blockers (ARBs), commonly used for the treatment of hypertension and related end-organ damage, augments AT$_2$-mediated beneficial effects of RAAS signaling in fat tissue. Some of the ARBs, when applied to animals fed a high-fat diet (HFD), were able to improve metabolic characteristic of those animals. With regard to adipose tissue metabolism, ARBs were reported to improve insulin signaling and restore insulin sensitivity, decrease sympathetic activity, and induce beneficial adipose tissue remodeling (reviewed by Kintscher et al.[13]). Importantly, beneficial metabolic actions of ARBs are also linked with their putative anti-inflammatory actions in adipose tissue, significantly reducing fat-tissue-derived systemic and local inflammation. Although the molecular mechanism underlying those anti-inflammatory effects is not completely understood, inhibition of the proinflammatory pathway in adipose tissue was directly linked with the increase in systemic glucose tolerance and insulin sensitivity and led to augmented insulin-mediated glucose uptake in skeletal muscle and adipocytes.

The Protective Arm of the Renin–Angiotensin System (RAS). http://dx.doi.org/10.1016/B978-0-12-801364-9.00016-X

Some of the beneficial metabolic effects of the ARBs, as in the case of telmisartan, losartan metabolite EXP3179, and irbesartan, are linked rather with their ability to activate peroxisome proliferator-activated receptor gamma (PPAR-gamma) in adipose tissue[14–16] or PPAR-alpha in the liver.[17] PPAR-gamma is regarded as a master regulator of the lipid and glucose homeostasis, and its activation leads not only to increased adipogenesis and elevated insulin sensitivity in white adipose tissue (WAT), skeletal muscle, and the liver but also—due to the interaction between PPAR-gamma signaling and NF-kappaB signaling—to the inhibition of proinflammatory pathways and attenuation of local proinflammatory reactions in WAT. A recent study performed by Zidek and colleagues indicated that most of the beneficial metabolic effects of telmisartan are directly linked with the activation of PPAR-gamma in adipose tissue,[18] as the adipose tissue-specific PPAR-gamma-deficient mice showed impaired metabolic benefit of the telmisartan treatment in comparison to controlled mice, when fed with HFD. Also, the data from Rong et al. indicated that telmisartan when applied to the AT_1R-deficient mice still provides some metabolic benefit in HFD-feeding experiments.[19]

Importantly, only a few of the ARBs are able to induce PPAR-gamma activation *in vitro*, and the molecular mechanism underlying the beneficial metabolic effects of this group of hypertensive drugs remains elusive. For example, valsartan is regarded as an ARB with very low PPAR-gamma activating potential.[20] However, when applied to diabetic KKAy mice, valsartan was demonstrated to attenuate hyperglycemia and hyperinsulinemia and improve glucose tolerance in those animals.[15] The beneficial metabolic action of valsartan relied upon enhanced insulin-mediated glucose uptake and increased activation of the insulin receptor/PI3K/Glut4 pathway in skeletal muscle of those mice. These data indicate that ARBs exert antidiabetogenic actions likely independent from PPAR-gamma via classical RAAS inhibition.

In summary, inhibition of AT_1R in obesity leads to beneficial metabolic effects. Some of those effects in a distinct ARB subgroup seem to rely on ARB-dependent activation of PPAR-gamma.

METABOLIC FUNCTION OF AT_2R: THE LESSONS FROM AT_2R-DEFICIENT MICE

Metabolic phenotyping of AT_2R-deficient mice pointed toward a metabolic function of this receptor *in vivo*, although available results seemed to be somehow controversial.[21,22] Experiments performed by Shiuchi and collaborators[15] demonstrated that AT_2R deficiency in mice strongly abrogated insulin-mediated glucose uptake in WAT, when compared to WT littermate animals. Intriguingly, the glucose uptake measured in the soleus muscle (skeletal muscle) of those mice was not altered by AT_2R deficiency, which can be explained by a very low expression level of AT_2R in this tissue.[15] In addition, AT_2R-KO mice showed a lower locomotor activity, when compared to WT littermate animals.[21] In contrast, metabolic phenotyping experiments of AT_2R-deficient mice performed by Yvan-Charvet and collaborators[23] indicated that those animals are characterized by adipose tissue hypertrophy and hyperplasia and decreased Ang II-dependent glucose and FA uptake measured in adipocytes. Surprisingly—when fed with HFD—those mice seemed to be protected against diet-induced obesity, mostly due to the preserved insulin sensitivity, when compared to HFD-fed WT mice.

In summary, the existent discrepancies from previous studies relied most likely on differences in the genetic background of AT_2R-deficient mice[21,22] and in applied diet and experimental protocols. Overall, the published results on the metabolic phenotype of global AT_2R-deficient animals seem to be difficult to interpret. There is a growing need to generate tissue-specific AT_2R-KO animals, which allow the better distinction among metabolic functions of the AT_2R in different metabolically important tissues and organs.

METABOLIC FUNCTION OF AT_2R: EXPERIMENTS WITH DIRECT AT_2R STIMULATION

The metabolic contribution of AT_2R was postulated previously. Importantly, major parts of earlier studies on the metabolic role of AT_2R using pharmacological interventions are complex to interpret, since specific and selective AT_2R agonists were not available. Due to the lack of specific and selective AT_2R agonists, most of the published data are based on complex or indirect experimental approaches such as stimulation by Ang II under concomitant AT_1R blockade and use of AT_2R antagonists or AT_2R-deficient mice.[24] Also, the commonly used AT_2R antagonist PD123319 lacks selectivity when applied *in vivo* in higher doses.[24]

Novel studies on the metabolic AT_2R function *in vivo* using a direct AT_2R agonist applied to KKAy mice were published recently by Ohshima and colleagues.[25] In this study, KKAy mice display moderate obesity, hyperphagia, hyperglycemia, and glucose intolerance. The mice were treated with the first nonpeptide AT_2R agonist compound 21 (C21).[26] Importantly, C21 applied to these mice over a 2-week period was able to reduce body weight and fat mass, to improve systemic glucose tolerance and insulin sensitivity, and to reduce plasma levels of the proinflammatory cytokines, such as IL-6, TNF-alpha, or MCP1. In addition to the beneficial metabolic effects of C21 on adipose tissue, application of C21 to those animals improved pancreatic beta-cell function and increased beta-cell number, when compared to control, untreated animals.

The most striking result from this study was the observation with regard to inflammation and oxidative stress. When applied to KKAy mice, C21 was able to markedly reduce oxidative stress and local/systemic inflammation of those animals. Interestingly, the beneficial effects were linked to enhanced adipogenesis resulting from the increased PPAR-gamma activity in adipose tissue of those mice. All experiments were controlled for PPAR-gamma specificity by simultaneous treatment of C21 and PPAR-gamma-specific inhibitor GW9662. Application of the inhibitory factor GW9662 completely abolished the beneficial metabolic effects of C21, supporting the idea that direct AT$_2$R stimulation ameliorates the metabolic phenotype of KKAy mice in a PPAR-gamma-dependent manner.

The second piece of evidence for C21-mediated metabolic actions comes from the study performed by Shum and colleagues on high-fat/high-fructose (HF/HF)-fed rats, supplemented with C21, losartan, or vehicle-treated, control animals.[27] In line with the results discussed previously,[25] C21, when applied to the animals, was able to improve insulin sensitivity and glucose tolerance, in a similar manner to the metabolic effects mediated by losartan treatment. Interestingly, C21—but not losartan—was able to reduce the plasma TG level, when compared to vehicle-treated control animals. In line with these results, the insulin-mediated inhibition of lipolysis, measured as the non-esterified fatty acids (NEFA) and TG release during clamp studies, was observed exclusively in C21-supplemented animals. In addition to its systemic effects, C21—when applied to the HF/HF-fed animals—showed surprising effects on adipose tissue metabolism. Stimulation of AT$_2$R with C21 *in vivo* promoted differentiation of preadipocytes, improved lipid storage capacity in subcutaneous adipose tissue, reduced the size of adipocytes, and improved overall fat-tissue-specific insulin sensitivity. The study was complemented by the elegant set of *in vitro* experiments performed on primary adipocytes, isolated from subcutaneous and gonadal adipose tissue. Those experiments indicated that AT$_2$R is necessary for the initial steps of adipogenesis, as the differentiation rate of AT$_2$R-deficient primary adipocytes was strongly attenuated, when compared to control WT cells. In line with the reduced adipogenesis, the adipose expression and activity of PPAR-gamma were strongly reduced in AT$_2$R-deficient cells.

The study published by Sabuhi and colleagues was performed in obese Zucker rats, supplemented with the AT$_2$R agonist CGP-42112A.[28] In line with the data discussed above,[25] direct AT$_2$R stimulation mediated by CGP-42112A decreased systemic inflammation and oxidative stress after a 2-week application time.

Besides its putative metabolic actions, specific activation of the AT$_2$R was recently shown to mediate a potent anti-inflammatory response *in vitro* and *in vivo*.[29–33] The studies performed on human fibroblasts treated with C21 indicated that the AT$_2$R-derived anti-inflammatory action relied on the inhibitory effect of AT$_2$R stimulation on NF-kappaB signaling mediated by augmented production of intracellular 11,12-epoxyeicosatrienoic acid (11,12EET).[33] The epoxygenase-EET pathway was recently demonstrated to regulate metabolic homeostasis, insulin sensitivity, and adiponectin production *in vivo*.[34]

In summary, direct AT$_2$R stimulation seems to exert beneficial antidiabetogenic effects observed *in vitro* and in different animal models (Figure 1). AT$_2$R activation induces a potent anti-inflammatory action in adipose tissue, presumably due to

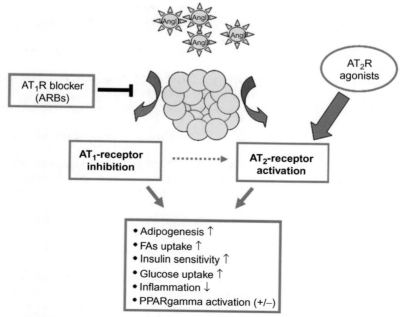

FIGURE 1   Metabolic function of AT$_2$R in adipose tissue. Ang II, angiotensin II.

the direct or indirect local PPAR-gamma regulation. The anti-inflammatory properties of direct AT_2R stimulation may be essential mechanisms for its beneficial metabolic actions, because inflammation in adipose tissue contributes to the pathogenesis of the obesity. From that point of view, AT_2R may be an attractive pharmacological target for the development of novel adipose-tissue-specific antidiabetic agents, with a potential to improve obesity-related metabolic disorders.

REFERENCES

1. Zimmet P, Alberti G, Kaufman F, Tajima N, Silink M, Arslanian S, et al. The metabolic syndrome in children and adolescents. *Lancet* 2007;**369**(9579):2059–61.
2. Hill JO, Wyatt HR, Peters JC. Energy balance and obesity. *Circulation* 2012;**126**(1):126–32.
3. Zimmermann R, Strauss JG, Haemmerle G, Schoiswohl G, Birner-Gruenberger R, Riederer M, et al. Fat mobilization in adipose tissue is promoted by adipose triglyceride lipase. *Science* 2004;**306**(5700):1383–6.
4. Zechner R, Kienesberger PC, Haemmerle G, Zimmermann R, Lass A. Adipose triglyceride lipase and the lipolytic catabolism of cellular fat stores. *J Lipid Res* 2009;**50**(1):3–21.
5. Lelliott C, Vidal-Puig AJ. Lipotoxicity, an imbalance between lipogenesis de novo and fatty acid oxidation. *Int J Obes Relat Metab Disord* 2004;**28**(Suppl. 4):S22–8.
6. Bluher M. Adipose tissue dysfunction in obesity. *Exp Clin Endocrinol Diabetes* 2009;**117**(6):241–50.
7. Hotamisligil GS. Inflammation and metabolic disorders. *Nature* 2006;**444**(7121):860–7.
8. Engeli S, Gorzelniak K, Kreutz R, Runkel N, Distler A, Sharma AM. Co-expression of renin-angiotensin system genes in human adipose tissue. *J Hypertens* 1999;**17**(4):555–60.
9. Engeli S, Negrel R, Sharma AM. Physiology and pathophysiology of the adipose tissue renin-angiotensin system. *Hypertension* 2000;**35**(6):1270–7.
10. Schling P, Loffler G. Effects of angiotensin II on adipose conversion and expression of genes of the renin-angiotensin system in human preadipocytes. *Horm Metab Res* 2001;**33**(4):189–95.
11. Goossens GH, Jocken JW, Blaak EE, Schiffers PM, Saris WH, van Baak MA. Endocrine role of the renin-angiotensin system in human adipose tissue and muscle: effect of beta-adrenergic stimulation. *Hypertension* 2007;**49**(3):542–7.
12. Janke J, Engeli S, Gorzelniak K, Luft FC, Sharma AM. Mature adipocytes inhibit in vitro differentiation of human preadipocytes via angiotensin type 1 receptors. *Diabetes* 2002;**51**(6):1699–707.
13. Kintscher U, Foryst-Ludwig A, Unger T. Inhibiting angiotensin type 1 receptors as a target for diabetes. *Expert Opin Ther Targets* 2008;**12**(10):1257–63.
14. Schupp M, Clemenz M, Gineste R, Witt H, Janke J, Helleboid S, et al. Molecular characterization of new selective peroxisome proliferator-activated receptor gamma modulators with angiotensin receptor blocking activity. *Diabetes* 2005;**54**(12):3442–52.
15. Shiuchi T, Iwai M, Li HS, Wu L, Min LJ, Li JM, et al. Angiotensin II type-1 receptor blocker valsartan enhances insulin sensitivity in skeletal muscles of diabetic mice. *Hypertension* 2004;**43**(5):1003–10.
16. Tomono Y, Iwai M, Inaba S, Mogi M, Horiuchi M. Blockade of AT1 receptor improves adipocyte differentiation in atherosclerotic and diabetic models. *Am J Hypertens* 2008;**21**(2):206–12.
17. Clemenz M, Frost N, Schupp M, Caron S, Foryst-Ludwig A, Bohm C, et al. Liver-specific peroxisome proliferator-activated receptor alpha target gene regulation by the angiotensin type 1 receptor blocker telmisartan. *Diabetes* 2008;**57**(5):1405–13.
18. Zidek V, Mlejnek P, Simakova M, Silhavy J, Landa V, Kazdova L, et al. Tissue-specific peroxisome proliferator activated receptor gamma expression and metabolic effects of telmisartan. *Am J Hypertens* 2013;**26**(6):829–35.
19. Rong X, Li Y, Ebihara K, Zhao M, Naowaboot J, Kusakabe T, et al. Angiotensin II type 1 receptor-independent beneficial effects of telmisartan on dietary-induced obesity, insulin resistance and fatty liver in mice. *Diabetologia* 2010;**53**(8):1727–31.
20. Benson SC, Pershadsingh HA, Ho CI, Chittiboyina A, Desai P, Pravenec M, et al. Identification of telmisartan as a unique angiotensin II receptor antagonist with selective PPARgamma-modulating activity. *Hypertension* 2004;**43**(5):993–1002.
21. Hein L, Barsh GS, Pratt RE, Dzau VJ, Kobilka BK. Behavioural and cardiovascular effects of disrupting the angiotensin II type-2 receptor in mice. *Nature* 1995;**377**(6551):744–7.
22. Ichiki T, Labosky PA, Shiota C, Okuyama S, Imagawa Y, Fogo A, et al. Effects on blood pressure and exploratory behaviour of mice lacking angiotensin II type-2 receptor. *Nature* 1995;**377**(6551):748–50.
23. Yvan-Charvet L, Even P, Bloch-Faure M, Guerre-Millo M, Moustaid-Moussa N, Ferre P, et al. Deletion of the angiotensin type 2 receptor (AT2R) reduces adipose cell size and protects from diet-induced obesity and insulin resistance. *Diabetes* 2005;**54**(4):991–9.
24. Steckelings UM, Rompe F, Kaschina E, Namsolleck P, Grzesiak A, Funke-Kaiser H, et al. The past, present and future of angiotensin II type 2 receptor stimulation. *J Renin Angiotensin Aldosterone Syst* 2010;**11**(1):67–73.
25. Ohshima K, Mogi M, Jing F, Iwanami J, Tsukuda K, Min LJ, et al. Direct angiotensin II type 2 receptor stimulation ameliorates insulin resistance in type 2 diabetes mice with PPARgamma activation. *PLoS One* 2012;**7**(11):e48387.
26. Georgsson J, Skold C, Botros M, Lindeberg G, Nyberg F, Karlen A, et al. Synthesis of a new class of druglike angiotensin II C-terminal mimics with affinity for the AT2 receptor. *J Med Chem* 2007;**50**(7):1711–15.
27. Shum M, Pinard S, Guimond MO, Labbe SM, Roberge C, Baillargeon JP, et al. Angiotensin II type 2 receptor promotes adipocyte differentiation and restores adipocyte size in high-fat/high-fructose diet-induced insulin resistance in rats. *Am J Physiol* 2013;**304**(2):E197–210.
28. Sabuhi R, Ali Q, Asghar M, Al-Zamily NR, Hussain T. Role of the angiotensin II AT2 receptor in inflammation and oxidative stress: opposing effects in lean and obese Zucker rats. *Am J Physiol Renal Physiol* 2011;**300**(3):F700–6.

29. Wu L, Iwai M, Li Z, Shiuchi T, Min LJ, Cui TX, et al. Regulation of inhibitory protein-kappaB and monocyte chemoattractant protein-1 by angiotensin II type 2 receptor-activated Src homology protein tyrosine phosphatase-1 in fetal vascular smooth muscle cells. *Mol Endocrinol* 2004;**18**(3):666–78.

30. Tani T, Ayuzawa R, Takagi T, Kanehira T, Maurya DK, Tamura M. Angiotensin II bi-directionally regulates cyclooxygenase-2 expression in intestinal epithelial cells. *Mol Cell Biochem* 2008;**315**(1-2):185–93.

31. Okada H, Inoue T, Kikuta T, Watanabe Y, Kanno Y, Ban S, et al. A possible anti-inflammatory role of angiotensin II type 2 receptor in immune-mediated glomerulonephritis during type 1 receptor blockade. *Am J Pathol* 2006;**169**(5):1577–89.

32. Horiuchi M, Hayashida W, Akishita M, Tamura K, Daviet L, Lehtonen JY, et al. Stimulation of different subtypes of angiotensin II receptors, AT1 and AT2 receptors, regulates STAT activation by negative crosstalk. *Circ Res* 1999;**84**(8):876–82.

33. Rompe F, Artuc M, Hallberg A, Alterman M, Stroder K, Thone-Reineke C, et al. Direct angiotensin II type 2 receptor stimulation acts anti-inflammatory through epoxyeicosatrienoic acid and inhibition of nuclear factor kappaB. *Hypertension* 2010;**55**(4):924–31.

34. Sodhi K, Inoue K, Gotlinger KH, Canestraro M, Vanella L, Kim DH, et al. Epoxyeicosatrienoic acid agonist rescues the metabolic syndrome phenotype of HO-2-null mice. *J Pharmacol Exp Ther* 2009;**331**(3):906–16.

Chapter 17

Sex Differences in AT$_2$R Expression and Action

Lucinda M. Hilliard*, Katrina M. Mirabito*, Robert E. Widdop†, Kate M. Denton*

*Department of Physiology, Monash University, Clayton Victoria 3800, Australia, †Department of Pharmacology, Monash University, Clayton Victoria 3800, Australia

BASIS FOR SEX DIFFERENCES IN FUNCTION

Biological differences between men and women are apparent in both health and disease. The primary sex-biasing elements are those encoded on the sex chromosomes that are inherently different in males and females. These genes as well as downstream factors, such as gonadal hormones, act directly on tissues to produce sex differences. It is well recognized that disease incidence and course can be markedly affected by sex chromosomal and hormonal influences. Evidence demonstrates that the (patho)physiological role of the renin–angiotensin system (RAS) is sexually dimorphic,[1–4] including, as outlined here, the expression and function of the angiotensin type 2 receptor (AT$_2$R).

AT$_2$R Gene Mapping

The AT$_2$R gene is located on the X-chromosome.[5] Therefore, XX females have two copies of the AT$_2$R gene as compared to the single copy in XY males. In general, the extra copy of the genes on the X-chromosome undergoes inactivation. However, this is not always the case as some of these genes have a dosage-dependent function in females by escaping X-inactivation.[6] It has also been suggested that inactivation or activation of genes on the X-chromosome may be epigenetically regulated and tissue-specific.[6] Thus, the location of the AT$_2$R on the X-chromosome suggests a greater role for the AT$_2$R in females, particularly in reproduction, as it is rich in genes associated with reproduction and as such warrants further investigation.

Single-nucleotide polymorphisms in the AT$_2$R gene have been associated with disease. One of the earliest studies to suggest that the function of the AT$_2$R was sexually dimorphic demonstrated that polymorphism of the AT$_2$R gene was linked to left ventricular mass index in women but not men with ventricular cardiomyopathy.[7] Since this report, several other studies have also demonstrated a sex bias in the linkage of AT$_2$R polymorphisms and disease. Significant positive associations with AT$_2$R polymorphism and hypertension, cardiac hypertrophy, urinary albumin excretion, and preeclamptic women have been observed in many but not all studies.[8,9] However, the effect size for each of these relationships was relatively small and therefore unlikely to have utility as a marker for future disease on its own.

Hormonal Regulation of AT$_2$R Expression

It is well established that major sex differences exist in the RAS, including the AT$_2$R, due to differential modulations of this system by sex hormones. Mounting evidence indicates that estrogen decreases the expression of the AT$_1$R in target tissue but has the opposite effect to increase AT$_2$R expression in the vasculature, heart, kidney, central and peripheral nervous systems, and adrenal gland.[1,3,10,11] In addition, during the high estrogenic state of pregnancy, AT$_2$R expression is reported to increase at some gestational ages, but not all, suggesting the AT$_2$R may have a temporal expression during pregnancy.[12,13] On the contrary, testosterone amplifies the expression of the AT$_1$R and decreases AT$_2$R expression.[14] In addition, estrogen administration to male rats has been demonstrated to enhance AT$_2$R expression.[15] This strongly suggests differential roles for these pathways between the sexes, with the balance shifted toward an increase in the AT$_2$R–AT$_1$R ratio in females of reproductive age as compared to males (Figure 1).

In addition, AT$_2$R expression has been reported to alter with age.[17] In males, AT$_2$R expression has been shown to increase in the vasculature of spontaneously hypertensive rats (SHR) or not change in the vasculature or kidneys in wild-type

The Protective Arm of the Renin–Angiotensin System (RAS). http://dx.doi.org/10.1016/B978-0-12-801364-9.00017-1

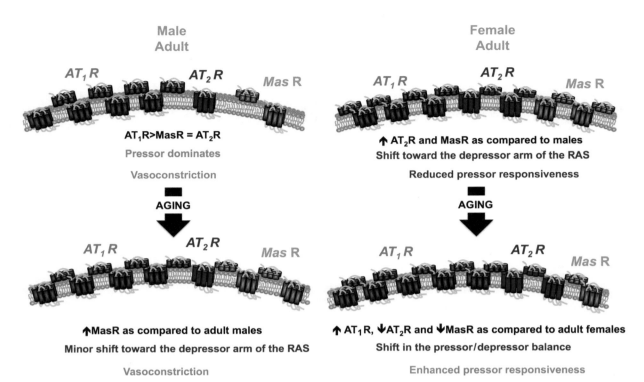

FIGURE 1   Schema of the effect of age and sex on the relative renal expression of angiotensin type 1 receptor (AT$_1$R), the angiotensin type 2 receptor (AT$_2$R), and the Mas receptor (Mas R) determined in adult (16-week-old) and aged (55-week-old) mice taken from Mirabito et al.[16] and the predicted impact on the response to angiotensin II. RAS, renin–angiotensin system.

mice with age.[16,18,19] In contrast, in female mice, renal AT$_2$R expression has been shown to decrease with age.[16,19] It is known that with age, the ratio of estrogen to testosterone increases in males but decreases in females.[20] These data suggest that the contribution of the AT$_2$R is modified by age in association with hormonal status and this may contribute to age-related differences in disease protection between the sexes, particularly in terms of cardiovascular disease (Figure 1). Beyond the scope of this chapter, but of interest, Mas receptor expression also varies with sex and age, contributing to the constrictor/vasodilator balance of the RAS[16] (Figure 1).

REGULATION OF ARTERIAL PRESSURE AND RENAL FUNCTION

The RAS is a major regulator of the cardiovascular system acting to control extracellular volume and arterial pressure. Summarized below is the evidence that the role of the AT$_2$R in the regulation of arterial pressure and renal function is enhanced in females.

Arterial Pressure

In males, though studies in isolated vessels have demonstrated the vasodilator potential of the AT$_2$R, direct AT$_2$R stimulation or blockade *in vivo* has not translated into a decrease in arterial pressure.[10,21] Reports of arterial pressure in AT$_2$R-knockout (KO) mice are divergent with a few studies reporting subtle increases in arterial pressure and others no change.[16,22,23] These disparities likely reflect differences in genetic background, the site of deletion, basal sodium intake that can influence the level of RAS activation, and methodological differences in arterial pressure measurement (radiotelemetry vs. tail-cuff).

However, studies have demonstrated an enhanced role for the AT$_2$R in the regulation of arterial pressure in females. Several studies have shown that the arterial pressure response to a pressor dose of angiotensin II is attenuated in females as compared to males, due to an AT$_2$R-mediated effect.[1,3,21] Remarkably, we showed that chronic infusion of angiotensin II at a low dose paradoxically decreased arterial pressure in female rats, at a dose that caused an increase in arterial pressure in male rats.[24] In subsequent studies, we demonstrated that the depressor effect of angiotensin II in females occurred via an AT$_2$R-mediated, estrogen-dependent mechanism.[24,25] More recently, we extended these findings into AT$_2$R-KO mice. While basal arterial pressure was not different between the genotypes, the chronic pressor response to angiotensin II was attenuated in female WT mice as compared to male WT and female AT$_2$R-KO mice.[26] These findings illustrate the dual nature of angiotensin II on arterial pressure control and support an enhanced role for the AT$_2$R in regulating arterial pressure in females.

Kidney

Long-term arterial pressure regulation is inextricably linked to the renal excretion of salt and water, such that derangements in renal tubular and/or hemodynamic function, which compromise the ability of the kidneys to maintain sodium and water homeostasis, are central to the pathophysiology of hypertension. Indeed, key RAS components are expressed throughout the kidney, and extensive evidence from our laboratory, and that of others, indicates that the RAS modulates renal excretory and hemodynamic function. We, and others, have detected differential expression of the AT$_2$R in the male and female kidney with a higher AT$_2$R/AT$_1$R ratio in females, suggesting that the AT$_2$R may play a sex-specific role in the regulation of renal function.[17] The AT$_2$R maintains autoregulation of renal blood flow and glomerular filtration rate at low renal perfusion pressures in females and protects against the vasoconstrictor effects of angiotensin II.[27,28] The sensitivity of the tubuloglomerular feedback mechanism to angiotensin II, which is an important regulator of vascular tone, is also reduced by the presence of the AT$_2$R in female but not male mice.[26] Finally, acute AT$_2$R stimulation using compound 21 (C21) evoked greater renal vasodilator responses in female rats than seen with equivalent graded doses in male rats.[28] Collectively, these data indicate that the AT$_2$R significantly influences renal function in females to a greater extent than males. However, notably in these studies, conducted in anesthetized rodents, acute AT$_2$R blockade with PD123319 or AT$_2$R stimulation with C21 did not affect arterial pressure.

Effective antihypertensive medications cause a sustained leftward shift in the pressure–natriuresis relationship, which takes 2–3 days to become fully realized in terms of a fall in arterial pressure. Studies in normotensive and hypertensive rats have shown that females demonstrate a protective leftward shift in the pressure–natriuresis curve such that they excrete the same amount of sodium as males at a lower arterial pressure.[29,30] In acute studies, we have demonstrated in normotensive Sprague-Dawley rats that AT$_2$R blockade with PD123319 shifted the pressure–natriuresis curve rightward, and conversely, direct AT$_2$R stimulation using C21 increased renal blood flow and natriuresis in both sexes.[28] Moreover, pharmacological AT$_2$R stimulation with C21 induced renal vasodilatory and natriuretic effects in the female SHR, albeit of smaller magnitude than that observed in female normotensive rats.[31] However, no significant renal vasodilatory or natriuretic response to C21 was observed in the male SHR. This difference in renal response to AT$_2$R stimulation between the normotensive and the hypertensive rat strains is associated with a lower AT$_2$R–AT$_1$R ratio in male and female SHR (Figure 2). Similar to Sprague-Dawley rats, we and others have found a greater AT$_2$R–AT$_1$R ratio in the kidney and vasculature of female than male SHR (Figure 2).[32] However, due to a greater AT$_1$R expression in SHR than Sprague-Dawley rats of both sexes, there is a difference in the relative AT$_2$R–AT$_1$R ratio (Figure 2). Thus, AT$_2$R stimulation in both normotensive and hypertensive strains of rats causes a potentially protective leftward shift in the pressure–natriuresis relationship. Together, this knowledge provides a strong rationale for subsequent studies into the long-term and sex-specific effects of pharmacological AT$_2$R stimulation on renal function and arterial pressure in order to establish whether AT$_2$R agonist therapy represents a novel therapeutic approach for hypertension, particularly in females.

EVIDENCE FOR A DIFFERENTIAL ROLE FOR THE AT$_2$R IN OTHER DISEASE CONDITIONS

The evidence for sex differences in the expression and function of the AT$_2$R in the regulation of arterial pressure and renal function as discussed is substantial. However, there is also evidence, more preliminary in nature (often only single reports), that suggests sex differences in the role of the AT$_2$R in other conditions. These observations need to be confirmed in future studies.

Heart and Vascular Disease

The AT$_2$R has been associated with improved cardiac function in ischemic injury models, and this has been linked to AT$_2$R-mediated effects on fibrosis and inflammation. AT$_1$R-mediated remodeling of both the heart and the vasculature (e.g., cellular growth, proliferation, and extracellular matrix deposition) is well recognized.[10] On the other hand, AT$_2$R has consistently been shown to play a protective role in cardiovascular hypertrophy and fibrosis, particularly in aged animals.[10] Likewise, AT$_2$R-KO mice generally exhibit vascular hypertrophy and increased cardiac fibrosis compared to wild-type littermates,[10] and greater AT$_2$R-mediated vasoprotective remodeling is seen in female wild types.[19] Thus, the possibility of sex-specific AT$_2$R distribution and vascular remodeling has important implications for the regulation of cardiovascular homeostasis.

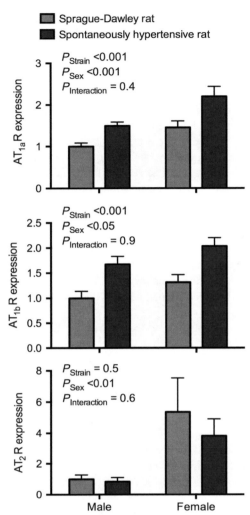

FIGURE 2 Relative renal gene expression of angiotensin type 1a receptor (AT$_{1a}$R), angiotensin type 1b receptor (AT$_{1b}$R), and angiotensin type 2 receptor (AT$_2$R) in age-matched adult male and female Sprague-Dawley and spontaneously hypertensive rats, as determined by real-time reverse transcription PCR. Rats were maintained on a sodium-controlled diet (2.5% NaCl). The kidneys were collected from female rats in estrus as determined by vaginal smear. All samples were run in triplicate using TaqMan gene expression assays. Relative expression was calculated using the comparative cycle of threshold fluorescence ($2^{-\Delta\Delta CT}$) method with 18S rRNA as the internal housekeeping gene. Data are presented as mean ± standard error of the mean and are expressed relative to the male Sprague-Dawley rat group. $n = 7$–8 per group. Data were analyzed using a two-way analysis of variance with factors: strain (Sprague-Dawley or spontaneously hypertensive rat), sex (male or female), and the interaction between strain and sex.

In addition, evidence suggests an enhanced role for the AT$_2$R in atherosclerosis,[33] diabetes,[34] and obesity-induced hypertension,[35] via its actions to reduce tissue fibrosis and inflammation and promote tissue repair.

Neurological Disorders

There is a clear sex bias in neurological disorders including stroke, ischemic injury, and pain syndromes.[36] In preceding chapters, evidence that the AT$_2$R protects against neural damage and cognitive impairment has been addressed. Again in studies that examined both sexes, a differential role for the AT$_2$R has been demonstrated. In AT$_2$R-KO mice, it was shown that in a model of cerebral ischemic injury, the AT$_2$R improved cognitive function in females but not males, whereas in males, the AT$_2$R reduced the infarct area in association with an increase in cerebral blood flow.[37] These findings await further study but suggest that therapeutic interventions may need to be tailored to suit the sexes.

Painful conditions occur more commonly in premenopausal women than men or postmenopausal women.[38] This increase in pain sensitivity has been linked to enhanced neurite outgrowth in the presence of estrogen.[39] Furthermore, AT$_2$R blockade with PD123319 reduced this neural sprouting.[39] Thus, estrogen-induced AT$_2$R upregulation may be linked to enhanced pain sensitivity in females, and the therapeutic potential of AT$_2$R inhibition for the treatment of pain is being investigated.

REPRODUCTION

Finally, the enhanced role of the AT$_2$R in females is also apparent in the normal cardiovascular and renal adaptations to pregnancy. During pregnancy, arterial pressure decreases despite activation of the RAS, and this is associated with a reduced pressor sensitivity to angiotensin II.[40] We, and others, have demonstrated that the AT$_2$R mediates the reduction in arterial pressure and reduced pressor response to angiotensin II during pregnancy.[12,41,42] Moreover, we have demonstrated that AT$_2$R deficiency causes an increase in arterial pressure in late gestation in mice in association with a phenotypic shift in renal T helper (Th) cell infiltration toward the Th1 phenotype, which increases renal proinflammatory cytokines and the Th1–Th2 balance.[12] This suggests that the AT$_2$R plays an important role in the regulation of arterial pressure during normal pregnancy and that defects in the expression or function of the AT$_2$R may contribute to the pathophysiology of pregnancy-induced hypertension.

CONCLUSION

Preclinical studies have demonstrated that the beneficial AT$_2$R-mediated effects on cardiovascular and renal function are enhanced in females, primarily in rodent models. However, while there is a paucity of data demonstrating similar effects in women, available data (polymorphisms and attenuated pressor responses to Ang II) are compatible with such a conclusion and warrant further investigation. In conclusion, the expression and therefore the actions of the AT$_2$R are amplified in females.

REFERENCES

1. Hilliard LM, Mirabito KM, Denton KM. Unmasking the potential of the angiotensin AT2 receptor as a therapeutic target in hypertension in men and women: what we know and what we still need to find out. *Clin Exp Pharmacol Physiol* 2013;**40**:542–50.
2. Komukai K, Mochizuki S, Yoshimura M. Gender and the renin-angiotensin-aldosterone system. *Fundam Clin Pharmacol* 2010;**24**:687–98.
3. Hilliard LM, Sampson AK, Brown RD, Denton KM. The "his and hers" of the renin-angiotensin system. *Curr Hypertens Rep* 2013;**15**:71–9.
4. Sullivan JC. Sex and the renin-angiotensin system: inequality between the sexes in response to RAS stimulation and inhibition. *Am J Physiol Regul Integr Comp Physiol* 2008;**294**:R1220–6.
5. Koike G, Winer ES, Horiuchi M, Brown DM, Szpirer C, Dzau VJ, et al. Cloning, characterization, and genetic mapping of the rat type 2 angiotensin II receptor gene. *Hypertension* 1995;**26**:998–1002.
6. Schulz EG, Heard E. Role and control of X chromosome dosage in mammalian development. *Curr Opin Genet Dev* 2013;**23**:109–15.
7. Deinum J, van Gool JM, Kofflard MJ, ten Cate FJ, Danser AH. Angiotensin II type 2 receptors and cardiac hypertrophy in women with hypertrophic cardiomyopathy. *Hypertension* 2001;**38**:1278–81.
8. Katsuya T, Morishita R. Gene polymorphism of angiotensin II type 1 and type 2 receptors. *Curr Pharm Des* 2013;**19**:2996–3001.
9. Rahimi Z, Rahimi Z, Aghaei A, Vaisi-Raygani A. AT2R–1332 G:A polymorphism and its interaction with AT1R 1166 A:C, ACE I/D and MMP-9–1562 C:T polymorphisms: risk factors for susceptibility to preeclampsia. *Gene* 2014;**538**:176–81.
10. Jones ES, Vinh A, McCarthy CA, Gaspari TA, Widdop RE. AT2 receptors: functional relevance in cardiovascular disease. *Pharmacol Ther* 2008;**120**:292–316.
11. Macova M, Armando I, Zhou J, Baiardi G, Tyurmin D, Larrayoz-Roldan IM, et al. Estrogen reduces aldosterone, upregulates adrenal angiotensin II AT(2) receptors and normalizes adrenomedullary Fra-2 in ovariectomized rats. *Neuroendocrinology* 2008;**88**(4):276–86.
12. Mirabito KM, Hilliard LM, Wei Z, Tikellis C, Widdop RE, Vinh A, et al. Role of inflammation and the angiotensin type 2 receptor in the regulation of arterial pressure during pregnancy in mice. *Hypertension* 2014;**64**(3):626–31.
13. Pulgar VM, Yamashiro H, Rose JC, Moore LG. Role of the AT2 receptor in modulating the angiotensin II contractile response of the uterine artery at mid-gestation. *J Renin Angiotensin Aldosterone Syst* 2011;**12**:176–83.
14. Sampson AK, Jennings GL, Chin-Dusting JP. Y are males so difficult to understand? A case where "X" does not mark the spot. *Hypertension* 2012;**59**:525–31.
15. Reckelhoff JF, Zhang H, Granger JP. Testosterone exacerbates hypertension and reduces pressure-natriuresis in male spontaneously hypertensive rats. *Hypertension* 1998;**31**:435–9.
16. Mirabito KM, Hilliard LM, Kett MM, Brown RD, Booth SC, Widdop RE, et al. Sex and age-related differences in the chronic pressure-natriuresis relationship: role of the angiotensin type 2 receptor. *Am J Physiol Renal Physiol* 2014;**307**(8):F901–7.
17. Sampson AK, Moritz KM, Denton KM. Postnatal ontogeny of angiotensin receptors and ACE2 in male and female rats. *Gender Med* 2012;**9**:21–32.
18. Bosnyak S, Widdop RE, Denton KM, Jones ES. Differential mechanisms of ang (1–7)-mediated vasodepressor effect in adult and aged candesartan-treated rats. *Int J Hypertens* 2012;**2012**:192567.
19. Okumura M, Iwai M, Ide A, Mogi M, Ito M, Horiuchi M. Sex difference in vascular injury and the vasoprotective effect of valsartan are related to differential AT2 receptor expression. *Hypertension* 2005;**46**:577–83.

20. Maric C. Sex, diabetes and the kidney. *Am J Physiol Renal Physiol* 2009;**296**:F680–8.

21. McCarthy CA, Widdop RE, Denton KM, Jones ES. Update on the angiotensin AT(2) receptor. *Curr Hypertens Rep* 2013;**15**:25–30.

22. Hein L, Barsh GS, Pratt RE, Dzau VJ, Kobilka BK. Behavioural and cardiovascular effects of disrupting the angiotensin II type-2 receptor in mice. *Nature* 1995;**377**:744–7.

23. Ichiki T, Labosky PA, Shiota C, Okuyama S, Imagawa Y, Fogo A, et al. Effects on blood pressure and exploratory behaviour of mice lacking angiotensin II type-2 receptor. *Nature* 1995;**377**:748–50.

24. Sampson AK, Hilliard LM, Moritz KM, Thomas MC, Tikellis C, Widdop RE, et al. The arterial depressor response to chronic low-dose angiotensin II infusion in female rats is estrogen dependent. *Am J Physiol Regul Integr Comp Physiol* 2012;**302**:R159–65.

25. Sampson AK, Moritz KM, Jones ES, Flower RL, Widdop RE, Denton KM. Enhanced angiotensin II type 2 receptor mechanisms mediate decreases in arterial pressure attributable to chronic low-dose angiotensin II in female rats. *Hypertension* 2008;**52**:666–71.

26. Brown RD, Hilliard LM, Head GA, Jones ES, Widdop RE, Denton KM. Sex differences in the pressor and tubuloglomerular feedback response to angiotensin II. *Hypertension* 2012;**59**:129–35.

27. Hilliard LM, Nematbakhsh M, Kett MM, Teichman E, Sampson AK, Widdop RE, et al. Gender differences in pressure-natriuresis and renal auto-regulation: role of the angiotensin type 2 receptor. *Hypertension* 2011;**57**:275–82.

28. Hilliard LM, Jones ES, Steckelings UM, Unger T, Widdop RE, Denton KM. Sex-specific influence of angiotensin type 2 receptor stimulation on renal function: a novel therapeutic target for hypertension. *Hypertension* 2012;**59**:409–14.

29. Khraibi AA. Renal interstitial hydrostatic pressure and sodium excretion in hypertension and pregnancy. *J Hypertens Suppl* 2002;**20**:S21–7.

30. Khraibi AA. Renal interstitial hydrostatic pressure and pressure natriuresis in pregnant rats. *Am J Physiol Renal Physiol* 2000;**279**:F353–7.

31. Hilliard LM, Chow CL, Mirabito KM, Steckelings UM, Unger T, Widdop RE, et al. Angiotensin type 2 receptor stimulation increases renal function in female, but not male, spontaneously hypertensive rats. *Hypertension* 2014;**64**(2):378–83.

32. Silva-Antonialli MM, Tostes RC, Fernandes L, Fior-Chadi DR, Akamine EH, Carvalho MH, et al. A lower ratio of AT1/AT2 receptors of angiotensin II is found in female than in male spontaneously hypertensive rats. *Cardiovasc Res* 2004;**62**:587–93.

33. Brosnihan KB, Hodgin JB, Smithies O, Maeda N, Gallagher P. Tissue-specific regulation of ACE/ACE2 and AT1/AT2 receptor gene expression by oestrogen in apolipoprotein E/oestrogen receptor-alpha knock-out mice. *Exp Physiol* 2008;**93**:658–64.

34. Pettersson-Fernholm K, Frojdo S, Fagerudd J, Thomas MC, Forsblom C, Wessman M, et al. The AT2 gene may have a gender-specific effect on kidney function and pulse pressure in type I diabetic patients. *Kidney Int* 2006;**69**:1880–4.

35. Samuel P, Khan MA, Nag S, Inagami T, Hussain T. Angiotensin AT(2) receptor contributes towards gender bias in weight gain. *PLoS One* 2013;**8**:e48425.

36. Beery AK, Zucker I. Sex bias in neuroscience and biomedical research. *Neurosci Biobehav Rev* 2011;**35**:565–72.

37. Sakata A, Mogi M, Iwanami J, Tsukuda K, Min LJ, Fujita T, et al. Sex-different effect of angiotensin II type 2 receptor on ischemic brain injury and cognitive function. *Brain Res* 2009;**1300**:14–23.

38. Berkley KJ, Zalcman SS, Simon VR. Sex and gender differences in pain and inflammation: a rapidly maturing field. *Am J Physiol Regul Integr Comp Physiol* 2006;**291**:R241–4.

39. Chakrabarty A, Blacklock A, Svojanovsky S, Smith PG. Estrogen elicits dorsal root ganglion axon sprouting via a renin-angiotensin system. *Endocrinology* 2008;**149**:3452–60.

40. Irani RA, Xia Y. The functional role of the renin-angiotensin system in pregnancy and preeclampsia. *Placenta* 2008;**29**:763–71.

41. Takeda-Matsubara Y, Iwai M, Cui TX, Shiuchi T, Liu HW, Okumura M, et al. Roles of angiotensin type 1 and 2 receptors in pregnancy-associated blood pressure change. *Am J Hypertens* 2004;**17**:684–9.

42. Chen K, Merrill DC, Rose JC. The importance of angiotensin II subtype receptors for blood pressure control during mouse pregnancy. *Reprod Sci* 2007;**14**:694–704.

Chapter 18

Discovery of Nonpeptide, Selective AT$_2$ Receptor Agonists

Mats Larhed, Rebecka Isaksson, Anders Hallberg

Department of Medicinal Chemistry, Division of Organic Pharmaceutical Chemistry, Uppsala University, Uppsala, Sweden

There are a large number of structurally diverse AT$_1$ receptor (AT$_1$R) antagonists in clinic, for example, losartan (**1**; AT$_1$R, $K_i = 16.4$ nM).[1,2] There are, however, neither AT$_2$ receptor (AT$_2$R) agonists nor AT$_2$R antagonists in clinic, although EMA401 (**2**; AT$_1$R, $K_i = 408$ nM, and AT$_2$R, $K_i = 39.5$ nM), reported to be a selective AT$_2$R antagonist, is now undergoing clinical trials.[3,4] Selective AT$_2$R ligands, proposed to act as agonists, have been known for a long time and have been used extensively as research tools, for example, CGP-42112A (**3**; AT$_1$R, $K_i = 568$ nM, and AT$_2$R, $K_i = 0.73$ nM) and (*p*-amino-Phe^6)-Ang II (**4**; AT$_2$R, $K_i = 0.7$ nM), but those agonists are all peptides.[1,5,6] Short peptides exhibit in general a very low bioavailability and are therefore not useful as drugs. Consequently, metabolically stable, drug-like AT$_2$R agonists with fair oral bioavailability are desired both as research tools and possibly as potential therapeutics.

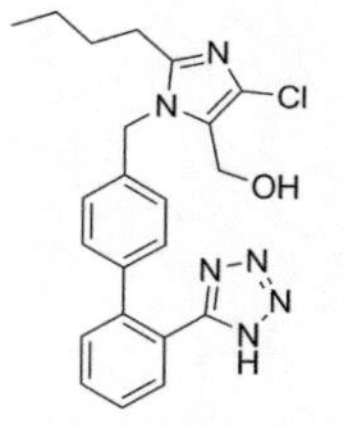

Losartan (1)
K_i (nM)
AT$_1$R: 16.4
AT$_2$R: >10,000
Antagonist (AT$_1$R)

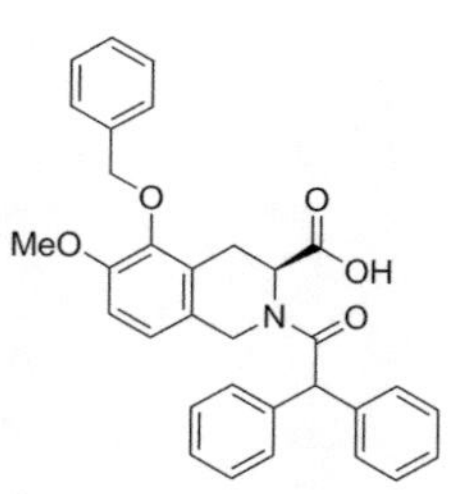

EMA401 (2)
K_i (nM)
AT$_1$R: 408 ± 335
AT$_2$R: 39.5 ± 5.2
Antagonist (AT$_2$R)

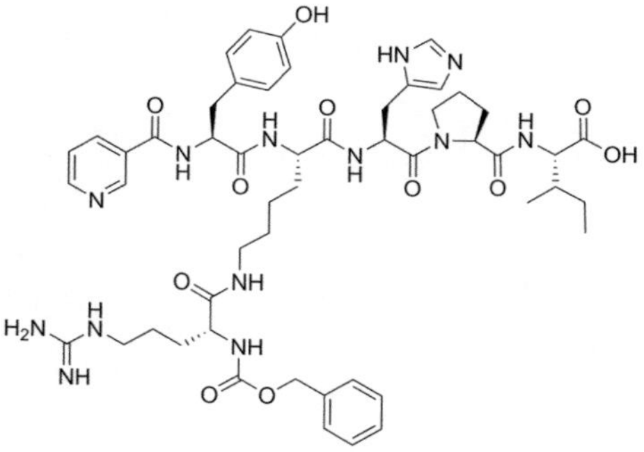

CGP-42112A (3)
K_i (nM)
AT$_1$R: 568
AT$_2$R: 0.73
Agonist (AT$_2$R)

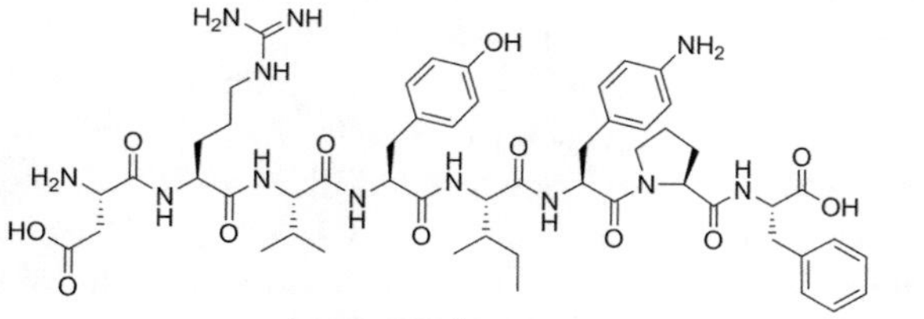

(*p*-amino-Phe^6)Ang II (4)
K_i (nM)
AT$_1$R: >10,000
AT$_2$R: 0.7

CHART 1

The Protective Arm of the Renin–Angiotensin System (RAS). http://dx.doi.org/10.1016/B978-0-12-801364-9.00018-3

Some 10–15 years ago, we embarked on a drug discovery program aimed at identifying drug-like AT$_2$R agonists. We applied two different approaches: (a) to start from nonpeptide, nonselective AT$_1$/AT$_2$ receptor ligands and (b) to start from and transform the nonselective native parent peptide angiotensin II (Ang II) to selective drug-like ligands exhibiting agonism at the AT$_2$R.

SELECTIVE DRUG-LIKE AT$_2$R AGONISTS FROM NONPEPTIDE, NONSELECTIVE AT$_1$/AT$_2$ RECEPTOR LIGANDS

The first nonpeptide AT$_1$R agonist, L-162,313 (**5**; AT$_1$R, IC$_{50}$ = 1.1 nM, and AT$_2$R, IC$_{50}$ = 2.0 nM), which is nonselective, was reported in 1994.[7,8] Nonpeptide agonists of peptide receptors were rare 20 years ago, being confined mainly to agonists of the opioid receptors. The biphenyl derivative L-162,782 (**6**; AT$_1$R, IC$_{50}$ = 2.1 nM, and AT$_2$R, IC$_{50}$ = 0.7 nM) is a nonselective AT$_1$R agonist structurally similar to L-162,313 (**5**). The removal of a methyl group from the side chain of L-162,782 (**6**) resulted in L-162,389 (**7**; AT$_1$R, IC$_{50}$ = 2.1 nM, and AT$_2$R, IC$_{50}$ = 3.8 nM) found to act as an AT$_1$R antagonist *in vivo*.[9–11] Thus, subtle molecular alterations determine the agonist/antagonist properties of these ligands at the AT$_1$R. The compounds **5**, **6**, and **7** are all binding to both the AT$_1$R and the AT$_2$R with similar affinities. Notably, replacing the alkyl group with an *m*-methoxybenzyl group afforded a selective AT$_1$R agonist, L-163,491 (**8**; AT$_1$R, IC$_{50}$ = 1.4 nM, and AT$_2$R, IC$_{50}$ = 100 nM).[12] Thus, it seems that a larger side chain might render ligands with lower preference for the AT$_2$R.

The AT$_1$R agonist L-162,313 (**5**), a proved partial AT$_1$R agonist with an AT$_1$R K_i value of 3.9 nM and an AT$_2$R K_i value of 2.8 nM, was found to act as an agonist also at the AT$_2$R, as deduced from data from a secretor model in rat.[13] Duodenal mucosal alkaline secretion in rats has been shown previously to be inhibited by AT$_1$R activation and increased after AT$_2$R stimulation.[14,15] The secretion could be blocked by the AT$_2$R-selective antagonist PD 123,319.[16]

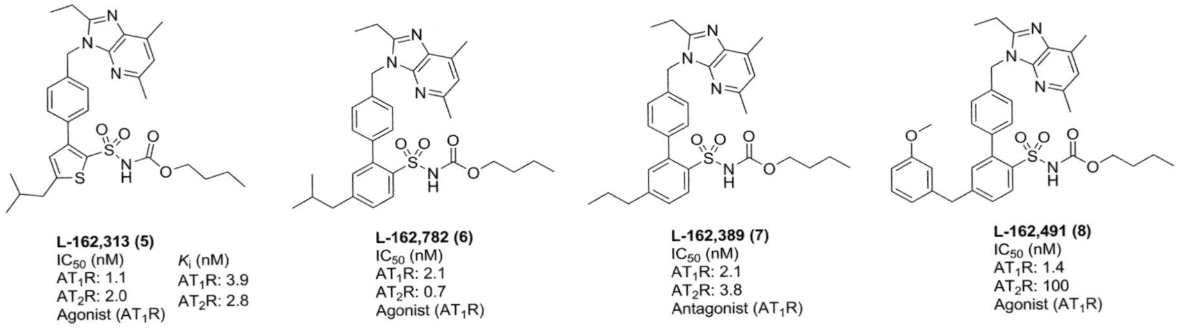

L-162,313 (5)
IC$_{50}$ (nM) K_i (nM)
AT$_1$R: 1.1   AT$_1$R: 3.9
AT$_2$R: 2.0   AT$_2$R: 2.8
Agonist (AT$_1$R)

L-162,782 (6)
IC$_{50}$ (nM)
AT$_1$R: 2.1
AT$_2$R: 0.7
Agonist (AT$_1$R)

L-162,389 (7)
IC$_{50}$ (nM)
AT$_1$R: 2.1
AT$_2$R: 3.8
Antagonist (AT$_1$R)

L-162,491 (8)
IC$_{50}$ (nM)
AT$_1$R: 1.4
AT$_2$R: 100
Agonist (AT$_1$R)

CHART 2

In efforts to improve receptor selectivity, a series of structural alterations of the butyloxy group of L-162,313 (**5**) were assessed.[13] All variations investigated were deleterious with respect to AT$_2$R affinity. Lower preferences for the AT$_1$R were also encountered by these alterations but less pronounced. Kevin et al.[17] reported that even the small structural modifications of the isobutyl group in the 5-position tended to reduce AT$_2$R affinity while retaining AT$_1$R affinity, which partly seems to apply also to the sulfonylcarbamate portion of L-162,313 (**5**).

A series of substituted quinazolinones were made, where the bicyclic imidazopyridine ring system in **5** had been replaced,[13] since quinazolinone fragments are found in several selective AT$_2$R ligands.[18–24] One of the most potent and selective compounds synthesized in the series, the acetyl amide **9** (AT$_1$R, K_i = 728 nM, and AT$_2$R, K_i = 17 nM), was subjected to *in vivo* studies in the secretor model, but **9** did not act as an AT$_2$R agonist. Thus, it did not seem productive to further increase the size of the nitrogen heterocycle part of the molecule. However, by stepwise minimizing the heterocyclic system, very high AT$_2$/AT$_1$ receptor selectivity could be achieved.[25] The nonselective ligand **5** that exhibits an AT$_2$R K_i value of 2.8 nM could be converted to the receptor-selective compound **10** with a K_i value of 7.1 nM. Further simplifications of the structure furnished **11**, **12**, **13**, and **14** with K_i values of 3.2, 4.0, 0.5, and 0.4 nM, respectively. Migration of the methylene linker between the benzene ring and the imidazole heterocycle of the high-affinity compound **14** resulted in the inactive compound **15**.

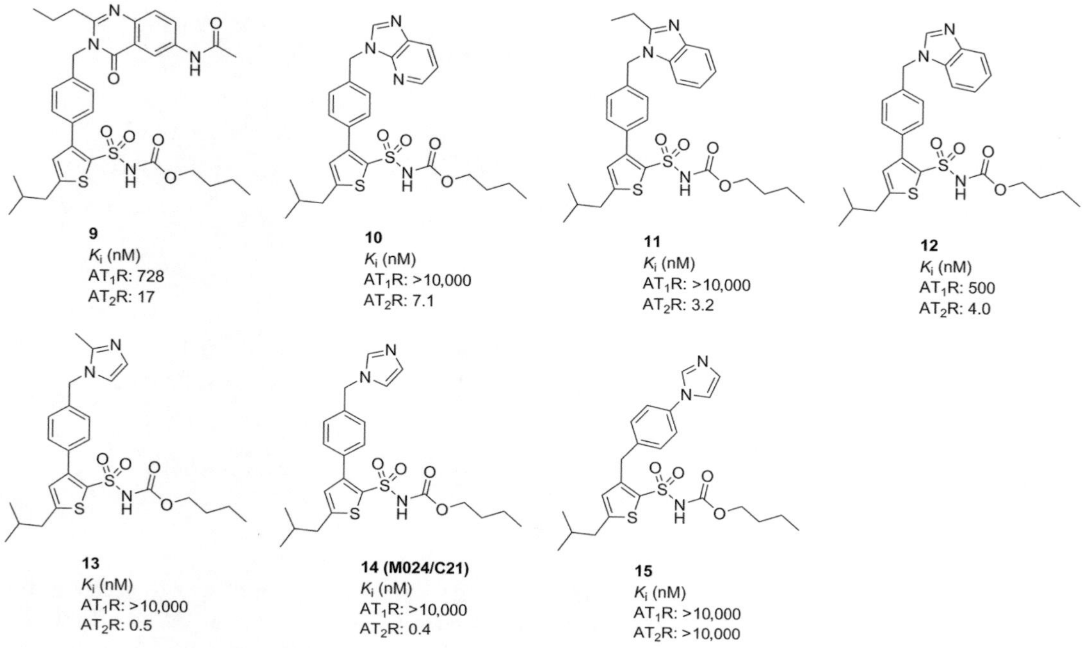

9
K_i (nM)
AT$_1$R: 728
AT$_2$R: 17

10
K_i (nM)
AT$_1$R: >10,000
AT$_2$R: 7.1

11
K_i (nM)
AT$_1$R: >10,000
AT$_2$R: 3.2

12
K_i (nM)
AT$_1$R: 500
AT$_2$R: 4.0

13
K_i (nM)
AT$_1$R: >10,000
AT$_2$R: 0.5

14 (M024/C21)
K_i (nM)
AT$_1$R: >10,000
AT$_2$R: 0.4

15
K_i (nM)
AT$_1$R: >10,000
AT$_2$R: >10,000

CHART 3

Compound **14** was studied in more detail and was found in a NG108-15 cell assay to induce neurite outgrowth,[25] one of the first steps of neuronal differentiation, an effect that both Ang II and CGP-42112A show.[26] In addition, **14** like Ang II stimulates MAPK, and as previously shown, this sustained activation of the p42/p44^{MAPK} pathway is essential to promote neurite outgrowth.[27] Incubation with the AT$_2$R antagonist PD 123,319 abolishes the effect of **14** indicating that the latter acts as an agonist at the AT$_2$R.

The duodenal mucosal alkaline secretion increases after stimulation of the AT$_2$R located in the duodenal mucosa/submucosa. The imidazole derivative **14** markedly increased the mucosal alkaline secretion either when infused intravenously (0.3 mg/(kg h)) or given topically in the luminal perfusate (100 µM). In the presence of PD 123,319 (0.3 mg/kg *i.v. bolus*), secretion was absent or greatly reduced. Increasing i.v. doses of **14** demonstrated a dose–response relationship, further supporting the agonistic nature of **14** *in vivo*. Compound **14**, named M024/C21 or more frequently C21, with a bioavailability of 20–30% after oral administration and a half-life estimated at 4 h in rat, and which was discovered in our laboratory, is the first selective, nonpeptide AT$_2$R agonist reported to our knowledge.[25]

A large number of alterations have been made in the 2- and 5-positions of the thiophene core structure of the imidazole derivative **14**, and it has been proven that the imidazole ring structure is a powerful determinant for AT$_2$R selectivity.[28] Thus, the majority of the alterations in the sulfonylcarbamate side chain in the 2-position furnished ligands with AT$_2$R affinity. Neither of the AT$_2$R ligands prepared—the sulfonylcarbamates **16** ($K_i = 10$ nM), **17** ($K_i = 37$ nM), and **18** ($K_i = 31$ nM); the sulfonylamides **19** ($K_i = 34$ nM) and **20** ($K_i = 1.0$ nM); the acylsulfonamide **21** ($K_i = 79$ nM); and the sulfonylureas **22** ($K_i = 13$ nM) and **23** ($K_i = 4$ nM)—nor the tetrazole derivative **24** ($K_i = 189$ nM) exhibited any affinity to the AT$_1$R. It is worth noting that the tetrazole ring structure (like the sulfonylcarbamate function) often serves as a carboxylic acid bioisostere and is a commonly used element in AT$_1$R antagonists, as with sartans in clinic (e.g., losartan (**1**), valsartan, and candesartan).[29] Considering AT$_2$R affinities, none of the modifications of the sulfonylcarbamate side chain performed produced AT$_2$R binders more powerful than **14**.[28]

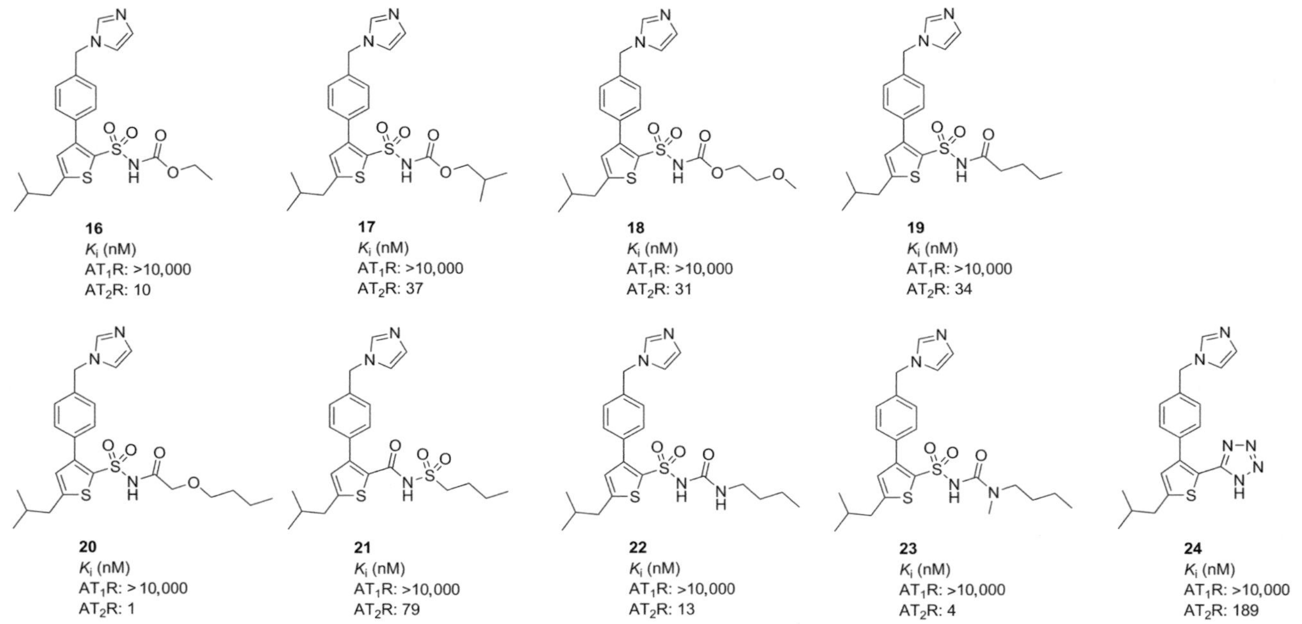

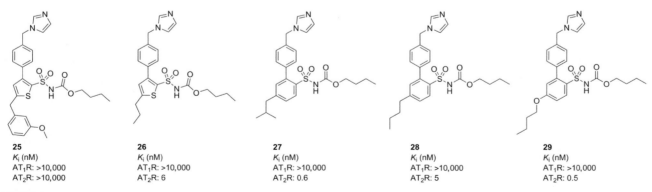

CHART 4

By replacing the branched alkyl chain of **14** with an *m*-methoxybenzyl group, in efforts to create a molecule with affinity to the AT_1R (cf. the structure of **8**), compound **25** was obtained. This compound had no affinity to neither the AT_1R nor the AT_2R. As discussed previously, the ligand **6** is an AT_1R agonist, while compound **7** is an AT_1R antagonist *in vivo*. Removal of the analogous methyl group of the AT_2R agonist **14** provides the *n*-propyl derivative **26**, which exhibiting a ten-fold lower binding affinity to the AT_2R is still an agonist at the AT_2R. Thus, a deletion of a methyl group from the branched side chain was not transforming these ligands from agonists to antagonists as was the case among the AT_1R ligands (cf. **6** and **7**). Replacement of the thiophene nucleus of **14** for a benzene ring did not in this series seem to affect the structure–activity relationship significantly. As an example, the isobutyl compound **27** exhibits a K_i value of 0.6 nM and the *n*-butyl compound **28** a K_i value of 5.0 nM. Notably, the butyloxyphenyl derivative **29** with a K_i of 0.5 nM demonstrated a pronounced agonistic effect as deduced from its capacity to induce neurite elongation in neuronal NG108-15 cells. Hence, the imidazole ring strongly discriminates between the two receptor subtypes, and ligands with this five-membered ring system favor binding to the AT_2R.[28]

CHART 5

As alternative to the imidazole ring structure in **14** that most likely accounted for the inhibition of, for example, the CYP 3A4 and 2C9 enzymes observed, a large series of alternative small aromatic and aliphatic ring systems were examined.[30] It was found that with a methyl group or a trifluoromethyl group in the 4-position of the imidazole core structure, although reducing the AT$_2$R affinity 50-fold and five-fold, respectively, a suppression of the CYP 450 inhibition occurred, as could be expected. Similar results were observed with pyrazoles, while with small aromatic heterocycles, such as thiazoles or oxazoles, considerable CYP inhibition was still encountered. However, with small nonaromatic ring systems, inhibition of the most common CYP enzymes was less pronounced. As an example, the 2-oxo-pyrrolidine derivative **30** that exhibited an AT$_2$R K_i value of 3.5 nM, at a 10 μM concentration of **30**, a CYP inhibition of 2B6 (33%), 2C9 (24%), 2C19 (25%), 2E1 (5%), 3A4 (28%), and 3A5 (25%) was observed. The AT$_2$R agonist **30** induced neurite outgrowth of NG108-15 cells after 3 days of treatment as Ang II and the AT$_2$R antagonist PD 123,319 virtually abolished this neurite elongation. Compound **30** exhibited a lower bioavailability than **14** in rat (unpublished results).

A large series of compounds where the imidazole ring was displaced with noncyclic moieties were prepared and evaluated with regard to AT$_2$/AT$_1$ receptor selectivity and affinity.[31] The compounds **31**, **32**, and **33** represent some of the derivatives with the highest binding affinities, with K_i values of 8.8, 2.8, and 7.8 nM, respectively. Compounds related to the amides **31** or **33** but with larger substituents as phenyl and benzyl groups or longer carbon chains attached to the amide nitrogen all demonstrated considerably lower AT$_2$R affinities, although still often K_i values below 100 nM. None of the compounds, regardless of substituents, showed any affinity to the AT$_1$R, which is notable considering the structural similarity in the amide part to, for example, valsartan, although the latter carries a carboxyl group, as several other sartans. As deduced from assessments in the neurite outgrowth cell assay, **31**, **32**, and **33** exert high agonistic effects.

It is worth noting that the 2,5-substituted furanyl derivative **34** exhibits a high affinity to the AT$_2$R, while the corresponding 2,4-substituted thiophene and 2,5-substituted pyridine compounds were inactive.[31] Furthermore, the furan derivative with a K_i of 9.1 nM acts as a potent AT$_2$R agonist in the NG108-15 cell assay.

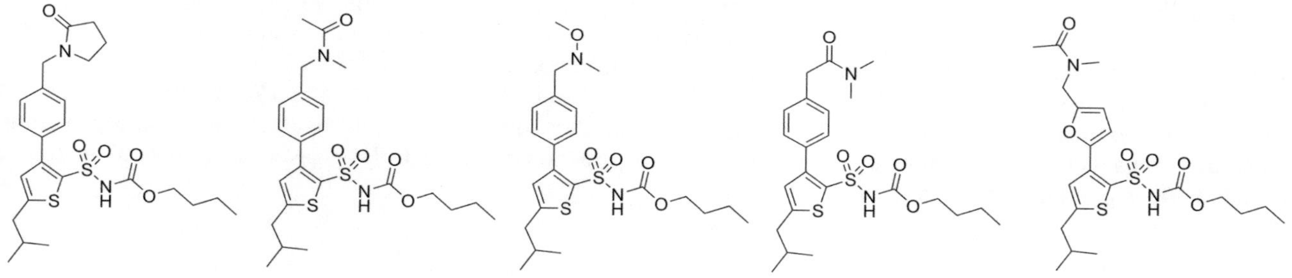

30
K_i (nM)
AT$_1$R: >10,000
AT$_2$R: 3.5

31
K_i (nM)
AT$_1$R: >10,000
AT$_2$R: 8.8

32
K_i (nM)
AT$_1$R: >10,000
AT$_2$R: 2.8

33
K_i (nM)
AT$_1$R: >10,000
AT$_2$R: 7.0

34
K_i (nM)
AT$_1$R: >10,000
AT$_2$R: 9.1

CHART 6

The deleterious effect of removal of the methylene linker in the imidazole series of compounds was not seen in the amide series.[32] Compound **35** with an AT$_2$R K_i value of 3.0 nM provides a good example. It is notable though that replacement of the thiophene ring for a benzene moiety in the latter series of compounds furnishes compounds with much lower affinities. Thus, with **35** as the reference, and with the isobutyl group in the 4-position of a benzene ring replacing the thiophene, a K_i of 145 nM was observed, and in the case where this side chain is attached to the 5-position of the benzene ring, a K_i of 122 nM was encountered. Furthermore, in this amide series, aromatic structures adjacent to the amide functionality, as in **36,** can be tolerated and moreover result frequently in very high affinities and AT$_2$/AT$_1$ receptor selectivities. Markedly, and somewhat surprisingly, the triaryl compound **37** that is AT$_2$R-selective with a K_i of 52.0 nM exhibits a better AT$_2$R affinity than the aforementioned analogs and biphenyl derivatives with isobutyl groups in either the 4- or the 5-positions (compare also **8** and **25**).

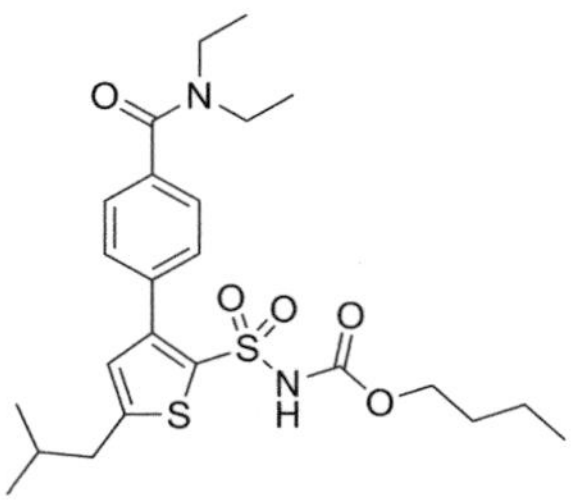

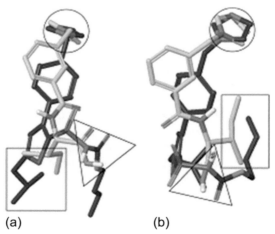

35
K_i (nM)
AT$_1$R: >10,000
AT$_2$R: 3.0 ± 0.3

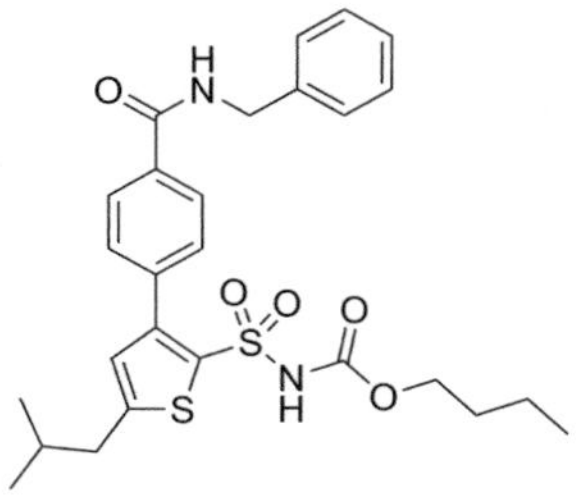

36
K_i (nM)
AT$_1$R: >10,000
AT$_2$R: 1.0 ± 0.08

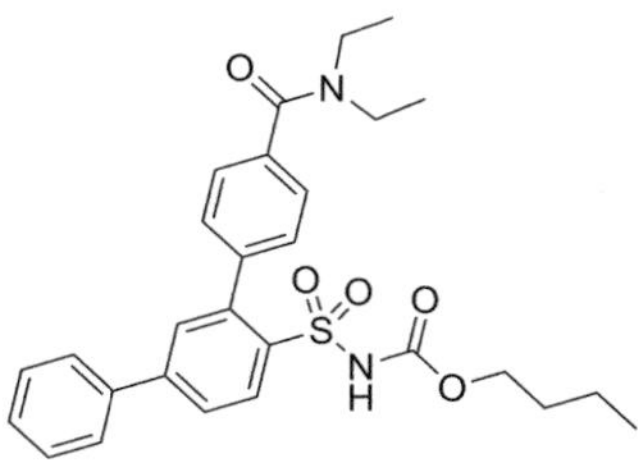

37
K_i (nM)
AT$_1$R: >10,000
AT$_2$R: 52.0 ± 4.3

CHART 7

SELECTIVE DRUG-LIKE AT$_2$R AGONISTS FROM THE NONSELECTIVE PARENT PEPTIDE ANG II

To successfully make low-molecular-weight drug-like nonpeptides by stepwise structural transformations of endogenous peptides is a challenge.[33] Recently, a number of bioactive AT$_2$R-selective analogs of similar size to Ang II, where the dipeptide fragment Tyr[4]-Ile[5] was replaced by various γ-turn mimetic scaffolds, were designed and prepared.[34–36] Furthermore, it was previously demonstrated, by de Gasparo et al. in 1991, that removal of three amino acids from the N-terminus of Ang II ([Val[5]]Ang II (4-8) and [Val[5]-Ile[8]]Ang II (4-8)) provided AT$_2$R selectivity.[37] More recently, we synthesized and evaluated the structurally similar compound **38** and the corresponding analog with Phe as in Ang II in the C-terminal. Notably, both of these pentapeptides acted as AT$_2$R agonists.[5] In addition, in the same study, it was demonstrated that replacement of the Tyr-Ile unit by a single aromatic core produced a ligand **39** with an impressive AT$_2$R affinity (K_i of 0.5 nM) and producing an agonistic activity in the NG108-15 cell model comparable to Ang II.[5] Further structural simplifications resulted in a series of low-molecular-weight compounds. Among those, the imidazole derivative **40** with a K_i of 16.6 nM is structurally very similar to **14** (M024/C21).[33] Both compounds contain a methylene imidazole fragment, a lipophilic side chain (**14** encompasses two lipophilic side chains), and an acidic group. The two compounds should be able to adopt similar binding modes and probably occupy the same regions in the active conformation of the AT$_2$R. Proposed pharmacophore models of **40** and of **14** are presented in Figure 1.

(a)   (b)

FIGURE 1 Models of **40** (gray carbons) and **14** (M024/C21) (black carbons) with the essential pharmacophore elements marked. (a) Model in which the Ile side chain of **40** and the *n*-butyl group of **14** are overlaid. (b) Model in which the Ile side chain of **40** and the isobutyl group of **14** are overlaid.

Notably, with low-molecular-weight peptidomimetic compounds (e.g., **40** and structural analogs), difference in receptor affinity between species (human versus animal) has been observed. Hence, the affinity to AT$_2$R in human embryonic kidney (HEK293-AT$_2$R) cell assay was significantly lower than in a cell membrane preparation from hamster or pig uterus. However, for the longer peptides and the nonpeptidic series (e.g., the M024/C21 class), this was not the case.[38–40]

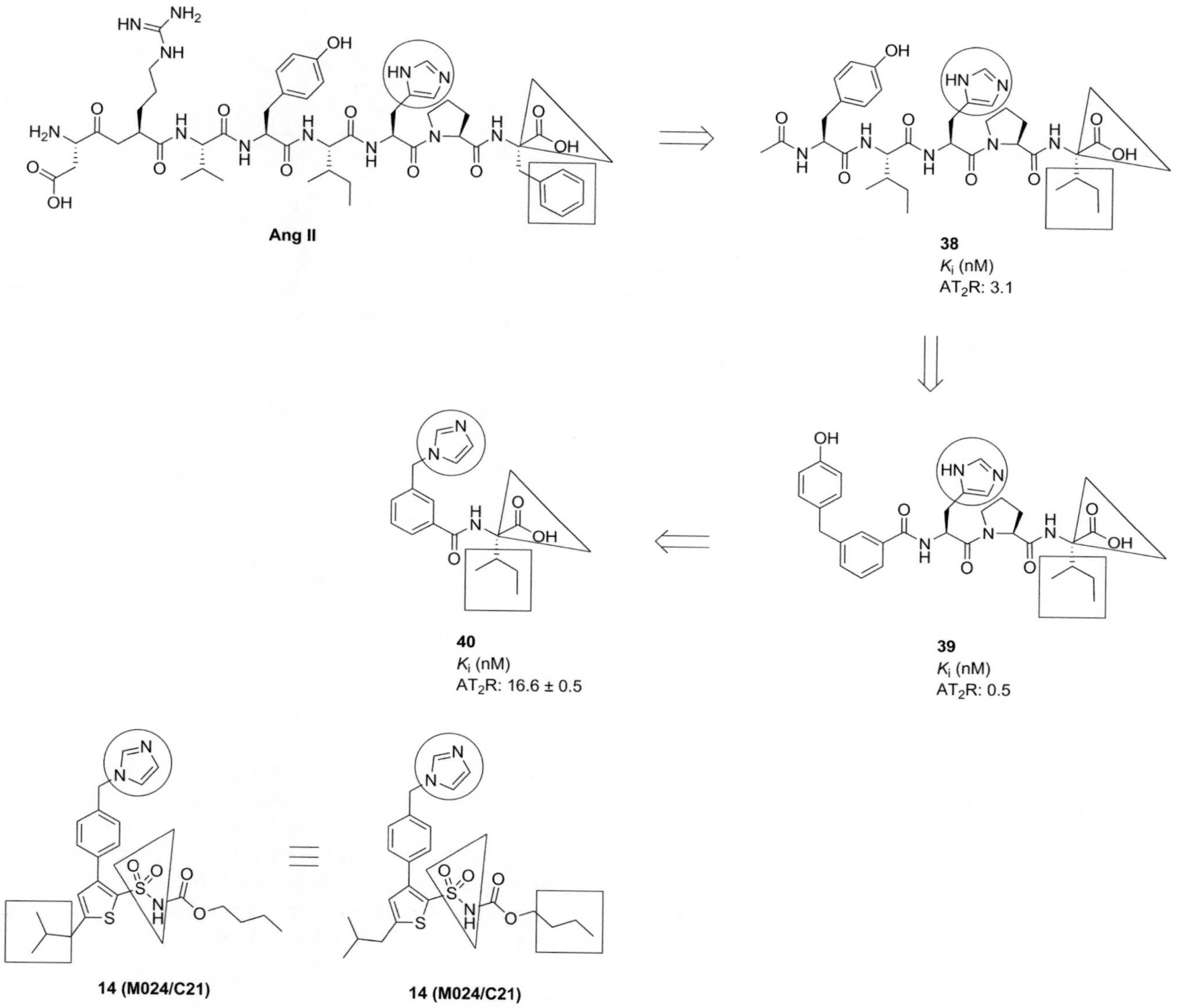

CHART 8

CONVERSION OF AT$_2$R AGONISTS TO AT$_2$R ANTAGONISTS

Can compounds be made that are structurally similar to, for example, M024/C21, that act as selective AT$_2$R antagonists? A characteristic feature of all selective AT$_2$R agonists discussed herein is the para substitution pattern of the phenyl ring in the biaryl scaffold. A large series of meta-substituted compounds was recently prepared. These derivatives were all AT$_2$/AT$_1$ receptor-selective but exhibited in general somewhat lower affinities to the AT$_2$R than the para-substituted counterparts.[41] Notably, some of these compounds with a substituent in the meta position functioned as antagonists at the AT$_2$R, but not all of them. Thus, the benzimidazole **41** and 2-chloroimidazole derivative **42** were antagonist, while the diketoester **43** induced neurite outgrowth in the NG108-15 cell assay and consequently served as an agonist.[41]

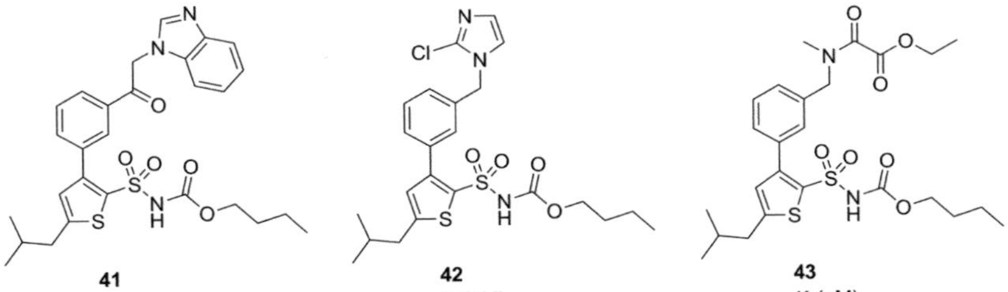

41 K_i (nM) AT_2R: 20

42 K_i (nM) AT_2R: 9.2

43 K_i (nM) AT_2R: 1.6

CHART 9

The migration of a methylene imidazole group from the para to the meta position illustrates how an agonist, compound **14** (M024/C21), via a minor modification can be converted to an antagonist, compound **44** (M132/C38).[41,42] The antagonist **44** was found to be considerably more effective as an antagonist in the neurite outgrowth model than PD 123,319, which is used in most laboratories as the standard selective AT_2R antagonist. The selective AT_2R antagonist **44** should exhibit a similar pharmacokinetic profile as **14**, thus facilitating adequate comparisons and studies in more complex animal models.

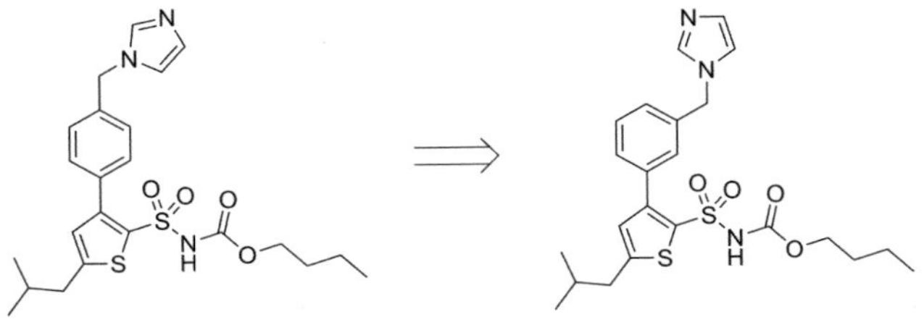

14 (M024/C21)
K_i (nM)
AT_1R: >10,000
AT_2R: 0.4
Agonist (AT_2R)

44
K_i (nM)
AT_2R: 19
Antagonist (AT_2R)

CHART 10

In summary, we have briefly discussed the structure–activity relationship of the first reported series of selective, nonpeptide AT_2R ligands. Drug-like selective AT_2R ligands with proper pharmacokinetic properties such as the agonist **14** (M024/C21)[25] and the antagonist **44** (M132/C38)[41] should serve as valuable tools in the assessment of the role of the AT_2R *in vivo*. The AT_2R-selective compound **14** (M024/C21) originates from the first reported nonselective, nonpeptide AT_1R agonist L-162,313 reported by Merck.[7,8] L-162,313 was found by us to also exert agonistic properties at the AT_2R.[13] The agonist **14** (M024/C21), one of more than 200 drug-like substances synthesized in our laboratory as selective AT_2R ligands, has attracted considerable attention and has recently been studied extensively in various *in vitro* and *in vivo* models.[43–59]

ACKNOWLEDGMENTS

We acknowledge Dr Mathias Alterman, Dr Charlotta Wallinder, Dr Jennie Georgsson, Dr Yiqian Wan, Dr A.K. Mahalingam, Dr Xiongyu Wu, Dr A.M.S. Murugaiah, and Dr Christian Sköld for their pivotal roles in this project. Furthermore, we would like to thank Professor Nicole Gallo-Payet, Professor Lars Fändriks, Professor Fred Nyberg, Professor Anders Karlén, Professor Ulrike Steckelings, and Professor Thomas Unger for stimulating fruitful and productive collaborations. We gratefully acknowledge the support from the Swedish Research Council and Kjell och Märta Beijer Foundation.

REFERENCES

1. Whitebread S, Mele M, Kamber B, de Gasparo M. Preliminary biochemical characterization of two angiotensin II receptor subtypes. *Biochem Biophys Res Commun* 1989;**163**:284–91.
2. Chiu AT, Herblin WF, McCall DE, Ardecky RJ, Carini DJ, Duncia JV, et al. Identification of angiotensin II receptor subtypes. *Biochem Biophys Res Commun* 1989;**165**:196–203.

3. Rice ASC, Dworkin RH, McCarthy TD, Anand P, Bountra C, McCloud PI, et al. EMA401, an orally administered highly selective angiotensin II type 2 receptor antagonist, as a novel treatment for postherpetic neuralgia: a randomised, double-blind, placebo-controlled phase 2 clinical trial. *Lancet* 2014;**6736**:1–11.

4. Smith MT, Wyse BD, Edwards SR. Small molecule angiotensin II type 2 receptor (AT2R) antagonists as novel analgesics for neuropathic pain: comparative pharmacokinetics, radioligand binding, and efficacy in rats. *Pain Med* 2013;**14**:692–705.

5. Georgsson J, Rosenström U, Wallinder C, Beaudry H, Plouffe B, Lindeberg G, et al. Short pseudopeptides containing turn scaffolds with high AT2 receptor affinity. *Bioorg Med Chem* 2006;**14**:5963–72.

6. Speth RC, Kim KH. Discrimination of two angiotensin II receptor subtypes with a selective agonist analogue of angiotensin II, *p*-aminophenylalanine6 angiotensin II. *Biochem Biophys Res Commun* 1990;**169**:997–1006.

7. Perlman S, Schambye HT, Rivero RA, Greenlee WJ, Hjorth SA, Schwartz TW. Non-peptide angiotensin agonist. *J Biol Chem* 1995;**270**:1493–6.

8. Kivlighn SD, Huckle WR, Zingaro GJ, Rivero RA, Lotti VJ, Chang RS, et al. Discovery of L-162,313: a nonpeptide that mimics the biological actions of angiotensin II. *Am J Physiol* 1995;**268**:R820–3.

9. Perlman S, Costa-Neto CM, Miyakawa AA, Schambye HT, Hjorth SA, Paiva AC, et al. Dual agonistic and antagonistic property of nonpeptide angiotensin AT1 ligands: susceptibility to receptor mutations. *Mol Pharmacol* 1997;**51**:301–11.

10. Rivero RA, Kevin NJ, Kivlighn SD, Zingaro GJ, Chang RS, Greenlee WJ. L-162,389: a potent orally active angiotensin II receptor antagonist with balanced affinity to both AT1 and AT2 receptor subtypes. *Bioorg Med Chem Lett* 1996;**6**:307–10.

11. Ashton WT, Chang LL, Flanagan KL, Mantlo NB, Ondeyka DL, Kim D, et al. AT1/AT2-balanced angiotensin II antagonists. *Eur J Med Chem* 1995;**30**:255–66.

12. Huckle WR, Kivlighn SD, Zingaro GJ, Kevin NJ, Rivero RA, Chang RSL, et al. Angiotensin II receptor-mediated activation of phosphoinositide hydrolysis and elevation of mean arterial pressure by a nonpeptide, L-163, 491. *Can J Physiol Pharmacol* 1994;**72**:543.

13. Wan Y, Wallinder C, Johansson B, Holm M, Mahalingam AK, Wu X, et al. First reported nonpeptide AT1 receptor agonist (L-162,313) acts as an AT2 receptor agonist in vivo. *J Med Chem* 2004;**47**:1536–46.

14. Johansson B, Holm M, Ewert S, Casselbrant A, Pettersson A, Fändriks L. Angiotensin II type 2 receptor-mediated duodenal mucosal alkaline secretion in the rat. *Am J Physiol Gastrointest Liver Physiol* 2001;**280**:G1254–60.

15. Johansson B, Holm M, Chen L, Pettersson A, Jönson C, Fändriks L. ANG II prolongs splanchnic nerve-mediated inhibition of duodenal mucosal alkaline secretion in the rat. *Am J Physiol* 1997;**273**:R942–6.

16. Blankley CJ, Hodges JC, Klutchko SR, Himmelsbach RJ, Chucholowski A, Connolly CJ, et al. Synthesis and structure-activity relationships of a novel series of non-peptide angiotensin II receptor binding inhibitors specific for the AT2 subtype. *J Med Chem* 1991;**34**:3248–60.

17. Kevin NJ, Rivero RA, Greenlee WJ, Chang RSL, Chen TB. Substituted phenylthiophene benzoylsulfonamides with potent binding affinity to angiotensin II AT1 and AT2 receptors. *Bioorg Med Chem Lett* 1994;**4**:189–94.

18. De Laszlo SE, Quagliato CS, Greenlee WJ, Patchett AA, Chang RSL, Lotti VJ, et al. A potent, orally active, balanced affinity angiotensin II AT1 antagonist and AT2 binding inhibitor. *J Med Chem* 1993;**36**:3207–10.

19. Glinka TW, de Laszlo SE, Tran J, Chang RS, Chen T-B, Lotti VJ, et al. L-161,638: a potent AT2selective quinazolinone angiotensin II binding inhibitor. *Bioorg Med Chem Lett* 1994;**4**:1479–84.

20. Glinka TW, de Laszlo SE, Siegl PKS, Chang RS, Kivlighn SD, Schorn TS, et al. A new class of balanced AT1/AT2 angiotensin II antagonists: quinazolinone AII antagonists with acylsulfonamide and sulfonylcarbamate acidic functionalities. *Bioorg Med Chem Lett* 1994;**4**:81–6.

21. Chakravarty PK, Strelitz RA, Chen T-B, Chang RSL, Lotti VJ, Zingaro GJ, et al. Quinazolinone biphenyl acylsulfonamides: a potent new class of angiotensin-II receptor antagonists. *Bioorg Med Chem Lett* 1994;**4**:75–80.

22. De Laszlo SE, Chang RS, Chen T-B, Faust KA, Greenlee WJ, Kivlighn SD, et al. The SAR of 6-(*N*-alkyl-*N*-acyl)-2-propyl-3-[(2′-tetrazol-5-yl) biphen-4-yl)methyl]-quinazolinones as balanced affinity antagonists of the human AT1 and AT2 receptors. *Bioorg Med Chem Lett* 1995;**5**:1359–64.

23. De Laszlo SE, Glinka TW, Greenlee WJ, Ball R, Nachbar RB, Prendergast K. The design, binding affinity prediction and synthesis of macrocyclic angiotensin II AT1 and AT2 receptor antagonists. *Bioorg Med Chem Lett* 1996;**6**:923–8.

24. Levin JI, Venkatesan AM, Chan PS, Baker JS, Francisco G, Bailey T, et al. 2,3,6-Substituted quinazolinones as angiotensin II receptor antagonists. *Bioorg Med Chem Lett* 1994;**4**:1135–40.

25. Wan Y, Wallinder C, Plouffe B, Beaudry H, Mahalingam AK, Wu X, et al. Design, synthesis, and biological evaluation of the first selective nonpeptide AT2 receptor agonist. *J Med Chem* 2004;**47**:5995–6008.

26. Laflamme L, de Gasparo M, Gallo JM, Payet MD, Gallo-Payet N. Angiotensin II induction of neurite outgrowth by AT2 receptors in NG108-15 cells. Effect counteracted by the AT1 receptors. *J Biol Chem* 1996;**271**:22729–35.

27. Gendron L, Laflamme L, Rivard N, Asselin C, Payet MD, Gallo-Payet N. Signals from the AT2 (angiotensin type 2) receptor of angiotensin II inhibit p21ras and activate MAPK (mitogen-activated protein kinase) to induce morphological neuronal differentiation in NG108-15 cells. *Mol Endocrinol* 1999;**13**:1615–26.

28. Wu X, Wan Y, Mahalingam AK, Murugaiah AM, Plouffe B, Botros M, et al. Selective angiotensin II AT2 receptor agonists: arylbenzylimidazole structure-activity relationships. *J Med Chem* 2006;**49**:7160–8.

29. Berellini G, Cruciani G, Mannhold R. Pharmacophore, drug metabolism, and pharmacokinetics models on non-peptide AT1, AT2, and AT1/AT2 angiotensin II receptor antagonists. *J Med Chem* 2005;**48**:4389–99.

30. Mahalingam AK, Wan Y, Murugaiah AMS, Wallinder C, Wu X, Plouffe B, et al. Selective angiotensin II AT(2) receptor agonists with reduced CYP 450 inhibition. *Bioorg Med Chem* 2010;**18**:4570–90.

31. Murugaiah AMS, Wallinder C, Mahalingam AK, Wu X, Wan Y, Plouffe B, et al. Selective angiotensin II AT(2) receptor agonists devoid of the imidazole ring system. *Bioorg Med Chem* 2007;**15**:7166–83.

32. Wallinder C, Botros M, Rosenström U, Guimond M-O, Beaudry H, Nyberg F, et al. Selective angiotensin II AT2 receptor agonists: benzamide structure-activity relationships. *Bioorg Med Chem* 2008;**16**:6841–9.

33. Georgsson J, Sköld C, Botros M, Lindeberg G, Nyberg F, Karlén A, et al. Synthesis of a new class of druglike angiotensin II C-terminal mimics with affinity for the AT 2 receptor. *J Med Chem* 2007;**50**:1711–15.

34. Rosenström U, Sköld C, Lindeberg G, Botros M, Nyberg F, Karlén A, et al. A selective AT2 receptor ligand with a gamma-turn-like mimetic replacing the amino acid residues 4–5 of angiotensin II. *J Med Chem* 2004;**47**:859–70.

35. Rosenström U, Sköld C, Plouffe B, Beaudry H, Lindeberg G, Botros M, et al. New selective AT2 receptor ligands encompassing a gamma-turn mimetic replacing the amino acid residues 4–5 of angiotensin II act as agonists. *J Med Chem* 2005;**48**:4009–24.

36. Georgsson J, Sköld C, Plouffe B, Lindeberg G, Botros M, Larhed M, et al. Angiotensin II pseudopeptides containing 1,3,5-trisubstituted benzene scaffolds with high AT2 receptor affinity. *J Med Chem* 2005;**48**:6620–31.

37. De Gasparo M, Whitebread S, Kamber B, Criscione L, Thomann H, Riniker B, et al. Effect of covalent dimer conjugates of angiotensin II on receptor affinity and activity in vitro. *J Recept Res* 1991;**11**:247–57.

38. Feng Y-H, Zhou L, Yan S, Douglas JG. Functional diversity of AT2 receptor orthologues in closely related species. *Kidney Int* 2005;**67**:1731–8.

39. Veron J-B, Joshi A, Wallinder C, Larhed M, Odell LR. Synthesis and evaluation of isoleucine derived angiotensin II AT2 receptor ligands. *Bioorg Med Chem Lett* 2014;**24**:476–9.

40. Behrends M, Wallinder C, Wieckowska A, Guimond M-O, Hallberg A, Gallo-Payet N, et al. N-Aryl isoleucine derivatives as angiotensin II AT 2 receptor ligands. *ChemistryOpen* 2014;**3**:65–75.

41. Murugaiah AMS, Wu X, Wallinder C, Mahalingam AK, Wan Y, Sköld C, et al. From the first selective non-peptide AT(2) receptor agonist to structurally related antagonists. *J Med Chem* 2012;**55**:2265–78.

42. Guimond M-O, Wallinder C, Alterman M, Hallberg A, Gallo-Payet N. Comparative functional properties of two structurally similar selective nonpeptide drug-like ligands for the angiotensin II type-2 (AT(2)) receptor. Effects on neurite outgrowth in NG108-15 cells. *Eur J Pharmacol* 2013;**699**:160–71.

43. Kaschina E, Grzesiak A, Li J, Foryst-Ludwig A, Timm M, Rompe F, et al. Angiotensin II type 2 receptor stimulation: a novel option of therapeutic interference with the renin-angiotensin system in myocardial infarction? *Circulation* 2008;**118**:2523–32.

44. Gelosa P, Pignieri A, Fändriks L, de Gasparo M, Hallberg A, Banfi C, et al. Stimulation of AT2 receptor exerts beneficial effects in stroke-prone rats: focus on renal damage. *J Hypertens* 2009;**27**:2444–51.

45. Rompe F, Artuc M, Hallberg A, Alterman M, Ströder K, Thöne-Reineke C, et al. Direct angiotensin II type 2 receptor stimulation acts anti-inflammatory through epoxyeicosatrienoic acid and inhibition of nuclear factor kappaB. *Hypertension* 2010;**55**:924–31.

46. Steckelings UM, Larhed M, Hallberg A, Widdop RE, Jones ES, Wallinder C, et al. Non-peptide AT2-receptor agonists. *Curr Opin Pharmacol* 2011;**11**:187–92.

47. Gallo-Payet N, Guimond M-O, Bilodeau L, Wallinder C, Alterman M, Hallberg A. Angiotensin II, a neuropeptide at the frontier between endocrinology and neuroscience: is there a link between the angiotensin II type 2 receptor and Alzheimer's disease? *Front Endocrinol (Lausanne)* 2011;**2**:17.

48. Foulquier S, Steckelings UM, Unger T. Impact of the AT(2) receptor agonist C21 on blood pressure and beyond. *Curr Hypertens Rep* 2012;**14**:403–9.

49. Ohshima K, Mogi M, Jing F, Iwanami J, Tsukuda K, Min L-J, et al. Direct angiotensin II type 2 receptor stimulation ameliorates insulin resistance in type 2 diabetes mice with PPARγ activation. *PLoS One* 2012;**7**:e48387.

50. Mogi M, Horiuchi M. Effect of angiotensin II type 2 receptor on stroke, cognitive impairment and neurodegenerative diseases. *Geriatr Gerontol Int* 2013;**13**:13–18.

51. McCarthy CA, Widdop RE, Denton KM, Jones ES. Update on the angiotensin AT(2) receptor. *Curr Hypertens Rep* 2013;**15**:25–30.

52. Shum M, Pinard S, Guimond M-O, Labbé SM, Roberge C, Baillargeon J-P, et al. Angiotensin II type 2 receptor promotes adipocyte differentiation and restores adipocyte size in high-fat/high-fructose diet-induced insulin resistance in rats. *Am J Physiol Endocrinol Metab* 2013;**304**:E197–210.

53. Namsolleck P, Boato F, Schwengel K, Paulis L, Matho KS, Geurts N, et al. AT2-receptor stimulation enhances axonal plasticity after spinal cord injury by upregulating BDNF expression. *Neurobiol Dis* 2013;**51**:177–91.

54. Guimond M-O, Battista M-C, Nikjouitavabi F, Carmel M, Barres V, Doueik AA, et al. Expression and role of the angiotensin II AT2 receptor in human prostate tissue: in search of a new therapeutic option for prostate cancer. *Prostate* 2013;**73**:1057–68.

55. Dhande I, Ali Q, Hussain T. Proximal tubule angiotensin AT2 receptors mediate an anti-inflammatory response via interleukin-10: role in renoprotection in obese rats. *Hypertension* 2013;**61**:1218–26.

56. Shao C, Yu L, Gao L. Activation of angiotensin type 2 receptors partially ameliorates streptozotocin-induced diabetes in male rats by islet protection. *Endocrinology* 2013;**155**:793–804.

57. Hrenák J, Arendášová K, Rajkovičová R, Aziriová S, Repová K, Krajčírovičová K, et al. Protective effect of captopril, olmesartan, melatonin and compound 21 on doxorubicin-induced nephrotoxicity in rats. *Physiol Res* 2013;**62**:S181–9.

58. Iwanami J, Mogi M, Tsukuda K, Jing F, Ohshima K, Wang X-L, et al. Possible synergistic effect of direct angiotensin II type 2 receptor stimulation by compound 21 with memantine on prevention of cognitive decline in type 2 diabetic mice. *Eur J Pharmacol* 2014;**724**:9–15.

59. Lauer D, Slavic S, Sommerfeld M, Thöne-Reineke C, Sharkovska Y, Hallberg A, et al. Angiotensin type 2 receptor stimulation ameliorates left ventricular fibrosis and dysfunction via regulation of tissue inhibitor of matrix metalloproteinase 1/matrix metalloproteinase 9 axis and transforming growth factor β1 in the rat heart. *Hypertension* 2014;**63**:e60–7.

Chapter 19

Angiotensin-Based Peptides as AT$_2$R Agonists

Emma S. Jones,#* Yan Wang,#* Mark Del Borgo,† Kate M. Denton,‡ Marie-Isabel Aguilar,† Robert E. Widdop*

*Department of Pharmacology, Monash University, Clayton Victoria 3800, Australia, †Department of Biochemistry and Molecular Biology, Monash University, Clayton Victoria 3800, Australia, ‡Department of Physiology, Monash University, Clayton Victoria 3800, Australia

INTRODUCTION

A recurrent criticism of peptides as therapeutic agents stems from the fact that they are susceptible to rapid breakdown by endogenous peptides particularly in the digestive tract, thus generally requiring intravenous or subcutaneous administration to reach sufficient plasma levels. However, there are many approaches to combat this drawback of peptide administration such as sustained-release depot injections,[1] chemical modification to prevent enzyme recognition and increase stability,[2] and use of "carrier" formulations to facilitate movement across biological membranes.[3,4] Indeed, several of these techniques have already been utilized in the context of angiotensin peptides with promising results, as will be discussed further in following sections.

Importantly, there are many characteristics of peptides that offer advantages in drug design that are well worth pursuing. For example, the endogenous ligands of drug targets are peptides themselves, which often show high affinity to their cognate receptor, albeit with little selectivity for specific receptor subtypes. However, this lack of selectivity can often be altered by relatively simple chemical modification of the peptide, thus transforming a high-affinity/low-selectivity agonist, into a potent and selective entity suitable for consideration as a lead compound for further development.[5] At the very least, "selective peptides" become valuable tools with which to probe AT$_2$R pathophysiology.

In the following sections, both endogenous and synthetic peptide AT$_2$R agonists will be discussed, with particular emphasis on recent development in the area of peptide agonists.

ENDOGENOUS ANGIOTENSIN PEPTIDES

Our understanding of angiotensin peptides has expanded dramatically over the last 10–15 years, moving from the straightforward concept of Ang II being the predominant mediator of the renin–angiotensin system (RAS) with other angiotensin peptides being primarily by-products of Ang II synthesis and metabolism to that of a system in which multiple angiotensin peptides have biological activity in their own right, thus providing an important physiological brake on overstimulation of AT$_1$R, via activation of AT$_2$R and other non-AT$_1$R binding sites. Indeed, current knowledge recognizes multiple angiotensin peptides that have protective effects via AT$_2$R stimulation, with rank order of binding affinity of peptides for the AT$_2$R being Ang III ≥ Ang II ≫ Ang IV > Ang 1-7 > Ang 1-9[6] (see Table 1).

Ang II

Considering the pivotal role of Ang II within the RAS, it is not surprising that this octapeptide shows high-affinity binding to AT$_2$R, which in fact is of very similar magnitude to that at AT$_1$R.[6] Indeed, before the development of selective AT$_2$R agonists, elucidation of AT$_2$R function was largely dependent on the assumption that any effects of either exogenous or endogenous Ang II during AT$_1$R blockade must be due to stimulation of unblocked AT$_2$R (for review, see Ref. [19]). However, since the development of much more selective AT$_2$R agonists, Ang II is now rarely used in this role to specifically probe AT$_2$R function.

# Both these authors contributed equally.

The Protective Arm of the Renin–Angiotensin System (RAS). http://dx.doi.org/10.1016/B978-0-12-801364-9.00019-5

TABLE 1 AT$_2$R Binding Affinity and Biological Stability of Peptide AT$_2$R Agonists

Peptide	AT$_2$R Binding Affinity	Biological Stability
Ang II	IC50 = 5.2 × 10^{-10} M in AT$_2$R-transfected HEK-293 cells[6] K_d = 1.1 × 10^{-10} M in AT$_2$R-transfected COS-3 cells[7]	$t_{1/2}$ = 4.4 min in human blood *in vitro*[8] $t_{1/2}$ = 14 s in circulation of rats[9] $t_{1/2}$ = 16 s in circulation of rats[10] $t_{1/2}$ = 28 min in 50% rat plasma[6]
Ang III	IC50 = 6.5 × 10^{-10} M in AT$_2$R-transfected HEK-293 cells[6] K_i = 0.15-0.36 M rat and human adrenal[11]	$t_{1/2}$ = 2.0 min in human blood *in vitro*[8] $t_{1/2}$ = 14 s in circulation of rats[10]
Ang IV	IC50 = 4.9 × 10^{-8} M in AT$_2$R-transfected HEK-293 cells[6]	$t_{1/2}$ = 2.4 min in human blood *in vitro*[8]
Ang 1-7	IC50 = 2.5 × 10^{-7} M AT$_2$R-transfected HEK-293 cells[6]	$t_{1/2}$ = 10 s in circulation of rats[12]
[cAng-(1-7)]	Not available	Not available
Ang 1-9	K_i = 3.3 × 10^{-6} M[13]	K_{cat}/K_m = 3.7 × 10^5 M(−1) × s(−1) as substrate for purified NEP *in vitro*[14]
CGP42112	K_i = 0.35-0.98 × 10^{-9} M in rat and human uterus and adrenal[11,15] IC50 = 2.3 × 10^{-10} M in AT$_2$R-transfected HEK-293 cells[6] IC50 = 4.0 × 10^{-9} M in AT$_2$R-transfected COS cells[16]	Not available
β-Ile Ang II	IC50 = 10.6 × 10^{-9} M in AT$_2$R-transfected HEK-293 cells[5]	$t_{1/2}$ = 295 min in 50% rat plasma[6]
[Y]^6-AII	K_i = 3.4 × 10^{-9} M in membranes from transiently transfected AT$_2$-HEK-293T cells[17]	Not available
LP2-3 (dKcAng-(1-7))	Not available	Not available
Novokinin	K_i = 7 × 10^{-6} M[18]	Not available

Stability in humans and rodents differs dramatically due to variation in enzymatic metabolism between species. IC50, half maximal inhibitory concentration; K_d, dissociation constant; K_i, inhibition constant; $t_{1/2}$, half-life; K_{cat}/K_m, specificity constant.

Ang III

Similarly to Ang II, Ang III has high affinity for the AT$_2$R, but due to relatively lower AT$_1$R binding, Ang III possesses greater AT$_2$R–AT$_1$R selectivity than Ang II.[6] As such, AT$_1$R-mediated effects are generally less than those of equimolar Ang II, whereas maximal AT$_2$R-mediated actions may be unmasked by concomitant AT$_1$R blockade.[20,21] Ang III is formed by cleavage of the N-terminal aspartate residue of Ang II by aminopeptidase A (APA), thus simultaneously metabolizing Ang II and producing a more AT$_2$R-selective peptide product. Indeed, it has been suggested that Ang III may in fact be the endogenous ligand for AT$_2$R. In seminal studies by Carey et al., Ang III infusion caused natriuresis, which was sensitive to PD123319, indicating AT$_2$R-mediated effects.[22] Moreover, this group demonstrated that *in vivo* inhibition of APA and thus prevention of endogenous Ang III production inhibited natriuresis, thus highlighting the necessity for the conversion of Ang II to Ang III for renal AT$_2$R function.[23,24] Similarly, other groups have suggested an important vasodilator role for Ang III-mediated AT$_2$R stimulation in the rat heart.[25–27]

Ang IV

Surprisingly, in spite of relatively good binding affinity at AT$_2$R (which is in fact greater than that of Ang 1-7[6]), there has been little direct research into the contribution of AT$_2$R stimulation to Ang IV effects. In ApoE-deficient mice fed with a high-fat diet for 8 weeks, there was significant endothelial dysfunction in aortic vessels due to reduced NO bioavailability, which was prevented by concurrent Ang IV treatment.[28] Interestingly, the protective effects of Ang IV in this model were attenuated by simultaneous treatment with either the AT$_4$R antagonist, divalinal Ang IV, or PD123319, indicating involvement of both AT$_4$R and AT$_2$R.[28] Similarly, Ang IV was also vasoprotective when treatment was initiated after the development of atherosclerotic lesions, effects that were sensitive to AT$_2$R inhibition.[29]

Ang 1-7

Interest in Ang 1-7 as a cardioprotective peptide gained significant momentum with the discovery of angiotensin converting enzyme 2 (ACE2), which is involved in Ang 1-7 production from Ang II or Ang I. Ang 1-7 has highest affinity for MasR and it exerts a number of cardiovascular-protective effects via this mechanism.[30] However, Ang 1-7 also shows significant binding at AT$_2$R,[6] and AT$_2$R-mediated effects of Ang 1-7 have also been observed *in vivo*. We have shown that Ang-(1-7) reduces blood pressure in both normotensive and hypertensive adult rats (during low-level AT$_1$R inhibition), an effect that was blocked by PD123319, but not by the MasR antagonist, A-779.[31] Interestingly, this effect was preserved in aged normotensive rats, but in these animals, the vasodepressor effect was via both the AT$_2$R and the Mas receptor and corresponded with an upregulation of MasR with senescence.[32] In isolated mouse hearts exposed to the AT$_2$R antagonist PD123319, Ang-(1-7) increased perfusion pressure, an effect not observed with Ang-(1-7) infusion alone and that was independent of both AT$_1$R and MasR.[33]

In terms of therapeutic potential, several groups have developed techniques to enable oral delivery of Ang 1-7. Ang-(1-7) has been incorporated into a cyclodextrin (CyD) solution [Ang-(1-7)-CyD], which is a commonly used method to enhance drug solubility, stability, and absorption across biological barriers.[34] In an alternative approach, stabilization of Ang 1-7 by introduction of thioether bridges resulted in a peptide cAng-(1-7) with sufficient protease resistance to enable oral or pulmonary delivery, with therapeutic plasma concentrations reached via both routes of administration.[2]

Ang 1-9

Originally, Ang 1-9 was thought to be biologically inactive, contributing indirectly to counterregulate actions of Ang II by competing with Ang I for the ACE active site, thus reducing Ang II and increasing Ang 1-7 levels.[35] However, there are more recent suggestions that Ang 1-9 may exert direct biological effects in the cardiovascular system, as advocated by Ocaranza et al.[36] Ang 1-9 was antihypertrophic in neonatal and adult cardiomyocytes, an effect inhibited by the AT$_2$R antagonist PD123319, whereas those of Ang 1-7 were only inhibited in the presence of the MasR antagonist A-779.[37] Likewise, Ang 1-9 reduced cardiac fibrosis in spontaneously hypertensive rat-stroke prone (SHRSP) by 50%, whereas cotreatment with PD123319 attenuated the antifibrotic effects of Ang 1-9.[38] However, this group also showed that Ang 1-9 binds with only moderate affinity at AT$_2$R (~100-fold lower than Ang II for the AT$_2$R),[39] questioning whether concentrations reached *in vivo* would in fact achieve levels required for AT$_2$R stimulation. Furthermore, Ang 1-9 is known to inhibit cardiac ACE, thus reducing Ang II formation,[13] and also to potentiate bradykinin signaling at concentrations at least ten-fold lower than its pK_i for AT$_2$R,[40] which could contribute to the effects reported. Additionally, the authors' interpretation is dependent on the AT$_2$R selectivity of PD123319, which has recently been reported to also be a competitive antagonist at the newly identified Mas-related G protein-coupled receptor D,[7] although this could be a criticism of many studies inferring AT$_2$R involvement solely based on use of PD123319.

SYNTHETIC PEPTIDES

Design of synthetic peptides has invariably commenced with the endogenous peptide ligand Ang II and subsequent modifications to enhance AT$_2$R selectivity by maintaining AT$_2$R affinity while reducing AT$_1$R binding. Using this approach, investigation into AT$_2$R ligand-receptor structure–activity relationships has identified several interactions, which are critical for AT$_2$R selectivity. It has become clear that the C-terminal end of Ang II interacts with the inner half of the third transmembrane domain of AT$_2$R,[41–43] and thus, structural analogs require conservation of this interaction for effective binding.[44] On the other hand, the N-terminal portion of Ang II is not critical for AT$_2$R binding,[45–47] which is consistent with enhanced AT$_2$R binding of Ang III compared to Ang II, and demonstration that the high AT$_2$R binding affinity of CGP42112 is independent of amino terminal interaction.[41,44,48,49]

CGP42112

Until the recent development of the nonpeptide agonist, compound 21 (C21), CGP42112 had long been considered the "gold standard" AT$_2$R agonist. Although CGP42112 has been criticized for the lack of selectivity, CGP42112 behaves as a full agonist both *in vitro* and *in vivo*,[15,16] and AT$_1$R binding actually only occurs at relatively high concentrations (i.e., $>1 \times 10^{-5}$ M).[6] Moreover, CGP42112 still remains the highest-affinity AT$_2$R agonist to date ($K_i = 2 \times 10^{-10}$ M) and shows ~ten-fold greater AT$_2$R–AT$_1$R selectivity when directly compared to that of C21 in the same binding assay.[6]

β-Amino-Acid–Substituted Angiotensin Peptides

Given that Ang II activates both AT_1R and AT_2R, there is obvious merit in performing modifications to the native effector molecule Ang II to produce AT_2R ligands. β-Amino acids are similar in structure to naturally occurring α-amino acids except they contain an "extra" carbon atom, that is, β-amino acid substitution results in an amino acid with an identical side chain (R group) but containing an additional methylene group (CH_2) in the peptide backbone. We have previously performed a β-amino acid scan using the octapeptide Ang II as the template.[50] This study identified compounds with remarkable AT_2R selectivity including β-Tyr[4]-Ang II and β-Ile[5]-Ang II, which exhibited negligible AT_1R binding and were >1000-fold selective for AT_2R compared to AT_1R.[5] Furthermore, β-Tyr[4]-Ang II and β-Ile[5]-Ang II both evoked AT_2R-mediated vasorelaxation of mouse thoracic aortas via a nitric-oxide-dependent mechanism involving AT_2R (blocked by the NOS inhibitor L-NAME and AT_2R antagonist PD123319, respectively), and β-Ile[5]-Ang II lowered blood pressure in conscious spontaneously hypertensive rat (SHR) against a background of AT_1R blockade, as we have reported for CGP42112[15] and C21.[51] In addition, using an *in vitro* plasma stability assay, we determined that the single substitution in β-Ile[5]-Ang II increased half-life to approximately ten times greater than that of native Ang II.

Given that the approach of β-substitution of Ang II provided such striking effects on AT_2R selectivity, we have recently synthesized β-scans of a number of Ang peptides including Ang III. Several analogs exhibited marked AT_2R selectivity and caused vasorelaxation of precontracted mouse aortas that was highly correlated with AT_2R affinity (unpublished data). Thus, our strategy of using novel β-substituted Ang analogs that are cheap and straightforward to synthesize is likely to yield highly selective lead compounds that, importantly, do not exhibit AT_1R effects at high doses, unlike CGP42112 and C21.[51–53]

[Y]⁶-AII

[Y]⁶-AII is a recently developed AT_2R agonist based on the peptide sequence of Ang II, with modifications to improve $AT_2R–AT_1R$ selectivity. Peptide analogs [Y]⁶-AII and [4-OPO₃H₂-F]⁶-AII were synthesized and used to probe the binding pocket of AT_2R and to discriminate between differently charged entities at position 6. It was found that [Y]⁶-AII analog (with an electron-donating group at position 6) results in K_i values for AT_2R one order of magnitude larger than the [4-OPO₃H₂-F]⁶-AII ligand (electron-deficient aromatic residue), confirming the importance of hydrostatic interactions between the C-terminal portions of agonists within the AT_2R binding pocket. Importantly, [Y]⁶-AII displayed high binding affinity at AT_2R ($K_i = 3$ nM), with 18,000-fold selectivity for AT_2R compared to AT_1R.[54] [Y]⁶-AII induced neurite outgrowth in PC12 cells, presumably via AT_2R stimulation, but functional effects are yet to be tested *in vivo*.

LP2-3 (dKcAng-(1-7))

LP2-3 is a recently reported AT_2R agonist developed by Lanthio Pharma, whose structure is based on a cyclized form of Ang 1-7 with an additional NH_2-terminal D-lysine (to reduce susceptibility to enzymatic degradation) and substitution of Tyr[4] by D-cysteine (to prevent AT_1R interaction). LP2-3 was administered to neonatal rat pups, which were exposed to hyperoxic conditions as an experimental model of chronic lung disease. LP2-3 reduced indexes of cardiopulmonary remodeling including alveolar septal thickness, arterial medial wall thickness, macrophage influx, and right ventricular hypertrophy.[17] However, whether or not LP2-3 effects were selective for AT_2R or due to off-target effects is yet to be established. AT_2R selectivity of response was not confirmed, and inhibition of NO production, which would usually be anticipated to at least dampen AT_2R signaling, did not modify the LP2-3 response. Moreover, data regarding pharmacological characterization of LP2-3 have not yet been made available, other than a reference to unpublished data claiming that LP2-3 is fully resistant to ACE and induces Erk phosphorylation in human bronchial epithelial cells, which was inhibited by the AT_2R antagonist PD123319.[17] Clearly, confirmation of binding affinity and selectivity is required before LP2-3 can be conclusively considered an AT_2R agonist.

Novokinin

Novokinin is a small peptide that was designed based on ovokinin, a vasorelaxing peptide present in egg albumin. The sequence of novokinin (Arg-Phe-Lys-K-Phe-Trp) has clear similarity to that of Ang II; thus, it is not surprising that the peptide displays affinity for AT_2R ($K_i = 7 \times 10^{-6}$ M).[18,55] Novokinin caused vasodilation in mesenteric arteries at high concentrations (>10^{-5} M) and reduced blood pressure in SHR by ~10 mmHg, both effects that were blocked by PD123319 and are consistent with a moderate-affinity AT_2R agonist.[56] Analogs of ovokinin were synthesized and screened for stability in a pancreatic cell assay. It was found that incorporation of Trp at the C-terminus of the peptide afforded novokinin >100-fold resistance to enzymatic degradation, and consequently, the peptide displayed ~30% oral availability. In addition, the hypotensive effect of oral novokinin was absent in AT$_2$R-deficient mice, confirming that AT$_2$R selectivity is retained following ingestion.[56] Novokinin has since been incorporated by genetic manipulation into rice, and oral consumption of these seeds by rodents was found to modestly reduce blood pressure.[57]

CONCLUSIONS

From the preceding discussion, it is clear that a number of highly selective AT$_2$R peptide ligands have been developed using strategies based on modifications to the effector peptide Ang II. In particular, chemical modifications or amino acid substitutions to the C-terminal portion of Ang II have resulted in ligands exhibiting affinity that was several orders of magnitude greater at AT$_2$R than AT$_1$R. However, relatively few of these peptides have been tested for functional activity, which is critical in order to develop a greater range of compounds by which to probe both AT$_2$R agonist and AT$_2$R antagonist functions.

ACKNOWLEDGMENTS

Studies from the authors' laboratories were supported in part by grants from the National Health and Medical Research Council of Australia and the National Heart Foundation of Australia.

REFERENCES

1. Amiram M, Luginbuhl KM, Li X, Feinglos MN, Chilkoti A. A depot-forming glucagon-like peptide-1 fusion protein reduces blood glucose for five days with a single injection. *J Control Release* 2013;**172**(1):144–51.
2. de Vries L, Reitzema-Klein CE, Meter-Arkema A, van Dam A, Rink R, Moll GN, et al. Oral and pulmonary delivery of thioether-bridged angiotensin-(1–7). *Peptides* 2010;**31**(5):893–8.
3. Santos CF, Santos SH, Ferreira AV, Botion LM, Santos RA, Campagnole-Santos MJ. Association of an oral formulation of angiotensin-(1–7) with atenolol improves lipid metabolism in hypertensive rats. *Peptides* 2013;**43**:155–9.
4. Santos SH, Andrade JM, Fernandes LR, Sinisterra RD, Sousa FB, Feltenberger JD, et al. Oral angiotensin-(1–7) prevented obesity and hepatic inflammation by inhibition of resistin/TLR4/MAPK/NF-kappaB in rats fed with high-fat diet. *Peptides* 2013;**46**:47–52.
5. Jones ES, Del Borgo MP, Kirsch JF, Clayton D, Bosnyak S, Welungoda I, et al. A single β-amino acid substitution to angiotensin II confers AT2 receptor selectivity and vascular function. *Hypertension* 2011;**57**(3):570–6.
6. Bosnyak S, Jones ES, Christopolous A, Aguilar MI, Thomas WG, Widdop RE. Relative affinity of angiotensin peptides and novel ligands at AT1 and AT2 receptors. *Clin Sci* 2011;**121**:297–303.
7. Lautner RQ, Villela DC, Fraga-Silva RA, Silva N, Verano-Braga T, Costa-Fraga F, et al. Discovery and characterization of alamandine: a novel component of the renin-angiotensin system. *Circ Res* 2013;**112**(8):1104–11.
8. Chapman BJ, Brooks DP, Munday KA. Half-life of angiotensin II in the conscious and barbiturate-anaesthetized rat. *Br J Anaesth* 1980;**52**(4):389–93.
9. Al-Merani SA, Brooks DP, Chapman BJ, Munday KA. The half-lives of angiotensin II, angiotensin II-amide, angiotensin III, Sar1-Ala8-angiotensin II and renin in the circulatory system of the rat. *J Physiol* 1978;**278**:471–90.
10. Whitebread S, Mele M, Kamber B, de Gasparo M. Preliminary biochemical characterization of two angiotensin II receptor subtypes. *Biochem Biophys Res Commun* 1989;**163**(1):284–91.
11. Yamada K, Iyer SN, Chappell MC, Ganten D, Ferrario CM. Converting enzyme determines plasma clearance of angiotensin-(1–7). *Hypertension* 1998;**32**(3):496–502.
12. Rice GI, Thomas DA, Grant PJ, Turner AJ, Hooper NM. Evaluation of angiotensin-converting enzyme (ACE), its homologue ACE2 and neprilysin in angiotensin peptide metabolism. *Biochem J* 2004;**383**(Pt 1):45–51.
13. Kokkonen JO, Saarinen J, Kovanen PT. Regulation of local angiotensin II formation in the human heart in the presence of interstitial fluid. Inhibition of chymase by protease inhibitors of interstitial fluid and of angiotensin-converting enzyme by Ang-(1–9) formed by heart carboxypeptidase A-like activity. *Circulation* 1997;**95**(6):1455–63.
14. Semple PF, Boyd AS, Dawes PM, Morton JJ. Angiotensin II and its heptapeptide (2–8), hexapeptide (3–8), and pentapeptide (4–8) metabolites in arterial and venous blood of man. *Circ Res* 1976;**39**(5):671–8.
15. Barber MN, Sampey DB, Widdop RE. AT(2) receptor stimulation enhances antihypertensive effect of AT(1) receptor antagonist in hypertensive rats. *Hypertension* 1999;**34**(5):1112–16.
16. Brechler V, Jones PW, Levens N, et al. Agonist and antagonist properties of angiotensin analogs at the AT2 receptor in PC12W cells. *Regul Pept* 1993;**44**:207–13.
17. Wagenaar GT, Laghmani ElH, Fidder M, Sengers RM, de Visser YP, de Vries L, et al. Agonists of MAS oncogene and angiotensin II type 2 receptors attenuate cardiopulmonary disease in rats with neonatal hyperoxia-induced lung injury. *Am J Physiol* 2013;**305**(5):L341–51.
18. Yamada Y, Yamauchi D, Yokoo M, Ohinata K, Usui H, Yoshikawa M. A potent hypotensive peptide, novokinin, induces relaxation by AT2- and IP-receptor-dependent mechanism in the mesenteric artery from SHRs. *Biosci Biotechnol Biochem* 2008;**72**(1):257–9.
19. Widdop RE, Jones ES, Hannan RE, Gaspari TA. Angiotensin AT2 receptors: cardiovascular hope or hype? *Br J Pharmacol* 2003;**140**:809–24.
20. Walters PE, Gaspari TA, Widdop RE. Ang III acts as a vasodepressor agent via the AT(2) receptor. *Hypertension* 2003;**42**(3):149.

21. Scheuer DA, Perrone MH. Angiotensin type 2 receptors mediate depressor phase of biphasic pressure response to angiotensin. *Am J Physiol* 1993;**264**(5 Pt 2):R917–23.

22. Padia SH, Howell NL, Siragy HM, Carey RM. Renal angiotensin type 2 receptors mediate natriuresis via angiotensin III in the angiotensin II type 1 receptor-blocked rat. *Hypertension* 2006;**47**(3):537–44.

23. Padia SH, Kemp BA, Howell NL, Fournie-Zaluski MC, Roques BP, Carey RM. Conversion of renal angiotensin II to angiotensin III is critical for AT2 receptor-mediated natriuresis in rats. *Hypertension* 2008;**51**(2):460–5.

24. Padia SH, Kemp BA, Howell NL, Siragy HM, Fournie-Zaluski M-C, Roques BP, et al. Intrarenal aminopeptidase N inhibition augments natriuretic responses to angiotensin III in angiotensin type 1 receptor-blocked rats. *Hypertension* 2007;**49**(3):625–30.

25. van Esch JH, Oosterveer CR, Batenburg WW, van Veghel R, Jan Danser AH. Effects of angiotensin II and its metabolites in the rat coronary vascular bed: is angiotensin III the preferred ligand of the angiotensin AT2 receptor? *Eur J Pharmacol* 2008;**588**(2–3):286–93.

26. Park BM, Gao S, Cha SA, Park BH, Kim SH. Cardioprotective effects of angiotensin III against ischemic injury via the AT2 receptor and KATP channels. *Physiol Rep* 2013;**1**(6):e00151.

27. Park BM, Oh YB, Gao S, Cha SA, Kang KP, Kim SH. Angiotensin III stimulates high stretch-induced ANP secretion via angiotensin type 2 receptor. *Peptides* 2013;**42**:131–7.

28. Vinh A, Widdop RE, Drummond GR, Gaspari TA. Chronic angiotensin IV treatment reverses endothelial dysfunction in ApoE-deficient mice. *Cardiovasc Res* 2008;**77**(1):178–87.

29. Vinh A, Widdop RE, Chai SY, Gaspari TA. Angiotensin IV-evoked vasoprotection is conserved in advanced atheroma. *Atherosclerosis* 2008;**200**(1):37–44.

30. Ohshima K, Mogi M, Nakaoka H, Iwanami J, Min LJ, Kanno H, et al. Possible role of angiotensin-converting enzyme 2 and activation of angiotensin II type 2 receptor by angiotensin-(1–7) in improvement of vascular remodeling by angiotensin II type 1 receptor blockade. *Hypertension* 2014;**63**(3):e53–9.

31. Walters PE, Gaspari TA, Widdop RE. Angiotensin-(1–7) acts as a vasodepressor agent via angiotensin II type 2 receptors in conscious rats. *Hypertension* 2005;**45**(5):960–6.

32. Bosnyak S, Widdop RE, Denton KM, Jones ES. Differential mechanisms of Ang (1–7)-mediated vasodepressor effect in adult and aged candesartan-treated rats. *Int J Hypertens* 2012;**2012**:192567.

33. Castro CH, Santos RA, Ferreira AJ, Bader M, Alenina N, Almeida AP. Evidence for a functional interaction of the angiotensin-(1–7) receptor Mas with AT1 and AT2 receptors in the mouse heart. *Hypertension* 2005;**46**(4):937–42.

34. Feltenberger JD, Andrade JM, Paraiso A, Barros LO, Filho AB, Sinisterra RD, et al. Oral formulation of angiotensin-(1–7) improves lipid metabolism and prevents high-fat diet-induced hepatic steatosis and inflammation in mice. *Hypertension* 2013;**62**(2):324–30.

35. McKinney CA, Fattah C, Loughrey CM, Milligan G, Nicklin SA. Angiotensin-(1–7) and angiotensin-(1–9): function in cardiac and vascular remodelling. *Clin Sci* 2014;**126**(12):815–27.

36. Ocaranza MP, Moya J, Barrientos V, Alzamora R, Hevia D, Morales C, et al. Angiotensin-(1–9) reverses experimental hypertension and cardiovascular damage by inhibition of the angiotensin converting enzyme/Ang II axis. *J Hypertens* 2014;**32**(4):771–83.

37. Ocaranza MP, Lavandero S, Jalil JE, Moya J, Pinto M, Novoa U, et al. Angiotensin-(1–9) regulates cardiac hypertrophy in vivo and in vitro. *J Hypertens* 2010;**28**(5):1054–64.

38. Flores-Munoz M, Work LM, Douglas K, Denby L, Dominiczak AF, Graham D, et al. Angiotensin-(1–9) attenuates cardiac fibrosis in the stroke-prone spontaneously hypertensive rat via the angiotensin type 2 receptor. *Hypertension* 2012;**59**(2):300–7.

39. Flores-Munoz M, Smith NJ, Haggerty C, Milligan G, Nicklin SA. Angiotensin1-9 antagonises pro-hypertrophic signalling in cardiomyocytes via the angiotensin type 2 receptor. *J Physiol* 2011;**589**(4):939–51.

40. Erdos EG, Jackman HL, Brovkovych V, Tan F, Deddish PA. Products of angiotensin I hydrolysis by human cardiac enzymes potentiate bradykinin. *J Mol Cell Cardiol* 2002;**34**(12):1569–76.

41. Prokop JW, Santos RA, Milsted A. Differential mechanisms of activation of the Ang peptide receptors AT1, AT2, and MAS: using in silico techniques to differentiate the three receptors. *PLoS One* 2013;**8**(6):e65307.

42. Servant G, Laporte SA, Leduc R, Escher E, Guillemette G. Identification of angiotensin II-binding domains in the rat AT2 receptor with photolabile angiotensin analogs. *J Biol Chem* 1997;**272**(13):8653–9.

43. Wallinder C, Botros M, Rosenstrom U, Guimond MO, Beaudry H, Nyberg F, et al. Selective angiotensin II AT2 receptor agonists: benzamide structure-activity relationships. *Bioorg Med Chem* 2008;**16**(14):6841–9.

44. Behrends M, Wallinder C, Wieckowska A, Guimond MO, Hallberg A, Gallo-Payet N, et al. *N*-Aryl isoleucine derivatives as angiotensin II AT2 receptor ligands. *ChemistryOpen* 2014;**3**(2):65–75.

45. Bouley R, Perodin J, Plante H, Rihakova L, Bernier SG, Maletinska L, et al. N- and C-terminal structure-activity study of angiotensin II on the angiotensin AT2 receptor. *Eur J Pharmacol* 1998;**343**(2–3):323–31.

46. Regoli D, Rioux F, Park WK, Choi C. Role of the N-terminal amino acid for the biological activities of angiotensin and inhibitory analogues. *Can J Physiol* 1974;**52**(1):39–49.

47. Rosenstrom U, Skold C, Lindeberg G, Botros M, Nyberg F, Hallberg A, et al. Synthesis and AT2 receptor-binding properties of angiotensin II analogue. *J Pept Res* 2004;**64**(5):194–201.

48. Yee DK, Heerding JN, Krichavsky MZ, Fluharty SJ. Role of the amino terminus in ligand binding for the angiotensin II type 2 receptor. *Brain Res Mol Brain Res* 1998;**57**(2):325–9.

49. Hines J, Heerding JN, Fluharty SJ, Yee DK. Identification of angiotensin II type 2 (AT2) receptor domains mediating high-affinity CGP 42112A binding and receptor activation. *J Pharmacol Exp Ther* 2001;**298**(2):665–73.

50. Jones ES, Del Borgo MP, Kirsch JF, Clayton D, Bosnyak S, Welungoda I, et al. A single beta-amino acid substitution to angiotensin II confers AT(2) receptor selectivity and vascular function. *Hypertension* 2011;**57**(3):570–6.

51. Bosnyak S, Welungoda IK, Hallberg A, Alterman M, Widdop RE, Jones ES. Stimulation of angiotensin AT2 receptors by the non-peptide agonist, compound 21, evokes vasodepressor effects in conscious spontaneously hypertensive rats. *Br J Pharmacol* 2010;**159**(3):709–16.

52. Macari D, Bottari S, Whitebread S, De Gasparo M, Levens N. Renal actions of the selective angiotensin AT2 receptor ligands CGP 42112B and PD 123319 in the sodium-depleted rat. *Eur J Pharmacol* 1993;**249**(1):85–93.

53. Macari D, Whitebread S, Cumin F, De Gasparo M, Levens N. Renal actions of the angiotensin AT2 receptor ligands CGP 42112 and PD 123319 after blockade of the renin-angiotensin system. *Eur J Pharmacol* 1994;**259**(1):27–36.

54. Magnani F, Pappas CG, Crook T, Magafa V, Cordopatis P, Ishiguro S, et al. Electronic sculpting of ligand-GPCR subtype selectivity: the case of angiotensin II. *ACS Chem Biol* 2014;**9**(7):1420–5.

55. Yamada Y, Yamauchi D, Usui H, Zhao H, Yokoo M, Ohinata K, et al. Hypotensive activity of novokinin, a potent analogue of ovokinin(2–7), is mediated by angiotensin AT(2) receptor and prostaglandin IP receptor. *Peptides* 2008;**29**(3):412–18.

56. Yoshikawa M, Ohinata K, Yamada Y. The pharmacological effects of novokinin; a designed peptide agonist of the angiotensin AT2 receptor. *Curr Pharm Des* 2013;**19**(17):3009–12.

57. Wakasa Y, Zhao H, Hirose S, Yamauchi D, Yamada Y, Yang L, et al. Antihypertensive activity of transgenic rice seed containing an 18-repeat novokinin peptide localized in the nucleolus of endosperm cells. *Plant Biotechnol J* 2011;**9**(7):729–35.

Chapter 20

Potential Clinical Application of Angiotensin 2 Receptor Agonists

Nephtali Marina,* Björn Dahlöf,† Bryan Williams‡

*Department of Clinical Pharmacology and Experimental Therapeutics, University College London, London, UK, †Department of Molecular and Clinical Medicine, Institute of Medicine, Sahlgrenska Academy, Gothenburg, Sweden, ‡Institute of Cardiovascular Science, University College London, London, UK

INTRODUCTION

The role of the renin–angiotensin system (RAS) plays a fundamental role, not only in cardiovascular homeostasis and water/electrolyte balance but also in the pathophysiology of conditions associated with, for example, chronic inflammation and fibrosis via activation of the angiotensin AT_1 receptor (AT_1R). Pharmacological blockade of the RAS by either renin inhibitors, angiotensin-converting enzyme (ACE) inhibitors, or AT_1R blockers (ARBs) has become an important therapeutic target in the management of hypertension, myocardial infarction, heart failure, and diabetic nephropathy.

In recent years, the protective RAS system composed of the AT_2 receptor (AT_2R) and the ACE2-angiotensin-(1-7) (Ang-[1-7])-MAS axis has gained significant attention due to the beneficial effects associated with its activation as outlined in previous chapters. In particular, the AT_2R signaling pathway has become a major focus for the design, synthesis, and patenting of a unique collection of small nonpeptide molecules, which selectively stimulate the AT_2R (the so-called AT_2R agonists), including (1) CGP42112A, the first AT_2R compound described as an AT_2R agonist[1]; (2) compound 21 (C21; Vicore Pharma, Gothenburg, Sweden, www.vicorepharma.com),[2] which emerged from an extensive medicinal chemistry program, chosen from a large number of patent-protected AT_2 agonists; (3) the cyclic Ang II derivative LP2-3 (Lanthio Pharma, Groningen, the Netherlands, www.lanthiopep.nl)[3]; and (4) a group of Ang II derivatives (β-amino acid substituted Ang II).[4]

PRECLINICAL TESTING

Bringing AT_2R agonists to the market as clinically effective drugs requires a series of research stages and regulatory approvals. So far, C21 is the only small molecule to have reached the final stage of preclinical development and is expected to enter clinical testing in 2015. C21 has been tested for its safety and effectiveness in several animal models of cardiovascular disease, ischemia, neurodegeneration, metabolic disease, and inflammation. The results from these preclinical studies have been summarized in previous chapters, and they have revealed that there is sufficient potential for C21 to proceed to testing in human clinical trials. The compound has particularly shown great potential for the treatment of diseases (Table 1) where the most distinct disease-modifying actions have been anti-inflammation, antiproliferation, antifibrosis, neuroregeneration, and neuroreparation and metabolic benefits and cardiovascular protection.[5] The chemical and pharmaceutical properties of C21 render it well suited for both oral administration and intravenous administration. The likelihood of adverse effects is small since the AT_2R in animals and man is, in most cases, not continuously expressed, but rather, its expression is localized to areas requiring tissue reparation, temporally related to the injury and healing, fibrosis, and growth and repair processes. Thus, the receptor is upregulated when there is tissue damage and stimulated via its endogenous ligand, angiotensin II. Inhibition of the AT_1R is well established as a drug target to limit these deleterious processes. However, there is potential to go beyond AT_1 antagonism and directly influence fibrosis, inflammation, etc., via direct and potent stimulation of the AT_2R. The documented broad action of C21 on fibrosis and inflammation in experimental animal models of disease states highlights the potential effectiveness of this drug when disease processes are active. The fact that the AT_2R is not widely expressed in the absence of such disease means that such a drug is unlikely to have any AT_2R-mediated effects when disease is not present or to organs not affected by disease. This points to the likelihood of an excellent safety profile (unless there are any untoward off-target effects of the molecule) and has important implications for clinical use.

The Protective Arm of the Renin–Angiotensin System (RAS). http://dx.doi.org/10.1016/B978-0-12-801364-9.00020-1

TABLE 1 A Broad Therapeutic Platform for C21

Cardiovascular	Neuroprotection	Anti-inflammation/ Antifibrosis	Metabolic	Dermatologic
Post MI (remodeling → HF) Heart failure (REF and PEF) Diabetes end organ damage Chronic kidney disease Aortic stiffness Aortic aneurysm/MS Atherosclerosis Hypertension (combo ARB) Hypertensive end organ damage	Spinal cord injury Stroke Multiple sclerosis Dementia	Fibrosis (lung, liver, and kidney) Scleroderma Rheumatoid arthritis Autoimmune myocarditis ARDS	Metabolic syndrome Diabetic control High TG	Psoriasis Eczema Wound healing Keloid

HF, heart failure; REF, reduced ejection fraction; PEF, preserved ejection fraction; MS, Marfan syndrome; ARB, angiotensin receptor blocker; ARDS, acute respiratory distress syndrome; TG, triglycerides.

INVESTIGATIONAL NEW DRUG APPLICATION

Before an AT_2R agonist can enter human trials, an Investigational New Drug (IND) application must be filed with the relevant authorities such as the Food and Drug Administration in the United States or the EMEA in Europe or the MHRA in the United Kingdom. The IND application must contain information in three key areas:

- *Animal pharmacology and toxicology studies.* Preclinical data must be shown to demonstrate whether the product is reasonably safe for initial testing in humans. The main objectives for nonclinical testing are safety (toxicity), tolerance (local and systemic), biodistribution, persistence (duration of exposure), carcinogenesis, reproducibility, biological activity (potency) *in vivo* and/or *in vitro*, dose definition including rationale for starting dose in man, route of administration, and schedule and study duration to monitor for toxicity.
- *Chemistry and manufacturing information.* This information is assessed to ensure that the company can consistently produce and supply active batches of the drug according to GMP. The company is expected to present information related to the chemical structure, manufacturing methods, stability and controls used for manufacturing the drug substance, and the drug product.
- *Clinical protocols and investigator information.* The proposed clinical studies must be accompanied by detailed protocols to assess whether the initial-phase trials may expose the subjects to unnecessary risks. Information on the qualifications of clinical investigators who supervise the organization and administration of the experimental compound must be presented in order to assess whether they are qualified to perform their clinical trial duties. Furthermore, the sponsor must present commitments to obtain informed consent from the participants, to obtain review of the study by an Institutional Review Board, and to adhere to the IND regulations.

CLINICAL TRIALS

The first phase I trial conducted in humans usually aims to establish the safety of the drug and reveal any toxicities associated with the drug, the maximum tolerated dose, dose-limiting toxicity, and pharmacokinetics. The need to establish a dose that is suitable for further evaluation in phase II clinical trials is of critical importance.

Given the broad spectrum of diseases that could potentially benefit from AT_2 signaling and the proven efficacy of AT_2R stimulation in terms of a unique combination of anti-inflammatory, antifibrotic, and neuroregeneration effects,[5] it is challenging to choose a specific therapeutic indication. Many "classical" indications have been recently proposed for the first clinical trial with AT_2R agonists, including post-myocardial infarction ventricular remodeling, heart failure especially with preserved ejection fraction, stroke, spinal cord injury, rheumatoid arthritis, diseases with a strong fibrosis component, and dermatologic applications (Table 1). However, when conducting phase I studies to assess safety, it is ideal if a signal of efficacy can also be recorded so that it is clear that safety is being evaluated at a dose that is producing a biological effect. Herein lies a major challenge with AT_2R stimulation. If safety is to be evaluated as is usual in phase I studies, that is, in normal healthy volunteers, then it would be difficult to demonstrate an efficacy signal because there are unlikely to be many AT_2Rs expressed in tissues in the absence of disease. However, the effect on natriuresis and maybe some metabolic parameters in slightly overweight healthy individuals might be measurable.[6]

One way to circumvent this problem is to test the safety of this compound in clinical disease states for which there is no effective treatment. If no effective therapy exists, then there is an obvious unmet medical need and a new treatment effective in that condition would be of potential benefit for the patient. This approach allows safety and toxicity to be tested alongside the measurement of a biomarker indicative of drug activity against the underlying disease. Importantly, however, such studies are usually of a short duration, as they are still primarily studies of safety and toxicity, and as such, the biomarker of drug activity against the underlying disease should ideally be measurable over a short period of exposure, that is, usually 3 months but occasionally 6 months. In effect, this kind of early-phase trial is combining phase I and early-phase II studies in patients with clear unmet medical need. However, to justify such a study, there must be robust preclinical data to demonstrate that the AT$_2$R agonist has the potential to be more effective than existing therapies in favorably impacting on the disease process.

POTENTIAL THERAPEUTIC INDICATIONS FOR AT$_2$R AGONISTS

Orphan Indications

In this section, we present two unmet medical needs that may be used as models to propel the use of AT$_2$R agonists, that is, Marfan syndrome (MFS) and idiopathic pulmonary fibrosis (IPF).

Marfan Syndrome

MFS is an autosomal dominant connective tissue disorder caused by a mutation in fibrillin-1 that is associated with structural dysfunction in the aortic wall and biochemical changes including overexpression of TGF-β. Patients with MFS are at an increased risk of sudden death due to aortic root aneurysm and aortic rupture. Clinical management is limited primarily to prophylactic aortic root replacement and afterload reduction agents with drugs that reduce blood pressure and stress on the aortic and mitral valves and the aortic root. However, these drugs have little or no effect on the relentless progression of aortic dilatation.

Recent preclinical studies and clinical trials have shown that the selective ARB losartan can significantly reduce the rate of aortic enlargement in murine models and in patients with MFS.[7,8] The beneficial effect of the ARB losartan was recently evaluated in patients with MFS.[7] The results suggest that the addition of losartan to standard care patients with MFS reduces the rate of aortic dilatation but does not prevent it. Recent studies have shown that protective effects of losartan require intact expression of AT$_2$Rs to be effective, which suggests that AT$_2$ signaling may confer all or additional protective effects against cardiovascular abnormalities in MFS.[8] This suggests that an AT$_2$R agonist, either alone or in combination with losartan, could have the potential to further reduce the progression of aortic root dilatation in MFS or eliminate it.

Idiopathic Pulmonary Fibrosis

IPF is a chronic, progressive, and ultimately fatal condition, characterized by fibrosis of the lung interstitium. The pathogenesis includes an abnormal "wound healing" leading to restrictive fibrosis of the lung. Disease progression is poorly responding to current pharmacological therapy, and the median survival of patients with IPF is approximately 3 years from the time of diagnosis. There has been no clear-cut effect on mortality with the currently available therapy.[9] A number of preclinical studies have shown that AT$_2$R stimulation has the potential to reduce acute and chronic tissue fibrosis in various organs, including reverse lung fibrosis in several models.

Reflecting on AT$_2$R agonist preclinical data, target validation, target selection, and the design of the ideal early-phase clinical trial. Inflammation and fibrosis are ubiquitous processes in many disease states, both acute and chronic. Their treatment is a challenge because both the inflammation cascade and fibrosis cascade are important aspects of the acute healing process. Reasonable anti-inflammatory agents exist, and new biologics have added to the armory for chronic disease states such as rheumatoid arthritis. There is, however, a major unmet need for the treatment of progressive fibrosis. There are compelling data from preclinical studies of a variety of disease states in different tissues that AT$_2$ stimulation has antifibrotic effects. So, the next question is what is the target? IPF is usually a relentless and fatal disease that is without an effective therapy. Recent studies have suggested that pirfenidone can delay the progressive loss of lung forced vital capacity but the impact is modest and the drug has some debilitating side effects. Furthermore, the mechanism of action of pirfenidone is not understood. This is in stark contrast to AT$_2$R agonists such as C21, for which the mechanism of action is much better understood, and from this, one would predict that significant side effects are unlikely to occur. If AT$_2$R stimulation adds to the impressive data on fibrosis by producing compelling preclinical data in IPF, there would be a very real potential to move into early-phase clinical trials of patients with IPF as an orphan disease indication. The early-phase trial would combine safety and toxicity measures with biological and imaging markers of efficacy against the fibrotic process. The key here is that human proof of concept could be established quickly, leading to a pivotal outcome study in larger numbers of patients. The decision to focus on disease indications for the phase I clinical trial (rather than healthy human volunteers) is directly related to the biology of the AT$_2$R. It would be difficult to evaluate both the safety and efficacy of a highly specific and selective agonist of a receptor in a population of people (i.e., healthy individuals) in whom the receptor is not abundantly expressed. This has steered the discussion toward an orphan indication for the phase I human trial, that is, a disease in which there is no really effective treatment but for which there is a strong preclinical rationale for the drug and acceptable safety and preclinical toxicology data.

Other Indications

In the next section, we discuss some examples of other clinical conditions in which stimulation of AT$_2$R might provide significant therapeutic benefit.

Diabetic nephropathy: The best renal protection seems to be achieved by a combination of ARB and AT$_2$R antagonists, resulting in good BP control and enhanced antiproteinuric effect. Such therapy will potentially prevent renal fibrosis and reduce the risk for end-stage renal disease in many diabetic patients.[10]

Natriuresis: The natriuretic action of C21 uniquely takes place in the renal proximal tubule, where no other available diuretic exerts its action. This effect is also seen in healthy kidneys. Additive and/or synergistic effects can therefore be expected clinically during combination therapy with other diuretic agents and might also contribute to a small antihypertensive effect.[6]

Cardiovascular protection: AT$_2$Rs are upregulated and contribute to Ang II-induced vasodilation in resistance arteries of hypertensive diabetic patients treated with ARBs and may mediate, in part, vascular actions of these drugs in high-cardiovascular-risk patients. Thus, combination therapy with C21 can be expected to provide enhanced cardiovascular protection in several groups of ARB-treated patients and maybe also contribute to an antihypertensive effect in the long run.[11] Furthermore, in a classical model of ligated coronary artery, C21 was more effective than candesartan in reducing infarct size and improving ejection fraction after 1 week.[12] After 6 weeks in the same model, the AT$_2$R agonist improved systolic and diastolic function and improved ejection fraction in parallel with reduced fibrosis better than control.[13] The improvement of diastolic function coupled with a very strong antifibrotic effect making heart failure with preserved ejection fraction a very attractive clinical indication.

Stroke: Positive results in animal stroke studies are often difficult to interpret in terms of clinical relevance. However, positive findings have been reported with C21 from five independent research centers, all using different stroke models where C21 has been given at stroke induction[5] (see also previous chapters). This markedly increases the likelihood for beneficial effects of C21 also in patients with stroke. Additionally, it seems that several of the actions of C21 are important, for example, improved circulation in the penumbra, anti-inflammation and neuroprotection, and maybe neuroregeneration in the longer term.

Spinal cord injury: AT$_2$R stimulation has been shown to combine in one molecule what is usually the sole mechanism of action of different drugs, that is, neuroprotection, neuroregeneration, anti-inflammation, and antiapoptosis.[14] Therefore, AT$_2$R stimulation can be regarded as a novel, putative therapeutic approach for the improvement of neurological outcome after spinal cord injury.

Dementia: AT$_2$R stimulation has been shown to prevent cognitive decline in a mouse model of Alzheimer's disease, and these effects seemed to be mediated, at least in part, by an increase in cerebrovascular flow, enhancement in hippocampal field-excitatory postsynaptic potential, and hippocampal neurite outgrowth.[15] Thus, AT$_2$R stimulation can be used as a novel therapy to prevent Alzheimer's disease associated cognitive decline.

Type II diabetes/metabolic syndrome: There is evidence that AT$_2$R stimulation accompanied with a PPARγ antagonist attenuates insulin resistance in a mouse model of type 2 diabetes. This improved metabolic profile is achieved at least in part by combined adipocyte dysfunction improvement and β-cell protection.[16,17]

CONCLUSION

The protective RAS is an impressive therapeutic platform with a unique mechanism of action representing a new class of drugs likely to result in novel therapeutic interventions for several indications and patient categories. In addition, availability of AT$_2$R agonists will also serve as a potent pharmacological tool to better understand pathophysiological and repair mechanisms involving the protective RAS in various clinical conditions.

REFERENCES

1. Whitebread S, Mele M, Kamber B, de Gasparo M. Preliminary biochemical characterization of two angiotensin II receptor subtypes. *Biochem Biophys Res Commun* 1989;**163**:284–91.
2. Wan Y, Wallinder C, Plouffe B, Beaudry H, Mahalingam AK, Wu X, et al. Design, synthesis, and biological evaluation of the first selective nonpeptide AT2 receptor agonist. *J Med Chem* 2004;**47**:5995–6008.
3. Wagenaar GT, Laghmani el H, Fidder M, Sengers RM, de Visser YP, de Vries L, et al. Agonists of MAS oncogene and angiotensin II type 2 receptors attenuate cardiopulmonary disease in rats with neonatal hyperoxia-induced lung injury. *Am J Physiol Lung Cell Mol Physiol* 2013;**305**:L341–51.
4. Jones ES, Del Borgo MP, Kirsch JF, Clayton D, Bosnyak S, Welungoda I, et al. A single beta-amino acid substitution to angiotensin II confers AT2 receptor selectivity and vascular function. *Hypertension* 2011;**57**:570–6.
5. Namsolleck P, Recarti C, Foulquier S, Steckelings UM, Unger T. AT(2) receptor and tissue injury: therapeutic implications. *Curr Hypertens Rep* 2014;**16**:416.
6. Kemp BA, Howell NL, Gildea JJ, Keller SR, Padia SH, Carey RM. AT$_2$ receptor activation induces natriuresis and lowers blood pressure. *Circ Res* 2014;**115**:388–99.
7. Groenink M, den Hartog AW, Franken R, Radonic T, de Waard V, Timmermans J, et al. Losartan reduces aortic dilatation rate in adults with Marfan syndrome: a randomized controlled trial. *Eur Heart J* 2013;**34**:3491–500.
8. Habashi JP, Doyle JJ, Holm TM, Aziz H, Schoenhoff F, Bedja D, et al. Angiotensin II type 2 receptor signaling attenuates aortic aneurysm in mice through ERK antagonism. *Science* 2011;**332**:361–5.
9. King Jr TE, Pardo A, Selman M. Idiopathic pulmonary fibrosis. *Lancet* 2011;**378**:1949–61.
10. Castoldi G, di Gioia CR, Bombardi C, Maestroni S, Carletti R, Steckelings UM, et al. Prevention of diabetic nephropathy by compound 21, selective agonist of the angiotensin type 2 receptors, in zucker diabetic fatty rats. *Am J Physiol Renal Physiol* 2014;**307**(10):F1123–31.
11. Rehman A, Leibowitz A, Yamamoto N, Rautureau Y, Paradis P, Schiffrin EL. Angiotensin type 2 receptor agonist compound 21 reduces vascular injury and myocardial fibrosis in stroke-prone spontaneously hypertensive rats. *Hypertension* 2012;**59**:291–9.
12. Kaschina E, Grzesiak A, Li J, Foryst-Ludwig A, Timm M, Rompe F, et al. Angiotensin II type 2 receptor stimulation: a novel option of therapeutic interference with the renin-angiotensin system in myocardial infarction? *Circulation* 2008;**118**:2523–32.
13. Lauer D, Slavic S, Sommerfeld M, Thöne-Reineke C, Sharkovska Y, Hallberg A, et al. Angiotensin type 2 receptor stimulation ameliorates left ventricular fibrosis and dysfunction via regulation of tissue inhibitor of matrix metalloproteinase 1/matrix metalloproteinase 9 axis and transforming growth factor β1 in the rat heart. *Hypertension* 2014;**63**:e60–7.
14. Namsolleck P, Boato F, Schwengel K, Paulis L, Matho KS, Geurts N, et al. AT2-receptor stimulation enhances axonal plasticity after spinal cord injury by upregulating BDNF expression. *Neurobiol Dis* 2013;**51**:177–91.
15. Jing F, Mogi M, Sakata A, Iwanami J, Tsukuda K, Ohshima K, et al. Direct stimulation of angiotensin II type 2 receptor enhances spatial memory. *J Cereb Blood Flow Metab* 2012;**32**(2):248–55.
16. Ohshima K, Mogi M, Jing F, Iwanami J, Tsukuda K, Min LJ, et al. Direct angiotensin II type 2 receptor stimulation ameliorates insulin resistance in type 2 diabetes mice with PPARγ activation. *PLoS One* 2012;**11**:e48387.
17. Shao C, Yu L, Gao L. Activation of angiotensin type 2 receptors partially ameliorates streptozotocin-induced diabetes in male rats by islet protection. *Endocrinology* 2014;**155**:793–804.

Chapter 21

Angiotensin-(1-7) and Mas: A Brief History

Natalia Alenina*,†,‡, Robson Augusto Souza dos Santos§

*Max-Delbrueck-Center for Moleculare Medicine (MDC), Berlin-Buch, Germany, †Federal University of Minas Gerais, Belo Horizonte, Brazil, ‡Instituto Nacional de Ciência e Tecnologia em, NanoBiofarmacêutica (NanoBIOFAR), Belo Horizonte, Brazil, §Departamento de Fisiologia e Biofísica—ICB, Universidade Federal de Minas Gerais, Belo Horizonte, Minas Gerais, Brazil

INTRODUCTION

Angiotensin-(1-7) is a vasoactive peptide of the renin–angiotensin system (RAS), which is generated mainly by angiotensin-converting enzyme 2 (ACE2) and exerts its actions via activation of its receptor Mas. The Ang-(1-7)/ACE2/Mas axis is nowadays considered to be a main mechanism, which counterbalances the vasoconstrictive actions of classical RAS, which includes renin, ACE, ANG II, and its receptors AT_1 and AT_2 (Figure 1). Whereas the classical RAS has been known for more than 100 years,[1] the protective arm of the RAS was relatively recently discovered. Both Mas[2] and Ang-(1-7)[3,4] were first described almost 30 years ago; however, it took an additional 15 years until the interaction of these components was revealed.[5] The third component, ACE2, was the latest to be discovered, in 2000.[6] Interestingly, besides carboxypeptidase activity, ACE2 turned out to have totally different functions and was shown to mediate the adsorption of large amino acids in the gut[7] and to be the receptor for the human severe acute respiratory syndrome virus.[8] Here, we will shortly describe the story of Mas and Ang-(1-7), which was full of errors and uncertainty at the beginning, until the interrelationship between the two was unveiled in 2003.

DISCOVERY OF MAS: IS IT A PROTO-ONCOGENE?

The Mas gene was first described in 1986.[2] Intriguingly, the first three main discoveries about this gene turned out to be erroneous, mostly due to the fact that the quantity of collected data and the general knowledge at that time obscured the interpretation of the obtained results and led to inappropriate conclusions.

The human Mas gene was originally isolated from DNA of a human epidermoid carcinoma cell line due to its ability to transform NIH3T3 cells upon transfection[2] and therefore was called a proto-oncogene. Since computer analysis of the Mas amino acid sequence suggested that the protein belongs to the class of G protein-coupled receptors (GPCRs) with seven transmembrane domains,[2] this provided the first direct evidence for an oncogenic activity of a GPCR. Though such tumorigenicity was confirmed in independent experiments,[9,10] the oncogenic potential of Mas was challenged. In these experiments, the transfected cells or the tertiary tumor in nude mice contained amplified Mas sequences characterized by rearrangements in 5′- and 3′-noncoding regions, such as an insertion of human centromeric alpha-satellite repeat DNA.[10] However, the original tumor DNA, used in the first round of transfection, was neither rearranged nor amplified or mutated in the Mas coding sequence and therefore cannot be considered as the driving cause for tumor development. Presumably, the *Mas* 5′-noncoding region represents a hot spot of recombination, and the rearrangement of the 5′-noncoding sequence, which occurred during transfection, was responsible for the activation of the *Mas* gene in the tumorigenicity assay. Thus, *Mas* is not an oncogene, but can transform cells, when artificially overexpressed.

FUNCTIONS OF MAS: IS IT AN ANG II RECEPTOR?

To investigate the functions of Mas, Jackson et al.[11] expressed Mas transiently in *Xenopus* oocytes and stably in a mammalian cell line. Under voltage-clamp conditions, oocytes injected with *Mas* RNA exhibited a dose-dependent induction of an inward current in response to the Ang I, II, and III, whereas in the transfected cells, stimulation of Mas with Ang II

The Protective Arm of the Renin–Angiotensin System (RAS). http://dx.doi.org/10.1016/B978-0-12-801364-9.00021-3

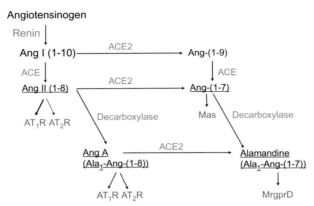

FIGURE 1 Renin-angiotensin system (RAS). The system consists of several components. First, the protein angiotensinogen is cleaved by renin to the decapeptide, angiotensin I (Ang I, Ang (1-10)). Subsequent cleavage of Ang I by ACE leads to the formation of the octapeptide, Ang II (Ang-(1-8)), which can activate its receptors, AT_1 and AT_2. This part of the system is the most studied one and is called the "classical RAS" or ACE/Ang II/AT_1 axis. However, Ang II is not the only active peptide of the system. Another important component is Ang-(1-7), a heptapeptide, which is mainly produced by the enzyme ACE2 and acts on its receptor Mas. The axis is called ACE2/Ang-(1-7)/Mas and exerts cardioprotective actions. Two further vasoactive peptides of RAS were recently identified: Ang A and alamandine.[61,62] Both are produced by yet unknown aspartate-decarboxylating enzymes from Ang II and Ang-(1-7), respectively. While Ang A binds AT_1 and AT_2 receptors, alamandine was shown to interact with MrgprD, a receptor of the Mas-related gene (Mrg) family. Enzymatic pathways (blue), peptides (black), and receptors (magenta) in the RAS.

and III led to the mobilization of intracellular Ca^{2+} and to the initiation of DNA synthesis. Based on these results, Mas was proposed to be a functional Ang II receptor—a molecule whose identification had been pending for years. Although a number of further studies were in agreement with this assumption,[12–15] the activation of inward currents by Ang II in *Mas* mRNA-injected oocytes was not inhibited by Ang antagonists.[11] Moreover, Ambroz et al.[16] could show that the intracellular Ca^{2+} increase in *Mas*-transfected cells after Ang II treatment was only observed in cells already expressing endogenously Ang II receptors. Therefore, doubts arose as to whether the *Mas* gene product per se is an Ang II receptor. Moreover, cloning of the Ang II receptor AT_1 in 1991[17,18] did not favor the original hypothesis of Ang II being a ligand for Mas. The later identification of Ang-(1-7) as Mas agonist in 2003 and of the direct interaction between Mas and AT_1 receptors in 2005[19,20] partly explained the original observations of Jackson et al.[11] in *Xenopus* oocytes and clarified that Mas is not an Ang II receptor per se, but a modulator of AT_1R signaling.

MAS, ITS ANTISENSE RNA, AND IMPRINTING

A third intriguing story about Mas was published in 1994. Mas was reported to be maternally imprinted in mice during embryonic development and for some organs such as the tongue and heart also in adults[21] and in human breast tissue.[22] In genomic imprinting, one of the two parental alleles of an autosomal gene is silenced epigenetically by a *cis*-acting mechanism. The *Mas* gene is located in close proximity to the imprinted *Igf2r* gene in the mouse and human genomes.[23,24] Imprinting of the maternally expressed *Igf2r* gene is controlled by an intronic imprint control element that contains the promoter of the long noncoding RNA, Airn (antisense *Igf2r* RNA noncoding), which overlaps the silenced paternal *Igf2r* promoter and partially the *Mas* gene in an antisense orientation.[25,26] Our work using Mas-deficient mice and RNase protection assay clearly demonstrated that Mas is biallelically expressed.[27] Thus, due to the lack of strand selectivity in the RT-PCR assays used by Villar and Pedersen[21] and Miller et al.,[22] the maternally imprinted RNA detected by them was most probably not the coding mRNA but the antisense RNA of the *Mas* gene as part of Airn. Altogether, these data demonstrate that Airn but not the *Mas* mRNA is monoallelically expressed in mouse and human.

MAS EXPRESSION

The distribution of *Mas* expression in different organs of rodents was extensively investigated. The highest expression was found in the brain and testis. In the rat and mouse brain, Mas transcripts are localized not only in the hippocampus and cerebral cortex, in particular in the dentate gyrus, the CA3 and CA4 areas of the hippocampus, the olfactory tubercle, the pyriform cortex, and the olfactory bulb, but also at lower levels all over the neocortex and especially in the frontal lobe.[28,29] In the rodent testis, Mas expression is not detectable in newborn animals, but starts a few weeks after birth and continuously increases during puberty.[30,31] On the cellular level, Mas is confined to Leydig and Sertoli cells, with a clear preference for

Leydig cells.[30] Mas expression was also discovered in other tissues of mice and rats such as the heart, kidney, lung, liver, spleen, tongue, and skeletal muscle.[21,31–33] This ubiquitous low-level presence of Mas mRNA may partly be due to its expression in the endothelial layer of vessels in different organs, as has been shown for the brain,[34] heart,[32] and corpus cavernosum,[35] supporting an important role of this protein in the function of the endothelium. Mas expression was also detected in cardiomyocytes[36] and more recently in cardiac fibroblasts[37] and in the sinoatrial node.[38] Moreover, Mas is present in several tissues involved in glucose and lipid metabolism including the pancreas,[39,40] liver,[41] adipose tissue,[42,43] and skeletal muscle.[44,45]

MAS FUNCTIONS: MAS-DEFICIENT MICE

Despite this well-described expression pattern of Mas, early studies failed to identify its functions (see above). Some progress was made after mouse deficient for Mas was generated in 1998.[46] These studies showed the importance of Mas for the anxiety-related behavior and indicated Mas as the first GPCR involved in the modulation of long-term potentiation in the dentate gyrus of male mice, whereas spatial learning was not affected in these animals. Interestingly, these behavior phenotypes turned out to be gender-specific and were not detected in female *Mas*-deficient mice.[47] Elucidation of cardiovascular parameters in these mice, which were on the mixed genetic background at that time, did not reveal significant differences in heart rate or blood pressure between knockout and control mice (later, it was shown that *Mas* deficiency on the FVB/N background leads to the elevation in blood pressure, whereas *Mas*-deficient mice on the C57BL/6 background are normotensive[48,49]). Interestingly, female mice showed a reduction of heart rate variability, and knockout animals of both genders showed an increased sympathetic tone.[50] However, the molecular bases of its actions remained obscure until 2003, when finally, the ligand for Mas was identified.

DISCOVERY OF ANG-(1-7)

Based on the lack of demonstrable pressor effect in structure–activity relationship studies, the heptapeptide Ang-(1-7) was initially considered inactive (reviewed in Ref. [51]). One decade later, this concept was reinforced by the demonstration that it also lacks one of the most classical actions of Ang II: induction of drinking behavior.[52] In keeping with this concept, enzymes capable of forming Ang-(1-7) were called angiotensinases.[53–55] In 1988, we observed that 125I-Ang-(1-7) was the principal product of the 125I-Ang I by micropunches of brain stem homogenates.[3] The formation of Ang-(1-7) was independent of ACE activity. In the same year, the first biological action of Ang-(1-7) in the brain, release of vasopressin from the rat hypothalamo-neurohypophyseal system, was reported.[4] It should be pointed out that 2 years before, Kono et al.[56] described the first biological action of Ang-(1-7) in human. However, the pressor effect described in this study with a high dose of the heptapeptide probably was due to the stimulation of AT_1 receptors.[56]

CLOSING THE GAP: MAS IS AN ANG-(1-7) RECEPTOR

The fact that Ang-(1-7) was equipotent to Ang II for release of vasopressin from hypothalamo-neurohypophyseal explants,[4] contrasting with the lack of its effect on drinking behavior,[52] was the first evidence for the existence of a distinct receptor for Ang-(1-7). Moreover, Ang-(1-7) was reported to release nitric oxide, induce diuresis, and have a vasodilatory effect, favoring a lowering of blood pressure.[56,57] These vascular and baroreflex actions of Ang-(1-7), counteracting the effects of Ang II, suggested that Ang-(1-7) mediates its effects through a novel non-AT_1/AT_2 receptor subtype. Finally, the description of a selective antagonist for Ang-(1-7) in 1994[58–60] clearly indicated the existence of a receptor for this heptapeptide.

However, only in 2003, more definitive evidence for a specific binding site for Ang-(1-7) was obtained with the demonstration that Mas is a receptor for the heptapeptide.[5] In this study, specific binding of 125I-Ang-(1-7) to Mas-transfected cells was reported. Moreover, the specific binding of 125I-Ang-(1-7) to kidney sections was abolished by genetic deletion of Mas. In addition, *Mas*-deficient mice completely lack the antidiuretic action of Ang-(1-7) after an acute water load, and Mas-deficient aortas lost their Ang-(1-7)-induced relaxation response. These findings provided a clear molecular basis for the physiological actions of this biologically active peptide. At this point, an orphan receptor met an orphan peptide filling an important gap in our understanding of the RAS.

REFERENCES

1. Tigerstedt R, Bergman PG. Niere und Kreislauf. *Skand Arch Physiol* 1989;**8**:223–71.
2. Young D, Waitches G, Birchmeier C, Fasano O, Wigler M. Isolation and characterization of a new cellular oncogene encoding a protein with multiple potential transmembrane domains. *Cell* 1986;**45**:711–19.

3. Santos RA, Brosnihan KB, Chappell MC, Pesquero J, Chernicky CL, Greene LJ, et al. Converting enzyme activity and angiotensin metabolism in the dog brainstem. *Hypertension* 1988;**11**(2 Pt 2):I153–7.

4. Schiavone MT, Santos RA, Brosnihan KB, Khosla MC, Ferrario CM. Release of vasopressin from the rat hypothalamo-neurohypophysial system by angiotensin-(1–7) heptapeptide. *Proc Natl Acad Sci U S A* 1988;**85**(11):4095–8.

5. Santos RA, Simoes e Silva AC, Maric C, Silva DMR, Machado RP, de Buhr I, et al. Angiotensin-(1–7) is an endogenous ligand for the G-protein coupled receptor Mas. *Proc Natl Acad Sci U S A* 2003;**100**:8258–63.

6. Donoghue M, Hsieh F, Baronas E, Godbout K, Gosselin M, Stagliano N, et al. A novel angiotensin-converting enzyme-related carboxypeptidase (ACE2) converts angiotensin I to angiotensin 1–9. *Circ Res* 2000;**87**(5):E1–9.

7. Singer D, Camargo SM, Ramadan T, Schäfer M, Mariotta L, Herzog B, et al. Defective intestinal amino acid absorption in Ace2 null mice. *Am J Physiol Gastrointest Liver Physiol* 2012;**303**(6):G686–95.

8. Li W, Moore MJ, Vasilieva N, Sui J, Wong SK, Berne MA, et al. Angiotensin-converting enzyme 2 is a functional receptor for the SARS coronavirus. *Nature* 2003;**426**(6965):450–4.

9. Janssen JWG, Steenvoorden ACM, Schmidtberger M, Bartram CR. Activation of the Mas oncogene during transfection of monoblastic cell line DNA. *Leukemia* 1988;**2**:318–20.

10. van`t Veer LJ, van der Feltz MJM, van der Berg-Bakker CAM, Cheng NC, Hermens RPMG, et al. Activation of the *Mas* oncogene involves coupling to human alphoid sequences. *Oncogene* 1993;**8**:2673–81.

11. Jackson TR, Blair AC, Marshall J, Goedert M, Hanley MR. The Mas oncogene encodes an angiotensin receptor. *Nature* 1988;**335**:437–40.

12. Andrawis NS, Dzau VJ, Pratt RE. Autocrine stimulation of Mas oncogene leads to altered growth control. *Cell Biol Int Rep* 1992;**16**:547–56.

13. Jackson TR, Hanley MR. Tumor promoter 12-*O*-tetradecanoylphorbol 13-acetate inhibits mas/angiotensin receptor-stimulated inositol phosphate production and intracellular Ca²⁺ elevation in the 401L-C3 neuronal cell line. *FEBS Lett* 1989;**251**(1–2):27–30.

14. McGillis JP, Sudduth-Klinger J, Harrowe G, Mitsuhashi M, Payan DG. Transient expression of the angiotensin II receptor: a rapid and functional analysis of a calcium-mobilizing seven-transmembrane domain receptor in COS-7 cells. *Biochem Biophys Res Commun* 1989;**165**:935–41.

15. Poyner DR, Hawkins PT, Benton HP, Hanley MR. Changes in inositol lipids and phosphates after stimulation of the Mas-transfected NG115-401L-C3 cell line by mitogenic and non-mitogenic stimuli. *Biochem J* 1990;**271**:605–11.

16. Ambroz C, Clark AJL, Catt KJ. The Mas oncogene enhances angiotensin-induced [Ca2+]i responses in cells with pre-existing angiotensin II receptors. *Biochim Biophys Acta* 1991;**1133**:107–11.

17. Murphy TJ, Alexander RW, Griendling KK, Runge MS, Bernstein KE. Isolation of a CDNA encoding the vascular type-1 angiotensin II receptor. *Nature* 1991;**351**:233–6.

18. Sasaki K, Yamano Y, Kostenis Bardhan S, Iwai N, Murray JJ, Hasegawa M, et al. Cloning and expression of a complementary DNA encoding a bovine adrenal angiotensin II type-1 receptor. *Nature* 1991;**351**:230–2.

19. Kostenis E, Milligan G, Christopoulos A, Sanchez-Ferrer CF, Heringer-Walther S, Sexton PM, et al. G-protein-coupled receptor Mas is a physiological antagonist of the angiotensin II type 1 receptor. *Circulation* 2005;**111**:1806–13.

20. Santos EL, Reis RI, Silva RG, Shimuta SI, Pecher C, Bascands JL, et al. Functional rescue of a defective angiotensin II AT1 receptor mutant by the Mas protooncogene. *Regul Pept* 2007;**141**(1–3):159–67.

21. Villar AJ, Pedersen RA. Parental imprinting of the *Mas* protooncogene in mouse. *Nat Genet* 1994;**8**:373–9.

22. Miller N, McCann AH, O'Connell D, Pedersen IS, Spiers V, Gorey T, et al. The Mas proto-oncogene is imprinted in human breast tissue. *Genomics* 1997;**46**:509–12.

23. Barlow DP, Stoger R, Herrmann BG, Saito K, Schweifer N. The mouse insulin-like growth factor type-2 receptor is imprinted and closely linked to the Tme locus. *Nature* 1991;**349**:84–7.

24. Schweifer N, Valk PJM, Delwel R, Cox R, Francis F, Meierwert S, et al. Characterization of the C3 YAC contig from proximal mouse chromosome 17 and analysis of allelic expression of genes flanking the imprinted Igf2r gene. *Genomics* 1997;**43**:285–97.

25. Lyle R, Watanabe D, te Vruchte D, Lerchner W, Smrzka OW, Wutz A, et al. The imprinted antisense RNA at the Igf2r locus overlaps but does not imprint Mas1. *Nat Genet* 2000;**25**:19–21.

26. Wutz A, Smrzka OW, Schweifer N, Schellander K, Wagner EF, Barlow DP. Imprinted expression of the Igf2r gene depends on an intronic CpG island. *Nature* 1997;**389**:745–9.

27. Alenina N, Bader M, Walther T. Imprinting of the murine Mas protooncogene is restricted to its antisense RNA. *Biochem Biophys Res Commun* 2002;**290**:1072–8.

28. Bunnemann B, Fuxe K, Metzger R, Mullins J, Jackson TR, Hanley MR, et al. Autoradiographic localization of Mas proto-oncogene MRNA in adult rat brain using in situ hybridization. *Neurosci Lett* 1990;**114**:147–53.

29. Young D, O'Neill K, Jessell T, Wigler M. Characterization of the rat Mas oncogene and its high-level expression in the hippocampus and cerebral cortex of rat brain. *Proc Natl Acad Sci U S A* 1988;**85**:5339–42.

30. Alenina N, Baranova TV, Smirnov E, Bader M, Lippoldt A, Patkin EL, et al. Cell-type specific expression of the *Mas* protooncogene in testis. *J Histochem Cytochem* 2002;**59**:691–6.

31. Metzger R, Bader M, Ludwig T, Berberich C, Bunnemann B, Ganten D. Expression of the mouse and rat *Mas* proto-oncogene in the brain and peripheral tissues. *FEBS Lett* 1995;**357**:27–32.

32. Alenina N, Xu P, Rentzsch B, Patkin EL, Bader M. Genetically altered animal models for Mas and angiotensin-(1–7). *Exp Physiol* 2008;**93**:528–37.

33. Ferrario CM, Jessup J, Chappell MC, Averill DB, Brosnihan KB, Tallant EA, et al. Effect of angiotensin-converting enzyme inhibition and angiotensin II receptor blockers on cardiac angiotensin-converting enzyme 2. *Circulation* 2005;**111**:2605–10.

34. Kumar M, Grammas P, Giacomelli F, Wiener J. Selective expression of C-*Mas* proto-oncogene in rat cerebral endothelial cells. *NeuroReport* 1996;**8**:93–6.

35. Goncalves ACC, Leite R, Silva RAF, Pinheiro SVB, Sampaio WO, Reis AB, et al. The vasodilator angiotensin-(1–7)-Mas axis plays an essential role in erectile function. *Am J Physiol* 2007;**293**:2588–96.

36. Tallant EA, Ferrario CM, Gallagher PE. Angiotensin-(1–7) inhibits growth of cardiac myocytes through activation of the Mas receptor. *Am J Physiol Heart Circ Physiol* 2005;**289**:H1560–6.

37. Iwata M, Cowling RT, Yeo SJ, Greenberg B. Targeting the ACE2-Ang-(1–7) pathway in cardiac fibroblasts to treat cardiac remodeling and heart failure. *J Mol Cell Cardiol* 2011;**51**:542–7.

38. Ferreira AJ, Moraes PL, Foureaux G, Andrade AB, Santos RA, Almeida AP. The angiotensin-(1–7)/Mas receptor axis is expressed in sinoatrial node cells of rats. *J Histochem Cytochem* 2011;**59**:761–8.

39. Bindom SM, Lazartigues E. The sweeter side of ACE2: physiological evidence for a role in diabetes. *Mol Cell Endocrinol* 2009;**302**:193–202.

40. Frantz ED, Crespo-Mascarenhas C, Barreto-Vianna AR, Aguila MB, Mandarim-de-Lacerda CA. Renin-angiotensin system blockers protect pancreatic islets against diet-induced obesity and insulin resistance in mice. *PLoS ONE* 2013;**8**:e67192.

41. Herath CB, Warner FJ, Lubel JS, Dean RG, Jia Z, Lew RA, et al. Upregulation of hepatic angiotensin-converting enzyme 2 (ACE2) and angiotensin-(1–7) levels in experimental biliary fibrosis. *J Hepatol* 2007;**47**:387–95.

42. Liu C, Lv XH, Li HX, Cao X, Zhang F, Wang L, et al. Angiotensin-(1–7) suppresses oxidative stress and improves glucose uptake via Mas receptor in adipocytes. *Acta Diabetol* 2012;**49**:291–9.

43. Santos SHS, Fernandes LR, Mario EG, Ferreira AVM, Porto LCJ, Alvarez-Leite JI, et al. Mas deficiency in FVB/N in mice produces marked changes in lipid and glycemic metabolism. *Diabetes* 2008;**57**:340–7.

44. Acuna MJ, Pessina P, Olguin H, Cabrera D, Vio CP, Bader M, et al. Restoration of muscle strength in dystrophic muscle by angiotensin-1-7 through inhibition of TGF-beta signalling. *Hum Mol Genet* 2013;**23**(5):1237–49.

45. Prasannarong M, Santos FR, Henriksen EJ. ANG-(1–7) reduces ANG II-induced insulin resistance by enhancing Akt phosphorylation via a Mas receptor-dependent mechanism in rat skeletal muscle. *Biochem Biophys Res Commun* 2012;**426**:369–73.

46. Walther T, Balschun D, Voigt JP, Fink H, Zuschratter W, Birchmeier C, et al. Sustained long term potentiation and anxiety in mice lacking the Mas protooncogene. *J Biol Chem* 1998;**273**:11867–73.

47. Walther T, Voigt JP, Fink H, Bader M. Sex specific behavioural alterations in Mas-deficient mice. *Behav Brain Res* 2000;**107**:105–9.

48. Xu P, Costa-Goncalves AC, Todiras M, Rabelo LA, Sampaio WO, Moura MM, et al. Endothelial dysfunction and elevated blood pressure in Mas gene-deleted mice. *Hypertension* 2008;**51**(2):574–80.

49. Rabelo LA, Xu P, Todiras M, Sampaio WO, Buttgereit J, Bader M, et al. Ablation of angiotensin (1–7) receptor Mas in C57Bl/6 mice causes endothelial dysfunction. *J Am Soc Hypertens* 2008;**2**(6):418–24.

50. Walther T, Wessel N, Kang N, Sander A, Tschöpe C, Malberg H, et al. Altered heart rate and blood pressure variability in mice lacking the Mas protooncogene. *Braz J Med Biol Res* 2000;**33**:1–9.

51. Page IH, Bumbus FM. Angiotensin. *Physiol Rev* 1961;**41**:331–90.

52. Fitzsimons JT. The effect on drinking of peptide precursors and of shorter chain peptide fragments of angiotensin II injected into the rat's diencephalon. *J Physiol* 1972;**214**(2):295–303.

53. Koida M, Walter R. Post-proline cleaving enzyme. Purification of this endopeptidase by affinity chromatography. *J Biol Chem* 1976;**251**(23):7593–9.

54. Yang HYT, Erdös EG, Chiang TS. New enzymatic route for the inactivation of angiotensin. *Nature* 1968;**218**:1224–6.

55. Greene LJ, Spadaro AC, Martins AR, Perussi De Jesus WD, Camargo AC. Brain endo-oligopeptidase B: a post-proline cleaving enzyme that inactivates angiotensin I and II. *Hypertension* 1982;**4**(2):178–84.

56. Kono T, Taniguchi A, Imura H, Oseko F, Khosla MC. Biological activities of angiotensin II-(1-6)-hexapeptide and angiotensin II-(1-7)-heptapeptide in man. *Life Sci* 1986;**38**(16):1515–19.

57. Santos RA, Campagnole-Santos MJ, Andrade SP. Angiotensin-(1-7): an update. *Regul Pept* 2000;**91**(1–3):45–62.

58. Santos RA, Campagnole-Santos MJ, Baracho NC, Fontes MA, Silva LC, Neves LA, et al. Characterization of a new angiotensin antagonist selective for angiotensin-(1–7): evidence that the actions of angiotensin-(1–7) are mediated by specific angiotensin receptors. *Brain Res Bull* 1994;**35**(4):293–8.

59. Fontes MA, Silva LC, Campagnole-Santos MJ, Khosla MC, Guertzenstein PG, Santos RA. Evidence that angiotensin-(1–7) plays a role in the central control of blood pressure at the ventro-lateral medulla acting through specific receptors. *Brain Res* 1984;**665**(1):175–80.

60. Ambuhl P, Felix D, Khosla MC. [7-D-ALA]-angiotensin-(1–7): selective antagonism of angiotensin-(1–7) in the rat paraventricular nucleus. *Brain Res Bull* 1994;**35**(4):289–91.

61. Jankowski V, Vanholder R, van der Giet M, Tölle M, Karadogan S, Gobom J, et al. Mass-spectrometric identification of a novel angiotensin peptide in human plasma. *Arterioscler Thromb Vasc Biol* 2007;**27**(2):297–302.

62. Lautner RQ, Villela DC, Fraga-Silva RA, Silva N, Verano-Braga T, Costa-Fraga F, et al. Discovery and characterization of alamandine: a novel component of the renin-angiotensin system. *Circ Res* 2013;**112**(8):1104–11.

Chapter 22

Animal Models with a Genetic Alteration of the ACE2/Ang-(1-7)/Mas Axis

Luiza A. Rabelo,* Valéria Nunes-Souza,* Michael Bader†

*Max Delbrück Center for Molecular Medicine, Berlin, Germany; Laboratório de Reatividade Cardiovascular (LRC), Setor de Fisiologia e Farmacologia, ICBS, Universidade Federal de Alagoas, Maceió, Al, Brazil; National Institute of Science and Technology in NanoBiopharmaceutics (N-BIOFAR), Brazil, †Max Delbrück Center for Molecular Medicine (MDC), Berlin, Germany; Charité Medical Faculty, Berlin, Germany; Universidade Federal de Minas Gerais, Belo Horizonte, Brazil; Institute for Biology, University of Lübeck, Germany

INTRODUCTION

The study of hormone systems involved in cardiovascular and metabolic diseases, such as the renin–angiotensin system (RAS), can only be performed in whole organisms due to the complex interplay of different organs, which determines cardiovascular physiology. In contrast to pharmacological interventions in these systems, which often lack specificity, the targeted genetic alteration of the expression of single-hormone system components is the most straightforward method to analyze their functions in cardiovascular and metabolic homeostasis and disorders. Accordingly, the generation of transgenic and knockout (KO) animals was widely used to study the role of RAS components in cardiovascular control and in the pathogenesis of diseases.[1,2] The aim of this chapter is to describe the animal models generated by transgenic technology for the functional analysis of the protective axis of the RAS, consisting of angiotensin-converting enzyme 2 (ACE2), Ang-(1-7), and Mas.

TRANSGENIC ANIMAL TECHNOLOGY: A BRIEF UPDATE

In biomedical research, the use of rats and mice has become a major tool, considering the easiness of breeding, growth, and maintenance and the similarity with human organisms in most cardiovascular and metabolic systems.[3] Currently, the use of transgenic technologies to access animal physiology is routine, but the pioneering work was performed in the 1980s. The first successful genetic modifications of a mouse were achieved by Gordon and colleagues.[4] This group demonstrated that foreign genes can be integrated into the mouse genome by transfer of DNA constructs into the pronuclei of zygotes. By the use of specific promoters, the investigator can direct the expression of the transgene into specific tissues. About 10 years later, the same technology was also established for the rat, interestingly first targeting the RAS component renin.[5]

Since 1986, the elimination of gene expression is also possible by homologous recombination-mediated targeted gene KO in embryonic stem (ES) cells.[6–8] To this purpose, firstly, a targeting vector has to be designed and constructed that contains parts of DNA sequences homologous to the gene of interest and the intended mutation. Following the transfection of this vector into ES cells, a few cells will incorporate it into the endogenous gene via homologous recombination and can be selected by appropriate methods. These successfully targeted ES cells are injected into host blastocysts, and a chimeric animal is obtained. This animal carries the mutation in part of its cells and will eventually pass it on to its offspring, which will be heterozygous or homozygous (KO) for the mutant gene. The KO animals will present the physiological phenotype caused by the absence of the gene product.[6] In general, this method is still the most commonly used for the targeted alteration of the mouse genome. It took more than 20 years for this technology to also became available for the rat by the discovery of a method to establish germ-line-competent ES cells from this species in 2008,[9] and a few KO rat models have been developed since using ES cells.[10,11] However, recently, novel methods for the targeted alteration of genes in the mouse and rat genome have become available, which will probably replace the relatively complicated ES cell method in the near future. They are based on nucleases that are targeted to a certain site in the genome by different methods.[12] Zinc finger (ZFN) and TALE nucleases use protein domains that specifically recognize a DNA sequence of choice, while the CRISPR/Cas9 system uses a guide RNA that binds DNA by specific base pairing. Since only one animal model for the RAS has been described yet produced by ZFN technology, the renin-KO rat,[13] and since the Mas-KO rat has obviously been generated using ZFNs (http://rgd.mcw.edu/rgdweb/report/strain/main.html?id=5131952; access 25.04.2014) but is not yet published, we will not go into more detail on these novel technologies.

The Protective Arm of the Renin–Angiotensin System (RAS). http://dx.doi.org/10.1016/B978-0-12-801364-9.00022-5

TRANSGENIC AND KO RODENT MODELS OF THE ACE2/ANG-(1-7)/MAS AXIS

Our group and others have developed several transgenic and KO rat and mouse models with genetic deletion and/or overexpression of components of the ACE2/Ang-(1-7)/Mas axis. Some of these models produced pleiotropic phenotypes depending on the genetic background of the strain they were generated in.

In the following, we will list these models and summarize the insights into the physiology of the protective RAS axis gained by their analysis (see also Table 1).

ACE2 Models

ACE2 KO Mice

The ACE2 gene is located on the X chromosome, and thus, heterozygous deletion (ACE2$^{-/y}$) already results in the complete absence of the enzyme in male animals. Deletion of ACE2 in mouse leads to several cardiac abnormalities. In an elegant study, Crackower and colleagues provided the first *in vivo* evidence[14] supporting the hypothesis that the loss of ACE2 promotes heart dysfunction.[60] However, later studies on the function of ACE2 in the heart resulted in contradictory observations.[61] Two independent groups showed that the baseline cardiac function and morphology appeared normal in their ACE2-deficient mouse lines.[16,18] Nevertheless, ACE2$^{-/y}$ and even heterozygous female ACE2$^{+/-}$ mice are more susceptible to pressure overload or diabetes-induced cardiac injury.[18-20] Moreover, the lack of ACE2 also exacerbated diabetic and shock-induced kidney injury.[25-27]

There is also still a debate whether the deletion of ACE2 changes basic blood pressure in mice. Most likely, this effect depends strongly on the genetic background of the mice analyzed: 129 mice show hardly any effect, while C57BL/6 or FVB/N ACE2$^{-/y}$ mice are clearly hypertensive[61] (our unpublished results). Nevertheless, C57BL/6 ACE2$^{-/y}$ mice displayed high blood pressure during pregnancy and reduced weight gain and gave birth to smaller pups.[17] Besides an upregulation of oxidative stress in the brain and consequently of the sympathetic nervous system,[41] the cause for the hypertensive effect of ACE2 deletion may be an endothelial dysfunction as evidenced by an impaired acetylcholine-induced aortic vasodilatation.[15]

ACE2 deletion in apolipoprotein E (ApoE) KO mice, a classical model for atherosclerosis, worsened plaque formation and vascular inflammation.[21,22] In low-density lipoprotein receptor KO mice, fed a high-fat diet, ACE2 deletion also aggravated atherosclerosis.[62] Moreover, loss of ACE2 led to increased arterial neointima formation in response to endovascular injury in the femoral artery accompanied by an overexpression of inflammation-related genes.[22]

Several studies have shown an important role of ACE2 in metabolism. C57BL/6 ACE2$^{-/y}$ mice show impaired glucose homeostasis at different ages.[23,24] Also, ACE2 gene deletion aggravated liver fibrosis in models of chronic hepatic injury.[63]

ACE2$^{-/y}$ mice displayed aggravated pathologies in the acute respiratory distress syndrome[28] and in bleomycin-induced lung injury[29] rendering ACE2 an important target for inflammatory lung diseases.

ACE2$^{-/y}$ animals were also instrumental for the surprising finding that the protein is not only an enzyme but also a trafficking molecule in the gut being responsible for the functional expression of the amino acid transporter SLC6A19.[30,31] ACE2$^{-/y}$ mice, therefore, show reduced levels of large amino acids, such as tryptophan, in the circulation, altered gut microbiota, and intestinal inflammation.[32]

Human ACE2 Overexpression in Mouse

Besides being an enzyme and trafficking molecule, ACE2 is also the receptor for the human severe acute respiratory syndrome (SARS) coronavirus. In order to study this function, transgenic mouse models have been generated by several groups that overexpress human ACE2 using the mouse ACE2 promoter,[33] the cytomegalovirus promoter,[36,37] or the cytokeratin 18 promoter specific for the airway and other epithelia.[34,35] As expected, human ACE2 expression led to an increased susceptibility of the transgenic mice to SARS virus infection. In the kidney, ACE2 overexpression protected the mice from shock-induced injury.[26]

Human ACE2 Overexpression in the Mouse Brain

The transgenic mouse overexpressing human ACE2 in the brain using the synapsin promoter has confirmed the important actions of central Ang II in the pathogenesis of cardiovascular diseases and the protective role of Ang-(1-7). The animals are protected from hypertension induced by low peripheral infusions of Ang II[40] and DOCA-salt treatment.[64] Moreover, they show an amelioration of cardiac hypertrophy induced by AngII,[65] of chronic heart failure induced by coronary ligation,[66] and of stroke induced by middle cerebral artery occlusion.[42,67] In most cases, an increase in Ang-(1-7) and a decrease of Ang II in the brain both influencing the levels of local NO and the autonomic nervous system were shown to be instrumental.

TABLE 1 Transgenic Rodent Models of the ACE2/Ang-(1-7)/Mas Axis

Component of the ACE2/Ang-(1–7)/Mas Axis	Promoter	Species	Effect on Expression	Phenotypes/References
ACE2	-	Mouse	Knockout	Cardiac[14] and endothelial dysfunction[15] (our unpublished data) Increased blood pressure (depending on genetic background)[16] Increased blood pressure during pregnancy[17] Increased susceptibility to cardiac damage[18–20] Aggravated atherosclerosis[21,22] Disturbed glucose homeostasis[23,24] Aggravated kidney[25–27] and lung injury[28,29] Amino acid uptake deficiency in the gut[30–32]
ACE2 (human)	ACE2 (mouse)	Mouse	General overexpression	Increased susceptibility to SARS virus infection[33]
ACE2 (human)	Cytokeratin 18	Mouse	Overexpression in epithelia	Increased susceptibility to SARS virus infection[34,35]
ACE2 (human)	CMV	Mouse	General overexpression	Increased susceptibility to SARS virus infection[36,37]
ACE2 (human)	αMHC	Mouse	Overexpression in heart	Increase in ventricular tachycardia and sudden death[38]
ACE2 (human)	Nephrin	Mouse	Overexpression in podocytes	Ameliorated nephropathy induced by diabetes[39]
ACE2 (human)	Synapsin	Mouse	Overexpression in brain	Attenuated neurogenic hypertension[40] Decreased oxidative stress and sympathetic activity[40,41] Protection from brain ischemic injury[42]
ACE2 (human)	SMMHC	Rat (SHRSP)	Overexpression in smooth muscle	Ameliorated hypertension, reduction in oxidative stress[43]
Mas	-	Mouse	Knockout	Increased anxiety (sex-dependent)[44,45] Cardiac[46] and endothelial dysfunction[47,48] Increased vascular and systemic oxidative stress[47,48] Erectile dysfunction[49] Nephropathy[50] Metabolic dysfunction[51]
Mas (rat)	Opsin	Mouse	Overexpression in retina	Degeneration of photoreceptors[52]
PRCP	-	Mouse	Knockout	Decrease in body weight, hypertension, vascular dysfunction[53] (Schadock et al., unpublished results)
Ang-(1–7)	CMV	Rat	Overexpression in testis	Improved cardiac and endothelial function,[54] improved lipid and glycolytic profile[55,56] Protection from heart hypertrophy[57]
Ang-(1–7)	αMHC	Mouse	Overexpression in heart	Protection from heart hypertrophy[58]
Ang-(1–7)	αMHC	Rat	Overexpression in heart	Improved cardiac function and protection from heart hypertrophy[59]

PRCP, prolylcarboxypeptidase; SHRSP, spontaneously hypertensive stroke-prone rat; SMMHC, smooth muscle myosin heavy chain; CMV, cytomegalovirus; αMHC, α-cardiac myosin heavy chain.

Human ACE2 Overexpression in the Mouse Heart

Transgenic mice overexpressing human ACE2 in the heart surprisingly showed an increase in ventricular tachycardia and sudden death, which was due to a dysregulation of connexins.[38] The underlying mechanism and the involved peptides could not be elucidated.

Human ACE2 Overexpression in Mouse Podocytes

When human ACE2 was overexpressed in podocytes of mice using the nephrin promoter, the nephropathy induced by diabetes was ameliorated.[39] The authors suggest that this is due to reduced renal Ang II levels leading to a reduced expression of TGF-beta, but they could also not exclude a protective effect of Ang-(1-7).

Human ACE2 Overexpression in the Rat Vascular Smooth Muscle

ACE2 is highly expressed in the endothelium and smooth muscle cells (SMC), and its expression is reduced in the spontaneously hypertensive stroke-prone rat (SHRSP). When human ACE2 was overexpressed in vascular SMC of transgenic SHRSP, endothelial dysfunction and hypertension were ameliorated, which was accompanied by a reduction in oxidative stress linked to a decrease in Ang II and/or an increase in Ang-(1-7).[43]

Mas Models

Mas-KO Mice

In 1998, we generated Mas-deficient (Mas$^{-/-}$) mice on the mixed 129×C57BL/6 background and showed that these animals were healthy in appearance and grew normally and exhibited normal Ang II plasma levels.[44] However, male (but not female[45]) Mas$^{-/-}$ mice displayed increased anxiety on the elevated-plus maze. We also showed not only that long-term potentiation was markedly increased in the hippocampal CA1 region of Mas$^{-/-}$ mice[44] but also that object recognition memory is impaired.[68] Collectively, these data support a role of Mas in behavior.

C57BL/6 Mas$^{-/-}$ mice show a marked cardiac dysfunction both *in vitro*[69] and *in vivo*[46] accompanied by an increase in extracellular matrix proteins, such as collagen I, collagen III, and fibronectin.[46] Furthermore, Mas$^{-/-}$ animals exhibit vascular oxidative stress, endothelial dysfunction, and high blood pressure at least on the FVB/N genetic background.[47,48] Consistently, endothelial function is also impaired in isolated Mas$^{-/-}$ vessels.[70] Consequently, Mas$^{-/-}$ mice exhibit a pronounced decrease in blood flow and a marked increase in resistance in different vascular beds.[71] These functional changes in both regional and systemic hemodynamics in Mas$^{-/-}$ mice suggest that the Ang-(1-7)/Mas axis plays an important role in vascular regulation. A dysregulation of the vascular function in the *corpus cavernosum* is probably also the reason for the erectile dysfunction observed in Mas$^{-/-}$ mice.[49]

Pinheiro et al.[50] showed an imbalance in renal function in Mas$^{-/-}$ mice: reduced urine volume and fractional sodium excretion and increased glomerular filtration rate and proteinuria. Surprisingly, a proinflammatory role for Ang-(1-7) and Mas was reported in a model of unilateral ureteral obstruction in mice[72] in contrast to the anti-inflammatory effects of Ang-(1-7) and Mas in other models of kidney nephropathy.[73] Certainly, additional studies are needed to clarify the role of Ang-(1-7) and Mas in the kidney. The anti-inflammatory actions of Mas were also recently confirmed in a lipopolysaccharide (LPS)-induced endotoxic shock model.[74]

Santos and colleagues[51] have revealed the effects of Mas deficiency on lipid and glucose metabolism. They used Mas$^{-/-}$ mice on the FVB/N background and demonstrated that loss of Mas increases the risk of metabolic complications by causing several features of the metabolic syndrome, such as type 2 diabetes mellitus, hypertension, dyslipidemia, and nonalcoholic fatty liver disease. Mas deletion decreased the responsiveness of adipocytes to insulin accompanied by a decreased expression of PPARγ in adipose tissue.[75] Moreover, Silva et al. recently showed that Mas deletion in ApoE$^{-/-}$ mice leads to dyslipidemia and liver steatosis.[76]

Mas Overexpression in Retina

When rat Mas is overexpressed in the retina of transgenic mice using the opsin promoter, degeneration of photoreceptors is the consequence, which is probably induced by proliferative signaling pathways activated in these cells due to the constitutively active Mas protein.[77]

Prolylcarboxypeptidase KO Mice

Prolylcarboxypeptidase (PRCP, EC 3.4.16.2) is another enzyme that can generate Ang-(1-7) from Ang II, but it is also not specific for angiotensin peptides. PRCP-KO mice show elevated levels of α-melanocyte-stimulating hormone (α-MSH), an anorexigenic neuropeptide, in the hypothalamus. The phenotype of these animals is characterized by a decrease in body weight and body length under normal diet, accompanied by decreased white adipose tissue.[53] The lean phenotype is also observed after high-fat diet-induced obesity (Schadock et al., unpublished results).

Moreover, PRCP-KO mice display hypertension and vascular dysfunction probably due to an increase in reactive oxygen species and uncoupled eNOS.[52] In addition, PRCP also regulates angiogenesis and vascular repair.[78] How much of these cardiovascular actions are due to alterations in Ang II or Ang-(1-7) levels remains to be elucidated.

Ang-(1-7) Models

Based on an elegant system to generate RAS peptides by release from an artificial protein, which is processed by furin during secretion,[58,79] several transgenic animal models with altered Ang-(1-7) levels were generated.

Transgenic Rats Overexpressing Ang-(1-7)

The transgenic rat TGR(A1-7)3292 expresses the Ang-(1-7)-producing protein mainly in the testis.[80] In this model, Ang-(1-7) levels are chronically elevated in plasma and testis ~2.5-fold and ~4.5-fold, respectively. Surprisingly, this chronic increase in Ang-(1-7) did not alter basal blood pressure levels measured by telemetry. However, TGR(A1-7)3292 rats displayed a marked reduction in isoproterenol-induced heart hypertrophy and an improvement of postischemic systolic function.[54] Botelho-Santos and colleagues[57] showed changes in systemic and regional hemodynamic parameters in these rats, resulting in an increase of vascular conductance in several tissues and a decreased total peripheral resistance. Furthermore, the transgenic rats showed a significant increase in stroke volume and cardiac index[57] and a reduction in basal urinary flow, leading to increased urinary osmolality and osmolal clearance.[80] Furthermore, chronic elevation of circulating Ang-(1-7) levels considerably improved the lipid and glycolytic profile and lowered the fat mass accompanied by a decrease in triglycerides and cholesterol in plasma and an improved glucose tolerance and insulin sensitivity and reduced gluconeogenesis in the liver.[55,56]

Transgenic Mice and Rats Overexpressing Ang-(1-7) in the Heart

Transgenic mice carrying the Ang-(1-7)-releasing construct under the control of the alpha-cardiac myosin heavy-chain promoter exhibit about an eightfold increase of the peptide in the heart and show a normal basic cardiac function but are protected from hypertensive cardiac hypertrophy induced by Ang II infusion[58] but not from myocardial infarction.[81] Transgenic rats generated with the same construct showed a slightly improved resting cardiac function and were also protected from hypertrophy in this case induced by isoproterenol.[59]

CONCLUSIONS AND OUTLOOK

Transgenic and KO rodent models were pivotal for our understanding of the protective functions of the novel RAS axis, ACE2/Ang-(1-7)/Mas. Transgenic overexpression of the components of this axis in general led to ameliorated cardiac and vascular damage in disease states and to an improved metabolic profile. KO models for ACE2 and Mas, however, show aggravated cardiovascular pathologies and a metabolic-syndrome-like state. In particular, the local production of Ang-(1-7) in the vascular wall, in the heart, and in the brain was found to be of high physiological relevance by the use of transgenic animals overexpressing ACE2 or Ang-(1-7) in these tissues. Inducible and cell-type-specific KO models for Mas and ACE2 will be helpful in the future to deepen our understanding of the ACE2/Ang-(1-7)/Mas axis.

REFERENCES

1. Bader M. Mouse knockout models of hypertension. *Method Mol Med* 2005;**108**:17–32.
2. Alenina N, Xu P, Rentzsch B, Patkin EL, Bader M. Genetically altered animal models for mas and angiotensin-(1–7). *Exp Physiol* 2008;**93**:528–37.
3. Vaquer G, Riviere F, Mavris M, Bignami F, Llinares-Garcia J, Westermark K, et al. Animal models for metabolic, neuromuscular and ophthalmological rare diseases. *Nat Rev Drug Discov* 2013;**12**:287–305.
4. Gordon JW, Scangos GA, Plotkin DJ, Barbosa JA, Ruddle FH. Genetic transformation of mouse embryos by microinjection of purified DNA. *Proc Natl Acad Sci U S A* 1980;**77**:7380–4.
5. Mullins JJ, Peters J, Ganten D. Fulminant hypertension in transgenic rats harbouring the mouse ren-2 gene. *Nature* 1990;**344**:541–4.
6. Ramirez-Solis R, Bradley A. Advances in the use of embryonic stem cell technology. *Curr Opin Biotechnol* 1994;**5**:528–33.
7. Thomas KR, Capecchi MR. Introduction of homologous DNA sequences into mammalian cells induces mutations in the cognate gene. *Nature* 1986;**324**:34–8.

8. Doetschman T, Gregg RG, Maeda N, Hooper ML, Melton DW, Thompson S, et al. Targetted correction of a mutant hprt gene in mouse embryonic stem cells. *Nature* 1987;**330**:576–8.

9. Buehr M, Meek S, Blair K, Yang J, Ure J, Silva J, et al. Capture of authentic embryonic stem cells from rat blastocysts. *Cell* 2008;**135**:1287–98.

10. Tong C, Li P, Wu NL, Yan Y, Ying QL. Production of p53 gene knockout rats by homologous recombination in embryonic stem cells. *Nature* 2010;**467**:211–13.

11. Meek S, Buehr M, Sutherland L, Thomson A, Mullins JJ, Smith AJ, et al. Efficient gene targeting by homologous recombination in rat embryonic stem cells. *PLoS One* 2010;**5**:e14225.

12. Gaj T, Gersbach CA, Barbas 3rd CF. Zfn, talen, and crispr/cas-based methods for genome engineering. *Trends Biotechnol* 2013;**31**:397–405.

13. Moreno C, Hoffman M, Stodola TJ, Didier DN, Lazar J, Geurts AM, et al. Creation and characterization of a renin knockout rat. *Hypertension* 2011;**57**:614–19.

14. Crackower MA, Sarao R, Oudit GY, Yagil C, Kozieradzki I, Scanga SE, et al. Angiotensin-converting enzyme 2 is an essential regulator of heart function. *Nature* 2002;**417**:822–8.

15. Lovren F, Pan Y, Quan A, Teoh H, Wang G, Shukla PC, et al. Angiotensin converting enzyme-2 confers endothelial protection and attenuates atherosclerosis. *Am J Physiol Heart Circ Physiol* 2008;**295**:H1377–84.

16. Gurley SB, Allred A, Le TH, Griffiths R, Mao L, Philip N, et al. Altered blood pressure responses and normal cardiac phenotype in ace2-null mice. *J Clin Invest* 2006;**116**:2218–25.

17. Bharadwaj MS, Strawn WB, Groban L, Yamaleyeva LM, Chappell MC, Horta C, et al. Angiotensin-converting enzyme 2 deficiency is associated with impaired gestational weight gain and fetal growth restriction. *Hypertension* 2011;**58**:852–8.

18. Yamamoto K, Ohishi M, Katsuya T, Ito N, Ikushima M, Kaibe M, et al. Deletion of angiotensin-converting enzyme 2 accelerates pressure overload-induced cardiac dysfunction by increasing local angiotensin II. *Hypertension* 2006;**47**:718–26.

19. Patel VB, Bodiga S, Fan D, Das SK, Wang Z, Wang W, et al. Cardioprotective effects mediated by angiotensin II type 1 receptor blockade and enhancing angiotensin 1–7 in experimental heart failure in angiotensin-converting enzyme 2-null mice. *Hypertension* 2012;**59**:1195–203.

20. Wang W, Patel VB, Parajuli N, Fan D, Basu R, Wang Z, et al. Heterozygote loss of ace2 is sufficient to increase the susceptibility to heart disease. *J Mol Med* 2014;**92**(8):847–58.

21. Thomas MC, Pickering RJ, Tsorotes D, Koitka A, Sheehy K, Bernardi S, et al. Genetic ace2 deficiency accentuates vascular inflammation and atherosclerosis in the apoe knockout mouse. *Circ Res* 2010;**107**:888–97.

22. Sahara M, Ikutomi M, Morita T, Minami Y, Nakajima T, Hirata Y, et al. Deletion of angiotensin-converting enzyme 2 promotes the development of atherosclerosis and arterial neointima formation. *Cardiovasc Res* 2014;**101**:236–46.

23. Niu MJ, Yang JK, Lin SS, Ji XJ, Guo LM. Loss of angiotensin-converting enzyme 2 leads to impaired glucose homeostasis in mice. *Endocrine* 2008;**34**:56–61.

24. Takeda M, Yamamoto K, Takemura Y, Takeshita H, Hongyo K, Kawai T, et al. Loss of ace2 exaggerates high-calorie diet-induced insulin resistance by reduction of glut4 in mice. *Diabetes* 2013;**62**:223–33.

25. Wong DW, Oudit GY, Reich H, Kassiri Z, Zhou J, Liu QC, et al. Loss of angiotensin-converting enzyme-2 (ace2) accelerates diabetic kidney injury. *Am J Pathol* 2007;**171**:438–51.

26. Yang XH, Wang YH, Wang JJ, Liu YC, Deng W, Qin C, et al. Role of angiotensin-converting enzyme (ace and ace2) imbalance on tourniquet-induced remote kidney injury in a mouse hindlimb ischemia-reperfusion model. *Peptides* 2012;**36**:60–70.

27. Shiota A, Yamamoto K, Ohishi M, Tatara Y, Ohnishi M, Maekawa Y, et al. Loss of ace2 accelerates time-dependent glomerular and tubulointerstitial damage in streptozotocin-induced diabetic mice. *Hypertens Res* 2010;**33**:298–307.

28. Imai Y, Kuba K, Rao S, Huan Y, Guo F, Guan B, et al. Angiotensin-converting enzyme 2 protects from severe acute lung failure. *Nature* 2005;**436**:112–16.

29. Rey-Parra GJ, Vadivel A, Coltan L, Hall A, Eaton F, Schuster M, et al. Angiotensin converting enzyme 2 abrogates bleomycin-induced lung injury. *J Mol Med* 2012;**90**:637–47.

30. Singer D, Camargo SM, Ramadan T, Schafer M, Mariotta L, Herzog B, et al. Defective intestinal amino acid absorption in ace2 null mice. *Am J Physiol Gastr Liver Physiol* 2012;**303**:G686–95.

31. Camargo SM, Singer D, Makrides V, Huggel K, Pos KM, Wagner CA, et al. Tissue-specific amino acid transporter partners ace2 and collectrin differentially interact with hartnup mutations. *Gastroenterology* 2009;**136**:872–82.

32. Hashimoto T, Perlot T, Rehman A, Trichereau J, Ishiguro H, Paolino M, et al. Ace2 links amino acid malnutrition to microbial ecology and intestinal inflammation. *Nature* 2012;**487**:477–81.

33. Yang XH, Deng W, Tong Z, Liu YX, Zhang LF, Zhu H, et al. Mice transgenic for human angiotensin-converting enzyme 2 provide a model for sars coronavirus infection. *Comp Med* 2007;**57**:450–9.

34. McCray PB, Pewe L, Wohlford-Lenane C, Hickey M, Manzel L, Shi L, et al. Lethal infection of k18-hace2 mice infected with severe acute respiratory syndrome coronavirus. *J Virol* 2007;**81**:813–21.

35. Netland J, Meyerholz DK, Moore S, Cassell M, Perlman S. Severe acute respiratory syndrome coronavirus infection causes neuronal death in the absence of encephalitis in mice transgenic for human ace2. *J Virol* 2008;**82**:7264–75.

36. Tseng CT, Huang C, Newman P, Wang N, Narayanan K, Watts DM, et al. Severe acute respiratory syndrome coronavirus infection of mice transgenic for the human angiotensin-converting enzyme 2 virus receptor. *J Virol* 2007;**81**:1162–73.

37. Yoshikawa N, Yoshikawa T, Hill T, Huang C, Watts DM, Makino S, et al. Differential virological and immunological outcome of severe acute respiratory syndrome coronavirus infection in susceptible and resistant transgenic mice expressing human angiotensin-converting enzyme 2. *J Virol* 2009;**83**:5451–65.

38. Donoghue M, Wakimoto H, Maguire CT, Acton S, Hales P, Stagliano N, et al. Heart block, ventricular tachycardia, and sudden death in ace2 transgenic mice with downregulated connexins. *J Mol Cell Cardiol* 2003;**35**:1043–53.

39. Nadarajah R, Milagres R, Dilauro M, Gutsol A, Xiao F, Zimpelmann J, et al. Podocyte-specific overexpression of human angiotensin-converting enzyme 2 attenuates diabetic nephropathy in mice. *Kidney Int* 2012;**82**:292–303.

40. Feng Y, Xia H, Cai Y, Halabi CM, Becker LK, Santos RA, et al. Brain-selective overexpression of human angiotensin-converting enzyme type 2 attenuates neurogenic hypertension. *Circ Res* 2010;**106**:373–82.

41. Xia H, Suda S, Bindom S, Feng Y, Gurley SB, Seth D, et al. Ace2-mediated reduction of oxidative stress in the central nervous system is associated with improvement of autonomic function. *PLoS One* 2011;**6**:e22682.

42. Zheng JL, Li GZ, Chen SZ, Wang JJ, Olson JE, Xia HJ, et al. Angiotensin converting enzyme 2/ang-(1–7)/mas axis protects brain from ischemic injury with a tendency of age-dependence. *CNS Neurosci Ther* 2014;**20**:452–9.

43. Rentzsch B, Todiras M, Iliescu R, Popova E, Campos LA, Oliveira ML, et al. Transgenic angiotensin-converting enzyme 2 overexpression in vessels of shrsp rats reduces blood pressure and improves endothelial function. *Hypertension* 2008;**52**:967–73.

44. Walther T, Balschun D, Voigt JP, Fink H, Zuschratter W, Birchmeier C, et al. Sustained long term potentiation and anxiety in mice lacking the mas protooncogene. *J Biol Chem* 1998;**273**:11867–73.

45. Walther T, Voigt JP, Fink H, Bader M. Sex specific behavioural alterations in mas-deficient mice. *Behav Brain Res* 2000;**107**:105–9.

46. Santos RA, Castro CH, Gava E, Pinheiro SV, Almeida AP, Paula RD, et al. Impairment of in vitro and in vivo heart function in angiotensin-(1–7) receptor mas knockout mice. *Hypertension* 2006;**47**:996–1002.

47. Xu P, Costa-Goncalves AC, Todiras M, Rabelo LA, Sampaio WO, Moura MM, et al. Endothelial dysfunction and elevated blood pressure in mas gene-deleted mice. *Hypertension* 2008;**51**:574–80.

48. Rabelo LA, Xu P, Todiras M, Sampaio WO, Buttgereit J, Bader M, et al. Ablation of angiotensin (1–7) receptor mas in c57bl/6 mice causes endothelial dysfunction. *J Am Soc Hypertens* 2008;**2**:418–24.

49. da Costa Goncalves AC, Leite R, Fraga-Silva RA, Pinheiro SV, Reis AB, Reis FM, et al. Evidence that the vasodilator angiotensin-(1–7)-mas axis plays an important role in erectile function. *Am J Physiol Heart Circ Physiol* 2007;**293**:H2588–96.

50. Pinheiro SV, Ferreira AJ, Kitten GT, da Silveira KD, da Silva DA, Santos SH, et al. Genetic deletion of the angiotensin-(1–7) receptor mas leads to glomerular hyperfiltration and microalbuminuria. *Kidney Int* 2009;**75**:1184–93.

51. Santos SH, Fernandes LR, Mario EG, Ferreira AV, Porto LC, Alvarez-Leite JI, et al. Mas deficiency in fvb/n mice produces marked changes in lipid and glycemic metabolism. *Diabetes* 2008;**57**:340–7.

52. Adams GN, LaRusch GA, Stavrou E, Zhou Y, Nieman MT, Jacobs GH, et al. Murine prolylcarboxypeptidase depletion induces vascular dysfunction with hypertension and faster arterial thrombosis. *Blood* 2011;**117**:3929–37.

53. Wallingford N, Perroud B, Gao Q, Coppola A, Gyengesi E, Liu ZW, et al. Prolylcarboxypeptidase regulates food intake by inactivating alpha-MSH in rodents. *J Clin Invest* 2009;**119**:2291–303.

54. Santos RA, Ferreira AJ, Nadu AP, Braga AN, de Almeida AP, Campagnole-Santos MJ, et al. Expression of an angiotensin-(1–7)-producing fusion protein produces cardioprotective effects in rats. *Physiol Genomics* 2004;**17**:292–9.

55. Santos SH, Braga JF, Mario EG, Porto LC, Rodrigues-Machado Mda G, Murari A, et al. Improved lipid and glucose metabolism in transgenic rats with increased circulating angiotensin-(1–7). *Arterioscler Thromb Vasc Biol* 2010;**30**:953–61.

56. Bilman V, Mares-Guia L, Nadu AP, Bader M, Campagnole-Santos MJ, Santos RA, et al. Decreased hepatic gluconeogenesis in transgenic rats with increased circulating angiotensin-(1–7). *Peptides* 2012;**37**:247–51.

57. Botelho-Santos GA, Sampaio WO, Reudelhuber TL, Bader M, Campagnole-Santos MJ, Souza dos Santos RA. Expression of an angiotensin-(1–7)-producing fusion protein in rats induced marked changes in regional vascular resistance. *Am J Phys Heart Circ Physiol* 2007;**292**:H2485–90.

58. Mercure C, Yogi A, Callera GE, Aranha AB, Bader M, Ferreira AJ, et al. Angiotensin(1–7) blunts hypertensive cardiac remodeling by a direct effect on the heart. *Circ Res* 2008;**103**:1319–26.

59. Ferreira AJ, Castro CH, Guatimosim S, Almeida PW, Gomes ER, Dias-Peixoto MF, et al. Attenuation of isoproterenol-induced cardiac fibrosis in transgenic rats harboring an angiotensin-(1–7)-producing fusion protein in the heart. *Ther Adv Cardiovasc Dis* 2010;**4**:83–96.

60. Oudit GY, Kassiri Z, Patel MP, Chappell M, Butany J, Backx PH, et al. Angiotensin II-mediated oxidative stress and inflammation mediate the age-dependent cardiomyopathy in ace2 null mice. *Cardiovasc Res* 2007;**75**:29–39.

61. Gurley SB, Coffman TM. Angiotensin-converting enzyme 2 gene targeting studies in mice: mixed messages. *Exp Physiol* 2008;**93**:538–42.

62. Thatcher SE, Zhang X, Howatt DA, Lu H, Gurley SB, Daugherty A, et al. Angiotensin-converting enzyme 2 deficiency in whole body or bone marrow-derived cells increases atherosclerosis in low-density lipoprotein receptor−/− mice. *Arterioscler Thromb Vasc Biol* 2011;**31**:758–65.

63. Osterreicher CH, Taura K, De Minicis S, Seki E, Penz-Osterreicher M, Kodama Y, et al. Angiotensin-converting-enzyme 2 inhibits liver fibrosis in mice. *Hepatology* 2009;**50**:929–38.

64. Xia H, Sriramula S, Chhabra KH, Lazartigues E. Brain angiotensin-converting enzyme type 2 shedding contributes to the development of neurogenic hypertension. *Circ Res* 2013;**113**:1087–96.

65. Feng Y, Hans C, McIlwain E, Varner KJ, Lazartigues E. Angiotensin-converting enzyme 2 over-expression in the central nervous system reduces angiotensin-II-mediated cardiac hypertrophy. *PLoS One* 2012;**7**:e48910.

66. Xiao L, Gao L, Lazartigues E, Zucker IH. Brain-selective overexpression of angiotensin-converting enzyme 2 attenuates sympathetic nerve activity and enhances baroreflex function in chronic heart failure. *Hypertension* 2011;**58**:1057–65.

67. Chen J, Zhao Y, Chen S, Wang J, Xiao X, Ma X, et al. Neuronal over-expression of ace2 protects brain from ischemia-induced damage. *Neuropharmacology* 2014;**79**:550–8.

68. Lazaroni TL, Raslan AC, Fontes WR, de Oliveira ML, Bader M, Alenina N, et al. Angiotensin-(1–7)/Mas axis integrity is required for the expression of object recognition memory. *Neurobiol Learn Mem* 2012;**97**:113–23.

69. Castro CH, Santos RA, Ferreira AJ, Bader M, Alenina N, Almeida AP. Effects of genetic deletion of angiotensin-(1–7) receptor mas on cardiac function during ischemia/reperfusion in the isolated perfused mouse heart. *Life Sci* 2006;**80**:264–8.

70. Peiro C, Vallejo S, Gembardt F, Azcutia V, Heringer-Walther S, Rodriguez-Manas L, et al. Endothelial dysfunction through genetic deletion or inhibition of the g protein-coupled receptor Mas: a new target to improve endothelial function. *J Hypertens* 2007;**25**:2421–5.

71. Botelho-Santos GA, Bader M, Alenina N, Santos RA. Altered regional blood flow distribution in mas-deficient mice. *Ther Adv Cardiovasc Dis* 2012;**6**:201–11.

72. Esteban V, Heringer-Walther S, Sterner-Kock A, de Bruin R, van den Engel S, Wang Y, et al. Angiotensin-(1–7) and the g protein-coupled receptor Mas are key players in renal inflammation. *PLoS One* 2009;**4**:e5406.

73. Silveira KD, Barroso LC, Vieira AT, Cisalpino D, Lima CX, Bader M, et al. Beneficial effects of the activation of the angiotensin-(1–7) mas receptor in a murine model of adriamycin-induced nephropathy. *PLoS One* 2013;**8**:e66082.

74. Souza LL, Duchene J, Todiras M, Azevedo LC, Costa-Neto CM, Alenina N, et al. Receptor mas protects mice against hypothermia and mortality induced by endotoxemia. *Shock* 2014;**41**:331–6.

75. Mario EG, Santos SH, Ferreira AV, Bader M, Santos RA, Botion LM. Angiotensin-(1–7) mas-receptor deficiency decreases peroxisome proliferator-activated receptor gamma expression in adipocytes. *Peptides* 2012;**33**:174–7.

76. Silva AR, Aguilar EC, Alvarez-Leite JI, da Silva RF, Arantes RM, Bader M, et al. Mas receptor deficiency is associated with worsening of lipid profile and severe hepatic steatosis in apoe-knockout mice. *Am J Physiol Regul Integr Comp Physiol* 2013;**305**:R1323–30.

77. Xu X, Quiambao AB, Roveri L, Pardue MT, Marx JL, Rohlich P, et al. Degeneration of cone photoreceptors induced by expression of the mas1 protooncogene. *Exp Neurol* 2000;**163**:207–19.

78. Adams GN, Stavrou EX, Fang C, Merkulova A, Alaiti MA, Nakajima K, et al. Prolylcarboxypeptidase promotes angiogenesis and vascular repair. *Blood* 2013;**122**:1522–31.

79. Methot D, LaPointe MC, Touyz RM, Yang XP, Carretero OA, Deschepper CF, et al. Tissue targeting of angiotensin peptides. *J Biol Chem* 1997;**272**:12994–9.

80. Ferreira AJ, Pinheiro SV, Castro CH, Silva GA, Silva AC, Almeida AP, et al. Renal function in transgenic rats expressing an angiotensin-(1–7)-producing fusion protein. *Regul Pept* 2006;**137**:128–33.

81. Wang Y, Qian C, Roks AJ, Westermann D, Schumacher SM, Escher F, et al. Circulating rather than cardiac angiotensin-(1–7) stimulates cardioprotection after myocardial infarction. *Circ Heart Fail* 2010;**3**:286–93.

Chapter 23

Mas Signaling: Resolved and Unresolved Issues

Augusto C. Montezano, Aurelie Nguyen, Rhian M. Touyz

Institute of Cardiovascular & Medical Sciences, University of Glasgow, Glasgow, UK

INTRODUCTION

Mas is the receptor to which Ang-(1-7) binds to mediate its cellular effects. The Ang-(1-7)-Mas axis of the renin–angiotensin system is functionally active in many organs including the heart, blood vessels, kidneys, adipose tissue, and brain and plays an important role in cardiovascular, renal, and neuronal regulation and influences lipid and glucose metabolism.[1,2] Mas has a widespread distribution in the adult human (Figure 1). Mas-mediated effects of Ang-(1-7) are opposite to those of Ang II and counteract most of the injurious actions of the Ang II/AT_1R axis[3,4] (Figure 2). Mas is a G protein-coupled receptor (GPCR) that has a widespread distribution and signals through multiple pathways to induce vasorelaxation and anti-inflammatory, antiproliferative, and antifibrotic effects.[5] Mas is encoded by the *Mas* gene, originally described as a proto-oncogene, since it induced tumorigenicity in nude mice.[6–8] The human Mas gene is found on chromosome 6 and when activated by DNA from carcinoma cell lines, promotes cell growth. However, experimental evidence using transgenic mice overexpressing Mas showed that the Mas gene per se has no oncogenic activity.

Little is known about the structure and regulation of the Mas gene. Mas protein shares many features with the GPCR subfamily of hormone receptor proteins.[7] It has seven hydrophobic transmembrane domains, while the NH2- and COOH-terminal domains are hydrophilic. Although Mas was originally discovered in 1986,[6] it was only in 2003 that Ang-(1-7) was discovered as its ligand, establishing the ACE2/Ang-(1-7)/Mas axis as a new arm of the renin–angiotensin system.[9] To date, whether there are other Mas ligands and the endogenous regulation of Ang-(1-7)-Mas are still unclear. Unlike the other GPCRs in the renin–angiotensin system, particularly the AT_1R and AT_2R, where the signaling pathways have been well characterized, the downstream effector molecules and signaling cascades of Mas still need to be better characterized, although activations of Akt, endothelial nitric oxide synthase (eNOS), MAP kinases, RhoA, cAMP/PKA, and protein tyrosine phosphatases[10–12] have been implicated. Mas has also been shown to inhibit Ang II/AT_1R signaling by blocking Ang II signaling and through direct interaction between AT_1R and Mas.[10,13]

EXPRESSION OF MAS

Mas is expressed at the mRNA and protein levels in many organs including the brain, heart, vasculature, kidneys, eyes, adipose tissue, and testes as assessed by reverse transcriptase-polymerase chain reaction, RNase protection assay, Western blotting, and immunofluorescence[14–17] (Figure 1). In the central nervous system, Mas expression occurs at postnatal day 1 in rats. In rodent testes, Mas becomes evident 2–5 weeks after birth and continuously increases into adulthood.[18] Mas associates with mature Leydig cells and may be a marker and functionally involved in adult Leydig cells. Newborns do not seem to express testicular Mas. In the kidney, Mas appears primarily in the renal cortex.[17,19] In the heart, low levels are present in cardiomyocytes.[16] In the vascular system, Mas is abundantly expressed in the endothelium, where it may play a role in endothelial function.[20]

MAS AS A GPCR

Mas belongs to the family of GPCRs also known as seven-transmembrane domain receptors.[21] Over 800 GPCRs have been identified in humans, with the endogenous or natural exogenous ligands of more than 100 GPCRs still being unknown. Because most clinically important drugs target GPCRs, there is much interest in identifying the endogenous ligands of

The Protective Arm of the Renin–Angiotensin System (RAS). http://dx.doi.org/10.1016/B978-0-12-801364-9.00023-7

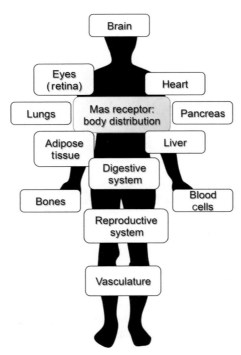

FIGURE 1   Mas receptor expression in human adults. Mas is ubiquitously expressed with high abundance particularly in the brain and testes. The functional significance of this remains unclear.

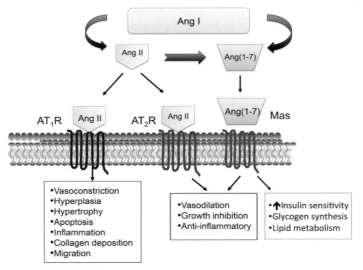

FIGURE 2   Counterregulatory effects of Ang-(1-7) and Ang II. Ang-(1-7), derived from Ang II, binds to the Mas receptor or the AT_2R, which leads to vasorelaxation, anti-inflammatory effects, antifibrotic actions, and improved metabolic profile. These downstream functional effects are opposite to those induced by Ang II, which signals through the AT_1R. As such, the Ang-(1-7)/Mas axis has been considered as the protective arm of the renin–angiotensin system.

these "orphan" receptors. For a long time, Mas was considered an orphan receptor, since it was only in 2003 that Santos identified Ang-(1-7) as the endogenous ligand.[9] Classically, intracellular signaling by these receptors is mediated by one or more members of the heterotrimeric G protein family. The G protein α subunit cycles between an inactive GDP-bound form and an active GTP-bound form, where the inactive α subunit is bound to the receptor and to a βγ heterodimer. Upon ligand stimulation, the receptor stimulates GDP–GTP exchange to promote the formation of the GTP-bound α subunit, which then dissociates from both the receptor and the βγ dimer. Both the Gα and βγ subunits then mediate activation of downstream effectors, including activation (Gαs) or inhibition (Gαi) of adenylyl cyclase or activation of phospholipase

C (Gαq or Gα11).[22] GPCRs also activate many other downstream signaling molecules including protein kinase C (PKC), mitogen-activated protein kinases (MAPKs), protein tyrosine phosphatases, tyrosine kinases, and NADPH oxidases. Mas has been shown to activate phospholipase C, indicating that Mas couples to the Gαq/11 family of heterotrimeric G proteins.[23] However, the signal transduction pathways activated by Mas, and those that cause cellular transformation, still remain largely unknown, although growing evidence indicates that many pathways may be involved. GPCRs are themselves regulated by agonist-induced receptor phosphorylation leading to uncoupling from cognate G proteins (desensitization) and subsequent receptor internalization through association with β-arrestin.[24] GPCRs can heterodimerize with receptor subtypes of other GPCRs. Such heterodimerization may be particularly important for Mas, since it may interact with other receptors of the renin–angiotensin system, such as the AT_1R,[25] and possibly with the bradykinin 2 receptor (BK_2R).[26]

Although intracellular signaling by Ang-(1-7)/Mas is still not well characterized, in general, actions of Ang-(1-7) seem to oppose those of Ang II/AT_1R signaling.[3,27,28] Interestingly, although both Mas and AT_1R are GPCRs, which classically increase intracellular free Ca^{2+} concentration ($[Ca^{2+}]_i$), Mas does not appear to signal through increased $[Ca^{2+}]_i$, although in pathological conditions, such as in ischemia-reperfusion injury and cardiac hypertrophy, Ang-(1-7) may influence Ca^{2+} transients.[29–31] However, in general, Ang-(1-7) does not seem to modulate $[Ca^{2+}]_i$.

EARLY DISCOVERY OF MAS AND ITS SIGNALING PATHWAYS

Soon after Mas was discovered in 1986 by Young et al.,[6] attempts were made to elucidate the signaling pathways underlying Mas transformation. Zohn et al.[32] demonstrated that the foci of transformed NIH 3T3 cells caused by Mas were similar to those caused by activated Rho and Rac proteins. They demonstrated that similar to activated Rac1, Mas cooperated with activated Raf and caused synergistic transformation of NIH 3T3 cells. This was associated with activation of actin stress fibers and enhanced membrane ruffling, and in porcine aortic endothelial cells, lamellipodia were formed. Mas activated JNK (JUN N-terminal kinase) and p38 MAPK, but not ERK, and stimulated transcription from common DNA promoter elements: NF-kappaB, serum response factor (SRF), Jun/ATF-2, and the cyclin D1 promoter. Mas transformation and signaling (SRF and cyclin D1, but not NF-kappaB activation) were blocked by dominant negative Rac1, suggesting that Mas effects are mediated in part by activation of Rac-dependent signaling pathways.[32] Interestingly, since these initial studies, there have been very few, if any, studies to examine the role of Rac1 in Mas signaling and to our knowledge, nothing is known about Rac1 in Ang-(1-7)/Mas signaling.

IDENTIFYING ANG-(1-7) TARGETS THROUGH PROTEOMICS

There is an urgent need to comprehensively identify proteins involved in Ang-(1-7) signaling. To address this, Verano-Braga et al. performed a mass spectrometry-based time-resolved quantitative phosphoproteome study of human aortic endothelial cells exposed to treated Ang-(1-7).[33] Over 1000 unique phosphosites on 699 different proteins with 99% certainty of correct peptide identification and phosphorylation site localization were identified. Phosphorylation levels were significantly changed on 121 sites on 79 proteins by Ang-(1-7). In particular, proteins of interest that were identified included those related to antiproliferative activity of Ang-(1-7), such as forkhead box protein O1 (FOXO1), mitogen-activated protein kinase 1 (MAPK1), and proline-rich AKT1 substrate 1 (AKT1S1), among others. Ang-(1-7) also induced changes in the phosphorylation status of several known downstream effectors of the insulin signaling, indicating an important role of Ang-(1-7) in glucose homeostasis. These proteomic-based findings support existing data that have characterized signaling pathways in cell-based and whole animal systems. Highlighted below are major signaling pathways that have best been linked to Ang-(1-7)/Mas signaling (Figure 3).

ANG-(1-7)/MAS SIGNALING THROUGH AKT

Akt (also called protein kinase B) is a serine/threonine kinase that plays a critical role in cell survival, growth, proliferation, angiogenesis, migration, apoptosis, autophagy, and lipid and glycogen metabolism.[34] Akt is activated by many factors including cytokines, TGF-β, advanced glycation end products and Ang-(1-7).[34–37] Ang-(1-7) stimulates phosphorylation of the regulatory sites Ser473 and Thr308, effects that are blocked by the Ang-(1-7)/Mas antagonist A-779, indicating the importance of Mas in this process.[38] The key role of Akt in Mas signaling was recently confirmed by phosphoproteomic analysis.[33] Activation of phosphatidylinositol 3-kinase (PI3K) through Mas has been demonstrated *in vitro* in human endothelial cells and mouse cardiomyocytes and *in vivo* in rodent liver, heart, adipose tissue, brain, and skeletal muscle.[38–40] Akt has numerous downstream targets including glycogen synthase kinase-3 (GSK-3), eNOS, mTOR, FOXO1, and AS160, all of which have been shown to be variably phosphorylated in response to Ang-(1-7).[41,42]

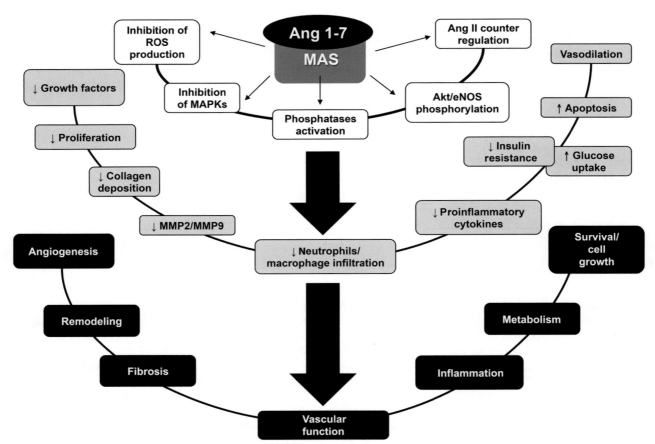

FIGURE 3 Diagram demonstrating signaling through Mas. Ang-(1-7) binding to Mas results in activation of multiple signaling pathways that impact on many cell types and systems. In general, Mas effects are vasoprotective and improve glucose and lipid metabolism.

Akt phosphorylation by Ang-(1-7)/Mas may be important in glucose and lipid metabolism through its effects on GSK-3 and insulin signaling[43] (Figure 4). *In vivo* studies demonstrated that in rats with diabetic cardiomyopathy and diabetic nephropathy, Ang-(1-7) treatment significantly reduces dyslipidemia in a Mas/Akt-dependent way.[44] In addition, transgenic animals with increased plasma levels of Ang-(1-7) had decreased triglyceride and cholesterol levels.[45] Ang-(1-7) signals through many pathways that are also activated by insulin, including PI3K/Akt. Through such cross talk, Ang-(1-7) has been shown to improve insulin sensitivity.[43,46]

Molecular mechanisms underlying effects on glucose and lipid metabolism by Ang-(1-7)/Mas involve PI3K/Akt in adipose tissue, liver, and skeletal muscle as indicated in *in vitro* studies.[47,48] Ang-(1-7) stimulates phosphorylation of GSK-3β, a proline-directed serine/threonine kinase that regulates glycogen metabolism.[43,49] Ang-(1-7) also stimulates glucose transport, processes that are associated with decreased activation of NADPH oxidase and reactive oxygen species (ROS) production in adipocytes.[43] In the liver, Ang-(1-7) increases glucose uptake and decreases glycogen synthesis with associated increased expression of glucose transporters, insulin receptor substrates, and decreased expression of enzymes for glycogen synthesis.[50] These processes are mediated through PI3K/Akt/insulin receptor substrate (IRS) signaling and activation of JNK. Such events may improve hepatic insulin sensitivity in the liver, which could be an important strategy in the treatment of insulin resistance and diabetes.

Other molecular mechanisms whereby Ang-(1-7)/Mas signaling may impact on insulin resistance and lipid metabolism relate to effects on pancreatic islet microvascular endothelial cells.[51] Pancreatic islets have a dense capillary network, with each β cell being surrounded by at least one islet endothelial cell, which may provide signals for islet cell development and are important for adult β-cell proliferation. Endothelial dysfunction was observed in insulin resistance and prediabetic populations. Islet microvascular endothelial cells (MS-1) are cells treated with the lipid palmitate-induced apoptosis, an effect that was prevented by Ang-(1-7).[51] Palmitate decreased the phosphorylation of AKT and eNOS, while Ang-(1-7) increased the phosphorylation of these kinases. Ang-(1-7) also inhibited the palmitate-induced ROS production and attenuated activation of the apoptosis-related signaling molecules JNK and p38.[51] These findings suggest that

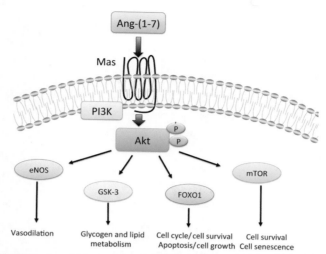

FIGURE 4 Mas signaling through Akt. Activation of Mas leads to phosphorylation of Akt, which induces activation of downstream targets including eNOS, GSK-3, FOXO1, and mTOR, among others.

Ang-(1-7) reduces palmitate-induced islet endothelial cell apoptosis by modulating AKT/eNOS/NO signaling and activating JNK and p38 MAPK.

Another downstream target of Mas-stimulated Akt is FOXO transcription factors that regulate genes controlling cell growth and glucose and lipid metabolism. Akt phosphorylates FOXO1 on T24, S256, and S319 and FOXO4 on three equivalent sites.[52] Phosphorylation of FOXO by Akt occurs in the nucleus, and when phosphorylated on T24 and S256, they are bound by 14-3-3 proteins, which displace FOXO transcription factors from target genes. Through this mechanism, Akt blocks FOXO-mediated transcription of target genes that regulate apoptosis, cell cycle, and metabolic processes. In human endothelial cells, Ang-(1-7) stimulates the dephosphorylation of FOXO1. In A549 lung adenocarcinoma cells, Ang-(1-7) increased the localization of FORKO1 in the nucleus, effects mediated through Mas.[53] Exact functions of Mas-regulated FOXO transcription factors in different cell types are still unclear, although there is some evidence from *in vivo* studies that this pathway may play a role in regulating metabolic processes. Mice receiving a high-fat diet together with Ang-(1-7) exhibited reduced adipose mass, improved insulin sensitivity, and glucose tolerance as well as lower plasma levels of fasting glucose and lipids, processes associated with increased expression of GLUT4 and increased activation of Akt/FOXO1/PPARγ.[53,54] These effects were shown to be Mas-mediated because Mas antagonists blocked Ang-(1-7) actions in primary culture adipocytes.

ANG-(1-7)/MAS SIGNALING AND NITRIC OXIDE PRODUCTION

A major downstream effector of Ang-(1-7) is nitric oxide (NO), which is a Mas-inhibitable process (Figure 5). In human endothelial cells that constitutively express Mas, and in Chinese hamster ovary (CHO) cells transfected with Mas, Ang-(1-7) induces activation of a PI3K/Akt signaling pathway that leads to increased expression and activation of eNOS.[37,38] We showed that Ang-(1-7) stimulates phosphorylation of eNOS on Ser1177, the active site, and reduces phosphorylation of the inactive site Thr495, with consequent increased NOS activation and increased NO production.[38] In endothelial cells, this promotes vasodilation and protects against vascular injury through its anti-inflammatory and anticoagulant effects. At the whole vessel level, Ang-(1-7) and its analogs, AVE 0991 and CGEN-856S, induce endothelium-dependent vasorelaxation effects that are blocked by A-779, the Mas antagonist.[37,38,55] In support of the obligatory role of Mas in these processes are studies showing that in Mas-knockout mice, Ang-(1-7)-induced endothelium-dependent vasorelaxation is absent.[56] Ang-(1-7) can also produce NO indirectly through cross talk between the Mas receptor and the bradykinin receptor (BK$_2$R).[57] In addition to stimulating NOS in endothelial cells, Ang-(1-7) stimulates production of NO in platelets[58] and thereby prevents platelet activation and coagulation. Vascular smooth muscle cells and cardiac cells also produce NO in response to Ang-(1-7)/Mas.[59] Through these mechanisms, Ang-(1-7) influences vascular tone and cardiac contraction.

In the brain, Ang-(1-7)/Mas stimulates production of NO through direct and indirect mechanisms. Ang-(1-7)/Mas directly induces phosphorylation and activation of neuronal NOS, which increases NO generation leading to activation of voltage-gated outward K$^+$ currents.[60] Mas also increases brain NO through bradykinin-mediated processes. Ang-(1-7) binding to Mas stimulates bradykinin production, which through BK$_2$R promotes NO formation.[61]

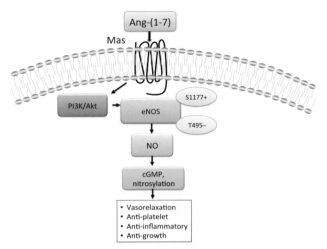

FIGURE 5 Mas activation signals through eNOS in endothelial cells. Activation of Mas by Ang-(1-7) leads to phosphorylation of eNOS, through Akt, which results in increased production of nitric oxide (NO), which promotes normal endothelial function. eNOS is phosphorylated at its stimulatory site (Ser1177) and its inhibitory site (Thr495).

MAS AND MAPK SIGNALING

MAP kinases are key regulators of multiple cell functions. There are many family members of which ERK1/2, p38 MAPK, JNK, and ERK5 are the best characterized. These kinases regulate multiple cell functions including cell growth, apoptosis, inflammation, fibrosis, and migration.[62] In the cardiovascular and renal systems, under physiological conditions, MAPKs are critically involved in maintaining vascular and renal functions, and in pathological conditions, they play a major role in vascular remodeling, renal fibrosis, and cardiac hypertrophy.[62,63] Whereas activation of the Ang II/AT$_1$R axis promotes activation of MAPKs, many studies have shown that Ang-(1-7) inhibits these pathways, thereby protecting against cardiovascular and renal injuries. Molecular mechanisms whereby Ang-(1-7) regulates MAPK signaling occur through indirect and direct mechanisms (Figure 4). Ang-(1-7) can modulate MAPK activation directly by stimulating activation of protein kinase phosphatases, such as SHP-2, which dephosphorylate MAPKs, thereby reducing activation of these kinases.[38,64,65] On the other hand, Ang-(1-7) can antagonize AT$_1$R-mediated activation of MAPKs.[66] In vascular smooth muscle cells, Ang-(1-7) attenuates Ang II-induced activation of PKC and reduces phosphorylation of ERK1/2.[66] We showed in Mas-expressing CHO cells that Ang-(1-7) reduces Ang II-stimulated activation of c-Src and ERK1/2 and that production of reactive oxygen species is inhibited.[37,38] These processes were associated with increased activation of SHP-2, which negatively regulates ERK1/2.[38] Similar results were observed in human endothelial cells that express Mas.[38] A-779 inhibited these Ang-(1-7) effects, confirming that responses are mediated through the Mas receptor. To further support a role for Mas in Ang-(1-7) modulation of MAP kinases, in cardiomyocytes, oligonucleotides against Mas block the MAPK inhibitory effects of Ang-(1-7).[64,65]

Ang-(1-7) antagonistic actions on MAPK signaling have also been shown in cultured rat aortic smooth muscle cells, where Ang-(1-7) inhibited Ang II signaling through prostacyclin-mediated production of cAMP and activation of cAMP-dependent protein kinase.[67] In proximal tubular cells, Ang II-stimulated phosphorylation of ERK1/2, p38 MAPK, and JNK was also attenuated by Ang-(1-7), an effect that was inhibited by A-779, the Mas antagonist.[68] Growing evidence indicates that vascular growth signaling is associated with Ang II-mediated transactivation of growth factor receptors in vascular cells. Ang-(1-7) has been shown to attenuate epidermal growth factor receptor (EGFR) transactivation by Ang II.[69] In vivo studies confirm a modulatory role of Ang-(1-7)/Mas on MAPK signaling. In atherosclerotic ApoE−/− mice, AVE 0991, an Ang-(1-7) peptidomimetic, influenced renal MAPK protein expression.[70]

In general, Ang-(1-7)/Mas reduces cell growth by downregulating MAPK growth signaling. However, not all cells exhibit this profile as in renal cells, Ang-(1-7) seems to upregulate MAPKs. In rat mesangial cells, Ang-(1-7) stimulated activation of ERK1/2 in a time- and concentration-dependent manner.[71] Pretreatment with A-779 but not with the AT$_1$R antagonist losartan or the AT$_2$R antagonist PD 123319 abrogated ERK1/2 phosphorylation. Ang-(1-7) also increased intracellular cAMP levels and activated protein kinase A (PKA). Inhibition of adenylyl cyclase and PKA blunted Ang-(1-7)-induced ERK1/2 activation.[71] Inhibition of NADPH oxidase and EGFR did not alter Ang-(1-7)-induced phosphorylation of ERK1/2. These data suggest that Ang-(1-7) stimulates activation of ERK1/2 in a cAMP/PKA-dependent manner and is independent of NADPH oxidase and the EGFR. Other studies showed that Ang-(1-7) and Ang II alone induce phosphorylation of

ERK1/2 and stimulate cell proliferation and extracellular matrix synthesis; however, when given in combination, Ang-(1-7) counteracted Ang II-induced effects.[71]

In cultured human mesangial cells, Ang-(1-7) increased phosphorylation of p38 MAPK, ERK1/2, and JNK, effects that were blocked by A-779.[72] Ang-(1-7) stimulated the release of arachidonic acid in mesangial cells. This was blocked by SB202190, a p38 MAPK inhibitor, and A-779, indicating the importance of Mas and p38 MAPK in Ang-(1-7)-mediated production of arachidonic acid. Functionally, these effects were associated with enhanced DNA synthesis and increased production of transforming growth factor-beta1 (TGF-beta1), fibronectin, and collagen IV, suggesting that in human mesangial cells, unlike in other cell types, Ang-(1-7) exerts growth-stimulatory effects.[72] How these effects impact on renal structure and function *in vivo* still remains unclear.[73]

Unlike mesangial cells, Ang-(1-7) does not seem to directly activate MAP kinases in other renal cells. In primary cultures of rat proximal tubular cells, Ang II stimulated phosphorylation of ERK1/2, p38 MAPK, and JNK and increased production of TGF-beta1, while Ang-(1-7) had no impact on these processes.[68,74] However, when cells are exposed to Ang II and Ang-(1-7) in combination, Ang-(1-7) abrogated Ang II-stimulated phosphorylation of p38 MAPK, ERK1/2, and JNK and partially reduced TGF-beta1 synthesis.[74] The Ang-(1-7) receptor antagonist, D-Ala7-Ang-(1-7), reversed Ang-(1-7) actions. Unlike mesangial cells, Ang-(1-7) had no significant effect on cyclic 3',5'-adenosine monophosphate production. Taken together, it is apparent that Ang-(1-7) inhibits MAPK signaling by Ang II in both mesangial and proximal tubule cells, effects that may protect against Ang II-induced renal injury. Reasons why Ang-(1-7) stimulates MAPKs in mesangial cells while it has no effect in proximal tubule cells are unclear but may reflect cell-specific processes.

Renoprotective effects of Ang-(1-7) are not specific for Ang II. In renal proximal tubule cells, glucose-induced activation of p38 MAPK and profibrotic signaling were attenuated by Ang-(1-7) through processes that involve activation of the protein tyrosine phosphatase SHP-1. In diabetic nephropathy, Ang-(1-7) may partly counteract the profibrotic effects of high glucose by stimulating SHP-1 and inhibiting p38 MAPK signaling.[75]

In the brain, Mas signaling involves PI3K/Akt, ERK1/2, and cyclooxygenase-2.[76] PI3K and ERK1/2 activation leads to changes in gene transcription of the norepinephrine transporter, with consequent reduced norepinephrine activity and decreased blood pressure. In the paraventricular nucleus and rostral ventrolateral medulla, Ang-(1-7)/Mas signals through cAMP/PKA, which mediates enhanced sympathetic outflow and cardiac sympathetic afferent reflexes, processes that involve NADPH oxidase-derived reactive oxygen species and which are associated with elevated blood pressure.[77] Thus, depending on the area of Mas stimulation in the brain, functional effects may differ.[78]

MAS AND RHOA

Early studies showed that Mas transformation requires Rac1 and RhoA because downregulation of RhoA using a dominant negative strategy resulted in impaired Mas signaling and transformation.[21,79] A few recent studies have shown that RhoA plays a role in Ang-(1-7) signaling, particularly by attenuating Ang II effects. In human airway smooth muscle cells[80] and in cardiomyocytes,[81] Ang II-induced activation of MAPKs and RhoA was attenuated by Ang-(1-7), effects that were blocked by A-779.[80] These data imply a role for RhoA in the protective actions of Ang-(1-7), although there is a paucity of data in the field and more definitive studies are still needed.

BIOLOGICAL SIGNIFICANCE OF MAS SIGNALING: INSIGHTS FROM MAS-KNOCKOUT MICE

Following its discovery in the late 1990s, the direct function of Mas remained a mystery. With the creation of Mas-knockout mice by Walther et al. in 1998, it became evident that Mas is an Ang-(1-7) receptor and that without a functioning Mas receptor, Ang-(1-7) actions are absent.[80] Findings from Mas-knockout mice demonstrated that Mas plays a role in behavior and cardiovascular function because Mas-deficient mice display increased anxiety[81,82] and when crossed on an FVB/N genetic background are hypertensive.[81] Also, mice that are Mas-deficient exhibit exaggerated renal hypertension, which is improved when oxidative stress is reduced.[83] Interestingly, despite the abundance of Mas in the brain and testes, ablation of the Mas gene had no significant impact on brain or testicular morphology or function.

UNRESOLVED ISSUES

Although there is now extensive evidence supporting Mas as the endogenous receptor for Ang-(1-7), with agonists and peptidomimetics already under development for potential clinical use, our knowledge about Mas signaling is still deficient and comprehensive characterization of intracellular signaling pathways is needed. Specific questions that need to be addressed include the following: (1) Does Mas signal through different pathways in different cell types? (2) Is Mas regulated by internalization through β-arrestin? (3) Are there other endogenous ligands besides Ang-(1-7) that bind to Mas? and (4) What are the physiological functions of Mas activation in different tissues and organs?

In addition to these questions, the significance of the Ang-(1-7)/Mas system in humans still remains unclear. While the expression pattern of Mas has been comprehensively characterized in various organ systems, in mice, rats, and other animals, there is still a paucity of information regarding the distribution pattern in humans. Moreover, the functional significance of the widespread expression is unclear. This is confirmed by the normal phenotype in Mas-KO mice. It should also be highlighted that Ang-(1-7) may be more promiscuous than previously thought, because recent evidence indicates that in addition to binding to Mas, it binds to Mas-related GPCR, hMRGPRD, and AT_2R.[84] The functional significance of this is unclear and it is unknown whether Mas, hMRGPRD, and ACE2 activation by Ang-(1-7) engages similar signaling pathways. Moreover, some effects of Ang-(1-7) occur through the AT_1R,[84] such as activation of the Janus kinase (JAK)/STAT pathway and phosphorylation in the rat heart, where Ang-(1-7) induces phosphorylation of JAK2, STAT3, and STAT5 through a Mas-independent, losartan-dependent manner.[75,84]

CONCLUSIONS

Mas is a GPCR that is activated by its ligand Ang-(1-7). Mas is ubiquitously expressed and signals through multiple pathways, including Akt, NOS-NO, MAP kinases, RhoA, and cAMP/PKA. Although many Mas-sensitive signaling molecules have been identified, there is still a paucity of information on the in-depth regulation and characterization of signal transduction by Mas. Ang-(1-7)/Mas signaling seems to oppose that of Ang II/AT_1R and to have cardio- and renal-protective effects. However, it is still unclear exactly how Mas signals in different systems, and the *in vivo* significance of differential Mas signaling is unclear, since most signaling studies were performed in cell culture-based models. Moreover, it is unknown whether there are endogenous ligands besides Ang-(1-7). Finally, it is very intriguing why more in-depth studies have not focused on elucidating the specific G proteins that couple to Mas, especially since Mas is a GPCR. With the exciting new phosphoproteomic studies, which have recently identified in a target-specific manner Ang-(1-7)/Mas-regulated proteins, and with the availability of Mas-knockout mice, the time is ripe to fill the gaps in better understanding the upstream regulators and downstream targets of Mas. This is especially important as there is growing interest in developing Mas agonists for use in the clinical arena.

ACKNOWLEDGMENTS

Publications from the author's laboratory were funded by the Canadian Institutes of Health Research and the British Heart Foundation.
There are no conflicts of interest to declare.

REFERENCES

1. Passos-Silva DG, Verano-Braga T, Santos RA. Angiotensin-(1–7): beyond the cardio-renal actions. *Clin Sci (Lond)* 2013;**124**:443–5.
2. Tirupula KC, Desnoyer R, Speth RC, Karnik SS. Atypical signaling and functional desensitization response of MAS receptor to peptide ligands. *PLoS One* 2014;**9**:e10352.
3. Iwai M, Horiuchi M. Devil and angel in the renin–angiotensin system: ACE–angiotensin II–AT1 receptor axis vs. ACE2–angiotensin-(1–7)–Mas receptor axis. *Hypertens Res* 2009;**32**:533–6.
4. Chappell MC, Marshall AC, Alzayadneh EM, Shaltout HA, Diz DI. Update on the angiotensin converting enzyme 2-angiotensin (1–7)-MAS receptor axis: fetal programing, sex differences, and intracellular pathways. *Front Endocrinol (Lausanne)* 2014;**4**:201–10.
5. Allahdadi K, Pessoa DC, Costa-Fraga FP, Fraga-Silva RA, Cojocaru G, Cohen Y. Vascular relaxation, antihypertensive effect, and cardioprotection of a novel peptide agonist of the MAS receptor. *Hypertension* 2010;**56**:112–20.
6. Young D, Waitches G, Birchmeier C, Fasano O, Wigler M. Isolation and characterization of a new cellular oncogene encoding a protein with multiple potential transmembrane domains. *Cell* 1986;**45**:711–19.
7. Young D, O'Neill K, Jessell T, Wigler M. Characterization of the rat mas oncogene and its high-level expression in the hippocampus and cerebral cortex of rat brain. *Proc Natl Acad Sci U S A* 1988;**85**:5339–42.
8. Becker LK, Etelvino GM, Walther T, Santos RA, Campagnole-Santos MJ. Immunofluorescence localization of the receptor Mas in cardiovascular-related areas of the rat brain. *Am J Physiol Heart Circ Physiol* 2007;**293**:H1416–24.
9. Santos RA, Simoes e Silva AC, Maric C, Silva DM, Machado RP, de Buhr I, et al. Angiotensin-(1–7) is an endogenous ligand for the G protein-coupled receptor Mas. *Proc Natl Acad Sci U S A* 2003;**100**:8258–63.
10. Dias-Peixoto MF, Santos RA, Gomes ER, Alves MN, Almeida PW, Greco L, et al. Molecular mechanisms involved in the angiotensin-(1–7)/Mas signaling pathway in cardiomyocytes. *Hypertension* 2008;**52**:542–8.
11. Xue H, Zhou L, Yuan P, Wang Z, Ni J, Yao T, et al. Counteraction between angiotensin II and angiotensin-(1–7) via activating angiotensin type I and Mas receptor on rat renal mesangial cells. *Regul Pept* 2012;**177**:12–20.

12. Than A, Leow MK, Chen P. Control of adipogenesis by the autocrine interplays between angiotensin (1–7)/mas receptor and angiotensin II/AT1 receptor signaling pathways. *J Biol Chem* 2013;**288**:15520–31.

13. Tao X, Fan J, Kao G, Zhang X, Su L, Yin Y, et al. Angiotensin-(1–7) attenuates angiotensin II-induced signalling associated with activation of a tyrosine phosphatase in Sprague–Dawley rats cardiac fibroblasts. *Biol Cell* 2014;**106**:182–92.

14. Santos SH, Andrade JM. Angiotensin 1–7: a peptide for preventing and treating metabolic syndrome. *Peptides* 2014;**59C**:34–41.

15. Martin KA, Grant SG, Hockfield S. The mas proto-oncogene is developmentally regulated in the rat central nervous system. *Dev Brain Res* 1992;**68**:75–82.

16. Villar AJ, Pedersen RA. Parental imprinting of the Mas protooncogene in mouse. *Nat Genet* 1994;**8**:373–9.

17. Metzger R, Bader M, Ludwig T, Berberich C, Bunnemann B, Ganten D. Expression of the mouse and rat mas proto-oncogene in the brain and peripheral tissues. *FEBS Lett* 1995;**357**:27–32.

18. Alenina N, Bader M, Walther T. Cell-type specific expression of the Mas protooncogene in testis. *J Histochem Cytochem* 2002;**59**:691–6.

19. Ferreira AJ, Pinheiro SVB, Castro CH, Silva GAB, Simoes e Silva AC, Almeida AP, et al. Renal functions in transgenic rats expressing an angiotensin-(1–7)-producing fusion protein. *Regul Pept* 2006;**137**:128–33.

20. McKinney CA, Fattah C, Loughrey CM, Milligan G, Nicklin SA. Angiotensin-(1–7) and angiotensin-(1–9): function in cardiac and vascular remodelling. *Clin Sci (Lond)* 2014;**126**:815–27.

21. Zhou L, Bohn LM. Functional selectivity of GPCR signaling in animals. *Curr Opin Cell Biol* 2014;**27**:102–8.

22. Wang D, Chen T, Zhou X, Couture R, Hong Y. Activation of Mas oncogene-related gene (Mrg) C receptors enhances morphine-induced analgesia through modulation of coupling of μ-opioid receptor to Gi-protein in rat spinal dorsal horn. *Neuroscience* 2013;**253**:455–64.

23. Canals M, Jenkins L, Kellett E, Milligan G. Up-regulation of the angiotensin II type 1 receptor by the MAS proto-oncogene is due to constitutive activation of Gq/G11 by MAS. *J Biol Chem* 2006;**281**:16757–67.

24. Kang DS, Tian X, Benovic JL. Role of β-arrestins and arrestin domain-containing proteins in G protein-coupled receptor trafficking. *Curr Opin Cell Biol* 2014;**27**:63–7.

25. Castro CH, Santos RA, Ferreira AJ, Bader M, Alenina N, Almeida AP. Evidence for a functional interaction of the angiotensin-(1–7) receptor Mas with AT1 and AT2 receptors in the mouse heart. *Hypertension* 2005;**46**:937–42.

26. Carvalho MB, Duarte FV, Faria-Silva R, Fauler B, da Mata Machado LT, de Paula RD, et al. Evidence for Mas-mediated bradykinin potentiation by the angiotensin-(1–7) nonpeptide mimic AVE 0991 in normotensive rats. *Hypertension* 2007;**50**:762–7.

27. Patel KP, Schultz HD. Angiotensin peptides and nitric oxide in cardiovascular disease. *Antioxid Redox Signal* 2013;**19**:1121–32.

28. Santos EL, Reis RI, Silva RG, Shimuta SI, Pecher C, Bascands JL, et al. Functional rescue of a defective angiotensin II AT1 receptor mutant by the Mas protooncogene. *Regul Pept* 2007;**141**:159–67.

29. Wang L, Luo D, Liao X, He J, Liu C, Yang C, et al. Ang-(1–7) offers cytoprotection against ischemia-reperfusion injury by restoring intracellular calcium homeostasis. *J Cardiovasc Pharmacol* 2014;**63**:259–65.

30. Castelo-Branco RC, Leite-Delova DC, de Mello-Aires M. Dose-dependent effects of angiotensin-(1–7) on the NHE3 exchanger and [Ca(2+)](i) in in vivo proximal tubules. *Am J Physiol Renal Physiol* 2013;**304**:F1258–65.

31. Shah A, Gul R, Yuan K, Gao S, Oh YB, Kim UH, et al. Angiotensin-(1–7) stimulates high atrial pacing-induced ANP secretion via Mas/PI3-kinase/Akt axis and Na+/H+ exchanger. *Am J Physiol Heart Circ Physiol* 2010;**298**:H1365–74.

32. Zohn IE, Symons M, Chrzanowska-Wodnicka M, Westwick JK, Der CJ. Mas oncogene signaling and transformation require the small GTP-binding protein Rac. *Mol Cell Biol* 1998;**18**:1225–35.

33. Verano-Braga T, Schwämmle V, Sylvester M, Passos-Silva DG, Peluso AA, Etelvino GM, et al. Time-resolved quantitative phosphoproteomics: new insights into angiotensin-(1–7) signaling networks in human endothelial cells. *J Proteome Res* 2012;**11**:3370–81.

34. Manning BD, Cantley LC. AKT/PKB signaling: navigating downstream. *Cell* 2007;**129**:1261–74.

35. Fu Z, Zhao L, Aylor KW, Carey RM, Barrett EJ, Liu Z. Angiotensin-(1–7) recruits muscle microvasculature and enhances insulin's metabolic action via mas receptor. *Hypertension* 2014;**63**:1219–27.

36. Cheng WH, Lu PJ, Hsiao M, Hsiao CH, Ho WY, Cheng PW, et al. Renin activates PI3K-Akt-eNOS signalling through the angiotensin AT₁ and Mas receptors to modulate central blood pressure control in the nucleus tractus solitarii. *Br J Pharmacol* 2012;**166**:2024–35.

37. Sampaio WO, Henrique de Castro C, Santos RA, Schiffrin EL, Touyz RM. Angiotensin-(1–7) counterregulates angiotensin II signaling in human endothelial cells. *Hypertension* 2007;**50**:1093–8.

38. Sampaio WO, Souza dos Santos RA, Faria-Silva R, da Mata Machado LT, Schiffrin EL, Touyz RM. Angiotensin-(1–7) through receptor Mas mediates endothelial nitric oxide synthase activation via Akt-dependent pathways. *Hypertension* 2007;**49**:185–92.

39. Tassone EJ, Sciacqua A, Andreozzi F, Presta I, Perticone M, Carnevale D, et al. Angiotensin (1–7) counteracts the negative effect of angiotensin II on insulin signalling in HUVECs. *Cardiovasc Res* 2013;**99**:129–36.

40. Muñoz MC, Giani JF, Burghi V, Mayer MA, Carranza A, Taira CA, et al. The Mas receptor mediates modulation of insulin signaling by angiotensin-(1–7). *Regul Pept* 2012;**177**:1–11.

41. Hafizi S, Wang X, Chester AH, Yacoub MH, Proud CG. ANG II activates effectors of mTOR via PI3-K signaling in human coronary smooth muscle cells. *Am J Physiol Heart Circ Physiol* 2004;**287**:H1232–8.

42. Muñoz MC, Giani JF, Dominici FP. Angiotensin-(1–7) stimulates the phosphorylation of Akt in rat extracardiac tissues in vivo via receptor Mas. *Regul Pept* 2010;**161**:1–7.

43. Dominici FP, Burghi V, Muñoz MC, Giani JF. Modulation of the action of insulin by angiotensin-(1–7). *Clin Sci* 2014;**126**:613–30.

44. Mori J, Patel VB, Abo Alrob O, Basu R, Altamimi T, Desaulniers J, et al. Angiotensin 1–7 ameliorates diabetic cardiomyopathy and diastolic dysfunction in db/db mice by reducing lipotoxicity and inflammation. *Circ Heart Fail* 2014;**7**:327–39.

45. Santos SHS, Braga JF, Mario EG. Improved lipid and glucose metabolism in transgenic rats with increased circulating angiotensin-(1–7). *Arterioscler Thromb Vasc Biol* 2010;**30**:953–61.

46. Santos SH, Giani JF, Burghi V, Miquet JG, Qadri F, Braga JF, et al. Oral administration of angiotensin-(1–7) ameliorates type 2 diabetes in rats. *J Mol Med (Berl)* 2014;**92**:255–65.

47. Prasannarong M, Santos FR, Henriksen EJ. ANG-(1–7) reduces ANG II-induced insulin resistance by enhancing Akt phosphorylation via a Mas receptor-dependent mechanism in rat skeletal muscle. *Biochem Biophys Res Commun* 2012;**426**:369–73.

48. Santos SHS, Fernandes LR, Mario EG. Mas deficiency in FVB/N mice produces marked changes in lipid and glycemic metabolism. *Diabetes* 2008;**57**:340–7.

49. Henriksen EJ. Dysregulation of glycogen synthase kinase-3 in skeletal muscle and the etiology of insulin resistance and type 2 diabetes. *Curr Diabetes Rev* 2010;**6**:285–93.

50. Cao X, Yang FY, Xin Z, Xie RR, Yang JK. The ACE2/Ang-(1–7)/Mas axis can inhibit hepatic insulin resistance. *Mol Cell Endocrinol* 2014;**393**:30–8.

51. Yuan L, Lu CL, Wang Y, Li Y. Li XY Ang (1–7) protects islet endothelial cells from palmitate-induced apoptosis by AKT, eNOS, p38 MAPK, and JNK pathways. *J Diabetes Res* 2014;**2014**:391476.

52. Huo X, Liu S, Shao T, Hua H, Kong Q, Wang J, et al. GSK3 positively regulates type I insulin-like growth factor receptor through forkhead transcription factors FOXO1/3/4. *J Biol Chem* 2014;**289**:24759–70, pii: jbc.M114.580738.

53. Andrade JM, Lemos FD, da Fonseca PS, Millán RD, de Sousa FB, Guimarães AL, et al. Proteomic white adipose tissue analysis of obese mice fed with a high-fat diet and treated with oral angiotensin-(1–7). *Peptides* 2014;**60C**:56–62.

54. Oliveira Andrade JM, Paraíso AF, Garcia ZM, Ferreira AV, Sinisterra RD, Sousa FB, et al. Cross talk between angiotensin-(1–7)/Mas axis and sirtuins in adipose tissue and metabolism of high-fat feed mice. *Peptides* 2014;**55**:158–65.

55. Savergnini SQ, Beiman M, Lautner RQ, de Paula-Carvalho V, Allahdadi K, Pessoa DC, et al. Vascular relaxation, antihypertensive effect, and cardioprotection of a novel peptide agonist of the MAS receptor. *Hypertension* 2010;**56**:112–20.

56. Lemos VS, Silva DM, Walther T, Alenina N, Bader M, Santos RA. The endothelium-dependent vasodilator effect of the nonpeptide Ang(1–7) mimic AVE 0991 is abolished in the aorta of mas-knockout mice. *J Cardiovasc Pharmacol* 2005;**46**:274–9.

57. Lyngsø C, Erikstrup N, Hansen JL. Functional interactions between 7TM receptors in the renin-angiotensin system—dimerization or crosstalk? *Mol Cell Endocrinol* 2009;**302**:203–12.

58. Fang C, Stavrou E, Schmaier AA, Grobe N, Morris M, Chen A, et al. Angiotensin 1–7 and Mas decrease thrombosis in Bdkrb2−/− mice by increasing NO and prostacyclin to reduce platelet spreading and glycoprotein VI activation. *Blood* 2013;**121**:3023–32.

59. Sampaio WO, Nascimento AA, Santos RA. Systemic and regional hemodynamic effects of angiotensin-(1–7) in rats. *Am J Physiol* 2003;**284**:H1985–94.

60. Staschewski J, Kulisch C, Albrecht D. Different isoforms of nitric oxide synthase are involved in angiotensin-(1–7)-mediated plasticity changes in the amygdala in a gender-dependent manner. *Neuroendocrinology* 2011;**94**:191–9.

61. Peiró C, Vallejo S, Gembardt F, Palacios E, Novella S, Azcutia V, et al. Complete blockade of the vasorelaxant effects of angiotensin-(1–7) and bradykinin in murine microvessels by antagonists of the receptor Mas. *J Physiol* 2013;**591**:2275–85.

62. Bogatcheva NV, Dudek SM, Garcia JG, Verin AD. Mitogen-activated protein kinases in endothelial pathophysiology. *J Investig Med* 2003;**51**:341–52.

63. Nguyen Dinh Cat A, Montezano AC, Burger D, Touyz RM. Angiotensin II, NADPH oxidase, and redox signaling in the vasculature. *Antioxid Redox Signal* 2013;**19**:1110–20.

64. Tallant EA, Ferrario CM, Gallagher PE. Angiotensin-(1–7) inhibits growth of cardiac myocytes through activation of the Mas receptor. *Am J Physiol Heart Circ Physiol* 2005;**289**:H1560–6.

65. Tallant EA, Clark MA. Molecular mechanisms of inhibition of vascular growth by angiotensin-(1–7). *Hypertension* 2003;**42**:574–9.

66. Zhu Z, Zhong J, Zhu S, Liu D, Van Der Giet M, Tepel M. Angiotensin-(1–7) inhibits angiotensin II-induced signal transduction. *J Cardiovasc Pharmacol* 2002;**40**:693–700.

67. Muthalif MM, Benter IF, Uddin MR, Harper JL, Malik KU. Signal transduction mechanisms involved in angiotensin-(1–7)-stimulated arachidonic acid release and prostanoid synthesis in rabbit aortic smooth muscle cells. *J Pharmacol Exp Ther* 1998;**284**:388–98.

68. Su Z, Zimpelmann J, Burns KD. Angiotensin-(1–7) inhibits angiotensin II-stimulated phosphorylation of MAP kinases in proximal tubular cells. *Kidney Int* 2006;**69**:2212–18.

69. Akhtar S, Yousif MH, Dhaunsi GS, Chandrasekhar B, Al-Farsi O, Benter IF. Angiotensin-(1–7) inhibits epidermal growth factor receptor transactivation via a Mas receptor-dependent pathway. *Br J Pharm* 2012;**165**:1390–400.

70. Suski M, Olszanecki R, Stachowicz A, Madej J, Bujak-Giżycka B, Okoń K, et al. The influence of angiotensin-(1–7) Mas receptor agonist (AVE 0991) on mitochondrial proteome in kidneys of apoE knockout mice. *Biochim Biophys Acta* 1834;**2013**:2463–9.

71. Liu GC, Oudit GY, Fang F, Zhou J, Scholey JW. Angiotensin-(1–7)-induced activation of ERK1/2 is cAMP/protein kinase A-dependent in glomerular mesangial cells. *Am J Physiol Renal Physiol* 2012;**302**:F784–90.

72. Zimpelmann J, Burns KD. Angiotensin-(1–7) activates growth-stimulatory pathways in human mesangial cells. *Am J Physiol Renal Physiol* 2009;**296**:F337–46.

73. Zimmerman D, Burns KD. Angiotensin-(1–7) in kidney disease: a review of the controversies. *Clin Sci (Lond)* 2012;**123**:333–46.

74. Gava E, Samad-Zadeh A, Zimpelmann J, Bahramifarid N, Kitten GT, Santos RA, et al. Angiotensin-(1–7) activates a tyrosine phosphatase and inhibits glucose-induced signalling in proximal tubular cells. *Nephrol Dial Transplant* 2009;**24**:1766–73.

75. Giani JF, Gironacci MM, Muñoz MC, Turyn D, Dominici FP. Angiotensin-(1–7) has a dual role on growth-promoting signalling pathways in rat heart in vivo by stimulating STAT3 and STAT5a/b phosphorylation and inhibiting angiotensin II-stimulated ERK1/2 and Rho kinase activity. *Exp Physiol* 2008;**93**:570–8.

76. Magierowski M, Jasnos K, Pawlik M, Krzysiek-Maczka G, Ptak-Belowska A, Olszanecki R, et al. Role of angiotensin-(1–7) in gastroprotection against stress-induced ulcerogenesis. The involvement of mas receptor, nitric oxide, prostaglandins, and sensory neuropeptides. *J Pharmacol Exp Ther* 2013;**347**:717–26.

77. Han Y, Sun HJ, Li P, Gao Q, Zhou YB, Zhang F, et al. Angiotensin-(1–7) in paraventricular nucleus modulates sympathetic activity and cardiac sympathetic afferent reflex in renovascular hypertensive rats. *PLoS One* 2012;**7**:e48966.

78. Gironacci MM, Cerniello FM, Longo Carbajosa NA, Goldstein J, Cerrato BD. Protective axis of the renin-angiotensin system in the brain. *Clin Sci (Lond)* 2014;**127**:295–306.

79. Qiu R-G, Chen J, McCormick F, Symons M. A role for Rho in Ras transformation. *Proc Natl Acad Sci U S A* 1995;**92**:11781–5.

80. Li N, Cai R, Niu Y, Shen B, Xu J, Cheng Y. Inhibition of angiotensin II-induced contraction of human airway smooth muscle cells by angiotensin-(1–7) via downregulation of the RhoA/ROCK2 signaling pathway. *Int J Mol Med* 2012;**30**:811–18.

81. Alenina N, Xu P, Rentzsch B, Patkin EL, Bader M. Genetically altered animal models for Mas and angiotensin-(1–7). *Exp Physiol* 2008;**93**:528–37.

82. Walther T, Balschun D, Voigt J-P, Fink H, Zuschratter W, Birchmeier C, et al. Sustained long-term potentiation and anxiety in mice lacking the Mas protooncogene. *J Biol Chem* 1998;**273**:11867–73.

83. Rakušan D, Bürgelová M, Vaněčková I, Vaňourková Z, Husková Z, Skaroupková P, et al. Knockout of angiotensin 1–7 receptor Mas worsens the course of two-kidney, one-clip Goldblatt hypertension: roles of nitric oxide deficiency and enhanced vascular responsiveness to angiotensin II. *Kidney Blood Press Res* 2010;**33**:476–88.

84. Giani JF, Gironacci MM, Muñoz MC, Peña C, Turyn D, Dominici FP. Angiotensin-(1–7) stimulates the phosphorylation of JAK2, IRS-1 and Akt in rat heart in vivo: role of the AT1 and Mas receptors. *Am J Physiol Heart Circ Physiol* 2007;**293**:H1154–63.

Chapter 24

Mas/AT$_2$ Cross Talk

Daniel C. Villela

Department of Basic Science - Faculty of Medicine, Federal University of Jequitinhonha and Mucuri Valleys, Diamantina, Brazil

Although our knowledge regarding the renin–angiotensin system (RAS) has more than a century of history,[1] certainly, there is still much more to unveil in a system that is conserved for millions of years throughout the vertebrates.[2,3] In the past 20 years, the classical concept of the RAS, where the octapeptide angiotensin II (Ang II), via the AT$_1$ receptor, is the main and only bioactive peptide from the system, has suffered significant conceptual changes, and a complex system with multifunctional peptides, enzymes, and receptors has been revealed.[4–6] One important branch of the system that has gained significant interest and acceptance is the protective arm of the RAS[6] (Figure 1) that includes the angiotensin AT$_2$ receptor (AT$_2$R); angiotensin-(1-7) (Ang-(1-7)); the angiotensin-converting enzyme 2 (ACE2) (mainly responsible for Ang-(1-7) synthesis); Mas, the receptor for Ang-(1-7); and the very recently discovered new member of the system, the heptapeptide alamandine that signals through the Mas-related G protein-coupled receptor D (MrgD), which has very similar actions to those observed by activation of the Mas and AT$_2$ receptors.[7]

Ang-(1-7) is mainly formed through the cleaving action of the ACE2 on Ang II, which cleaves the amino acid phenylalanine from the C-terminal end of Ang II. Ang-(1-7) can also be generated from angiotensin I through cleavage of a leucine from the C-terminal end by ACE2 generating Ang-(1-9) that subsequently can be hydrolyzed by ACE or neutral endopeptidase and form Ang-(1-7). As the catalytic efficiency of ACE2 is 400-fold higher with Ang II as a substrate than with angiotensin I[8], the direct cleavage from Ang II seems to be the more important way of Ang-(1-7) synthesis. It is also well established that Ang-(1-7) is the endogenous ligand for the Mas receptor.[5,9]

Ang II binds almost to the same affinity to AT$_1$R and the AT$_2$R, but since the basal expression of AT$_1$R is of much higher density than the AT$_2$R, stimulation with Ang II elicits AT$_1$R-mediated actions that blunt the AT$_2$R effects.[8] However, *in vivo/in vitro* studies performed by the administration of AT$_2$R-selective ligands in conjunction with recently developed transgenic animal models clearly established the protective role of the AT$_2$ receptor.[10–12]

Dynamic interactions between proteins are common events in biologic systems and much effort is now being made to investigate dimerization (physical interactions between proteins) of receptors.[13] Recent studies provide evidence of the existence of homodimers and heterodimers between the super families of the G protein-coupled receptors (GPCRs).[14] GPCRs are the most abundant type of receptors in the cardiovascular system,[14] and dimerizations are observed in physiological and pathophysiological states among receptors of the RAS.[15–18] An example of the importance of dimerizations among receptors of the RAS was demonstrated by AbdAlla et al., where heterodimerization between the AT$_1$ receptor and the B$_2$ receptor for bradykinin (BK) was related to preeclampsia.[15] It was also demonstrated that there is an increased AT$_1$/B$_2$ heterodimer in placentas from women with preeclampsia and that this interaction leads to increased responsiveness to Ang II.[15]

Both the Mas and AT$_2$ receptors interact separately with receptors of the RAS. Kostenis et al. observed that the receptor Mas forms dimers with the AT$_1$ receptor and that this interaction has an antagonistic effect to the AT$_1$ receptor, since the effects of inositol phosphate production and release of intracellular calcium mediated by Ang II were reduced to 50% after the AT$_1$/Mas coexpression.[18] Functional interactions between Ang-(1-7) and BK are also observed and suggest a possible dimerization between the Mas and the B$_2$ receptor for BK. It is believed that Ang-(1-7) increases the hypotensive effect of BK through a modulation of possible cross talk among receptors, rather than by preventing BK degradation by ACE.[19,20] The AT$_2$ receptor also interacts in an antagonistic way with the AT$_1$ receptor.[17] When AT$_2$R is coexpressed with the AT$_1$ receptor, the effects produced by Ang II via AT$_1$ are abolished. Also, the AT$_1$/AT$_2$ heterodimerization is reduced in biopsies from myometria of pregnant women. Interestingly, AT$_1$/AT$_2$ heterodimerization was also not affected by the administration of agonists or antagonists of these receptors.[17] Furthermore, the AT$_2$R is involved in BK production[21–23] and interacts in a synergistic way with the B2 receptor as a heterodimeric partner.[16] It has also been shown that the AT$_2$R can homodimerize.[24] Thus, dimerizations among GPCRs including those of the RAS are multifarious, and it seems possible that there may

The Protective Arm of the Renin–Angiotensin System (RAS). http://dx.doi.org/10.1016/B978-0-12-801364-9.00024-9

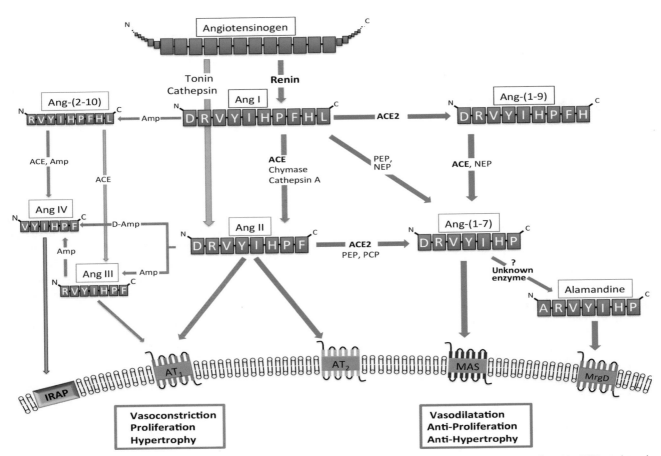

FIGURE 1 Representation of the renin–angiotensin system cascade. In red the classical cascade and in green the protective arm. ACE, angiotensin-converting enzyme; Ang, angiotensin; Amp, aminopeptidase; AT_1, Ang II type 1 receptor; AT_2, Ang II type 2 receptor; Mas, Ang-(1-7) receptor; MrgD, Mas-related G protein-coupled receptor D; D-Amp, dipeptidyl-aminopeptidase; IRAP, insulin-regulated aminopeptidase; PCP, prolyl carboxypeptidase; PEP, prolyl endopeptidase; NEP, neutral endopeptidase.

also be dimerization between the AT_2R and Mas receptor. In fact, studies show that these two receptors produce similar actions, such as inducing the release of nitric oxide,[25,26] inducing cardiac remodeling after infarction,[10,27] and eliciting antiproliferative effects.[28,29] In certain experimental conditions, MAS/AT_2 functional interactions are observed and suggest an involvement of AT_2 receptor-mediated action on Ang-(1-7), since the AT_2 receptor antagonist, PD 123319, blocks the effects produced by Ang-(1-7).[30–32] Controversially, other studies showed that Ang-(1-7) has a low affinity for the AT_2 receptor[33] and effects of Ang-(1-7) are not affected by AT_2 receptor antagonists.[34–37] This disparity may be explained by the fact that the antagonist PD 123319 may not be specific for the AT_2 receptor.[38,39] A recent study corroborates with this idea, as it was shown that PD 123319 blocked the binding of a newly discovered peptide of the RAS, alamandine, to the MrgD receptor. This same study also observed that PD 123319 blocked the vasorelaxant effect of alamandine in knockout animals for the AT_2 receptor.[7] The fact that Ang-(1-7) binds with low affinity to the AT_2R may explain why in some experiments PD 123319 partially blocks Ang-(1-7) effects. Nonetheless, this does not explain why in some assays the effect of Ang-(1-7) is completely blocked by PD 123319 and in others this antagonist has no effect at all. A MAS/AT_2 heterodimerization in selected tissues may explain these intriguing data. In this way, apart from the incongruence carried out by PD 123312, significant effort has been made to show that Mas and AT_2 can functionally and/or molecularly interact. Recent, but as yet unpublished, data signal strongly in this direction, where fluorescence resonance energy transfer (FRET) experiments show a very significant FRET efficiency for the MAS/AT_2 heterodimer.[40]

Initial studies indicated that the expression of MrgD was restricted to small-diameter nociceptive neurons,[41] although recent observations have identified its expression in different tissues such as the muscle, heart, and testis.[7,42,43] The MrgD receptor has also dimerizing properties[44] and may be considered as a possible candidate to interact with the Mas or AT_2 receptors. Here, we illustrate (Figure 2) the described and possible dimerizing partners regarding the protective receptors of the RAS, Mas, and AT_2.

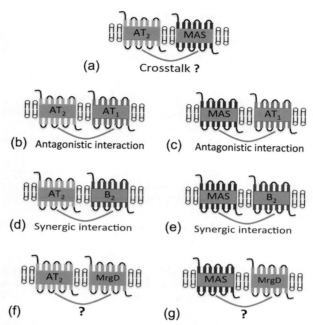

FIGURE 2 Prospective of dimerization partners within the protective arm receptors of the renin–angiotensin system, Mas and AT$_2$ receptors. (a) Mas and AT$_2$ possible cross talk.[30–32,40] (b) AT$_2$/AT$_1$ antagonistic interaction.[17] (c) Mas/AT$_1$ antagonistic interaction.[18] (d) AT$_2$/B$_2$ synergic interaction.[19,20] (e) Mas/B$_2$ synergic interaction.[19,20] (f) AT$_2$ and MrgD possible interaction. (g) Mas and MrgD possible interaction. Mas, Ang-(1-7) receptor; AT$_2$, Ang II type 2 receptor; AT$_1$, Ang II type 1 receptor; B$_2$, bradykinin type 2 receptor; MrgD, Mas-related G protein-coupled receptor D.

Despite that studies related to interactions between receptors of the RAS are still incipient and need more consistency in respect to the mechanism and function of these dimerizations, they expand the alternatives for new therapeutic avenues for diseases linked to the RAS, such as hypertension. Possibly, interactions between receptors of the RAS occur and are modulated to maintain homeostasis in certain physiological and pathophysiological situations.

REFERENCES

1. Phillips MI, Schmidt-Ott KM. The discovery of renin 100 years ago. *News Physiol Sci* 1999;**14**:271–4.
2. Fournier D, Luft FC, Bader M, Ganten D, Andrade-Navarro MA. Emergence and evolution of the renin–angiotensin–aldosterone system. *J Mol Med Berl* 2012;**90**:495–508.
3. Dell'Italia LJ, Ferrario CM. The never-ending story of angiotensin peptides: beyond angiotensin I and II. *Circ Res* 2013;**112**:1086–7.
4. Carey RM. Newly discovered components and actions of the renin–angiotensin system. *Hypertension* 2013;**62**:818–22.
5. Santos RAS, Ferreira AJ, Verano-Braga T, Bader M. Angiotensin-converting enzyme 2, angiotensin-(1–7) and Mas: new players of the renin–angiotensin system. *J Endocrinol* 2013;**216**:R1–17.
6. Sumners C, Horiuchi M, Widdop RE, McCarthy C, Unger T, Steckelings UM. Protective arms of the renin-angiotensin-system in neurological disease. *Clin Exp Pharmacol Physiol* 2013;**40**:580–8.
7. Lautner RQ, Villela DC, Fraga-Silva RA, Silva N, Verano-Braga T, Costa-Fraga F, et al. Discovery and characterization of alamandine: a novel component of the renin-angiotensin system. *Circ Res* 2013;**112**:1104–11.
8. Vickers C, Hales P, Kaushik V, Dick L, Gavin J, Tang J, et al. Hydrolysis of biological peptides by human angiotensin-converting enzyme-related carboxypeptidase. *J Biol Chem* 2002;**277**:14838–43.
9. Santos RAS, Simoes e Silva AC, Maric C, Silva DMR, Machado RP, de Buhr I, et al. Angiotensin-(1–7) is an endogenous ligand for the G protein-coupled receptor Mas. *Proc Natl Acad Sci U S A* 2003;**100**:8258–63.
10. Steckelings UM, Kaschina E, Unger T. The AT2 receptor—a matter of love and hate. *Peptides* 2005;**26**:1401–9.
11. Avila MD, Morgan JP, Yan X. Genetically modified mouse models used for studying the role of the AT2 receptor in cardiac hypertrophy and heart failure. *J Biomed Biotechnol* 2011;**2011**:141039.
12. Steckelings UM, Rompe F, Kaschina E, Namsolleck P, Grzesiak A, Funke-Kaiser H, et al. The past, present and future of angiotensin II type 2 receptor stimulation. *J Renin Angiotensin Aldosterone Syst* 2010;**11**:67–73.
13. Klemm JD, Schreiber SL, Crabtree GR. Dimerization as a regulatory mechanism in signal transduction. *Annu Rev Immunol* 1998;**16**:569–92.
14. Breitwieser GE. G protein-coupled receptor oligomerization implications for G protein activation and cell signaling. *Circ Res* 2004;**94**:17–27.
15. AbdAlla S, Lother H, el Massiery A, Quitterer U. Increased AT(1) receptor heterodimers in preeclampsia mediate enhanced angiotensin II responsiveness. *Nat Med* 2001;**7**:1003–9.

16. Abadir PM, Periasamy A, Carey RM, Siragy HM. Angiotensin II type 2 receptor-bradykinin B2 receptor functional heterodimerization. *Hypertension* 2006;**48**:316–22.

17. AbdAlla S, Lother H, Abdel-tawab AM, Quitterer U. The angiotensin II AT2 receptor is an AT1 receptor antagonist. *J Biol Chem* 2001;**276**:39721–6.

18. Kostenis E, Milligan G, Christopoulos A, Sanchez-Ferrer CF, Heringer-Walther S, Sexton PM, et al. G-protein-coupled receptor Mas is a physiological antagonist of the angiotensin II type 1 receptor. *Circulation* 2005;**111**:1806–13.

19. Fernandes L, Fortes ZB, Nigro D, Tostes RCA, Santos RAS, de Carvalho MHC. Potentiation of bradykinin by angiotensin-(1–7) on arterioles of spontaneously hypertensive rats studied in vivo. *Hypertension* 2001;**37**:703–9.

20. Carvalho MBL, Duarte FV, Faria-Silva R, Fauler B, da Mata Machado LT, de Paula RD, et al. Evidence for Mas-mediated bradykinin potentiation by the angiotensin-(1–7) nonpeptide mimic AVE 0991 in normotensive rats. *Hypertension* 2007;**50**:762–7.

21. Sosa-Canache B, Cierco M, Gutierrez CI, Israel A. Role of bradykinins and nitric oxide in the AT2 receptor-mediated hypotension. *J Hum Hypertens* 2000;**14**(Suppl 1):S40–6.

22. Siragy HM, Inagami T, Ichiki T, Carey RM. Sustained hypersensitivity to angiotensin II and its mechanism in mice lacking the subtype-2 (AT2) angiotensin receptor. *Proc Natl Acad Sci U S A* 1999;**96**:6506–10.

23. Abadir PM, Carey RM, Siragy HM. Angiotensin AT2 receptors directly stimulate renal nitric oxide in bradykinin B2-receptor-null mice. *Hypertension* 2003;**42**:600–4.

24. Miura S-I, Karnik SS, Saku K. Constitutively active homo-oligomeric angiotensin II type 2 receptor induces cell signaling independent of receptor conformation and ligand stimulation. *J Biol Chem* 2005;**280**:18237–44.

25. Siragy HM, Carey RM. The subtype 2 (AT2) angiotensin receptor mediates renal production of nitric oxide in conscious rats. *J Clin Invest* 1997;**100**:264–9.

26. Sampaio WO, Souza dos Santos RA, Faria-Silva R, da Mata Machado LT, Schiffrin EL, Touyz RM. Angiotensin-(1–7) through receptor Mas mediates endothelial nitric oxide synthase activation via Akt-dependent pathways. *Hypertension* 2007;**49**:185–92.

27. Santos RAS, Ferreira AJ, Simões E, Silva AC. Recent advances in the angiotensin-converting enzyme 2-angiotensin(1–7)-Mas axis. *Exp Physiol* 2008;**93**:519–27.

28. Stoll M, Steckelings UM, Paul M, Bottari SP, Metzger R, Unger T. The angiotensin AT2-receptor mediates inhibition of cell proliferation in coronary endothelial cells. *J Clin Invest* 1995;**95**:651–7.

29. Zhang F, Hu Y, Xu Q, Ye S. Different effects of angiotensin II and angiotensin-(1–7) on vascular smooth muscle cell proliferation and migration. *PLoS ONE* 2010;**5**:e12323.

30. Walters PE, Gaspari TA, Widdop RE. Angiotensin-(1–7) acts as a vasodepressor agent via angiotensin II type 2 receptors in conscious rats. *Hypertension* 2005;**45**:960–6.

31. Tesanovic S, Vinh A, Gaspari TA, Casley D, Widdop RE. Vasoprotective and atheroprotective effects of angiotensin (1–7) in apolipoprotein E-deficient mice. *Arterioscler Thromb Vasc Biol* 2010;**30**:1606–13.

32. de Castro CH, dos Santos RAS, Ferreira AJ, Bader M, Alenina N, de Almeida AP. Evidence for a functional interaction of the angiotensin-(1–7) receptor Mas with AT1 and AT2 receptors in the mouse heart. *Hypertension* 2005;**46**:937–42.

33. Bosnyak S, Jones ES, Christopoulos A, Aguilar M-I, Thomas WG, Widdop RE. Relative affinity of angiotensin peptides and novel ligands at AT1 and AT2 receptors. *Clin Sci Lond Engl* 2011;**121**:297–303.

34. Brosnihan KB, Li P, Ferrario CM. Angiotensin-(1–7) dilates canine coronary arteries through kinins and nitric oxide. *Hypertension* 1996;**27**:523–8.

35. Dharmani M, Mustafa MR, Achike FI, Sim M-K. Effects of angiotensin 1–7 on the actions of angiotensin II in the renal and mesenteric vasculature of hypertensive and streptozotocin-induced diabetic rats. *Eur J Pharmacol* 2007;**561**:144–50.

36. Silva DMR, Vianna HR, Cortes SF, Campagnole-Santos MJ, Santos RAS, Lemos VS. Evidence for a new angiotensin-(1–7) receptor subtype in the aorta of Sprague–Dawley rats. *Peptides* 2007;**28**:702–7.

37. Tallant EA, Ferrario CM, Gallagher PE. Angiotensin-(1–7) inhibits growth of cardiac myocytes through activation of the mas receptor. *Am J Physiol: Heart Circ Physiol* 2005;**289**:H1560–6.

38. Macari D, Bottari S, Whitebread S, De Gasparo M, Levens N. Renal actions of the selective angiotensin AT2 receptor ligands CGP 42112B and PD 123319 in the sodium-depleted rat. *Eur J Pharmacol* 1993;**249**:85–93.

39. Nossaman BD, Feng CJ, Kaye AD, Kadowitz PJ. Analysis of responses to ANG IV: effects of PD-123319 and DuP-753 in the pulmonary circulation of the rat. *Am J Physiol* 1995;**268**:L302–8.

40. Villela DC, Munter L-M, Multhaup G, Mayer M, Benz V, Namsolleck P, et al. 398 Evidence of a direct Mas-At2 receptor dimerization. *J Hypertens* 2012;**30**:e117.

41. Dong X, Han S, Zylka MJ, Simon MI, Anderson DJ. A diverse family of GPCRs expressed in specific subsets of nociceptive sensory neurons. *Cell* 2001;**106**:619–32.

42. Villela DC, Passos-Silva DG, Santos RAS. Alamandine: a new member of the angiotensin family. *Curr Opin Nephrol Hypertens* 2014;**23**:130–4.

43. Thorrez L, Laudadio I, Van Deun K, Quintens R, Hendrickx N, Granvik M, et al. Tissue-specific disallowance of housekeeping genes: the other face of cell differentiation. *Genome Res* 2011;**21**:95–105.

44. Milasta S, Pediani J, Appelbe S, Trim S, Wyatt M, Cox P, et al. Interactions between the Mas-related receptors MrgD and MrgE alter signalling and trafficking of MrgD. *Mol Pharmacol* 2006;**69**:479–91.

Chapter 25

ACE2 Cell Biology, Regulation, and Physiological Functions

Anthony J. Turner

School of Molecular & Cellular Biology, Faculty of Biological Sciences, University of Leeds, Leeds, UK

INTRODUCTION

The classical renin–angiotensin system (RAS) pathway regulating the cardiovascular system primarily through the hormone angiotensin II (Ang II) has evolved over a period of some 60 years into a complex network of angiotensin mediators and receptors influencing multiple physiological pathways at both intracellular and endocrine levels. In the simplified linear pathway from precursor protein (angiotensinogen) to the vasoconstrictor octapeptide, Ang II, two key proteolytic enzymes are involved: the aspartic protease renin cleaving the decapeptide Ang I from angiotensinogen and the metallopeptidase angiotensin-converting enzyme (ACE) converting Ang I to Ang II. Both of these proteases have served as primary targets for the development of antihypertensive drugs: for example, aliskiren for renin and captopril and its many successors for ACE. ACE also inactivates the vasodilator bradykinin, providing an intricate dual regulation of blood pressure. The discovery of Ang-(1-7) as a functional angiotensin mediator,[1] subsequently shown to counterbalance the action of Ang II,[2] set the scene for the hunt for the Ang-(1-7)-forming enzyme(s) and its cognate peptide receptor (Mas receptor[3]). Not until both of these entities were identified did the physiological significance of Ang-(1-7) become widely accepted.

ACE2: DISCOVERY AND BASIC BIOLOGY

ACE is a highly glycosylated, transmembrane protein existing in two differentially spliced forms: the two-domain somatic ACE (N- and C-domains) with similar but not identical substrate specificities, which is important for cardiovascular regulation, and the single-domain testicular form important for male fertility through metabolism of an unidentified peptide substrate. Somatic ACE also plays important roles in developmental processes, inflammation and immunity, and neurodegenerative disease (reviewed in Ref. [4]). For almost 50 years from the discovery of ACE, the existence of a human "ACE2" remained undiscovered and unpredicted notwithstanding the importance of ACE inhibitors to pharmacology and therapeutics, despite the known existence of ACE homologues in invertebrate species. For example, *Drosophila* expresses two ACE-like genes and catalytically active protein products (Ance and Acer) important in insect development and physiology.[5] It was on this basis that a search of human expressed sequence tags led us to identify, clone, and characterize a human ACE homologue, designated ACE2.[6] Shortly afterward and quite independently, Donoghue and colleagues identified the *ACE2* gene as one upregulated in a human heart failure cDNA library.[7] Despite the high sequence similarities between ACE and ACE2, there were some surprising differences in substrate specificity in that ACE2 functions exclusively as a carboxypeptidase removing a single C-terminal amino acid from Ang II generating Ang-(1-7) or, much less efficiently, from Ang I forming Ang-(1-9). In contrast, ACE principally acts as a carboxydipeptidase (peptidyldipeptidase) removing the C-terminal dipeptide from Ang I to form Ang II. ACE2 activity is also not affected by classical ACE inhibitors. Detailed structural and mutagenesis studies have revealed the important features leading to these specificity differences between ACE and ACE2.[8,9] It is these synergies in angiotensin metabolism, with ACE forming Ang II and ACE2 removing it by conversion to Ang-(1-7), that provides the counterbalancing of angiotensin metabolism by these two homologous enzymes providing metabolites that act through three distinct receptors, AT_1, AT_2, and Mas. ACE2, however, is not the only metabolic route to Ang-(1-7), which can also be formed directly from Ang I by neprilysin (NEP), and other peptidases (prolyl endopeptidase and thimet oligopeptidase) may also participate in Ang-(1-7) biosynthesis depending on tissue or cell type. However, kinetic analysis indicates that the most efficient pathway for Ang-(1-7) generation is directly from Ang II via ACE2.[10] A unique metallopeptidase may function to inactivate Ang-(1-7) to Ang-(1-4), at least in the brain.[11]

The Protective Arm of the Renin–Angiotensin System (RAS). http://dx.doi.org/10.1016/B978-0-12-801364-9.00025-0

ACE and ACE2 both serve a multiplicity of functions. Unlike ACE, ACE2 degrades [des-Arg9]-bradykinin but not bradykinin itself.[6] Since [des-Arg9]-bradykinin is the ligand for the B1 bradykinin receptor, ACE2 may function to turn off signaling at this receptor. Other substrates for ACE2, at least *in vitro*, include apelin-13/17, neurotensin (1-11), dynorphin A (1-13), and ghrelin.[12] There appears to be a close interplay between the inotropic and cardioprotective peptides apelin and ACE2 since ACE2 is downregulated in apelin-deficient mice and apelin acts as a positive regulator of ACE2 expression in failing hearts *in vivo*.[13] The recent discovery of a new player in the RAS, alamandine, has shown that this peptide can be formed from angiotensin A by ACE2 action and acts through a Mas-related gene receptor (MrgD).[14] Angiotensin A is itself formed by decarboxylation of the N-terminal aspartyl residue of Ang II to an alanyl residue.[15]

Roles for both ACE and ACE2 are emerging in Alzheimer's disease (AD), specifically the ability of both enzymes to hydrolyze the amyloid-β (Aβ) peptide. ACE can cleave Aβ40 at internal sites in the peptide reducing its aggregation.[16] It can also hydrolyze the more hydrophobic Aβ-(1-42) to the less neurotoxic form (Aβ40) with ACE inhibition enhancing Aβ deposition.[17] The larger Aβ43 species is also found in AD brain and Aβ43 is the earliest-depositing Aβ species in APP transgenic mouse brain.[18] We have now demonstrated[19] that ACE2 can efficiently hydrolyze Aβ43 to Aβ42 that is then further degraded to Aβ40 by ACE. The discovery that NEP was a major Aβ-degrading enzyme *in vivo* led to a cessation of developments of NEP inhibitors as novel antihypertensives because of the potential for promoting the onset of AD through enhanced Aβ accumulation and aggregation. This, then, begs the question as to whether ACE inhibitors, by promoting the accumulation of more toxic Aβ species in the brain, might also predispose to AD. To date, there are no epidemiological data to support this concept, and ACE inhibition with captopril even retards the development of signs of neurodegeneration in an animal model of AD.[20] This is most likely explained by the greater protective vascular effects on the brain of ACE inhibition, AD itself having a major vascular component in its etiology. Furthermore, ACE overexpression in myelomonocytes prevents Alzheimer's-like cognitive decline.[21] A comparison of the substrate specificities of ACE and ACE2 is provided in Table 1.

ACE2 also plays distinct biological roles independent of its enzymatic activity.[22] The ACE2 protein appears to be a chimera composed of an ACE-like domain fused to a collectrin-like domain. The collectrin protein is one that regulates renal amino acid transport and pancreatic insulin secretion. Likewise, ACE2 regulates transport of intestinal neutral amino acid transporters of the B⁰AT1 family to the plasma membrane and has been implicated in the pathology of Hartnup's disease, a disorder of amino acid homeostasis.[22] Through this process, ACE2 also appears to regulate intestinal inflammation and diarrhea, hence modulating the gut microbiome.[23] Other disease processes in which ACE2 is involved are infection and pathology induced by the severe acute respiratory syndrome (SARS) virus through its serendipitous function as the cell-surface receptor for the virus facilitating viral RNA entry in the lungs.[24] The consequent downregulation of surface ACE2 levels leads to increased local levels of Ang II, which probably contribute to the significant mortality rates resulting from the acute lung injury and fibrosis caused by SARS.[25,26] ACE2 may also be protective against liver fibrosis and other fibrotic diseases,[27] again through reduction in Ang II levels (or elevated Ang-(1-7) levels).

TABLE 1 Comparison of Substrate Specificities of ACE and ACE2 for Angiotensin, Bradykinin, and Amyloid-β (Aβ) Peptides

	ACE	ACE2
Ang I	AspArgValTyrIleHisProPhe↓HisLeu	AspArgValTyrIleHisProPheHis↓Leu
Ang II	Not cleaved	AspArgValTyrIleHisPro↓Phe
Ang-(1-9)	AspArgValTyrIleHisPro↓PheHis	Not cleaved
Ang-(1-7)	AspArgValTyrIle↓HisPro	Not cleaved
Ang A	Not cleaved	AlaArgValTyrIleHisPro↓Phe (alamandine)
BK	ArgProProGlyPheSerPro↓PheArg	Not cleaved
[des-Arg9]-BK	Not cleaved	ArgProProGlyPheSerPro↓Phe
Aβ42	Aβ-(1-42)→Aβ-(1-40)	Not cleaved
Aβ43	Aβ-(1-43)→Aβ-(1-41)	Aβ-(1-43)→Aβ-(1-42)

↓ indicates the site of enzyme cleavage. ACE primarily acts as a peptidyldipeptidase removing the C-terminal dipeptide in susceptible substrates. For some other peptide substrates (e.g., substance P), it can, however, act as an endopeptidase (not shown). ACE can also exhibit endopeptidase cleavage at a number of internal sites in both Aβ40 and Aβ42 (see Refs. [17,18] for further details). ACE2 converts Ang II to Ang-(1-7) approx 70-fold more efficiently than it does Ang I to Ang-(1-9), making it principally an Ang-(1-7)-forming enzyme under normal physiological conditions.[10] ACE2 hydrolysis of Ang A generates the peptide alamandine that acts through the Mas-related gene receptor MrgD14. ACE2 can also hydrolyze some other regulatory peptides, most notably apelin, which in turn regulates ACE2 levels, as does Ang II (see text).

ACE2 REGULATION

The discovery of ACE2 and its role in the RAS led to its rapidly becoming a focus as a novel cardiovascular target emphasized by studies of ACE2-null mice, revealing the enzyme as a key protective regulator of cardiovascular function.[28] Subsequent gene deletion models have not consistently reproduced the ACE2-null phenotype, perhaps reflecting their different genetic backgrounds, and it may be that ACE2 functions as a modulator of responses to injury rather than a primary mediator of cardiac phenotype (reviewed in Ref. [29]). The generally cardioprotective role of ACE2 has limited the development of ACE2 inhibitors since they are unlikely to be of therapeutic benefit in cardiovascular or other disease states where, instead, upregulation of ACE2 expression or activity is required. Nevertheless, given the knowledge of ACE2 structure, mechanism, and specificity, a number of ACE2 selective inhibitors of nM potency have been developed including MLN-4760 (GL1001), DX-600, and 416F2 (see Ref. [30] for a review). More interesting from a therapeutic perspective has been the rational development of ACE2 activator small molecules that lower blood pressure in animal models.[31] A structurally related antitrypanosomal drug (diminazene aceturate) also activates ACE2[32] and exhibits beneficial cardiovascular activity.[33] However, the biological effects of these compounds *in vivo* may be ACE2-independent so some caution is needed in the interpretation of these data at present.[34]

ACE2, like ACE, is shed into plasma in catalytically active form. The constitutive ACE-shedding enzyme remains to be unequivocally identified although in some stimulated conditions may involve the A disintegrin and metalloproteinase 9.[35] ACE2 shedding is mediated by ADAM17 (also known as TACE) both *in vitro* and *in vivo*.[36,37] Ang II induces ADAM17-mediated shedding of myocardial ACE2 providing a positive feedback mechanism in the RAS.[37] The cleavage of ACE2 at the plasma membrane, through binding of soluble ACE2 to integrins, can regulate integrin signaling modulating cell–extracellular matrix interactions and hence influence cardiac remodeling processes.[38] Furthermore, soluble circulating ACE2 may serve as a biomarker in hypertension and heart failure.[39]

While numerous agents have been shown to modulate ACE2 expression, including angiotensin peptides and some other peptide and steroid hormones, relatively little is known about the molecular details of transcriptional regulation of ACE2.[40] The regulatory element responsible for Ang II stimulation of human ACE2 gene expression in human cardiofibroblasts has, however, been identified in its promoter.[41] Hepatocyte nuclear factor (HNF) transcription factors, for example, HNF1α, upregulate ACE2 expression at least in pancreatic islets and HEK293 cells.[42,43] Cell energy stress including hypoxia, cytokine action, and AMP kinase activation lead to epigenetic control of ACE2 expression via the histone deacetylase SIRT1.[44] At the posttranscriptional level, a number of microRNA (miR) species have been reported to regulate ACE2 expression,

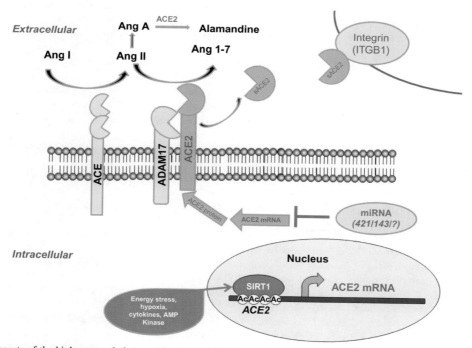

FIGURE 1 Some aspects of the biology, regulation, and function of ACE2 with particular reference to the protective arm of the RAS. Illustration of ACE2 regulation at the transcriptional/epigenetic,[43,44] posttranscriptional[45,46] (miRNA), and posttranslational[36,37] (plasma membrane shedding) levels as discussed in this chapter and its relevant angiotensin substrates in the RAS in conjunction with ACE, which interact with their specific receptors (AT1/2 and MrgD) (not shown). Shed ACE2 (sACE2) can interact with integrins modulating signaling as can membrane-bound ACE2 modulating cell–cell interactions.[38]

including miR-421 and miR-143.[45,46] All of the above transcriptional regulatory sites may provide mechanisms for future modulation of ACE2 levels as cardioprotective strategies. The metabolic and regulatory aspects of ACE2 function are summarized in Figure 1.

In summary, ACE2 is a multifunctional protein in health and disease, which serves as a counterregulatory component of the RAS functioning in a cardioprotective role. Hence, its transcriptional upregulation, activation of its catalytic activity, or administration of the recombinant protein[47] could well provide new strategies in hypertension and heart failure. Additionally, ACE2 modulation (and hence alteration of the circulating Ang II/Ang-(1-7) balance) may have relevance to diabetes, acute lung injury and fibrotic disease, and even dystrophic muscular conditions.[48] But much still remains to be explored in terms of the basic aspects of ACE2 cellular function and its regulation to be able to exploit these opportunities effectively and safely.

REFERENCES

1. Schiavone MT, Santos RA, Brosnihan KB, Khosla MC, Ferrario CM. Release of vasopressin from the rat hypothalamo-neurohypophysial system by angiotensin-(1–7) heptapeptide. *Proc Natl Acad Sci U S A* 1988;**85**:4095–8.

2. Ferrario CM, Chappell MC, Tallant EA, Brosnihan KB, Diz DI. Counterregulatory actions of angiotensin-(1–7). *Hypertension* 1997;**30**:535–41.

3. Santos RA, Simoes e Silva AC, Maric C, Silva DM, Machado RP, de Buhr I, et al. Angiotensin-(1–7) is an endogenous ligand for the G protein-coupled receptor Mas. *Proc Natl Acad Sci U S A* 2003;**100**:8258–63.

4. Gonzalez-Villalobos RA, Shen XZ, Bernstein EA, Janjulia T, Taylor B, Giani JF, et al. Rediscovering ACE: novel insights into the many roles of the angiotensin-converting enzyme. *J Mol Med (Berl)* 2013;**91**:1143–54.

5. Houard X, Williams TA, Michaud A, Dani P, Isaac RE, Shirras AD, et al. The Drosophila melanogaster-related angiotensin-I-converting enzymes Acer and Ance-distinct enzymic characteristics and alternative expression during pupal development. *Eur J Biochem* 1998;**257**:599–606.

6. Tipnis SR, Hooper NM, Hyde R, Karran E, Christie G, Turner AJ. A human homolog of angiotensin-converting enzyme. Cloning and functional expression as a captopril-insensitive carboxypeptidase. *J Biol Chem* 2000;**275**:33238–43.

7. Donoghue M, Hsieh F, Baronas E, Godbout K, Gosselin M, Stagliano N, et al. A novel angiotensin-converting enzyme-related carboxypeptidase (ACE2) converts angiotensin I to angiotensin 1–9. *Circ Res* 2000;**87**:E1–9.

8. Towler P, Staker B, Prasad SG, Menon S, Tang J, Parsons T, et al. ACE2 X-ray structures reveal a large hinge-bending motion important for inhibitor binding and catalysis. *J Biol Chem* 2004;**279**:17996–8007.

9. Rushworth CA, Guy JL, Turner AJ. Residues affecting the chloride regulation and substrate selectivity of the angiotensin-converting enzymes (ACE and ACE2) identified by site-directed mutagenesis. *FEBS J* 2008;**275**:6033–42.

10. Rice GI, Thomas DA, Grant PJ, Turner AJ, Hooper NM. Evaluation of angiotensin-converting enzyme (ACE), its homologue ACE2 and neprilysin in angiotensin peptide metabolism. *Biochem J* 2004;**383**:45–51.

11. Marshall AC, Pirro NT, Rose JC, Diz DI, Chappell MC. Evidence for an angiotensin-(1–7) neuropeptidase expressed in the brain medulla and CSF of sheep. *J Neurochem* 2014;**130**(2):313–23.

12. Vickers C, Hales P, Kaushik V, Dick L, Gavin J, Tang J, et al. Hydrolysis of biological peptides by human angiotensin-converting enzyme-related carboxypeptidase. *J Biol Chem* 2002;**277**:14838–43.

13. Sato T, Suzuki T, Watanabe H, Kadowaki A, Fukamizu A, Liu PP, et al. Apelin is a positive regulator of ACE2 in failing hearts. *J Clin Invest* 2013;**123**:5203–11.

14. Lautner RQ, Villela DC, Fraga-Silva RA, Silva N, Verano-Braga T, Costa-Fraga F, et al. Discovery and characterization of alamandine: a novel component of the renin-angiotensin system. *Circ Res* 2013;**112**:1104–11.

15. Jankowski V, Vanholder R, van der Giet M, Tölle M, Karadogan S, Gobom J, et al. Mass-spectrometric identification of a novel angiotensin peptide in human plasma. *Arterioscler Thromb Vasc Biol* 2007;**27**:297–302.

16. Hu J, Igarashi A, Kamata M, Nakagawa H. Angiotensin-converting enzyme degrades Alzheimer amyloid β-peptide (Aβ); retards Aβ aggregation, deposition, fibril formation; and inhibits cytotoxicity. *J Biol Chem* 2001;**276**:47863–8.

17. Zou K, Yamaguchi H, Akatsu H, Sakamoto T, Ko M, Mizoguchi K, et al. Angiotensin-converting enzyme converts amyloid β-protein 1–42 (Aβ (1–42)) to Aβ (1–40), and its inhibition enhances brain Aβ deposition. *J Neurosci* 2007;**7**:8628–35.

18. Zou K, Liu J, Watanabe A, Hiraga S, Liu S, Tanabe C, et al. Aβ43 is the earliest-depositing Aβ species in APP transgenic mouse brain and is converted to Aβ41 by two active domains of ACE. *Am J Pathol* 2013;**182**:2322–31.

19. Liu S, Liu J, Miura Y, Tanabe C, Maeda T, Terayama Y, et al. Conversion of Aβ43 to Aβ40 by the successive action of angiotensin-converting enzyme 2 and angiotensin-converting enzyme. *J Neurosci Res* 2014;**92**(9):1178–86.

20. AbdAlla S, Langer A, Fu X, Quitterer U. ACE inhibition with captopril retards the development of signs of neurodegeneration in an animal model of Alzheimer's disease. *Int J Mol Sci* 2013;**14**:16917–42.

21. Bernstein KE, Koronyo Y, Salumbides BC, Sheyn J, Pelissier L, Lopes DH, et al. Angiotensin-converting enzyme overexpression in myelomonocytes prevents Alzheimer's-like cognitive decline. *J Clin Invest* 2014;**124**:1000–12.

22. Kuba K, Imai Y, Penninger JM. Multiple functions of angiotensin-converting enzyme 2 and its relevance in cardiovascular diseases. *Circ J* 2013;**77**:301–8.

23. Hashimoto T, Perlot T, Rehman A, Trichereau J, Ishiguro H, Paolino M, et al. ACE2 links amino acid malnutrition to microbial ecology and intestinal inflammation. *Nature* 2012;**487**:477–81.

24. Li W, Moore MJ, Vasilieva N, Sui J, Wong SK, Berne MA, et al. Angiotensin-converting enzyme 2 is a functional receptor for the SARS coronavirus. *Nature* 2003;**426**:450–4.

25. Imai Y, Kuba K, Rao S, Huan Y, Guo F, Guan B, et al. Angiotensin-converting enzyme 2 protects from severe acute lung failure. *Nature* 2005;**436**:112–16.

26. Turner AJ, Hiscox JA, Hooper NM. ACE2: from vasopeptidase to SARS virus receptor. *Trends Pharmacol Sci* 2004;**25**:291–4.

27. Warner FJ, Lubel JS, McCaughan GW, Angus PW. Liver fibrosis: a balance of ACEs? *Clin Sci* 2007;**113**:109–18.

28. Crackower MA, Sarao R, Oudit GY, Yagil C, Kozieradzki I, Scanga SE, et al. Angiotensin-converting enzyme 2 is an essential regulator of heart function. *Nature* 2002;**417**:822–8.

29. Turner AJ. Angiotensin-converting enzyme 2. Cardioprotective player in the renin-angiotensin system? *Hypertension* 2008;**52**:816–17.

30. Clarke NE, Hooper NM, Turner AJ. Angiotensin-converting enzyme-2. In: Rawlings ND, Salvesen GS, editors. Handbook of proteolytic enzymes. 3rd ed. Elsevier; 2013. p. 499–504.

31. Hernández Prada JA, Ferreira AJ, Katovich MJ, Shenoy V, Qi Y, Santos RA, et al. Structure-based identification of small-molecule angiotensin-converting enzyme 2 activators as novel antihypertensive agents. *Hypertension* 2008;**51**:1312–17.

32. Kulemina LV, Ostrov DA. Prediction of off-target effects on angiotensin-converting enzyme. *J Biomol Screen* 2011;**16**:878–85.

33. Shenoy V, Gjymishka A, Jarajapu YP, Qi Y, Afzal A, Rigatto K, et al. Diminazene attenuates pulmonary hypertension and improves angiogenic progenitor cell functions in experimental models. *Am J Respir Crit Care Med* 2013;**187**:648–57.

34. Haber PK, Ye M, Wysocki J, Maier C, Haque SK, Batlle D. Angiotensin-converting enzyme 2-independent action of presumed angiotensin-converting enzyme 2 activators: studies in vivo, ex vivo, and in vitro. *Hypertension* 2014;**63**:774–82.

35. English WR, Corvol P, Murphy G. LPS activates ADAM9 dependent shedding of ACE from endothelial cells. *Biochem Biophys Res Commun* 2012;**421**:70–5.

36. Lambert DW, Yarski M, Warner FJ, Thornhill P, Parkin ET, Smith AI, et al. Tumor necrosis factor-alpha convertase (ADAM17) mediates regulated ectodomain shedding of the severe-acute respiratory syndrome-coronavirus (SARS-CoV) receptor, angiotensin-converting enzyme-2 (ACE2). *J Biol Chem* 2005;**280**:30113–19.

37. Patel VB, Clarke N, Wang Z, Fan D, Parajuli N, Basu R, et al. Angiotensin II induced proteolytic cleavage of myocardial ACE2 is mediated by TACE/ADAM-17: a positive feedback mechanism in the RAS. *J Mol Cell Cardiol* 2014;**66**:167–76.

38. Clarke NE, Fisher MJ, Porter KE, Lambert DW, Turner AJ. Angiotensin converting enzyme (ACE) and ACE2 bind integrins and ACE2 regulates integrin signalling. *PLoS One* 2012;**7**:e34747.

39. Uri K, Fagyas M, Mányiné Siket I, Kertész A, Csanádi Z, Sándorfi G, et al. New perspectives in the renin-angiotensin-aldosterone system (RAAS) IV: circulating ACE2 as a biomarker of systolic dysfunction in human hypertension and heart failure. *PLoS One* 2014;**9**:e87845.

40. Clarke NE, Turner AJ. Angiotensin-converting enzyme 2: the first decade. *Int J Hypertens* 2012;**2012**:307315.

41. Kuan TC, Yang TH, Wen CH, Chen MY, Lee IL, Lin CS. Identifying the regulatory element for human angiotensin-converting enzyme 2 (ACE2) expression in human cardiofibroblasts. *Peptides* 2011;**32**:1832–9.

42. Senkel S, Lucas B, Klein-Hitpass L, Ryffel GU. Identification of target genes of the transcription factor HNF1β and HNF1α in a human embryonic kidney cell line. *Biochim Biophys Acta* 2005;**1731**:179–90.

43. Pedersen KB, Chhabra KH, Nguyen VK, Xia H, Lazartigues E. The transcription factor HNF1α induces expression of angiotensin-converting enzyme 2 (ACE2) in pancreatic islets from evolutionarily conserved promoter motifs. *Biochim Biophys Acta* 1829;**2013**:1225–35.

44. Clarke NE, Belyaev ND, Lambert DW, Turner AJ. Epigenetic regulation of angiotensin-converting enzyme 2 (ACE2) by SIRT1 under conditions of cell energy stress. *Clin Sci* 2014;**126**:507–16.

45. Lambert DW, Lambert LA, Clarke NE, Hooper NM, Porter KE, Turner AJ. Angiotensin-converting enzyme-2 is subject to post-transcriptional regulation by microRNA-421. *Clin Sci* 2014;**127**(4):243–9.

46. Fernandes T, Hashimoto NY, Magalhães FC, Fernandes FB, Casarini DE, Carmona AK, et al. Aerobic exercise training-induced left ventricular hypertrophy involves regulatory MicroRNAs, decreased angiotensin-converting enzyme-angiotensin II, and synergistic regulation of angiotensin-converting enzyme 2-angiotensin (1–7). *Hypertension* 2011;**58**:182–9.

47. Oudit GY, Penninger JM. Recombinant human angiotensin-converting enzyme 2 as a new renin-angiotensin system peptidase for heart failure therapy. *Curr Heart Fail Rep* 2011;**8**:176–83.

48. Riquelme C, Acuña MJ, Torrejón J, Rebolledo D, Cabrera D, Santos RA, et al. ACE2 is augmented in dystrophic skeletal muscle and plays a role in decreasing associated fibrosis. *PLoS One* 2014;**9**:e93449.

Chapter 26

Mas in Myocardial Infarction and Congestive Heart Failure

Marcos Barrouin Melo

Department of Physiology and Pharmacology, UFMG (Federal University of Minas Gerais), Belo Horizonte, Minas Gerais, Brazil

MYOCARDIAL INFARCTION AND CONGESTIVE HEART FAILURE

Myocardial infarction (MI) and congestive heart failure (HF) are common occurrences in cardiovascular medicine and may represent an acute, possibly fatal, disease and the final stage of most cardiac diseases, respectively. The inflammatory and metabolic events following MI determine the cellular responses to injury. A number of inflammatory mediators are secreted and participate extensively in the MI pathophysiology. The responses depend on the duration and magnitude of coronary occlusion. The tissue injury and an increase in wall stress due to myocardial ischemia cause the activation of cellular signaling pathways related to remodeling. The remodeling process plays a key role in the progressive deterioration of cardiac function that leads to HF. There are conditions that can produce acute HF (malignant arrhythmias) without significant myocardial remodeling, but most heart diseases activate it during adaptation to disease, leading in their final stages to HF. Remodeling is morphologically characterized by cardiac hypertrophy and dilatation as well as conformational changes in the shape of the heart. Myocardial fibrosis is the most characteristic structural change in myocardial infarct and contributes to both systolic dysfunction and diastolic dysfunction.[1,2]

Left ventricular (LV) remodeling is directly related to future reduction of performance and faster worsening of the clinical conditions of patients. The functional impairment leads to changes in cardiac myocyte biology (volume), geometry, and architecture of the LV chamber caused by the nonmyocyte components of the myocardium.[3] A progressive loss of contractility is related to decreased α-MHC (myosin heavy chain) and increased β-MHC gene expression, loss of myofilaments in cardiomyocytes (necrosis, apoptosis, and autophagy), alterations in the extracellular matrix (ECM), disturbances in excitation–contraction coupling, and desensitization of β-adrenergic signaling.[2] ECM is important in keeping myocardial structure, interconnecting cardiomyocytes playing an essential role in cellular migration, proliferation, differentiation, and survival. This interstitial component is composed of collagens I and III that in physiological conditions maintain a balance between stiffness, resistance, and elasticity of the myocardium. ECM alterations observed in some heart disease could change the collagen proportion, and this leads to mechanical derangements in cardiac function by interfering substantially in the myocardium stiffness.[3] The fibrogenesis induced by remodeling generates a tissue formed by myofibroblasts, able to sustain the secretion of peptides (Ang II and its receptors), growth factors, and collagen fibrillar proteins. These substances will participate in cellular signaling that results in further fibrosis. The persistent presence of myofibroblasts causes a progressive adverse myocardial remodeling, worsening HF independent of its etiologic origin.[4,5]

RENIN–ANGIOTENSIN SYSTEM AND CARDIAC REMODELING

The classic renin–angiotensin system (RAS) is closely related to the cellular changes leading to an inappropriate adaptation of the heart to an abnormal increase in workload. Based on this, the manipulation of the RAS beyond ACE (angiotensin-converting enzyme) inhibitors and Ang II receptor blockers may be an important target in the treatment of heart diseases. Additionally, the Mas receptor stimulation by Ang-(1-7) has been shown to improve heart function in many models of disease including MI and HF.[6]

The description of ACE2 in the heart suggests its potential role in regulating cardiac function and it is proposed to have a protective role during pathological changes. This information is corroborated by the study of Crackower et al.[7] showing that ACE2-knockout mice develop severe impairment in myocardial contractility.[7,8]

The Protective Arm of the Renin–Angiotensin System (RAS). http://dx.doi.org/10.1016/B978-0-12-801364-9.00026-2

Conflicting results showing different patterns of ACE2 expression in MI models suggest that its effects may be possibly strain-dependent. In the failing human heart, it is described that ACE2 expression is increased. Ortiz-Pérez demonstrated an increase in ACE2 activity in the heart, right after myocardial infarct.[9] Ocaranza et al. described that at the early phase of MI, ACE2 activity in the plasma and left ventricle is increased in rats, while the plasma and LV ACE2 activities and mRNA levels are lower than those in controls at 8 weeks postinfarction.[10] Patients with MI and HF usually are predisposed to cardiac arrhythmias, and it is important to highlight the antiarrhythmogenic actions of the Ang-(1-7). Additionally, increased levels of Ang-(1-7) in the heart have been shown to modulate pathways related to fibrosis. This effect improves remodeling possibly through sodium channel changes. The cardioprotective actions of Mas activation were described also in a myocardial ischemia/reperfusion model reducing the incidences and the duration of reperfusion arrhythmias.[11] Ferreira et al. showed the cardiac effects of the compound AVE-0991, a Mas receptor agonist, in normal rats and a model of HF induced by left coronary artery ligation. AVE-0991 improved the cardiac function in normal rats through a mechanism involving the Mas by NO release and induced protective effects in HF, reducing the infarcted area.[12]

Ang II binds to AT_1R, stimulating the Gq pathway initiating a signaling sequence that leads to mitogen-activated protein kinase (MAPK) phosphorylation. The components of this pathway such as JNK and ERK promote the activation of survival or cell death (apoptosis) signaling. Also, due to Ang II effects, there is an increase of other substances related to maladaptive pathways such as TGF-β, aldosterone, and TNF-α, which promote collagen deposition and functional disruptions. According to Tao et al., Src homology 2-containing inositol phosphatase 1 (SHP-1), a redox-sensitive protein-tyrosine phosphatase, has negative effects in profibrotic, inflammatory, and hypertrophic signaling of Ang II. Ang-(1-7) blocked Ang II-stimulated phosphorylated extracellular signal-related kinase 1/2 (p-ERK1/2) and stimulated Src homology2-containing inositol phosphatase 1 (SHP-1). It was also described in this study that Ang-(1-7) via Mas receptor antagonized Ang II-induced increase in mRNA expressions of TGF-β1, types I and III collagen.[13]

Tallant et al. showed that the antiproliferative effect of Ang-(1-7) on cardiomyocytes is through the inhibition of MAPKs.[14] Santos et al. showed that Ang-(1-7) affects remodeling through actions over ECM, improving heart function.[15] There are evidences that the stimulation of Mas receptor modulates changes in ECM under disease conditions. Ang-(1-7) interacts with specific receptors on myofibroblasts in the heart, decreasing collagen synthesis, growth factor release, and fibroblast activation and eliciting antitrophic and antifibrotic effects capable of reverting Ang II remodeling actions.[16] Pan et al. demonstrated that the effects of Ang-(1-7) may be elicited by the regulation of matrix metalloproteinases (MMPs).[17]

Loot et al. demonstrated the beneficial effects of Ang-(1-7) infusion on an MI model. They observed a 40% reduction in the LV end-diastolic pressure, mostly preserved coronary flow, normal aortic endothelial function, and attenuation of myocyte hypertrophy.[6] Oudit et al. described that Ang-(1-7) is also capable of suppressing myocardial NADPH oxidase decreasing oxidative stress and remodeling.[18] According to De Mello et al., Ang-(1-7) has beneficial effects to a failing heart, causing hyperpolarization of the myocardial cells by activating the sodium pump, and may prevent cardiac arrhythmias.[19]

It was reported by Zhang et al.[20] in an ischemia/reperfusion model that exposing Mas receptors to high concentrations of agonists (not related to Ang-(1-7)) preferentially couples to the Gq pathway increasing PLC and intracellular Ca^{++}, causing vasoconstriction and decreasing coronary blood flow.

Recently, the protective effects of Ang-(1-7) were demonstrated in two models of HF, the treatment with isoproterenol and induced MI by ligature of the left descending coronary artery. Oral administration of Ang-(1-7) compound reduced significantly the lesion surface in both models, suggesting that the effects may be related to changes in the ECM and over the tissue repair signaling. A set of experiments in an MI model that induced a decrease in heart function showed the efficacy of Ang-(1-7) demonstrated through traditional echocardiographic measurements (fractional shortening and ejection fraction). Additionally, we applied a new technique to quantify velocity and displacement of the myocardium using strain analysis from high-resolution echocardiography. The results showed that rats under treatment with Ang-(1-7) suffered less decrement in heart function after MI.[21]

INFLAMMATORY PHASE

IL-1β expression increases after MI. Its importance in indicating cardiac myocyte injury and repair is known. IL-1β signaling is essential to activate apoptosis and hypertrophy. The influence of IL-1β on the expression and activity of MMPs is essential to the remodeling process. It is also a factor of myocardial depression, reducing contractility and triggering inflammatory pathways, stimulating the production of many chemokines related to leukocyte migration to the lesion area.[22–25]

Chemokine induction is a critical phase after MI and strongly supports tissue inflammatory activity. The stromal cell-derived factor 1 alpha (SDF-1 alpha) or CXCL12 is an inducible chemokine. Its expression increases in the first hours after the infarction due to ischemic condition, and it is considered a marker of angiogenic activity involved with cardiogenesis; it also modulates calcium homeostasis in cardiomyocytes.[26,27]

PROLIFERATIVE PHASE

TGF-β is a key modulator of tissue repair after MI, suppressing inflammation and contributing to ECM deposition to balance the transition between the end of inflammatory activity and fibrogenesis. Fibroblasts are the primary source of TGF-β in the heart, being required for the Ang II-induced cardiac hypertrophy. It elicits the conversion of fibroblasts to myofibroblasts, enhancing ECM protein synthesis.[28] It activates many signaling cascade components related to remodeling, including Smad, p38, ERK1/2, TAK1, and JNK.[29] Marques et al. showed a reduced increase in TGF-β expression in the MI/Ang-(1-7) treatment group.[21] Moreover, Mas activation induces the release of nitric oxide through eNOS stimulus.[30,31] eNOS could inhibit TGF-β signaling,[32] suggesting that this pathway explains the protective effects of Ang-(1-7) on cardiac function in infarcted rats.[21] TGF-β is considered a mediator of the proliferative phase after MI and participates in the final stages of myocardial healing (cicatricial phase—maturation). The TGF-β effects are critically linked to fibrosis production and persist during all steps of myocardial healing.

Biglycan is an ECM component, important for the modulation of structural arrangement of collagen fibers and for the formation of a stable matrix. The role of biglycans in remodeling is only partially understood, but its modulatory action is known over TGF-β. Biglycan has high affinity by TGF-β, modulating its profibrotic activity. There is a reciprocal modulatory activity between biglycan and TGF-β, controlling the adequate arrangement of collagen fiber deposition.[33] After 48 h of coronary ligation, the expression of biglycan increases, reaching the highest values on the seventh day, and the Ang-(1-7) treatment can abolish it completely. Ang-(1-7) also attenuates the expression of WISP-1, a growth factor related to profibrotic and hypertrophic actions in the heart in the MI model.[34] The effects of WISP-1 are mediated via AKT, stimulating fibroblast proliferation and collagen synthesis.[25]

Tenascin C (TNC) is a glycoprotein able to regulate cell behavior and matrix organization while remodeling, contributing to the formation of both types of fibrosis, reactive and replacement.[35] It is synthesized by cardiac fibroblasts under the influence of various stimuli. Cytokines, growth factors, hypoxia, tissue acidosis, Ang II, and mechanical stress can elicit TNC release. This makes TNC directly linked to myocardial injury, inflammation, and healing process. In humans, TNC is related to the higher incidence of remodeling and an unfavorable prognosis in HF acting as a marker for LV remodeling.[36,37]

MATURATION PHASE

Alterations in the proportion of collagens I and III determine increased myocardium stiffness leading to increased metabolic activity with increased oxygen consumption, worsening of function, and, at advanced conditions, congestive HF. The mediators described previously in this chapter carry out these alterations. As it is possible to note, Ang-(1-7) through Mas can block intracellular signaling leading to cardiovascular remodeling in disease conditions. The collagen I/III proportion increases in cardiovascular diseases and the collagen I deposition is more resistant to tension, thus producing deleterious effects in the right and left diastolic function due to decreased myocardial complacence.[38]

The overall effects of Mas receptor stimulation are described in Marques.[34] These effects are summarized in Figure 1. As it can be seen, Ang-(1-7) attenuates or abolishes the expression of many substances capable of producing remodeling.

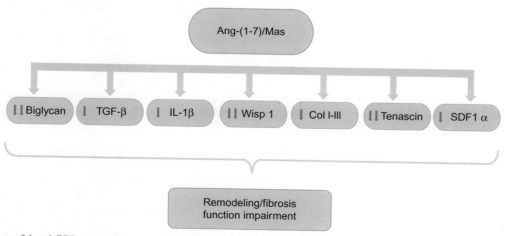

FIGURE 1   Effects of Ang 1-7/Mas on cardiac remodeling (↓Attenuate; ↓↓Abolish)

The information presented suggests that Ang-(1-7)/Mas axis stimulation has remarkable importance as a therapeutic possibility to treat MI and HF. The ongoing inflammatory pathways in cardiac diseases converge through many kinds of effectors to the development of fibrosis and worsening of heart function, by acting in specific times during myocardial injury. Mas receptor stimulation decreases the expression of many inflammatory/profibrotic components associated with myocardial remodeling, improving heart function.

REFERENCES

1. Iwata M, Cowling RT, Yeo SJ, Greenberg B. Targeting the ACE2-Ang-(1–7) pathway in cardiac fibroblasts to treat cardiac remodeling and heart failure. *J Mol Cell Cardiol* 2011;**51**:542–7.
2. Zipes DP, Libby P, Bonow RO, Braunwald E. Braunwald's Heart Disease: A Textbook of Cardiovascular Medicine, vols. 1 and 2. Elsevier Health Sciences; Saint Louis, MI. 2004.
3. Mann DL. Left ventricular size and shape: determinants of mechanical signal transduction pathways. *Heart Fail Rev* 2005;**10**:95–100.
4. Gava E, de Castro CH, Ferreira AJ, Colleta H, Melo MB, Alenina N, et al. Angiotensin-(1–7) receptor Mas is an essential modulator of extracellular matrix protein expression in the heart. *Regul Pept* 2012;**175**:30–42.
5. Weber KT, Sun Y, Bhattacharya SK, Ahokas RA, Gerling IC. Myofibroblast-mediated mechanisms of pathological remodelling of the heart. *Nat Rev Cardiol* 2013;**10**:15–26.
6. Loot AE, Roks AJM, Henning RH, Tio RA, Suurmeijer AJH, Boomsma F, et al. Angiotensin-(1–7) attenuates the development of heart failure after myocardial infarction in rats. *Circulation* 2002;**105**:1548–50.
7. Crackower MA, Sarao R, Oudit GY, Yagil C, Kozieradzki I, Scanga SE, et al. Angiotensin-converting enzyme 2 is an essential regulator of heart function. *Nature* 2002;**417**:822–8.
8. Santos RAS, Ferreira AJ, Verano-Braga T, Bader M. Angiotensin-converting enzyme 2, angiotensin-(1–7) and Mas: new players of the renin-angiotensin system. *J Endocrinol* 2013;**216**:R1–17.
9. Ortiz-Pérez JT, Riera M, Bosch X, De Caralt TM, Perea RJ, Pascual J, et al. Role of circulating angiotensin converting enzyme 2 in left ventricular remodeling following myocardial infarction: a prospective controlled study. *PLoS ONE* 2013;**8**:e61695.
10. Ocaranza MP, Godoy I, Jalil JE, Varas M, Collantes P, Pinto M, et al. Enalapril attenuates downregulation of angiotensin-converting enzyme 2 in the late phase of ventricular dysfunction in myocardial infarcted rat. *Hypertension* 2006;**48**:572–8.
11. Ferreira AJ, Santos RA, Almeida AP. Angiotensin-(1–7): cardioprotective effect in myocardial ischemia/reperfusion. *Hypertension* 2001;**38**:665–8.
12. Ferreira AJ, Jacoby BA, Araújo CAA, Macedo FAFF, Silva GAB, Almeida AP, et al. The nonpeptide angiotensin-(1–7) receptor Mas agonist AVE-0991 attenuates heart failure induced by myocardial infarction. *Am J Physiol Heart Circ Physiol* 2007;**292**:H1113–19.
13. Tao X, Fan J, Kao G, Zhang X, Su L, Yin Y, et al. Angiotensin-(1–7) attenuates angiotensin II-induced signaling associated with activation of a tyrosine phosphatase in Sprague–Dawley rats cardiac fibroblasts. *Biol Cell Auspices Eur Cell Biol Organ* 2014;**106**(6):182–92. doi:10.1111/boc.201400015, Published online first: 18 March 2014.
14. Tallant EA, Ferrario CM, Gallagher PE. Angiotensin-(1–7) inhibits growth of cardiac myocytes through activation of the mas receptor. *Am J Physiol Heart Circ Physiol* 2005;**289**:H1560–6.
15. Santos RAS, Castro CH, Gava E, Pinheiro SVB, Almeida AP, de Paula RD, et al. Impairment of in vitro and in vivo heart function in angiotensin-(1–7) receptor MAS knockout mice. *Hypertension* 2006;**47**:996–1002.
16. Iwata M, Cowling RT, Gurantz D, Moore C, Zhang S, Yuan JX-J, et al. Angiotensin-(1–7) binds to specific receptors on cardiac fibroblasts to initiate antifibrotic and antitrophic effects. *Am J Physiol Heart Circ Physiol* 2005;**289**:H2356–63.
17. Pan C-H, Wen C-H, Lin C-S. Interplay of angiotensin II and angiotensin(1–7) in the regulation of matrix metalloproteinases of human cardiocytes. *Exp Physiol* 2008;**93**:599–612.
18. Oudit GY, Liu GC, Zhong J, Basu R, Chow FL, Zhou J, et al. Human recombinant ACE2 reduces the progression of diabetic nephropathy. *Diabetes* 2010;**59**:529–38.
19. De Mello WC, Ferrario CM, Jessup JA. Beneficial versus harmful effects of angiotensin (1–7) on impulse propagation and cardiac arrhythmias in the failing heart. *J Renin-Angiotensin-Aldosterone Syst—JRAAS* 2007;**8**:74–80.
20. Zhang T, Li Z, Dang H, Chen R, Liaw C, Tran T-A, et al. Inhibition of Mas G-protein signaling improves coronary blood flow, reduces myocardial infarct size, and provides long-term cardioprotection. *Am J Physiol Heart Circ Physiol* 2012;**302**:H299–311.
21. Marques FD, Melo MB, Souza LE, Irigoyen MCC, Sinisterra RD, de Sousa FB, et al. Beneficial effects of long-term administration of an oral formulation of angiotensin-(1–7) in infarcted rats. *Int J Hypertens* 2012;**2012**:795452.
22. Palmer JN, Hartogensis WE, Patten M, Fortuin FD, Long CS. Interleukin-1 beta induces cardiac myocyte growth but inhibits cardiac fibroblast proliferation in culture. *J Clin Invest* 1995;**95**:2555–64.
23. Ing DJ, Zang J, Dzau VJ, Webster KA, Bishopric NH. Modulation of cytokine-induced cardiac myocyte apoptosis by nitric oxide, Bak, and Bcl-x. *Circ Res* 1999;**84**:21–33.
24. Deten A, Volz HC, Briest W, Zimmer H-G. Cardiac cytokine expression is upregulated in the acute phase after myocardial infarction. Experimental studies in rats. *Cardiovasc Res* 2002;**55**:329–40.
25. Colston JT, de la Rosa SD, Koehler M, Gonzales K, Mestril R, Freeman GL, et al. Wnt-induced secreted protein-1 is a prohypertrophic and profibrotic growth factor. *Am J Physiol Heart Circ Physiol* 2007;**293**:H1839–46.
26. Petit I, Jin D, Rafii S. The SDF-1-CXCR4 signaling pathway: a molecular hub modulating neo-angiogenesis. *Trends Immunol* 2007;**28**:299–307.

27. Zhuang Y, Chen X, Xu M, Zhang L, Xiang F. Chemokine stromal cell-derived factor 1/CXCL12 increases homing of mesenchymal stem cells to injured myocardium and neovascularization following myocardial infarction. *Chin Med J (Engl)* 2009;**122**:183–7.
28. Bujak M, Frangogiannis NG. The role of TGF-beta signaling in myocardial infarction and cardiac remodeling. *Cardiovasc Res* 2007;**74**:184–95.
29. Derynck R, Zhang YE. Smad-dependent and Smad-independent pathways in TGF-beta family signalling. *Nature* 2003;**425**:577–84.
30. Sampaio WO, Souza dos Santos RA, Faria-Silva R, da Mata Machado LT, Schiffrin EL, Touyz RM. Angiotensin-(1–7) through receptor Mas mediates endothelial nitric oxide synthase activation via Akt-dependent pathways. *Hypertension* 2007;**49**:185–92.
31. Dias-Peixoto MF, Santos RAS, Gomes ERM, Alves MNM, Almeida PWM, Greco L, et al. Molecular mechanisms involved in the angiotensin-(1–7)/Mas signaling pathway in cardiomyocytes. *Hypertension* 2008;**52**:542–8.
32. Chen L-L, Yin H, Huang J. Inhibition of TGF-beta1 signaling by eNOS gene transfer improves ventricular remodeling after myocardial infarction through angiogenesis and reduction of apoptosis. *Cardiovasc Pathol Off J Soc Cardiovasc Pathol* 2007;**16**:221–30.
33. Ahmed MS, Øie E, Vinge LE, Yndestad A, Andersen GGØ, Andersson Y, et al. Induction of myocardial biglycan in heart failure in rats—an extracellular matrix component targeted by AT(1) receptor antagonism. *Cardiovasc Res* 2003;**60**:557–68.
34. Marques FD. Mecanismos envolvidos no efeito cardioprotetor do composto de inclusão hidroxipropil beta-ciclodextrina/angiotensina-(1–7) em ratos. https://catalogobiblioteca.ufmg.br/pergamum/biblioteca/index.php; 2012 (accessed 25.04.14).
35. Imanaka-Yoshida K. Tenascin-C in cardiovascular tissue remodeling: from development to inflammation and repair. *Circ J Off J Jpn Circ Soc* 2012;**76**:2513–20.
36. Sato A, Aonuma K, Imanaka-Yoshida K, Yoshida T, Isobe M, Kawase D, et al. Serum tenascin-C might be a novel predictor of left ventricular remodeling and prognosis after acute myocardial infarction. *J Am Coll Cardiol* 2006;**47**:2319–25.
37. Terasaki F, Okamoto H, Onishi K, Sato A, Shimomura H, Tsukada B, et al. Higher serum tenascin-C levels reflect the severity of heart failure, left ventricular dysfunction and remodeling in patients with dilated cardiomyopathy. *Circ J Off J Jpn Circ Soc* 2007;**71**:327–30.
38. Wei S, Chow LT, Shum IO, Qin L, Sanderson JE. Left and right ventricular collagen type I/III ratios and remodeling post-myocardial infarction. *J Card Fail* 1999;**5**:117–26.

Chapter 27

Mas Receptor: Vascular and Blood Pressure Effects

Robson Augusto Souza dos Santos*, Walkyria Oliveira Sampaio[†,‡]

*Departamento de Fisiologia e Biofísica—ICB, Universidade Federal de Minas Gerais, Belo Horizonte, Minas Gerais, Brazil, †Department of Biological Sciences, Universidade de Itaúna, Itaúna, Minas Gerais, Brazil, ‡Department of Physiology, Universidade Federal de Minas Gerais, Belo Horizonte, Minas Gerais, Brazil

Vascular responses to angiotensin-(1-7) (Ang-(1-7)) significantly differ from those induced by angiotensin II (Ang II). In the blood vessels, Ang-(1-7)/Mas axis induces vasodilator, antiproliferative, and antithrombotic effects.[1] The blood vessels of human and several species generate Ang-(1-7) and express Mas receptor, in both the endothelial and vascular smooth muscle cells (VSMCs).[2-5] Vasodilation is the most well-characterized effect of Ang-(1-7). It has been shown that Ang-(1-7) produces relaxation of the aortic rings of Sprague-Dawley[6] and mRen-2 transgenic rats,[7] canine,[8] and porcine[9] coronary arteries; canine middle cerebral artery[10]; piglet pial arterioles[11]; feline systemic vasculature[12]; rabbit renal afferent arterioles[13]; and mesenteric microvessels of normotensive and hypertensive rats.[14,15] Furthermore, Ang-(1-7) potentiates the vasodilator effect of bradykinin in several vascular beds, including dog and rat coronary vessels,[8,16] rat kidney vessels, and mesenteric arterioles.[14,17]

The selectivity of the vasodilatory effect of Ang-(1-7)/Mas contrasts with the more widespread vasoconstriction described for Ang II/AT$_1$. In normotensive rats, Ang-(1-7) produced marked changes on regional blood flow, increasing vascular conductance in the mesenteric, cerebral, cutaneous, and renal territories. Furthermore, Ang-(1-7) simultaneously increases cardiac output (CO) by 30% and decreases the total peripheral resistance (TPR) by 26%. These opposite changes in TPR and CO lead to the absence of substantial changes in blood pressure.[18] Similarly, transgenic rats with a lifetime slight increase in circulating Ang-(1-7)/(TGRL-3292) present pronounced changes in regional blood flow, resulting in an increase in vascular conductance in the kidneys, lungs, adrenals, spleen, brain, testis, and brown fat tissue.[19] In contrast, Mas-deficient mice show an important increase in the vascular resistance of many territories such as the kidneys, lungs, adrenals, mesentery, spleen, and brown fat tissue. In this model, a parallel increase in the TPR and decrease in cardiac index were also observed.[20] Besides its actions on vascular tonus, Ang-(1-7) also inhibits vascular growth and has protective effects against thrombosis.[4,21] The antiproliferative potential was also observed in cardiac fibroblasts[22] and tumor cells.[23,24]

The main underlying event for many vascular actions of Ang-(1-7) is nitric oxide (NO) release. Human endothelial cells express the G protein-coupled receptor (GPCR) Mas, through which Ang-(1-7) mediates the activation of eNOS and NO production, via Akt-dependent pathways. Ang-(1-7) also promoted NO release in Mas-transfected Chinese hamster ovary cells through the phosphatidylinositol 3-kinase (PI3K)/Akt pathway.[3] Recently, it was demonstrated that Ang-(1-7) also is able to activate downstream components such as FOXO1 transcription factor, which is a well-known negative modulator of Akt signaling pathway.[25] These results highlight how the modulatory effect of Ang-(1-7) is complex and probably involves downstream effectors that, through fine-tuned mechanisms, coordinate negative feedback loops (Figure 1).

Regarding its effects on NO generation, Ang-(1-7) coordinately regulates Ser1177/Thr495 phosphorylation of eNOS.[3] In resting conditions, eNOS is phosphorylated on Thr495 and only weakly phosphorylated on Ser1177. The dual phosphorylation of Ser1177 and Thr495 determines the active state of eNOS in endothelial cells.[26] Ang-(1-7) increases the Ser1177 phosphorylation and simultaneously decreases Thr495 phosphorylation, and these posttranslational effects were blocked by the Mas antagonist A-779.[3]

In human endothelial cells, Ang-(1-7) also counterregulates Ang II signaling, blunting the phosphorylation of c-Src and ERK1/2 and the activation of NAD(P)H oxidase by Ang II. This modulatory effect is mediated by the phosphorylation of SHIP2, preventing Ang II-induced SHIP2 dephosphorylation and promoting interaction between SHIP2 and c-Src. The Ang-(1-7) antagonist, A-779, inhibited these actions, demonstrating that these effects are mediated through receptor Mas.[27] In keeping with this concept, the impairment of endothelium function in two different genetic backgrounds, C57Bl/6 and

The Protective Arm of the Renin–Angiotensin System (RAS). http://dx.doi.org/10.1016/B978-0-12-801364-9.00027-4

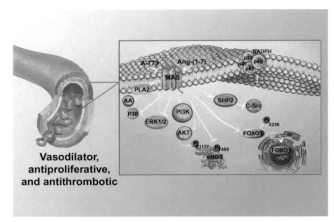

FIGURE 1   Vascular actions of Ang-(1-7). Ang-(1-7) coordinately regulates Ser1177/Thr495 phosphorylation of eNOS, increasing the eNOS Sr1177 phosphorylation, via PI3K/Akt pathway, and simultaneously decreasing Thr495 phosphorylation. Ang-(1-7) reduces Ang II-induced c-Src phosphorylation, which is crucial for NADPH oxidase activation. These effects are mediated via SHIP2, a negative regulator of Ang II-induced c-Src signaling. Ang-(1-7) also decreases phosphorylation of p38 and ERK1/2, well-characterized proliferative signaling pathways. FOXO1 is down-phosphorylated at S236 regulatory site after Ang-(1-7) treatment, inducing its nuclear translocation and transcriptional activities. GPCR Mas, G protein-coupled receptor Mas; NADPH oxidase, nicotinamide adenine dinucleotide phosphate oxidase; PLA2, phospholipase A2; AA, arachidonic acid; NRTK, nonreceptor tyrosine kinase (c-Src); FOXO1, forkhead box-containing protein O (FoxO) 1; eNOS, endothelial nitric oxide synthase; SHIP2, Src homology domain 2-containing inositol 5-phosphatase 2; PI3K, phosphatidylinositol 3-kinase; Akt, Ser/Thr protein kinase; p38, p38; MAPK, mitogen-activated protein kinase; ERK, extracellular signal-regulated kinase.

FVB/N, with Mas deficiency indicates a crucial role of Mas in endothelial function by its effects on the generation and metabolism of NO and ROS.[28,29] In the FVB/N background, endothelial dysfunction is associated with an increase in blood pressure,[28] whereas in C57BL/6 mice, no alteration of blood pressure was reported.[29] Furthermore, a worsening of 2-kidney, 1-clip Goldblatt hypertension was also observed in Mas-knockout mice.[30] Conversely, a short-term infusion of Ang-(1-7) improves endothelial function, significantly increasing the hypotensive effect of intra-arterial acetylcholine administration in normotensive rats.[31]

The Ang-(1-7)/Mas axis has also been recognized to have antithrombotic and antiproliferative effects. Mas-knockout mice possess increased venous thrombus size and short bleeding times.[32] Similar results were observed with the administration of an orally active form of Ang-(1-7), based on Ang-(1-7) inclusion in cyclodextrin.[33] The antithrombotic activity was also found in bradykinin B2 receptor-deleted (Bdkrb2−/−) mice, in which the Mas antagonist A-779 shortens the time to arterial thrombosis and the tail bleeding time. Mas-mediated NO release from the platelets and increased prostacyclin seem to mediate these antithrombogenic actions.[34]

The antiproliferative action of Ang-(1-7) in the vasculature is demonstrated in VSMC, in which Ang-(1-7) reduced Ang II-stimulated MAP kinase activities of ERK1/2 by an increase in prostacyclin (PGI$_2$).[35] A concurring result was observed in a rat stenting model, where Ang-(1-7) treatment produced a significant reduction in neointimal thickness, neointimal area, and percentage stenosis.[36] In rats with vascular calcification, Ang-(1-7) restored the decreased expression of lineage markers, including α-smooth muscle (SM) actin, SM22α, calponin, and smoothelin, in VSMCs and retarded the osteogenic transition of these cells by reducing the expression of bone-associated proteins.[37]

The actions of Ang-(1-7) in human vasculature still require further investigations to confirm the potent vasodilator effect described in rodents. Although the early research focused in human vessels demonstrated some controversies, the contrasting results obtained in humans could be due to methodological discrepancies, racial or vascular territory differences in the sensitivity to Ang-(1-7). The infusion of Ang-(1-7) in patients chronically treated with ACE inhibitors was without effect on forearm blood flow, whereas the infusion of bradykinin produced a noticeable vasodilation.[38] This observation was understood as evidence for a lack of relevance of Ang-(1-7) to the hemodynamic effects of ACE inhibitors. However, it was not considered that ACE inhibition significantly increases circulating levels of Ang-(1-7); therefore, the use of a Mas receptor antagonist rather than Ang-(1-7) would have been more suitable to evaluate the effect of this peptide on blood flow in this condition. A similar negative result was also found in the forearm blood flow of normotensive patients, in which Ang-(1-7) did not alter vasodilation produced by bradykinin infusion.[39] In contrast, Sasaki et al.[40] observed a dose-dependent vasodilation in the forearm circulation of normotensive subjects and patients with essential hypertension. Furthermore, Ueda et al.[41] similarly reported a dose-dependent potentiation of bradykinin vasodilation by Ang-(1-7) in normotensive forearm-resistant vessels of normotensive healthy men, which is in keeping with the well-known bradykinin-potentiating activity of

Ang-(1-7), described in animal models.[8,14,16,17] The attenuation of the vasoconstrictor effect of Ang II, but not norepinephrine, by Ang-(1-7) was described in the forearm of normotensive patients and in mammary arteries *in vitro*.[40,42] Ang-(1-7) also antagonized Ang II in renal vessels *in vitro*, but does not appear to have a pronounced effect in normal physiological regulation of renal vascular function *in vivo*.[43] More recently, van Twist et al.[44] observed a significant dose-dependent increase in blood flow to the kidney during intrarenal infusion of Ang-(1-7) in hypertensive patients. Interestingly, similar to the observation made in rat renal vessels,[45] this effect was weakened in patients on a low-salt diet, probably because of the fact that a low-salt diet leads to an increase in circulating angiotensin peptides, including Ang-(1-7).[46]

As discussed in previous chapters, the Mas-mediated effects of Ang-(1-7) are fairly numerous. Clearly, Ang-(1-7)/Mas axis acts as an important counterregulatory arm within RAS in the blood vessels. However, further studies are required to elucidate the complex interactions between the RAS receptors, peptides, and enzymes in the vasculature.

REFERENCES

1. Santos RA. Angiotensin-(1–7). *Hypertension* 2014;**63**.
2. Santos RA, Brosnihan KB, Jacobsen DW, DiCorleto PE, Ferrario CM. Production of angiotensin-(1–7) by human vascular endothelium. *Hypertension* 1992;**19**:II56–61.
3. Sampaio WO, Santos RAS, Faria-Silva R, Machado LTM, Schiffrin EL, Touyz RM. Angiotensin-(1–7) through receptor Mas mediates endothelial nitric oxide synthase activation via Akt-dependent pathways. *Hypertension* 2007;**49**:185–92.
4. Freeman EJ, Chisolm GM, Ferrario CM, Tallant EA. Angiotensin-(1–7) inhibits vascular smooth muscle cell growth. *Hypertension* 1996;**28**:104–8.
5. Xu P, Costa-Goncalves AC, Todiras M, Rabelo LA, Sampaio WO, Moura MM, et al. Endothelial dysfunction and elevated blood pressure in MAS gene-deleted mice. *Hypertension* 2008;**51**:574–80.
6. Tran Y, Forster C. Angiotensin-(1–7) and the rat aorta: modulation by endothelium. *J Cardiovasc Pharmacol* 1997;**30**:676–82.
7. Lemos VS, Cortes SF, Silva DM, Campgnole-Santos MJ, Santos RA. Angiotensin-(1–7) is involved in the endothelium-dependent modulation of phenylephrine-induced contraction in the aorta of m-Ren transgenic rats. *Br J Pharmacol* 2002;**135**:1743–8.
8. Brosnihan KB, Li P, Ferrario CM. Angiotensin-(1–7) dilates canine coronary arteries through kinins and nitric oxide. *Hypertension* 1996;**27**:523–8.
9. Pörsti I, Bara AT, Busse R, Hecker M. Release of oxide nitric by angiotensin-(1–7) from porcine coronary endothelium: implications for a novel angiotensin receptor. *Br J Pharmacol* 1994;**111**:652–4.
10. Feterik K, Smith L, Katusic ZS. Angiotensin-(1–7) causes endothelium-dependent relaxation in canine middle cerebral artery. *Brain Res* 2000;**873**:75–82.
11. Meng W, Busija DW. Comparative effects of angiotensin-(1–7) and angiotensin II on piglet pial arterioles. *Stroke* 1993;**24**:2041–5.
12. Osei SY, Ahima RS, Minkes RK, Weaver JP, Khosla MC, Kadowitz PJ. Differential responses to Ang-(1–7) in the feline mesenteric and hindquarter vascular beds. *Eur J Pharmacol* 1993;**234**:35–42.
13. Ren Y, Garvin JL, Carretero OA. Vasodilator action of angiotensin-(1–7) on isolated rabbit afferent arterioles. *Hypertension* 2002;**39**:799–802.
14. Fernandes L, Fortes ZB, Nigro D, Tostes RCA, Santos RAS, Carvalho MHC. Potentiation of bradykinin by angiotensin-(1–7) on arterioles of spontaneously hypertensive rats studies in vivo. *Hypertension* 2001;**37**:703–9.
15. Oliveira MA, Fortes ZB, Santos RAS, Khosla MC, Carvalho MHC. Synergistic effect of angiotensin-(1–7) on bradykinin arteriolar dilation in vivo. *Peptides* 1999;**20**:1195–201.
16. Almeida AP, Fabregas BC, Madureira MM, Santos RJ, Campagnole-Santos MJ, Santos RA. Angiotensin-(1–7) potentiates the coronary vasodilatory effect of bradykinin in the isolated rat heart. *Braz J Med Biol Res* 1999;**33**:709–13.
17. Santos RAS, Passaglio KT, Pesquero JB, Bader M, Simões e Silva AC. Interactions between kinins and angiotensin-(1–7) in kidney and blood vessels. *Hypertension* 2001;**38**:660–4.
18. Sampaio WO, Nascimento AA, Santos RA. Systemic and regional hemodynamic effects of angiotensin-(1–7) in rats. *Am J Physiol Heart Circ Physiol* 2003;**284**(6):H1985–94.
19. Botelho-Santos GA, Sampaio WO, Reudelhuber TL, Bader M, Campagnole-Santos MJ, Santos RAS. Expression of an angiotensin-(1–7)-producing fusion protein in rats induced marked changes in regional vascular resistance. *Am J Physiol Heart Circ Physiol* 2007;**292**(5):H2485–90.
20. Botelho-Santos GA, Bader M, Alenina N, Santos RA. Altered regional blood flow distribution in Mas-deficient mice. *Ther Adv Cardiovasc Dis* 2012;**6**(5):201–11.
21. Fraga-Silva RA, Da Silva DG, Montecucco F, Mach F, Stergiopulos N, Silva RF, et al. The angiotensin-converting enzyme 2/angiotensin-(1–7)/Mas receptor axis: a potential target for treating thrombotic diseases. *Thromb Haemost* 2012;**108**(6):1089–96.
22. McCollum LT, Gallagher PE, Tallant EA. Angiotensin-(1–7) abrogates mitogen-stimulated proliferation of cardiac fibroblasts. *Peptides* 2012;**34**:380–8.
23. Gallagher PE, Tallant EA. Inhibition of human lung cancer cell growth by angiotensin-(1–7). *Carcinogenesis* 2004;**25**:2045–52.
24. Ni L, Feng Y, Wan H, Ma Q, Fan L, Qian Y, et al. Angiotensin-(1–7) inhibits the migration and invasion of A549 human lung adenocarcinoma cells through inactivation of the PI3K/Akt and MAPK signaling pathways. *Oncol Rep* 2012;**27**:783–90.
25. Verano-Braga T, Schwämmle V, Sylvester M, Passos-Silva DG, Peluso AA, Etelvino GM, et al. Time-resolved quantitative phosphoproteomics: new insights into angiotensin-(1–7) signaling networks in human endothelial cells. *J Proteome Res* 2012;**11**(6):3370–81.
26. Fleming I, Fisslthaler B, Dimmeler S, Kemp BE, Busse R. Phosphorylation of Thr(495) regulates Ca(2+)/calmodulin-dependent endothelial nitric oxide synthase activity. *Circ Res* 2001;**88**:E68–75.

27. Sampaio WO, Castro CH, Santos RA, Schiffrin EL, Touyz RM. Angiotensin-(1–7) counterregulates angiotensin II signaling inhuman endothelial cells. *Hypertension* 2007;**50**:1093–8.

28. Xu P, Costa-Goncalves AC, Todiras M, Rabelo LA, Sampaio WO, Moura MM, et al. Endothelial dysfunction and elevated blood pressure in MAS gene-deleted mice. *Hypertension* 2008;**51**:574–80.

29. Rabelo LA, Xu P, Todiras M, Sampaio WO, Buttgereit J, Bader M, et al. Ablation of angiotensin-(1–7) receptor Mas in C57Bl/6 mice causes endothelial dysfunction. *J Am Soc Hypertens* 2008;**2**:418–24.

30. Rakušan D, Bürgelová M, Vaněčková I, Vaňourková Z, Husková Z, Skaroupková P, et al. Knockout of angiotensin-(1–7) receptor Mas worsens the course of two-kidney, one-clip Goldblatt hypertension: roles of nitric oxide deficiency and enhanced vascular responsiveness to angiotensin II. *Kidney Blood Press Res* 2010;**33**:476–88.

31. Faria-Silva R, Duarte FV, Santos RA. Short-term angiotensin(1–7) receptor MAS stimulation improves endothelial function in normotensive rats. *Hypertension* 2005;**46**:948–52.

32. Fraga-Silva RA, Pinheiro SVB, Gonçalves ACC, Alenina N, Bader M, Santos RA. The antithrombotic effect of angiotensin-(1–7) involves mas-mediated NO release from platelets. *Mol Med* 2008;**14**(1–2):28–35.

33. Fraga-Silva RA, Costa-Fraga FP, De Sousa FB, Alenina N, Bader M, Sinisterra RD, et al. An orally active formulation of angiotensin-(1–7) produces an antithrombotic effect. *Clinics* 2011;**66**(5):837–41.

34. Fang C, Stavrou E, Schmaier AA, Grobe N, Morris M, Chen A, et al. Angiotensin-(1–7) and Mas decrease thrombosis in Bdkrb2−/− mice by increasing NO and prostacyclin to reduce platelet spreading and glycoprotein VI activation. *Blood* 2013;**121**(15):3023–32.

35. Tallant EA, Clark MA. Molecular mechanisms of inhibition of vascular growth by angiotensin-(1–7). *Hypertension* 2003;**42**(4):574–9.

36. Langeveld B, vanGilst WH, Tio RA, Zijlstra F, Roks AJ. Angiotensin-(1–7) attenuates neointimal formation after stent implantation in the rat. *Hypertension* 2005;**45**(1):138–41.

37. Sui YB, Chang JR, Chen WJ, Zhao L, Zhang BH, Yu YR, et al. Angiotensin-(1–7) inhibits vascular calcification in rats. *Peptides* 2013;**42**:25–34.

38. Davie AP, McMurray JJ. Effect of angiotensin-(1–7) and bradykinin in patients with heart failure treated with an ACE inhibitor. *Hypertension* 1999;**34**:457–60.

39. Wilsdorf T, Gainer JV, Murphey LJ, Vaughan DE, Brown NJ. Angiotensin-(1–7) does not affect vasodilator or TPA responses to bradykinin human forearm. *Hypertension* 2001;**37**:1136–40.

40. Sasaki S, Higashi Y, Nakagawa K, Matsuura H, Kajiyama G. OshimaT. Effects of angiotensin-(1–7) on forearm circulation in normotensive subjects and patients with essential hypertension. *Hypertension* 2001;**38**:90–4.

41. Ueda S, Masumori-Maemoto S, Wada A, Ishii M, Brosnihan KB, Umemura S. Angiotensin(1–7) potentiates bradykinin-induced vasodilatation in man. *J Hypertens* 2001;**19**:2001–9.

42. Roks AJ, van Geel PP, Pinto YM, Buikema H, Henning RH, de Zeeuw D, et al. Angiotensin-(1–7) is a modulator of the human renin-angiotensin system. *Hypertension* 1999;**34**:296–301.

43. Roks AJ, Nijholt J, van Buiten A, van Gilst WH, de Zeeuw D, Henning RH. Low sodium diet inhibits the local counter-regulator effect of angiotensin-(1–7) on angiotensin II. *J Hypertens* 2004;**22**:2355–61.

44. van Twist DJ, Houben AJ, de Haan MW, Mostard GJ, Kroon AA, de Leeuw PW. Angiotensin-(1–7)-induced renal vasodilation in hypertensive humans is attenuated by low sodium intake and angiotensin II co-infusion. *Hypertension* 2013;**62**:789–93.

45. van der Wouden EA, Ochodnický P, van Dokkum RP, Roks AJ, Deelman LE, de Zeeuw D, et al. The role of angiotensin(1–7) in renal vasculature of the rat. *J Hypertens* 2006;**24**:1971–8.

46. Kocks MJ, Lely AT, Boomsma F, Jong PE, Navis G. Sodium status and angiotensin-converting enzyme inhibition: effects on plasma angiotensin-(1–7) in healthy man. *J Hypertens* 2005;**23**:597–602.

Chapter 28

Mas and Neuroprotection in Stroke

Douglas M. Bennion, Robert W. Regenhardt, Adam P. Mecca, Colin Sumners
Department of Physiology and Functional Genomics, University of Florida, Gainesville, Florida, USA

INTRODUCTION

As a cause of worldwide mortality, stroke ranks second only to ischemic heart disease. The majority of strokes (80–90%) are ischemic, resulting from a sudden interruption of blood flow, usually due to thromboembolism, that deprives a part of the brain of oxygen and nutrients. Within the core of the ischemic region, cell death and tissue necrosis rapidly develop. Outside this core exists tissue known as the penumbra that is somewhat buffered by collateral blood flow. In the hours and days following the ischemic event, the cells in this region offer a target for therapies that promote reperfusion and/or resolution pathways. A smaller percentage (10–20%) of strokes are classified as hemorrhagic, in which a loss of vessel integrity allows bleeding into the brain parenchyma. A large number of preclinical and human clinical trials have been conducted in an effort to discover novel stroke treatments, but the clinical unpredictability and multifactorial nature of the disease have hampered these efforts.

As one of the primary regulators of blood pressure, the renin–angiotensin system (RAS) has a profound impact in stroke. The original view of the RAS as a circulating endocrine system involving primarily the vasculature of the kidneys and lungs has been revised in recent decades to include the discovery of its ubiquitous expression in many tissues, including the brain, where both autocrine signaling and paracrine signaling contribute to its tissue-specific effects. The established physiological effects of the RAS in regulating blood pressure and fluid balance are mediated by angiotensin II (Ang II) binding to its type 1 receptor (AT_1R). Derangements in the RAS that result in chronic increases in circulating levels of Ang II and sustained overactivation of AT_1Rs contribute to increased risk for stroke and various cardiovascular diseases. Nowadays, the existence of the angiotensin-converting enzyme 2/angiotensin-(1-7)/receptor Mas [ACE2/Ang-(1-7)/Mas] axis of the RAS that exhibits anti-Ang II/AT_1R effects is well known. This chapter will explore evidence regarding the protective actions of this axis in stroke, the possible mechanisms to explain this protection, and the potential for targeting Ang-(1-7)/Mas using innovative translational treatments.

NEUROPROTECTION AND THE RAS

Neuroprotection is defined here as a therapy aimed at enhancing the brain's resilience to an injury, such as ischemia, in order to improve clinical outcome. This definition is not limited to actions only on neurons, but extends to include glia and other cells within the brain. This therapeutic approach represents the "holy grail" for drug discovery in the field of stroke research. Unfortunately, half a century of research has failed to translate many experimental treatments that elicit neuroprotection in animal models to human patients.[1] Despite this, there is much evidence that manipulating the brain RAS, in particular targeting the ACE2/Ang-(1-7)/Mas axis during stroke, has great therapeutic potential.

Much of the early work examining RAS manipulation in the treatment and prevention of stroke focused on ACE inhibitors and angiotensin type 1 receptor blockers (ARBs). In animal models, both treatments demonstrate therapeutic potential, which is supported by data from stroke prevention studies in human clinical trials.[2] Furthermore, these effects are independent of changes in blood pressure, and there is some debate as to whether the neuroprotective effects of ARBs are due to their blockade of the AT_1R or unopposed activation of the angiotensin type 2 receptor (AT_2R).[3] Activation of the ACE2/Ang-(1-7)/Mas axis has been demonstrated to have therapeutic potential for hypertension and related pathology[4] and so became a target for potential stroke therapies.

Importantly, all of the components of the ACE2/Ang-(1-7)/Mas axis occur within the brain.[4] In rodent models, immunoreactive Mas occurs throughout the brain and is present on neurons, microglia, and endothelial cells within the cerebral cortex and striatum.[5] Although cell culture experiments have suggested ACE2 to be expressed in glia,[6] mouse *in vivo* studies

The Protective Arm of the Renin–Angiotensin System (RAS). http://dx.doi.org/10.1016/B978-0-12-801364-9.00028-6

show that ACE2, at the mRNA and protein levels, is most predominantly expressed by neurons.[7] In a rat model of middle cerebral artery occlusion (MCAO), these RAS components were shown to increase in the hours and days following stroke.[8]

Our group has shown that experimentally manipulating the ACE2/Ang-(1-7)/Mas axis produces a neuroprotective effect in an endothelin-1 (ET-1)-induced transient MCAO rat model of ischemic stroke.[9] Central pretreatment with Ang-(1-7) decreased infarct size and neurological deficits when measured 3 days following stroke. These effects were mediated through a Mas-dependent pathway, as they were prevented by the coadministration of Mas antagonist A-779. The neuroprotective actions of centrally administered Ang-(1-7) were subsequently confirmed using a rat model of permanent MCAO.[10] In addition to ischemic stroke, activation of this axis also appears to have therapeutic potential in hemorrhagic stroke, where chronic central administration of Ang-(1-7) to stroke-prone spontaneously hypertensive rats (spSHR) fed with a high-salt diet increased survival, decreased the number of hemorrhages in the striatum, and improved lethargy and sensorimotor function.[11] Centrally administered Ang-(1-7) also reduced neurological deficits resulting from collagenase-induced hemorrhagic stroke in normotensive rats.[4] Alternative means of activating this axis have also proved effective. When transgenic mice with neural overexpression of ACE2 underwent MCAO, they exhibited decreased ischemic injury and neurological deficits. Interestingly, this effect was more pronounced in 8-month-old mice compared to 3-month-old mice.[12] In addition, human renin and angiotensinogen double-transgenic mice with neuronal ACE2 overexpression have decreased infarct volume, increased cerebral blood flow, increased neurological function, and increased microvascular density compared to the same double-transgenic mice with normal ACE2 expression.[13] Collectively, the above data provide compelling preclinical evidence that activation of the ACE2/Ang-(1-7)/Mas axis in the brain exerts neuroprotective effects in stroke. The effects of this system are robust; neuroprotection has been observed in several different models of ischemic and hemorrhagic stroke from different laboratories. Further, it is hypothesized that drug targets with multiple effects such as those described below for the ACE2/Ang-(1-7)/Mas axis, as opposed to a single mechanism of action, will prove more likely to translate.[14] For these and other reasons, it has been recently suggested that the ACE2/Ang-(1-7)/Mas axis in stroke represents an especially promising candidate for targeted stroke therapy.[15]

MECHANISMS OF PROTECTION

With the establishment of the protective benefits of the ACE2/Ang-(1-7)/Mas axis in stroke, there is increasing interest in understanding the mechanisms of neuroprotection. The following discussion explores several potential and nonmutually exclusive mechanisms. There is evidence that Ang-(1-7) has a multifactorial impact in reducing stroke damage, acting on the endothelium to increase cerebral blood flow and reduce inflammatory cell infiltration and on microglia and neurons to reduce oxidative stress, inflammation, and apoptosis. These actions are generally in direct opposition to the harmful effects of AngII/AT_1R activation.

Restoration and maintenance of cerebral perfusion are of primary importance in limiting stroke damage. Ang-(1-7) has an established role in stimulating nitric oxide (NO) release to increase vasodilation and regulate regional blood flow. During ischemic stroke, centrally administered Ang-(1-7) increased the release of NO and expression of endothelial NO synthase,[16] and in a model of neuronal ACE2 overexpression, blood flow to penumbral areas was increased and angiogenic factors were upregulated in a Mas-dependent fashion.[13] It is also likely that the observed vasodilatory effects of Ang-(1-7)/NO production involve interactions with the kallikrein-kinin system, as it has been demonstrated that central infusion of Ang-(1-7) induced increases in the levels of bradykinin and kinin B(1) and B(2) receptors in the ischemic cortex.[17] By contrast, in studies from our group, central infusion of Ang-(1-7) did not induce regional changes in cerebral blood flow during stroke that were detectable by laser Doppler flowmetry.[9] Nonetheless, the bulk of the evidence suggests that the neuroprotective action of Ang-(1-7)/Mas axis involves several vasodilatory pathways within the cerebrovasculature to maintain higher levels of perfusion to penumbral tissues.

Much of the damage in stroke that is thought be salvageable occurs days after the inciting injury, in response to the excessive inflammatory response in penumbral tissues. Smooth muscle cells, endothelial cells, astrocytes, and microglia are all induced to release proinflammatory cytokines. Microglial-mediated neurotoxicity is of particular importance in the progressive decline in penumbral viability in the days and weeks following a stroke. NF-κB (nuclear factor kappa-light-chain-enhancer of activated B cells) is a transcription factor that promotes activation of microglia to a proinflammatory state. Once activated, microglia induce neuron death via inflammatory signaling, specifically through tumor necrosis factor-α (TNF-α) signaling. In a study of permanent cerebral ischemia, Ang-(1-7) was shown to inhibit the NF-κB pathway and reduce proinflammatory cytokines via Mas signaling with concurrent reductions in infarct volume and improved neurological function in rats.[18] By inhibiting NF-κB in stroke, Ang-(1-7) may reduce inflammatory microglial activation and subsequent TNF-α-induced neurotoxicity.

Microglia help to orchestrate the inflammatory response that is known to drive much of the brain damage that results following stroke and thus represent an especially important site of action for stroke therapies such as those that target the

ACE2/Ang-(1-7)/Mas axis. We recently showed that centrally applied Ang-(1-7) attenuates the increased expression of the microglial marker CD11B and of proinflammatory cytokines within the cerebral cortex of rats 24h after ischemic stroke[5] and decreases the numbers of microglia in the striatum of spSHR rats.[11] In addition to limiting microglial activation by the NF-κB pathway, Mas signaling also impairs Ang II-induced increases in vascular cell adhesion protein-1 by inhibiting NF-κB nuclear translocation,[19] and Ang-(1-7) treatment reduces leukocyte chemotactic signaling, rolling and adhesion.[20] These results indicate a role for Ang-(1-7) in limiting inflammatory cell adhesion and subsequent invasion to further protect against stroke inflammatory injury, particularly so because the normally impermeable blood–brain barrier (BBB) is compromised during stroke leaving the brain especially vulnerable. Along with limiting inflammatory cell infiltration, further studies in this area may uncover a role for the ACE2/Ang-(1-7)/Mas axis in directly limiting BBB breakdown.

Besides reducing inflammation, the neuroprotective effects of Ang-(1-7) may also involve limiting neuronal cell death by decreasing oxidative stress. For example, Mas activation during permanent cerebral ischemia decreased oxidative stress and suppressed NF-κB activity,[18] whereas blockade of Mas increased oxidative stress levels, NF-κB activation, and proinflammatory signaling. The latter finding indicates a role for endogenously generated Ang-(1-7). Similar findings resulted from a study using a murine stroke model with neuronal ACE2 overexpression in which brain levels of NADPH oxidase subunits and reactive oxygen species were decreased, effects partially reversed by Mas blockade.[13] Another mechanism of neuronal death in stroke results from activation of the autophagy pathway, which might be inhibited to improve stroke outcomes. Long-term infusion of Ang-(1-7) into the brain of SHRs inhibits autophagic induction,[21] an effect that likely contributes to its protective actions. Although no evidence to date indicates a direct neurotrophic role for the ACE2/Ang-(1-7)/Mas axis, such as has been seen by activation of AT_2Rs to induce neurite outgrowth and neuron survival,[22] the likelihood that similar protective mechanisms can be induced via Mas signaling seems high, specifically since neurons themselves express Mas, and many Ang-(1-7)-induced protective effects are recapitulated by AT_2R agonism.

TRANSLATION

In view of the promising results from recent years of research on the neuroprotective ACE2/Ang-(1-7)/Mas axis, we now consider the topic of translating these findings into clinical relevance. The treatment of stroke provides several unique challenges. First, although there are known risk factors for stroke, ischemic events occur at unpredictable times, and patients are often unable to reach appropriate centers for medical care in a timely manner. This is particularly important if an intervention is targeting "at-risk" tissue of the penumbra where survival is time-dependent. Second, central nervous system (CNS) pharmacotherapy is complicated by the BBB, which effectively provides a physiological separation of the intraparenchymal brain from most drugs given through the systemic circulation. Preclinical models of disease and drug delivery are able to bypass these temporal and physiological barriers by standardizing stroke onset and preadministering drug directly into the CNS. Therefore, stroke treatments that target the ACE2/Ang-(1-7)/Mas axis will need to bypass these barriers. There are numerous candidate approaches that address the timing and route of administration of drug delivery to activate Mas in the CNS.

First, there are several options for delivery of stable nonpeptide molecules poststroke that activate the ACE2/Ang-(1-7)/Mas axis. One of the most promising, diminazene aceturate (DIZE), is a small-molecule antiprotozoal drug discovered to be a putative activator of ACE2.[23] Prestroke central administration of DIZE results in smaller infarct size and improved neurological function following stroke.[9] Because the BBB excludes many molecules and limits their activity centrally, less than 5% of drugs are active in the CNS.[24] The limitations of many small molecules to cross the BBB are due to a relatively narrow set of characteristics that seem to allow for BBB transport. For example, most of the small molecules that cross the BBB have a molecular mass less than 400–500 Da and are very lipophilic.[24] In addition, DIZE is approximately 515 Da and is also hydrophilic. In fact, there is direct evidence of the limited ability of DIZE delivered peripherally to cross the intact BBB.[25] However, DIZE may be bound by albumin in the blood, which has been shown to permeate the BBB in the early and late permeability phases of BBB opening after stroke.[4] In preliminary experiments, poststroke, peripheral administration of DIZE at 4, 24, and 48h after ET-1-induced MCAO reduced infarct size and significantly diminished neurological deficits.[26] The nonpeptide molecule DIZE is much more stable relative to Ang-(1-7), and stable nonpeptide drugs are more likely to reach peripheral therapeutic levels needed for ultimate BBB penetration after stroke. The nonpeptide Mas agonist, AVE 0991, which has been tested in models of cardioprotection, may be one such candidate for Ang-(1-7)/Mas activation during stroke.[27]

Alternatives to traditional drug therapy include viral-mediated methods of gene delivery and stem cell activation. Hematopoietic stem cells (HSCs) are mobilized from the bone marrow after stroke and home to sites of injury.[28] To facilitate delivery of a therapeutic molecule such as Ang-(1-7), HSCs harvested from donor bone marrow could be transduced with a lentiviral construct that constitutively produces a secretable form of Ang-(1-7). Furthermore, poststroke intravenous administration of nonvirally transduced HSCs is in itself protective, as demonstrated in mice following transient MCAO.[29] In addition, Ang-(1-7) stimulates hematopoietic progenitors *in vivo* and *in vitro* and transfusion of lentivirus-ACE2-primed EPCs reduces infarct volume and neurological deficits beyond that of control EPCs.[30,31] Therefore, targeting the Ang-(1-7)/ Mas axis using systemically administered Ang-(1-7)-secreting stem cells in the hours following stroke has potential as a neuroprotective therapy.

Continued study of translational approaches such as poststroke administration of stable nonpeptide molecules, methods to disrupt or pass through the BBB, and virally mediated gene delivery combined with stem cell therapy have the potential to identify a new set of angiotensin peptide-based stroke therapies.

CONCLUSION

New and innovative therapies are needed to counteract the devastating worldwide impact of stroke. In this chapter, we have explored the evidence that suggests that targeted activation of the ACE2/Ang-(1-7)/Mas axis of the brain RAS holds promise as one such stroke therapy. Accumulating data from animal studies have established the ACE2/Ang-(1-7)/Mas axis of the brain RAS as a neuroprotective target in stroke. Although more elucidative work is needed, it is apparent that vasodilatory, anti-inflammatory, antioxidative, and anti-AT$_1$R effects contribute to the mechanisms of Ang-(1-7)'s protective action in the ischemic brain. Efforts to translate preclinical findings are under way, and new strategies for noninvasively delivering stable agonists that cross BBB in the poststroke setting have the potential to validate previous studies using methods that more closely approximate the clinical setting.

REFERENCES

1. O'Collins VE, Macleod MR, Donnan GA, Horky LL, van der Worp BH, Howells DW. 1,026 experimental treatments in acute stroke. *Ann Neurol* 2006;**59**:467–77.
2. Ong HT, Ong LM, Ho JJ. Angiotensin-converting enzyme inhibitors (ACEIs) and angiotensin-receptor blockers (ARBs) in patients at high risk of cardiovascular events: a meta-analysis of 10 randomised placebo-controlled trials. *ISRN Cardiol* 2013;**2013**, Article ID 478597.
3. Li J, Culman J, Hortnagl H, Zhao Y, Gerova N, Timm M, et al. Angiotensin AT2 receptor protects against cerebral ischemia-induced neuronal injury. *FASB J* 2005;**19**:617–19.
4. Regenhardt RW, Bennion DM, Sumners C. Cerebroprotective action of angiotensin peptides in stroke. *Clin Sci (Lond)* 2014;**126**:195–205.
5. Regenhardt RW, Desland F, Mecca AP, Pioquinto DJ, Afzal A, Mocco J, et al. Anti-inflammatory effects of angiotensin-(1–7) in ischemic stroke. *Neuropharmacology* 2013;**71**:154–63.
6. Gallagher PE, Chappell MC, Ferrario CM, Tallant EA. Distinct roles for ANG II and ANG-(1–7) in the regulation of angiotensin-converting enzyme 2 in rat astrocytes. *Am J Physiol Cell Physiol* 2006;**290**:C420–6.
7. Doobay MF, Talman LS, Obr TD, Tian X, Davisson RL, Lazartigues E. Differential expression of neuronal ACE2 in transgenic mice with overexpression of the brain renin-angiotensin system. *Am J Physiol Regul Integr Comp Physiol* 2007;**292**:R373–81.
8. Lu J, Jiang T, Wu L, Gao L, Wang Y, Zhou F, et al. The expression of angiotensin-converting enzyme 2-angiotensin-(1–7)-Mas receptor axis are upregulated after acute cerebral ischemic stroke in rats. *Neuropeptides* 2013;**47**(5):289–95.
9. Mecca A, Regenhardt R, O'Connor T, Joseph J, Raizada M, Katovich M, et al. Cerebroprotection by angiotensin (1–7) in endothelin-1 induced ischemic stroke. *Exp Physiol* 2011;**96**(10):1084–96.
10. Jiang T, Gao L, Guo J, Lu J, Wang Y, Zhang Y. Suppressing inflammation by inhibiting the NF-kappaB pathway contributes to the neuroprotective effect of angiotensin-(1–7) in rats with permanent cerebral ischaemia. *Br J Pharmacol* 2012;**167**:1520–32.
11. Regenhardt RW, Mecca AP, Desland F, Ritucci-Chinni PF, Ludin JA, Greenstein D, et al. Centrally administered angiotensin-(1–7) increases the survival of stroke prone spontaneously hypertensive rats. *Exp Physiol* 2013;**99**(2):442–53.
12. Zheng JL, Li GZ, Chen SZ, Wang JJ, Olson JE, Xia HJ, et al. Angiotensin converting enzyme 2/ang-(1–7)/mas axis protects brain from ischemic injury with a tendency of age-dependence. *CNS Neurosci Ther* 2014;**20**(5):452–9.
13. Chen J, Zhao Y, Chen S, Wang J, Xiao X, Ma X, et al. Neuronal over-expression of ACE2 protects brain from ischemia-induced damage. *Neuropharmacology* 2014;**79**:550–8.
14. Fisher M. Characterizing the target of acute stroke therapy. *Stroke* 1997;**28**:866–72.
15. Pena Silva RA, Heistad DD. Promising neuroprotective effects of the angiotensin-(1–7)-angiotensin-converting enzyme 2-Mas axis in stroke. *Exp Physiol* 2014;**99**:342–3.
16. Zhang Y, Lu J, Shi J, Lin X, Dong J, Zhang S, et al. Central administration of angiotensin-(1–7) stimulates nitric oxide release and upregulates the endothelial nitric oxide synthase expression following focal cerebral ischemia/reperfusion in rats. *Neuropeptides* 2008;**42**:593–600.
17. Lu J, Zhang Y, Shi J. Effects of intracerebroventricular infusion of angiotensin-(1–7) on bradykinin formation and the kinin receptor expression after focal cerebral ischemia-reperfusion in rats. *Brain Res* 2008;**1219**:127–35.
18. Jiang T, Gao L, Guo J, Lu J, Wang Y, Zhang Y. Suppressing inflammation by inhibiting the NF-kappaB pathway contributes to the neuroprotective effect of angiotensin-(1–7) in rats with permanent cerebral ischaemia. *Br J Pharmacol* 2012;**167**:1520–32.

19. Zhang F, Ren J, Chan K, Chen H. Angiotensin-(1–7) regulates angiotensin II-induced VCAM-1 expression on vascular endothelial cells. *Biochem Biophys Res Commun* 2013;**430**:642–6.

20. da Silveira KD, Coelho FM, Vieira AT, Sachs D, Barroso LC, Costa VV, et al. Anti-inflammatory effects of the activation of the angiotensin-(1–7) receptor, MAS, in experimental models of arthritis. *J Immunol* 2010;**185**:5569–76.

21. Jiang T, Gao L, Shi J, Lu J, Wang Y, Zhang Y. Angiotensin-(1–7) modulates renin-angiotensin system associated with reducing oxidative stress and attenuating neuronal apoptosis in the brain of hypertensive rats. *Pharmacol Res* 2013;**67**:84–93.

22. Cote F, Laflamme L, Payet MD, Gallo-Payet N. Nitric oxide, a new second messenger involved in the action of angiotensin II on neuronal differentiation of NG108-15 cells. *Endocr Res* 1998;**24**:403–7.

23. Kulemina LV, Ostrov DA. Prediction of off-target effects on angiotensin-converting enzyme 2. *J Biomol Screen* 2011;**16**:878–85.

24. Pardridge WM. The blood–brain barrier: bottleneck in brain drug development. *NeuroRx* 2005;**2**:3–14.

25. Odika IE, Asuzu IU, Anika SM. The effects of hyperosmolar agents lithium chloride and sucrose on the brain concentration of diminazene aceturate in rats. *Acta Trop* 1995;**60**:119–25.

26. Bennion DM, Donnangelo L, Pioquinto D, Regenhardt R, Raizada M, Sumners C. Abstract TP111: activation of the brain renin-angiotensin system by translational approaches following stroke onset is neuroprotective in a rat model of ischemic stroke. *Stroke* 2013;**44**, ATP111.

27. Ferreira AJ, Jacoby BA, Araujo CA, Macedo FA, Silva GA, Almeida AP, et al. The nonpeptide angiotensin-(1–7) receptor mas agonist AVE-0991 attenuates heart failure induced by myocardial infarction. *Am J Physiol Heart Circ Physiol* 2007;**292**:H1113–19.

28. Kucia M, Jankowski K, Reca R, Wysoczynski M, Bandura L, Allendorf D, et al. CXCR4-SDF-1 signalling, locomotion, chemotaxis and adhesion. *J Mol Histol* 2004;**35**:233–45.

29. Schwarting S, Litwak S, Hao W, Bahr M, Weise J, Neumann H. Hematopoietic stem cells reduce postischemic inflammation and ameliorate ischemic brain injury. *Stroke* 2008;**39**:2867–75.

30. Heringer-Walther S, Eckert K, Schumacher SM, Uharek L, Wulf-Goldenberg A, Gembardt F, et al. Angiotensin-(1–7) stimulates hematopoietic progenitor cells in vitro and in vivo. *Haematologica* 2009;**94**:857–60.

31. Chen J, Xiao X, Chen S, Zhang C, Chen J, Yi D, et al. Angiotensin-converting enzyme 2 priming enhances the function of endothelial progenitor cells and their therapeutic efficacy. *Hypertension* 2013;**61**:681–9.

Chapter 29

Mas in the Kidney

Ana Cristina Simões Silva*,†, Sérgio Veloso Brant Pinheiro*

*Department of Pediatrics, Federal University of Minas Gerais—UFMG, Belo Horizonte, Minas Gerais, Brazil, †Interdisciplinary Laboratory of Medical Investigation—LIIM, Federal University of Minas Gerais—UFMG, Belo Horizonte, Minas Gerais, Brazil

INTRODUCTION

Over the past two decades, considerable advances have been made in our understanding of the renin–angiotensin system (RAS) and its roles in various disease states.[1] Besides the classical RAS axis formed by angiotensin-converting enzyme (ACE), angiotensin (Ang) II, and Ang type 1 receptor (AT_1), the biological relevance of another Ang fragment, the heptapeptide Ang-(1-7), was progressively recognized.[2,3] Ang-(1-7) acts through a specific G protein-coupled receptor, the Mas receptor,[4] and is mainly formed by the action of the ACE homologue enzyme, ACE2, which converts Ang II into Ang-(1-7).[5,6] In general, this alternative RAS axis formed by ACE2, Ang-(1-7), and Mas antagonizes the actions of the ACE/Ang II/AT_1 axis in many organs and systems. Therefore, the RAS can be envisioned as a dual-function system in which the final effects are primarily driven by the balance between both RAS axes.

In this chapter, we summarize recent findings related to the role of the ACE2/Ang-(1-7)/Mas axis in regulating renal function and in the physiopathology of experimental and human renal diseases.

ROLE OF ACE2/ANG-(1-7)/MAS IN RENAL PHYSIOLOGY

Many studies have addressed the complexity of renal actions of Ang-(1-7).[7–26] Differences between species, local and systemic concentrations of Ang-(1-7), nephron segment, level of RAS activation, and sodium and water status can be responsible for discrepant results concerning renal effects of Ang-(1-7). A diuretic/natriuretic action of Ang-(1-7) has been described in many *in vitro* and *in vivo* experimental studies, mostly by inhibition of sodium reabsorption at the proximal tubule.[7–15] Ang-(1-7) seems to be a potent inhibitor of Na-K-ATPase activity in the renal cortex [14] and in isolated convoluted proximal tubules.[15] In renal tubular epithelial cells, Ang-(1-7) inhibited transcellular flux of sodium through the activation of phospholipase A2.[8] *In vitro* studies also indicated that Ang-(1-7) modulates the stimulatory effect of Ang II on the Na-ATPase activity in the proximal tubule through an A779-sensitive receptor.[15] On the other hand, several studies demonstrated an antidiuretic/antinatriuretic effect induced by Ang-(1-7), especially in water-loaded animals.[16–26] The administration of Ang-(1-7) and the oral Mas receptor agonist, AVE 0991, has potent antidiuretic activity in water-loaded rats and mice via Mas receptor activation.[4,16,17,19,20,24,26] *In vitro*, Ang-(1-7) increased water transport in the inner medullary collecting duct through an interaction between Mas and the vasopressin type 2 receptor.[23] Accordingly, the administration of Mas receptor antagonists, A779 and D-Pro7, exerts diuretic effects associated with an increase in glomerular filtration rate and in water excretion.[16,19,21,22,25] Despite divergent data concerning tubular actions of Ang-(1-7), the studies support its role in the regulation of glomerular filtration and fluid handling.

Besides important tubular actions, Ang-(1-7) also contributes to renal hemodynamic regulation by opposing the hypertensive and fibrogenic effects of Ang II. When the RAS is inappropriately stimulated, high intrarenal Ang II levels, acting on AT_1 receptors, can lead to both systemic hypertension and glomerular capillary hypertension, which can induce hemodynamic injury to the vascular endothelium and glomerulus.[27–29] On the other hand, Ren et al.[30] reported that Ang-(1-7) induced dilatation of preconstricted renal afferent arterioles in rabbits, and Sampaio et al.[31] showed that an infusion of low concentrations of Ang-(1-7) increased renal blood flow in rats. Ang-(1-7) also attenuated the effect of Ang II-induced pressor responses and Ang II-enhanced nor epinephrine release to renal nerve stimulation in rat isolated kidney.[32]

Ang-(1-7) can also counteract the renal effects of Ang II by complex mechanisms as competing for the binding of Ang II to AT_1 receptors, modulating the signaling, and affecting the synthesis of AT_1 receptors.[33–38] In this regard, it was shown that prior exposure to Ang-(1-7) causes a modest decrease in the number of AT_1 receptors in the cortical tubulointerstitial

The Protective Arm of the Renin–Angiotensin System (RAS). http://dx.doi.org/10.1016/B978-0-12-801364-9.00029-8

area of the kidney.[38] Moreover, Kostenis et al.[39] showed that the Mas receptor hetero-oligomerizes with the AT_1 receptor and inhibits the intracellular Ca^{+2} mobilization effect of Ang II.

PHYSIOPATHOLOGICAL ROLE OF ACE2/ANG-(1-7)/MAS IN RENAL DISEASES

Experimental Studies

Ang-(1-7) may exert a protective role in the context of experimental models of renal diseases.[40–46] The infusion of Ang-(1-7) reduced glomerulosclerosis by opposing Ang II effects in experimental glomerulonephritis.[46] In adriamycin-induced nephropathy, the oral administration of the Mas agonist, the compound AVE 0991, improved renal function parameters, reduced urinary protein loss, and attenuated histological changes.[44] The administration of AVE 0991 had renoprotective effects in experimental acute renal injury, as seen by improvement of function, decreased tissue injury, prevention of local and remote leukocyte infiltration, and reduced release of the chemokine CXCL1.[40] The administration of Ang-(1-7) for 12 weeks in 5/6 nephrectomized male C57BL/6 mice reduced blood pressure, attenuated elevations in plasma urea and creatinine, and preserved cardiac function.[47] On the other hand, despite reducing blood pressure to the same extent as Ang-(1-7), the antihypertensive agent hydralazine was not able to improve renal function and avoid cardiac changes, thereby suggesting that renoprotection obtained with Ang-(1-7) was not mediated merely by the control of hypertension.[47]

Ang-(1-7) also produced beneficial effects in experimental diabetic nephropathy by reducing urinary protein excretion without affecting blood pressure in male adult STZ-diabetic rats compared with untreated diabetic animals.[48] The same research group treated diabetic spontaneous hypertensive rats with an identical dose of Ang-(1-7) and obtained for a second time a significant reduction in urinary protein excretion with no changes in mean arterial pressure.[49] Furthermore, Ang-(1-7) normalized the vascular responses to vasoconstrictors and prevented renal NOX-induced oxidative stress.[49] In an experimental model of type 2 diabetes, the KK-A^y/Ta mouse, the treatment with Ang-(1-7) opposed the Ang II-induced glomerular injury by reducing mesangial expansion, TGF-β, and fibronectin mRNA and NOX activity.[50] In Zucker diabetic fatty rats, Ang-(1-7) diminished triglyceridemia, proteinuria, and systolic blood pressure together with the restoration of creatinine clearance.[51] Additionally, Ang-(1-7) reduced renal fibrosis, attenuated renal oxidative stress, and decreased renal immunostaining of inflammatory markers to values similar to those displayed by control animals.[51] More recently, Mori et al.[52] reported that Ang-(1-7) treatment exerts renoprotective effects on diabetic nephropathy, associated with the reduction of oxidative stress, inflammation, fibrosis, and lipotoxicity.

Acquired or genetic ACE2 deficiency exacerbated renal damage and albuminuria in experimental models, possibly facilitating the damaging effects of Ang II.[53–58] Chronic administration of MLN-4760, an ACE2 inhibitor, produced albuminuria and matrix protein deposition in control or diabetic mice.[55] Renal expression of ACE2 was reduced in the renal cortex of mice that underwent subtotal nephrectomy and in a rat model of renal ischemia/reperfusion.[53,59] In a model of unilateral ureteral obstruction, the deletion of the ACE2 gene resulted in a four-fold increase in the ratio of intrarenal Ang II/Ang-(1-7), and these changes were associated with the development of progressive tubulointerstitial fibrosis and inflammation with high levels of TNF-α, IL-1β, and MCP-1.[42] Enhanced renal fibrosis and inflammation were attributed to marked increase in intrarenal Ang II signaling (AT_1/ERK1/2), TGF-β_1/Smad2/3, and NF-κB signaling pathways.[42] Dual RAS blockade normalized ACE2 expression and prevented hypertension, albuminuria, tubulointerstitial fibrosis, and tubular apoptosis in Akita angiotensinogen-transgenic mice.[60] Genetic deficiency of ACE2 activity in mice fosters oxidative stress via AT_1-dependent effect in the kidney.[61] In addition, these studies suggested that ACE2/Ang-(1-7)/Mas axis modulates oxidative stress, inflammation, and fibrosis at the renal tissue.

AT_1 and Mas receptors were codistributed in renal mesangial cells of rats and Ang-(1-7), through the binding to Mas, counteracted the stimulatory effects of Ang II on the ERK1/2 and TGF-β_1 pathways mediated by AT_1 receptors.[45] More recently, Ng et al.[62] showed that Mas receptor expression is reduced in the kidneys of rats with chronic kidney disease (CKD), and, in cultured human proximal tubular cells, indoxyl sulfate, a uremic toxin, downregulated renal expression of the Mas receptor and upregulated TGF-β_1.

On the other hand, few reports have suggested that Ang-(1-7) may exacerbate renal injury paradoxically in certain experimental conditions, suggesting that dose or route of administration, state of activation of the local RAS, cell-specific signaling, or non-Mas-mediated pathways may contribute to these deleterious responses.[63] Esteban et al.[64] reported that renal deficiency of Mas diminished renal damage in unilateral ureteral obstruction and in ischemia/reperfusion injury and that the infusion of Ang-(1-7) to wild-type mice elicited an inflammatory response. Velkoska et al.[65] verified that a 10-day infusion of Ang-(1-7) in rats with subtotal nephrectomy was associated with deleterious effects on blood pressure and heart function. Cell-specific signaling pathways associated with Ang-(1-7) in the kidney could play a role in the variable response. For instance, Ang-(1-7) displays growth-inhibitory properties and antagonizes the effects of Ang II in the proximal tubule,[66] whereas, in human mesangial cells, the heptapeptide seems to stimulate cell growth pathways by increasing arachidonic acid release and by MAPK phosphorylation.[67] Results obtained in rat mesangial cells are also divergent. While Liu et al.[68] reported that Ang-(1-7) stimulates ERK1/2 phosphorylation via Mas activation, Oudit and coworkers[69] showed that Ang-(1-7), acting via the Mas receptor, inhibits high-glucose-stimulated NOX activation. In addition, in primary cultures of mouse mesangial cells, Moon et al.[50] showed that Ang-(1-7) attenuated Ang II-induced MAPK phosphorylation and expression of TGF-β1, fibronectin, and collagen IV. It must be said, however, that the majority of studies suggest an overall renoprotective effect of administering Ang-(1-7) *in vivo*.

Clinical Studies

Ang-(1-7) can be measured in plasma and urine samples collected in healthy subjects and in patients with diverse clinical conditions.[70–76] The concentration of Ang-(1-7) may differ in plasma and urine samples of the same subject, since untreated adults with primary hypertension exhibited lower urinary levels of Ang-(1-7) than normotensive controls.[71] Significant differences in circulating levels of Ang II and Ang-(1-7) were detected in pediatric hypertensive patients.[74] Children with renovascular disease had plasma Ang II levels higher than plasma Ang-(1-7), whereas patients with primary hypertension had a selective elevation of plasma Ang-(1-7).[74] In pediatric patients with CKD, higher levels of Ang-(1-7) and Ang II were also detected in hypertensive patients when compared with normotensive patients at the same CKD stage.[75] Patients at end-stage renal disease presented an even more pronounced elevation of Ang-(1-7) levels, suggesting a deviation in RAS metabolism toward Ang-(1-7) synthesis.[75] Whether the elevation in plasma Ang-(1-7) provides a counterregulatory mechanism against Ang II-mediated vasoconstriction or nephron injury in primary hypertension and in CKD remains to be determined.

More recently, Mizuiri and coworkers[77] demonstrated that renal biopsies from patients with IgA nephropathy had significantly reduced glomerular and tubulointerstitial immunostaining for ACE2 compared with healthy controls. On the other hand, glomerular ACE staining was increased. These findings raise the possibility that an upward shift in the intrarenal ACE/ACE2 ratio favoring increased synthesis of Ang II and reduced Ang-(1-7) might lead to progressive nephron loss in this condition.[77] Circulating ACE2 activity was measured in kidney transplant patients and positively correlated with age, serum creatinine, and gamma-glutamyl transferase levels.[78]

It has been suggested that the beneficial effects of ACE inhibitors and of Ang receptor blockers may be due, at least in part, to an activation of the ACE2/Ang-(1-7)/Mas axis, since chronic treatment with these drugs increases plasma concentrations of Ang-(1-7).[70,72,75] Kocks et al.[72] showed that, during ACE inhibition, the administration of a low-sodium diet did not affect plasma levels of Ang II but induced a significant elevation in Ang-(1-7) concentration. Furthermore, in a murine model of adriamycin-induced nephropathy, renoprotective effects of losartan were blunted in mice with genetic deletion of the Mas receptor, indicating that Mas receptor activation is essential for renal actions of AT_1 receptor antagonism.[39] Very recently, Iwanami et al.[79] showed that the hypotensive and antihypertrophic effects of the newly developed AT_1 receptor blocker, azilsartan, may also involve activation of the ACE2/Ang-(1-7)/Mas axis. Another relevant aspect is the complex interaction between the Mas and AT_1 receptor. Kostenis et al.[39] showed that Mas can hetero-oligomerize with AT_1 and, by so doing, inhibit the actions of Ang II. Thus, Mas may act as a physiological antagonist of AT_1 receptor signaling.

CONCLUDING REMARKS

Experimental models of renal diseases suggest that the activation of the ACE2/Ang-(1-7)/Mas axis has a protective role. The few data provided by human studies also indicate a beneficial role for the activation of this alternative RAS axis. In addition, it has been hypothesized that the beneficial effects of ACEIs and ARBs might involve, at least in part, the elevation of plasma Ang-(1-7) levels. Further studies are clearly needed to elucidate the mechanisms by which both RAS axes modulate renal function and take part in the physiopathology of renal diseases. Nevertheless, current knowledge supports the possibility that drugs, which mimic or enhance the function of the ACE2/Ang-(1-7)/Mas axis, may be beneficial for the treatment of renal diseases.

REFERENCES

1. Simões e Silva AC, Flynn JT. The renin-angiotensin-aldosterone system in 2011: role in hypertension and chronic kidney disease. *Pediatr Nephrol* 2012;**27**:1835–45.
2. Santos RAS, Ferreira AJ, Simões e Silva AC. Recent advances in the angiotensin converting enzyme-angiotensin-(1–7)-Mas axis. *Exp Physiol* 2008;**93**:519–27.
3. Pinheiro SVB, Simões e Silva AC. Angiotensin converting enzyme 2, angiotensin-(1–7), and receptor Mas axis in the kidney. *Int J Hypertens* 2012;**2012**, Article ID 414128, 8 pp.

4. Santos RA, Simoes e Silva AC, Maric C, Silva DM, Machado RP, de Buhr I, et al. Angiotensin-(1–7) is an endogenous ligand for the G protein-coupled receptor Mas. *Proc Natl Acad Sci U S A* 2003;**100**:8258–63.

5. Tipinis SR, Hooper NM, Hyde R, Karran E, Christie G, Turner AJ. A human homolog of angiotensin-converting enzyme. Cloning and functional expression as a captopril-insensitive carboxypeptidase. *J Biol Chem* 2000;**275**:33238–43.

6. Donoghue M, Hsieh F, Baronas E, Godbout K, Gosselin M, Stagliano N, et al. A novel angiotensin-converting enzyme-related carboxypeptidase (ACE2) converts angiotensin I to angiotensin 1–9. *Circ Res* 2000;**87**:E1–9.

7. Ferrario CM, Jessup J, Gallagher PE, Averill DB, Brosnihan KB, Tallant EA, et al. Effects of renin-angiotensin system blockade on renal angiotensin-(1–7) forming enzymes and receptors. *Kidney Int* 2005;**68**:2189–96.

8. Andreatta-van Leyen S, Romero MF, Khosla MC, Douglas JG. Modulation of phospholipase A2 and sodium transport by angiotensin-(1–7). *Kidney Int* 1993;**44**:932–6.

9. Handa RK. Angiotensin-(1–7) can interact with the rat proximal tubule AT(4) receptor system. *Am J Physiol* 1999;**277**:F75–83.

10. Lara LS, De Carvalho T, Leao-Ferreira LR, Lopes AG, Caruso-Neves C. Modulation of the (Na(+)+K+)ATPase activity by angiotensin-(1–7) in MDCK cells. *Regul Pept* 2005;**129**:221–6.

11. Lara LS, Cavalcante F, Axelband F, De Souza AM, Lopes AG, Caruso-Neves C. Involvement of the Gi/o/cGMP/PKG pathway in the AT2-mediated inhibition of outer cortex proximal tubule Na+-ATPase by angiotensin-(1–7). *Biochem J* 2006;**395**:183–90.

12. Dellipizzi AM, Hilchley SD, Bell-Quilley CP. Natriuretic action of angiotensin-(1–7). *Br J Pharmacol* 1994;**111**:1–3.

13. Vallon V, Richter K, Heyne N, Osswald H. Effect of intratubular application of angiotensin-(1–7) on nephron function. *Kidney Blood Press Res* 1997;**20**:233–9.

14. Lopez O, Gironacci M, Rodriguez D, Pena C. Effect of angiotensin-(1–7) on ATPase activities in several tissues. *Regul Pept* 1998;**77**:135–9.

15. Bürgelová M, Kramer HJ, Teplan V, Velicková G, Vítko S, Heller J, et al. Intrarenal infusion of angiotensin-(1–7) modulates renal function responses to exogenous angiotensin II in the rat. *Kidney Blood Press Res* 2002;**25**:202–10.

16. Santos RAS, Campagnole-Santos MJ, Baracho NCV, Fontes MA, Silva LC, Neves LA, et al. Characterization of a new angiotensin antagonist selective for angiotensin-(1–7): evidence that the actions of angiotensin-(1–7) are mediated by specific angiotensin receptors. *Brain Res Bull* 1994;**35**:293–8.

17. Santos RAS, Baracho NC. Angiotensin-(1–7) is a potent antidiuretic peptide in rats. *Braz J Med Biol Res* 1992;**25**:651–4.

18. Garcia NH, Garvin JL. Angiotensin-(1–7) has a biphasic effect on fluid absorption in the proximal straight tubule. *J Am Soc Nephrol* 1994;**5**:1133–8.

19. Santos RAS, Simões e Silva AC, Magaldi AJ, Khosla MC, Cesar KR, Passaglio KT, et al. Evidence for a physiological role of angiotensin-(1–7) in the control of the hydroelectrolyte balance. *Hypertension* 1996;**27**:875–84.

20. Simões E Silva AC, Baracho NCV, Passaglio KT, Santos RAS. Renal actions of angiotensin-(1–7). *Braz J Med Biol Res* 1997;**30**:503–13.

21. Baracho NC, Simões e Silva AC, Khosla MC, Santos RAS. Effect of selective angiotensin antagonists on the antidiuresis produced by angiotensin-(1–7) in water-loaded rats. *Braz J Med Biol Res* 1998;**31**:1221–7.

22. Simões e Silva AC, Bello AP, Baracho NC, Khosla MC, Santos RA. Diuresis and natriuresis produced by long term administration of a selective angiotensin-(1–7) antagonist in normotensive and hypertensive rats. *Regul Pept* 1998;**74**:177–84.

23. Magaldi AJ, Cesar KR, Araujo M, Simões e Silva AC, Santos RA. Angiotensin-(1–7) stimulates water transport in rat inner medullary collecting duct: evidence for novel involvement of vasopressin V2 receptor. *Pflugers Arch* 2003;**447**:223–30.

24. Pinheiro SV, Simões e Silva AC, Sampaio WO, de Paula RD, Mendes EP, Bontempo ED, et al. The nonpeptide AVE 0991 is an angiotensin-(1–7) receptor Mas agonist in the mouse kidney. *Hypertension* 2004;**44**:490–6.

25. Santos RAS, Haibara AS, Campagnole-Santos MJ, Simões e Silva AC, de Paula RD, Pinheiro SV, et al. Characterization of a new selective antagonist for angiotensin-(1–7), D-pro7-angiotensin-(1–7). *Hypertension* 2003;**41**:737–43.

26. Ferreira AJ, Pinheiro SVB, Castro CH, Silva GA, Silva AC, Almeida AP, et al. Renal function in transgenic rats expressing an angiotensin-(1–7)-producing fusion protein. *Regul Pept* 2007;**137**:128–33.

27. Navar LG, Nishiyama A. Why are angiotensin concentrations so high in the kidney? *Curr Opin Nephrol Hypertens* 2004;**13**:107–15.

28. Brewster UC, Perazella MA. The renin-angiotensin-aldosterone system and the kidney: effects on kidney disease. *Am J Med* 2004;**116**:263–72.

29. Jaimes EA, Tian RX, Pearse D, Raij L. Up-regulation of glomerular COX-2 by angiotensin II: role of reactive oxygen species. *Kidney Int* 2005;**68**:2143–53.

30. Ren Y, Garvin JL, Carretero OA. Vasodilator action of angiotensin-(1–7) on isolated rabbit afferent arterioles. *Hypertension* 2002;**39**:799–802.

31. Sampaio WO, Nascimento AA, Santos RAS. Systemic and regional hemodynamic effects of angiotensin-(1–7) in rats. *Am J Physiol Heart Circ Physiol* 2003;**284**:H1985–94.

32. Stegbauer J, Oberhauser V, Vonend O, Rump LC. Angiotensin-(1–7) modulates vascular resistance and sympathetic neurotransmission in kidneys of spontaneously hypertensive rats. *Cardiovasc Res* 2004;**61**:352–9.

33. Roks AJ, Van Geel PP, Pinto YM, Buikema H, Henning RH, de Zeeuw D, et al. Angiotensin-(1–7) is a modulator of the human renin-angiotensin system. *Hypertension* 1999;**34**:296–301.

34. Mahon JM, Carr RD, Nicol AK, Henderson IW. Angiotensin-(1–7) is an antagonist at the type I angiotensin II receptor. *J Hypertens* 1994;**12**:1377–81.

35. Chansel D, Vandermeersch S, Oko A, Curat C, Ardaillou R. Effects of angiotensin IV and angiotensin-(1–7) on basal and angiotensin II-stimulated cytosolic Ca²⁺ in mesangial cells. *Eur J Pharmacol* 2001;**414**:165–75.

36. Oudot A, Vergely C, Ecarnot-Laubriet A, Rochette L. Pharmacological concentration of angiotensin-(1–7) activates NADPH oxidase after ischemia-reperfusion in rat heart through AT1 receptor stimulation. *Regul Pept* 2005;**127**:101–10.

37. Zhu Z, Zhong J, Zhu S, Liu D, Van Der Giet M, Tepel M. Angiotensin-(1–7) inhibits angiotensin II-induced signal transduction. *J Cardiovasc Pharmacol* 2002;**40**:693–700.

38. Clark MA, Tallant EA, Tommasi E, Bosch S, Diz DI. Angiotensin-(1–7) reduces renal angiotensin II receptors through a cyclooxygenase-dependent mechanism. *J Cardiovasc Pharmacol* 2003;**41**:276–83.

39. Kostenis E, Milligan G, Christopoulos A, Sanchez-Ferrer CF, Heringer-Walther S, Sexton PM, et al. G-protein-coupled receptor Mas is a physiological antagonist of the angiotensin II type 1 receptor. *Circulation* 2005;**111**:1806–13.

40. Barroso LC, Silveira KD, Lima CX, Borges V, Bader M, Rachid M, et al. Renoprotective effects of AVE0991, a nonpeptide Mas receptor agonist, in experimental acute renal injury. *Int J Hypertens* 2012;**2012**, Article ID 808726.

41. Giani JF, Munoz MC, Pons RA, Cao G, Toblli JE, Turyn D, et al. Angiotensin-(1–7) reduces proteinuria and diminishes structural damage in renal tissue of stroke-prone spontaneously hypertensive rats. *Am J Physiol Renal Physiol* 2011;**300**:F272–82.

42. Liu Z, Huang XR, Chen HY, Penninger JM, Lan HY. Loss of angiotensin-converting enzyme 2 enhances TGF-beta/Smad-mediated renal fibrosis and NF-kappaB-driven renal inflammation in a mouse model of obstructive nephropathy. *Lab Invest* 2012;**92**:650–61.

43. Pinheiro SV, Ferreira AJ, Kitten GT, da Silveira KD, da Silva DA, Santos SH, et al. Genetic deletion of the angiotensin-(1–7) receptor Mas leads to glomerular hyperfiltration and microalbuminuria. *Kidney Int* 2009;**75**:1184–93.

44. Silveira KD, Barroso LC, Vieira AT, Cisalpino D, Lima CX, Bader M, et al. Beneficial effects of the activation of the angiotensin-(1–7) Mas receptor in a murine model of adriamycin-induced nephropathy. *PLoS ONE* 2013;**8**, e66082.

45. Xue H, Zhou L, Yuan P, Wang Z, Ni J, Yao T, et al. Counteraction between angiotensin II and angiotensin-(1–7) via activating angiotensin type I and Mas receptor on rat renal mesangial cells. *Regul Pept* 2012;**177**:12–20.

46. Zhang J, Noble NA, Border WA, Huang Y. Infusion of angiotensin-(1–7) reduces glomerulosclerosis through counteracting angiotensin II in experimental glomerulonephritis. *Am J Physiol Renal Physiol* 2010;**298**:F579–88.

47. Li Y, Wu J, He Q, Shou Z, Zhang P, Pen W, et al. Angiotensin (1–7) prevent heart dysfunction and left ventricular remodeling caused by renal dysfunction in 5/6 nephrectomy mice. *Hypertens Res* 2009;**32**:369–74.

48. Benter IF, Yousif MH, Cojocel C, Al-Maghrebi M, Diz DI. Angiotensin-(1–7) prevents diabetes-induced cardiovascular dysfunction. *Am J Physiol Heart Circ Physiol* 2007;**292**:H666–72.

49. Benter IF, Yousif MH, Dhaunsi GS, Kaur J, Chappell MC, Diz DI. Angiotensin-(1–7) prevents activation of NADPH oxidase and renal vascular dysfunction in diabetic hypertensive rats. *Am J Nephrol* 2008;**28**:25–33.

50. Moon JY, Tanimoto M, Gohda T, Hagiwara S, Yamazaki T, Ohara I, et al. Attenuating effect of angiotensin-(1–7) on angiotensin II-mediated NAD(P)H oxidase activation in type 2 diabetic nephropathy of KK-A(y)/Ta mice. *Am J Physiol Renal Physiol* 2011;**300**:F1271–82.

51. Giani JF, Burghi V, Veiras LC, Tomat A, Munoz MC, Cao G, et al. Angiotensin-(1–7) attenuates diabetic nephropathy in Zucker diabetic fatty rats. *Am J Physiol Renal Physiol* 2012;**302**:F1606–15.

52. Mori J, Patel VB, Ramprasath T, Alrob OA, Desaulniers J, Scholey JW, et al. Angiotensin 1–7 mediates renoprotection against diabetic nephropathy by reducing oxidative stress, inflammation and lipotoxicity. *Am J Physiol Renal Physiol* 2014;**306**:F812–21.

53. Dilauro M, Zimpelmann J, Robertson SJ, Genest D, Burns KD. Effect of ACE2 and angiotensin-(1–7) in a mouse model of early chronic kidney disease. *Am J Physiol Renal Physiol* 2010;**298**:F1523–32.

54. Oudit GY, Herzenberg AM, Kassiri Z, Wong D, Reich H, Khokha R, et al. Loss of angiotensin-converting enzyme-2 leads to the late development of angiotensin II-dependent glomerulosclerosis. *Am J Pathol* 2006;**168**:1808–20.

55. Soler MJ, Wysocki J, Ye M, Lloveras J, Kanwar Y, Batlle D. ACE2 inhibition worsens glomerular injury in association with increased ACE expression in streptozotocin-induced diabetic mice. *Kidney Int* 2007;**72**:614–23.

56. Wong DW, Oudit GY, Reich H, Kassiri Z, Zhou J, Liu QC, et al. Loss of angiotensin-converting enzyme-2 (ACE2) accelerates diabetic kidney injury. *Am J Pathol* 2007;**171**:438–51.

57. Wysocki J, Ye M, Soler MJ, Gurley SB, Xiao HD, Bernstein KE, et al. ACE and ACE2 activity in diabetic mice. *Diabetes* 2007;**55**:2132–9.

58. Ye M, Wysocki J, William J, Soler MJ, Cokic I, Batlle D. Glomerular localization and expression of angiotensin-converting enzyme 2 and angiotensin-converting enzyme: implications for albuminuria in diabetes. *J Am Soc Nephrol* 2006;**17**:3067–75.

59. Silveira KD, Pompermayer Bosco KS, Diniz LR, Carmona AK, Cassali GD, Bruna-Romero O, et al. ACE2-angiotensin-(1–7)-Mas axis in renal ischaemia/reperfusion injury in rats. *Clin Sci (Lond)* 2010;**119**:385–94.

60. Lo CS, Liu F, Shi Y, Maachi H, Chenier I, Godin N, et al. Dual RAS blockade normalizes angiotensin-converting enzyme-2 expression and prevents hypertension and tubular apoptosis in Akita angiotensinogen-transgenic mice. *Am J Physiol Renal Physiol* 2012;**302**:F840–52.

61. Wysocki J, Ortiz-Melo DI, Mattocks NK, Xu K, Prescott J, Evora K, et al. ACE2 deficiency increases NADPH-mediated oxidative stress in the kidney. *Physiol Rep* 2014;**2**, e00264.

62. Ng HY, Yisireyili M, Saito S, Lee CT, Adelibieke Y, Nishijima F, et al. Indoxyl sulfate downregulates expression of Mas receptor via OAT3/AhR/Stat3 pathway in proximal tubular cells. *PLoS ONE* 2014;**9**, e91517.

63. Zimmerman D, Burns KD. Angiotensin-(1–7) in kidney disease: a review of the controversies. *Clin Sci (Lond)* 2012;**123**:333–46.

64. Esteban V, Heringer-Walther S, Sterner-Kock A, de Bruin R, van den Engel S, Wang Y, et al. Angiotensin-(1–7) and the g protein-coupled receptor MAS are key players in renal inflammation. *PLoS ONE* 2009;**4**, e5406.

65. Velkoska E, Dean RG, Griggs K, Burchill L, Burrell LM. Angiotensin-(1–7) infusion is associated with increased blood pressure and adverse cardiac remodelling in rats with subtotal nephrectomy. *Clin Sci (Lond)* 2011;**120**:335–45.

66. Su Z, Zimpelmann J, Burns KD. Angiotensin-(1–7) inhibits angiotensin II-stimulated phosphorylation of MAP kinases in proximal tubular cells. *Kidney Int* 2006;**69**:2212–18.

67. Zimpelmann J, Burns KD. Angiotensin-(1–7) activates growth-stimulatory pathways in human mesangial cells. *Am J Physiol Renal Physiol* 2009;**296**:F337–46.

68. Liu GC, Oudit GY, Fang F, Zhou J, Scholey JW. Angiotensin-(1–7)-induced activation of ERK1/2 is cAMP/protein kinase A-dependent in glomerular mesangial cells. *Am J Physiol Renal Physiol* 2012;**302**:F784–90.
69. Oudit GY, Liu GC, Zhong J, Basu R, Chow FL, Zhou J, et al. Human recombinant ACE2 reduces the progression of diabetic nephropathy. *Diabetes* 2010;**59**:529–38.
70. Azizi M, Menard J. Combined blockade of the renin-angiotensin system with angiotensin-converting enzyme inhibitors and angiotensin II type 1 receptor antagonists. *Circulation* 2004;**109**:2492–9.
71. Ferrario CM, Martell N, Yunis C, Flack JM, Chappell MC, Brosnihan KB, et al. Characterization of angiotensin-(1–7) in the urine of normal and essential hypertensive subjects. *Am J Hypertens* 1998;**11**:137–46.
72. Kocks MJ, Lely AT, Boomsma F, de Jong PE, Navis G. Sodium status and angiotensin-converting enzyme inhibition: effects on plasma angiotensin-(1–7) in healthy man. *J Hypertens* 2005;**23**:597–602.
73. Luque M, Martin P, Martell N, Fernandez C, Brosnihan KB, Ferrario CM. Effects of captopril related to increased levels of prostacyclin and angiotensin-(1–7) in essential hypertension. *J Hypertens* 1996;**14**:799–805.
74. Simoes e Silva AC, Diniz JS, Pereira RM, Pinheiro SV, Santos RA. Circulating renin angiotensin system in childhood chronic renal failure: marked increase of angiotensin-(1–7) in end-stage renal disease. *Pediatr Res* 2006;**60**:734–9.
75. Simoes e Silva AC, Diniz JS, Regueira Filho A, Santos RA. The renin angiotensin system in childhood hypertension: selective increase of angiotensin-(1–7) in essential hypertension. *J Pediatr* 2004;**145**:93–8.
76. Vilas-Boas WW, Ribeiro-Oliveira Jr A, Pereira RM, Ribeiro Rda C, Almeida J, Nadu AP, et al. Relationship between angiotensin-(1–7) and angiotensin II correlates with hemodynamic changes in human liver cirrhosis. *World J Gastroenterol* 2009;**15**:2512–19.
77. Mizuiri S, Hemmi H, Arita M, Aoki T, Ohashi Y, Miyagi M, et al. Increased ACE and decreased ACE2 expression in kidneys from patients with IgA nephropathy. *Nephron Clin Pract* 2011;**117**:c57–66.
78. Soler MJ, Riera M, Crespo M, Mir M, Márquez E, Pascula MJ, et al. Circulating angiotensin-converting enzyme 2 activity in kidney transplantation: a longitudinal pilot study. *Nephron Clin Pract* 2012;**121**:144–50.
79. Iwanami J, Mogi M, Tsukuda K, Wang XL, Nakaoka H, Ohshima K, et al. Role of angiotensin-converting enzyme 2/angiotensin-(1–7)/Mas axis in the hypotensive effect of azilsartan. *Hypertens Res* 2014;**37**(7):616–20.

Chapter 30

Mas and Inflammation

Lívia Corrêa Barroso*, Kátia Daniella Silveira‡, Mauro Martins Teixeira†, Ana Cristina Simões Silva‡

*Departamento de Bioquimica e Imunologia, UFMG, Belo Horizonte, Minas Gerais, Brazil, †Departamento de Bioquímica e Imunologia, Instituto de Ciências Biológicas, Belo Horizonte, Minas Gerais, Brazil, ‡Interdisciplinary Laboratory of Medical Investigation—LIIM, Federal University of Minas Gerais—UFMG, Belo Horizonte, Minas Gerais, Brazil

INTRODUCTION

The actions of the renin–angiotensin system (RAS) extend far beyond the regulation of the blood pressure and hydroelectrolyte balance.[1-3] The classical axis of the RAS, composed by angiotensin-converting enzyme (ACE), angiotensin (Ang) II, and Ang receptor type 1 (AT$_1$), is also known to have an important role in inflammation.[4,5] Many studies have shown that Ang II, acting via the AT$_1$ receptor, activates several cell functions and molecular signaling pathways related to tissue injury, inflammation, and fibrosis, including calcium mobilization, free radical generation, activation of protein kinases and nuclear transcription factors, recruitment of inflammatory cells, adhesion of monocytes and neutrophils to endothelial and mesangial cells, upregulation of adhesion molecules and stimulation of expression, and synthesis and release of cytokines.[6-11] On the other hand, the counterregulatory axis of the RAS, composed of angiotensin-converting enzyme 2 (ACE2), Ang-(1-7), and Mas receptor, may also influence inflammatory responses. Ang-(1-7), acting via receptor Mas, inhibits leukocyte migration, cytokine expression and release, macrophage activation, and fibrogenic pathways.[10,12-20]

In this chapter, we summarize recent findings on the anti-inflammatory and antifibrogenic role of ACE2/Ang-(1-7)/Mas axis in the context of experimental human diseases.

ACE2/ANG-(1-7)/MAS AXIS IN CYTOKINE REGULATION AND CELLULAR RECRUITMENT

Cytokine Regulation

The activation of the ACE2/Ang-(1-7)/Mas axis reduced the expression of proinflammatory and fibrogenic cytokines (tumor necrosis factor (TNF)-α, interleukin (IL)-1β, IL-6, monocyte chemotactic protein-1 (MCP-1), and transforming growth factor (TGF)-β) and increased the expression of the anti-inflammatory cytokine, IL-10, in a model of pulmonary hypertension.[21-23] In a murine model of asthma, Ang-(1-7) also attenuated leukocyte influx in airways spaces, perivascular and peribronchial inflammation, fibrosis, and goblet cell hyper-/metaplasia.[13] Using the same experimental model of asthma, Rodrigues-Machado et al. showed that an oral agonist of the Mas receptor, compound AVE 0991, prevented airway and pulmonary vascular remodeling, reduced IL-5 levels, and increased IL-10 levels in the bronchoalveolar fluid.[24]

In renal tissue, Barroso et al. showed that AVE 0991 reduced the circulating levels of the chemokine CXCL1 leading to less neutrophil accumulation in a murine model of renal ischemia and reperfusion.[12] More recently, Silveira et al. showed that oral administration of AVE 0991 improved renal function parameters and histological alterations and reduced urinary protein loss and renal levels of TGF-β in a model of adriamycin-induced nephropathy.[25] Ang(1-7) was also able to reduce oxidative stress; inflammatory markers, including IL-6, TNF-α, and endothelin (ED)-1; and renal fibrosis in Zucker diabetic fatty rats.[26]

Similarly, mice with genetic deletion of ACE2 had increased numbers of neutrophils, macrophages, and T cells and high levels of IL-1 and MCP-1.[27,28] In addition, the absence of ACE2 resulted in higher mRNA expression of MCP-1, IL-1β, and IL-6 and more aortic inflammation in response to Ang II infusion.[29] Accordingly, Souza et al. recently showed that Mas knockout mice produced higher levels of cytokines and exhibited higher mortality rates in response to lipopolysaccharide (LPS)-induced endotoxemia, whereas the treatment with Ang-(1-7) in LPS-induced wild-type animals inhibited the production of IL-6, IL-12, and CXCL1 and reduced hypothermia, thereby protecting mice from the fatal consequences of endotoxemia.[30]

The Protective Arm of the Renin–Angiotensin System (RAS). http://dx.doi.org/10.1016/B978-0-12-801364-9.00030-4

In the brain, infusion of Ang-(1-7) into the lateral ventricle of Sprague-Dawley rats subjected to permanent middle cerebral artery occlusion was associated with local decrease of oxidative stress, suppression of NF-κB activity, and reduction of proinflammatory cytokines.[14] Accordingly, it has been reported that the overexpression of ACE2 in paraventricular nucleus attenuated Ang II-induced expression of TNF-α, IL-1β, and IL-6.[17] In a rat model of cerebral ischemic stroke, Regenhardt et al.[20] showed that Ang-(1-7) reduces gene expression of IL-6, IL-1α, IL-1β, TNF-α, myeloperoxidase (MPO), MCP-1, chemokine receptor type 4 (CXCR4), and chemokine (C-X-C motif) ligand 12 (CXCL12).

Genetic ACE2 deficiency increased vascular inflammation and atherosclerosis in ApoE knockout mouse by increasing gene expression of vascular cell adhesion molecules (VCAM), cytokines, chemokines, and matrix metalloproteinases (MMPs).[18] Accordingly, Jin and colleagues showed that loss of ACE2 increases Ang II-induced mRNA expression of inflammatory cytokines and chemokines in the aorta.[15] Sukumaran and coworkers found that angiotensin receptor blockers (ARBs) increased ACE2, Ang-(1-7), and Mas expression in line with a reduction of proinflammatory cytokines (TNF-α, IFN-γ, IL-1β, and IL-6) and an elevation of the anti-inflammatory cytokine, IL-10, in a rat model of autoimmune myocarditis.[10,31] Al-Maghrebi et al. also showed that treatment with Ang-(1-7) prevents cardiac dysfunction in spontaneously hypertensive rats and reduces NF-κB activity and the expression of IL-1β, IL-6, Nalp12, and caspase 1 in response to acute ischemia-reperfusion injury.[32]

Cellular Recruitment and Function

Ang II contributes significantly to the process of leukocyte migration *in vivo* by modifying the interaction of leukocytes with endothelial cells, whereas experimental findings support an opposite effect of ACE2/Ang-(1-7)/Mas axis.[16,33–37] The administration of Ang-(1-7) or AVE 0991 decreased rolling and adhesion of leukocytes to the microvascular endothelium, neutrophil influx, hypernociception, and histological lesions at inflamed joints in a mouse model of antigen-induced arthritis.[16]

A few studies have reported that the ACE2/Ang-(1-7)/Mas axis decreases macrophage function.[18,19] Mas transcripts are upregulated in macrophages after LPS exposure, and Ang-(1-7) was able to regulate mRNA levels of proinflammatory cytokines (IL-6 and TNF-α) and to decrease the phosphorylation levels of Src kinases.[19] Macrophages isolated from the bone marrow of ACE2 knockout mice have increased expression of VCAM-1 and high levels of TNF-α, CCL2, and IL-6 following LPS administration.[18] Moreover, the overexpression of ACE2 inhibited the expression of MCP-1 induced by Ang II in macrophages, and this effect seemed to be mediated by increased levels of Ang-(1-7).[38]

ACE2/ANG-(1-7)/MAS AXIS AND FIBROGENIC PATHWAYS

Tissue Fibrosis

Fibrogenic and proliferative actions of the ACE/Ang II/AT$_1$ axis have been largely recognized, whereas many studies supported an antifibrogenic and antiproliferative role for the ACE2/Ang-(1-7)/Mas axis.[39–44]

Grobe et al. showed that the administration of Ang-(1-7) prevented cardiac fibrosis induced by Ang II in Sprague-Dawley rats.[45] In trained two-kidney one-clip hypertensive rats, the administration of Ang-(1-7) decreased fibrosis, and this effect was accompanied by upregulation of Mas, AT$_2$, and endothelial nitric oxide synthase phosphorylation in the heart.[46] ACE2-deficient mice, which have lower levels of Ang-(1-7), exhibited early cardiac hypertrophy[47] and adverse ventricular remodeling after myocardial infarction.[48,49] ACE2 deficiency also resulted in progressive cardiac fibrosis in models of aging and cardiac pressure overload,[49,50] while ACE2 overexpression reversed cardiac hypertrophy and fibrosis in mice.[51,52] Neonatal and adult Mas-deficient mice showed significantly higher levels of collagen types I and III and fibronectin and reduced levels of collagen IV in both the right ventricle and the AV valves.[53,54]

ACE2 mRNA expression and enzymatic activity were decreased in human idiopathic pulmonary fibrosis and in rat and mouse models of lung fibrosis induced by bleomycin.[55] Blockade of ACE2 in mice increased pulmonary levels of Ang II and enhanced bleomycin-induced lung fibrosis through a mechanism dependent on AT$_1$ receptors.[55] Similarly, the loss of ACE2 activity worsened liver fibrosis in chronic liver injury models,[56] and Mas deletion was associated with severe liver steatosis.[57]

TGF-β Signaling and Other Fibrogenic Pathways

The ACE2/Ang-(1-7)/Mas axis modifies the expression of TGF-β and of key components of the TGF-β pathway.[22,40,45,58,59] Ang-(1-7) decreased TGF-β mRNA levels in cultured cardiac fibroblasts,[40] reduced plasma levels of TGF-β in rat model of myocardial infarction,[45] and improved vascular remodeling via downregulation of TGF-β and inhibition of the Smad2 pathway.[59] Ang-(1-7) was able to counteract the signaling induced by Ang II in skeletal muscle by local decreasing of the expression and activity of TGF-β and by preventing early and late phosphorylation of p38 MAPK, Smad-2 phosphorylation, and Smad-4 nuclear translocation.[60] The administration of Ang-(1-7) protected damaged tissues in skeletal muscle of a Duchenne muscular dystrophy mouse model, improved muscle function, and decreased local fibrosis by the inhibition of TGF-β Smad 2/3 signaling.[28]

Ang-(1-7) also played a protective role in hepatic fibrosis since the pharmacological blockade of Mas receptor aggravated liver fibrosis with a significant elevation in TGF-β levels.[41] The oral formulation including Ang-(1-7) in hydroxypropyl β-cyclodextrin improved diastolic and systolic functions and reduced the expression of TGF-β and collagen type I in a rat model of myocardial infarction.[58] Intratracheal administration of lentiviruses expressing Ang-(1-7) or ACE2 induced overexpression of Ang-(1-7) in lungs and reduced lung fibrosis and TGF-β levels in rats given bleomycin.[22] Oral administration of Ang-(1-7) reduced fibrosis index, TGF-β expression, and Smad2/3 activation in lung injury induced by LPS.[61]

ACE2/Ang-(1-7)/Mas axis also regulates TGF-β-independent fibrogenic pathways. The molecular mechanisms that underlie the antiproliferative actions of Ang-(1-7) include stimulation of prostaglandins and cAMP production and inhibition of MAP kinases.[43] Ang-(1-7) caused inhibition of growth of human lung cells via reduction in serum-stimulated phosphorylation of ERK1 and ERK2.[39] In cultured vascular smooth muscle cells, Ang-(1-7) inhibited Ang II-activated protein kinase C and MAP kinases.[44] Similarly, in cultured human endothelial cells, Ang-(1-7) inhibited Ang II-stimulated phosphorylation of c-Src and ERK 1/2 and blunted Ang II-stimulated NADPH oxidase activity.[62] The treatment of cardiac fibroblasts with Ang-(1-7) reduced Ang II- or ET-1-stimulated increase in phospho-ERK1 and phospho-ERK2.[63] Moreover, Ang-(1-7) blocked the increase of cyclooxygenase 2 and prostaglandin synthase mRNAs in cardiac fibroblasts incubated with ET-1.[63] Treatment of cardiac fibroblasts with recombinant human ACE2 or ACE2 activators reduced collagen production and Ang II-mediated increase in phospho-MAP kinases.[64] Ang-(1-7) was able to inhibit the gene expression of IL-1β, IL-6, COX-2, and GFAP and activation of NF-κB and MAP kinases in primary cultures of rat astrocytes submitted to radiation.[65] Oral administration of Ang-(1-7) in rats receiving a high-fat diet prevented obesity and decreased liver proinflammatory cytokines (TNF-α and IL-6) by the downregulation of resistin/Toll-like receptor 4/NF-κB pathway.[66]

ACE2 deficiency increased tissue fibrosis, collagen accumulation, remodeling, hypertrophy, and macrophage inflammatory protein 1a (MIP1a) in the aorta and in cardiomyocytes of ACE2 knockout mice treated with Ang II. Accordingly, Mas-deficient mice also exhibited impairment of heart and renal function associated with changes in collagen expression toward a profibrotic profile.[67,68]

CONCLUDING REMARKS

Activation of the ACE2/Ang-(1-7)/Mas axis decreases inflammatory cell function and fibrogenesis in diverse models of human diseases. Further studies are clearly needed to elucidate the mechanisms by which this RAS axis modulates inflammation and fibrosis. Nevertheless, current knowledge supports the possibility that drugs that mimic or enhance the function of the ACE2/Ang-(1-7)/Mas axis may be beneficial for the treatment of human diseases with inflammatory and fibrotic components.

REFERENCES

1. Chappell MC. Nonclassical renin-angiotensin system and renal function. *Compr Physiol* 2012;**2**:2733–52.
2. Nabah YN, Mateo T, Estelles R, Mata M, Zagorski J, Sarau H, et al. Angiotensin II induces neutrophil accumulation in vivo through generation and release of CXC chemokines. *Circulation* 2004;**110**:3581–6.
3. Ferrario CM, Smith RD, Brosnihan B, Chappell MC, Campese VM, Vesterqvist O, et al. Effects of omapatrilat on the renin-angiotensin system in salt-sensitive hypertension. *Am J Hypertens* 2002;**15**:557–64.
4. Simoes e Silva AC, Silveira KD, Ferreira AJ, Teixeira MM. ACE2, angiotensin-(1–7) and Mas receptor axis in inflammation and fibrosis. *Br J Pharmacol* 2013;**169**:477–92.
5. Chappell MC. Angiotensin-converting enzyme 2 autoantibodies: further evidence for a role of the renin-angiotensin system in inflammation. *Arthritis Res Ther* 2010;**12**:128.
6. Silveira KD, Coelho FM, Vieira AT, Barroso LC, Queiroz-Junior CM, Costa VV, et al. Mechanisms of the anti-inflammatory actions of the angiotensin type 1 receptor antagonist losartan in experimental models of arthritis. *Peptides* 2013;**46**:53–63.
7. Capettini LS, Montecucco F, Mach F, Stergiopulos N, Santos RA, da Silva RF. Role of renin-angiotensin system in inflammation, immunity and aging. *Curr Pharm Des* 2012;**18**:963–70.
8. Ma TK, Kam KK, Yan BP, Lam YY. Renin-angiotensin-aldosterone system blockade for cardiovascular diseases: current status. *Br J Pharmacol* 2010;**160**:1273–92.
9. Ruiz-Ortega M, Lorenzo O, Suzuki Y, Ruperez M, Egido J. Proinflammatory actions of angiotensins. *Curr Opin Nephrol Hypertens* 2001;**10**:321–9.
10. Sukumaran V, Veeraveedu PT, Gurusamy N, Lakshmanan AP, Yamaguchi K, Ma M, et al. Telmisartan acts through the modulation of ACE-2/Ang 1–7/Mas receptor in rats with dilated cardiomyopathy induced by experimental autoimmune myocarditis. *Life Sci* 2012;**90**:289–300.

11. Suzuki Y, Ruiz-Ortega M, Lorenzo O, Ruperez M, Esteban V, Egido J. Inflammation and angiotensin II. *Int J Biochem Cell Biol* 2003;**35**:881–900.
12. Barroso LC, Silveira KD, Lima CX, Borges V, Bader M, Rachid M, et al. Renoprotective effects of AVE0991, a nonpeptide Mas receptor agonist, in experimental acute renal injury. *Int J Hypertens* 2012;**2012**, 808726.
13. El-Hashim AZ, Renno WM, Raghupathy R, Abduo HT, Akhtar S, Benter IF. Angiotensin-(1–7) inhibits allergic inflammation, via the MAS1 receptor, through suppression of ERK1/2- and NF-kappaB-dependent pathways. *Br J Pharmacol* 2012;**166**:1964–76.
14. Jiang T, Gao L, Guo J, Lu J, Wang Y, Zhang Y. Suppressing inflammation by inhibiting the NF-kappaB pathway contributes to the neuroprotective effect of angiotensin-(1–7) in rats with permanent cerebral ischaemia. *Br J Pharmacol* 2012;**167**:1520–32.
15. Jin HY, Song B, Oudit GY, Davidge ST, Yu HM, Jiang YY, et al. ACE2 deficiency enhances angiotensin II-mediated aortic profilin-1 expression, inflammation and peroxynitrite production. *PLoS ONE* 2012;**7**, e38502.
16. da Silveira KD, Coelho FM, Vieira AT, Sachs D, Barroso LC, Costa VV, et al. Anti-inflammatory effects of the activation of the angiotensin-(1–7) receptor, MAS, in experimental models of arthritis. *J Immunol* 2010;**185**:5569–76.
17. Sriramula S, Cardinale JP, Lazartigues E, Francis J. ACE2 overexpression in the paraventricular nucleus attenuates angiotensin II-induced hypertension. *Cardiovasc Res* 2011;**92**:401–8.
18. Thomas MC, Pickering RJ, Tsorotes D, Koitka A, Sheehy K, Bernardi S, et al. Genetic ACE2 deficiency accentuates vascular inflammation and atherosclerosis in the ApoE knockout mouse. *Circ Res* 2010;**107**:888–97.
19. Souza LL, Costa-Neto CM. Angiotensin-(1–7) decreases LPS-induced inflammatory response in macrophages. *J Cell Physiol* 2012;**227**:2117–22.
20. Regenhardt RW, Mecca AP, Desland F, Ritucci-Chinni PF, Ludin JA, Greenstein D, et al. Centrally administered angiotensin-(1–7) increases the survival of stroke-prone spontaneously hypertensive rats. *Exp Physiol* 2014;**99**:442–53.
21. Ferreira AJ, Shenoy V, Yamazato Y, Sriramula S, Francis J, Yuan L, et al. Evidence for angiotensin-converting enzyme 2 as a therapeutic target for the prevention of pulmonary hypertension. *Am J Respir Crit Care Med* 2009;**179**:1048–54.
22. Shenoy V, Ferreira AJ, Qi Y, Fraga-Silva RA, Diez-Freire C, Dooies A, et al. The angiotensin-converting enzyme 2/angiogenesis-(1–7)/Mas axis confers cardiopulmonary protection against lung fibrosis and pulmonary hypertension. *Am J Respir Crit Care Med* 2010;**182**:1065–72.
23. Yamazato Y, Ferreira AJ, Hong KH, Sriramula S, Francis J, Yamazato M, et al. Prevention of pulmonary hypertension by angiotensin-converting enzyme 2 gene transfer. *Hypertension* 2009;**54**:365–71.
24. Rodrigues-Machado MG, Magalhaes GS, Cardoso JA, Kangussu LM, Murari A, Caliari MV, et al. AVE 0991, a non-peptide mimic of angiotensin-(1–7) effects, attenuates pulmonary remodelling in a model of chronic asthma. *Br J Pharmacol* 2013;**170**:835–46.
25. Silveira KD, Barroso LC, Vieira AT, Cisalpino D, Lima CX, Bader M, et al. Beneficial effects of the activation of the angiotensin-(1–7) MAS receptor in a murine model of adriamycin-induced nephropathy. *PLoS ONE* 2013;**8**, e66082.
26. Giani JF, Burghi V, Veiras LC, Tomat A, Munoz MC, Cao G, et al. Angiotensin-(1–7) attenuates diabetic nephropathy in Zucker diabetic fatty rats. *Am J Physiol Renal Physiol* 2012;**302**:F1606–15.
27. Fang F, Liu GC, Zhou X, Yang S, Reich HN, Williams V, et al. Loss of ACE2 exacerbates murine renal ischemia-reperfusion injury. *PLoS ONE* 2013;**8**, e71433.
28. Acuña MJ, Pessina P, Olguin H, Cabrera D, Vio CP, Bader M, et al. Restoration of muscle strength in dystrophic muscle by angiotensin-1-7 through inhibition of TGF-β signalling. *Hum Mol Genet* 2014;**23**(5):1237–49.
29. Zhong JC, Yu XY, Lin QX, Li XH, Huang XZ, Xiao DZ, et al. Enhanced angiotensin converting enzyme 2 regulates the insulin/Akt signalling pathway by blockade of macrophage migration inhibitory factor expression. *Br J Pharmacol* 2008;**153**:66–74.
30. Souza LL, Duchene J, Todiras M, Azevedo LC, Costa-Neto CM, Alenina N, et al. Receptor MAS protects mice against hypothermia and mortality induced by endotoxemia. *Shock* 2014;**41**:331–6.
31. Sukumaran V, Veeraveedu PT, Gurusamy N, Yamaguchi K, Lakshmanan AP, Ma M, et al. Cardioprotective effects of telmisartan against heart failure in rats induced by experimental autoimmune myocarditis through the modulation of angiotensin-converting enzyme-2/angiotensin 1–7/mas receptor axis. *Int J Biol Sci* 2011;**7**:1077–92.
32. Al-Maghrebi M, Benter IF, Diz DI. Endogenous angiotensin-(1–7) reduces cardiac ischemia-induced dysfunction in diabetic hypertensive rats. *Pharmacol Res* 2009;**59**:263–8.
33. Alvarez A, Piqueras L, Bello R, Canet A, Moreno L, Kubes P, et al. Angiotensin II is involved in nitric oxide synthase and cyclo-oxygenase inhibition-induced leukocyte-endothelial cell interactions in vivo. *Br J Pharmacol* 2001;**132**:677–84.
34. Company C, Piqueras L, Naim Abu NY, Escudero P, Blanes JI, Jose PJ, et al. Contributions of ACE and mast cell chymase to endogenous angiotensin II generation and leucocyte recruitment in vivo. *Cardiovasc Res* 2011;**92**:48–56.
35. Mateo T, Abu Nabah YN, Abu Nabah YN, Abu TM, Mata M, Cerda-Nicolas M, et al. Angiotensin II-induced mononuclear leukocyte interactions with arteriolar and venular endothelium are mediated by the release of different CC chemokines. *J Immunol* 2006;**176**:5577–86.
36. Abu Nabah YN, Losada M, Estelles R, Mateo T, Company C, Piqueras L, et al. CXCR2 blockade impairs angiotensin II-induced CC chemokine synthesis and mononuclear leukocyte infiltration. *Arterioscler Thromb Vasc Biol* 2007;**27**:2370–6.
37. Piqueras L, Kubes P, Alvarez A, O'Connor E, Issekutz AC, Esplugues JV, et al. Angiotensin II induces leukocyte-endothelial cell interactions in vivo via AT(1) and AT(2) receptor-mediated P-selectin upregulation. *Circulation* 2000;**102**:2118–23.
38. Guo YJ, Li WH, Wu R, Xie Q, Cui LQ. ACE2 overexpression inhibits angiotensin II-induced monocyte chemoattractant protein-1 expression in macrophages. *Arch Med Res* 2008;**39**:149–54.
39. Gallagher PE, Tallant EA. Inhibition of human lung cancer cell growth by angiotensin-(1–7). *Carcinogenesis* 2004;**25**:2045–52.
40. Iwata M, Cowling RT, Gurantz D, Moore C, Zhang S, Yuan JX, et al. Angiotensin-(1–7) binds to specific receptors on cardiac fibroblasts to initiate antifibrotic and antitrophic effects. *Am J Physiol Heart Circ Physiol* 2005;**289**:H2356–63.

41. Pereira RM, dos Santos RA, Teixeira MM, Leite VH, Costa LP, da Costa Dias FL, et al. The renin-angiotensin system in a rat model of hepatic fibrosis: evidence for a protective role of angiotensin-(1–7). *J Hepatol* 2007;**46**:674–81.

42. Su Z, Zimpelmann J, Burns KD. Angiotensin-(1–7) inhibits angiotensin II-stimulated phosphorylation of MAP kinases in proximal tubular cells. *Kidney Int* 2006;**69**:2212–18.

43. Tallant EA, Clark MA. Molecular mechanisms of inhibition of vascular growth by angiotensin-(1–7). *Hypertension* 2003;**42**:574–9.

44. Tallant EA, Ferrario CM, Gallagher PE. Angiotensin-(1–7) inhibits growth of cardiac myocytes through activation of the mas receptor. *Am J Physiol Heart Circ Physiol* 2005;**289**:H1560–6.

45. Grobe JL, Mecca AP, Lingis M, Shenoy V, Bolton TA, Machado JM, et al. Prevention of angiotensin II-induced cardiac remodeling by angiotensin-(1–7). *Am J Physiol Heart Circ Physiol* 2007;**292**:H736–42.

46. Shah A, Oh YB, Lee SH, Lim JM, Kim SH. Angiotensin-(1–7) attenuates hypertension in exercise-trained renal hypertensive rats. *Am J Physiol Heart Circ Physiol* 2012;**302**:H2372–80.

47. Oudit GY, Kassiri Z, Patel MP, Chappell M, Butany J, Backx PH, et al. Angiotensin II-mediated oxidative stress and inflammation mediate the age-dependent cardiomyopathy in ACE2 null mice. *Cardiovasc Res* 2007;**75**:29–39.

48. Kassiri Z, Zhong J, Guo D, Basu R, Wang X, Liu PP, et al. Loss of angiotensin-converting enzyme 2 accelerates maladaptive left ventricular remodeling in response to myocardial infarction. *Circ Heart Fail* 2009;**2**:446–55.

49. Nakamura K, Koibuchi N, Nishimatsu H, Higashikuni Y, Hirata Y, Kugiyama K, et al. Candesartan ameliorates cardiac dysfunction observed in angiotensin-converting enzyme 2-deficient mice. *Hypertens Res* 2008;**31**:1953–61.

50. Yamamoto K, Ohishi M, Katsuya T, Ito N, Ikushima M, Kaibe M, et al. Deletion of angiotensin-converting enzyme 2 accelerates pressure overload-induced cardiac dysfunction by increasing local angiotensin II. *Hypertension* 2006;**47**:718–26.

51. Der SS, Grobe JL, Yuan L, Narielwala DR, Walter GA, Katovich MJ, et al. Cardiac overexpression of angiotensin converting enzyme 2 protects the heart from ischemia-induced pathophysiology. *Hypertension* 2008;**51**:712–18.

52. Huentelman MJ, Grobe JL, Vazquez J, Stewart JM, Mecca AP, Katovich MJ, et al. Protection from angiotensin II-induced cardiac hypertrophy and fibrosis by systemic lentiviral delivery of ACE2 in rats. *Exp Physiol* 2005;**90**:783–90.

53. Gava E, de Castro CH, Ferreira AJ, Colleta H, Melo MB, Alenina N, et al. Angiotensin-(1–7) receptor Mas is an essential modulator of extracellular matrix protein expression in the heart. *Regul Pept* 2012;**175**:30–42.

54. Uhal BD, Li X, Piasecki CC, Molina-Molina M. Angiotensin signalling in pulmonary fibrosis. *Int J Biochem Cell Biol* 2012;**44**:465–8.

55. Li X, Molina-Molina M, Abdul-Hafez A, Uhal V, Xaubet A, Uhal BD. Angiotensin converting enzyme-2 is protective but downregulated in human and experimental lung fibrosis. *Am J Physiol Lung Cell Mol Physiol* 2008;**295**:L178–85.

56. Osterreicher CH, Taura K, De Minicis S, Seki E, Penz-Osterreicher M, Kodama Y, et al. Angiotensin-converting-enzyme 2 inhibits liver fibrosis in mice. *Hepatology* 2009;**50**:929–38.

57. Silva AR, Aguilar EC, Alvarez-Leite JI, da Silva RF, Arantes RM, Bader M, et al. Mas receptor deficiency is associated with worsening of lipid profile and severe hepatic steatosis in ApoE-knockout mice. *Am J Physiol Regul Integr Comp Physiol* 2013;**305**:R1323–30.

58. Marques FD, Melo MB, Souza LE, Irigoyen MC, Sinisterra RD, De Sousa FB, et al. Beneficial effects of long-term administration of an oral formulation of angiotensin-(1–7) in infarcted rats. *Int J Hypertens* 2012;**2012**, 795452.

59. Zeng W, Chen W, Leng X, He JG, Ma H. Chronic angiotensin-(1–7) administration improves vascular remodeling after angioplasty through the regulation of the TGF-beta/Smad signaling pathway in rabbits. *Biochem Biophys Res Commun* 2009;**389**:138–44.

60. Morales MG, Abrigo J, Meneses C, Simon F, Cisternas F, Rivera JC, et al. Angiotensin 1-7/Mas-1 axis attenuates the expression and signaling of TGF-beta1 induced by Angiotensin II in skeletal muscle. *Clin Sci (Lond)* 2014;**127**(4):251–64.

61. Chen Q, Yang Y, Huang Y, Pan C, Liu L, Qiu H. Angiotensin-(1–7) attenuates lung fibrosis by way of Mas receptor in acute lung injury. *J Surg Res* 2013;**185**:740–7.

62. Sampaio WO, de Castro CH, Santos RA, Schiffrin EL, Touyz RM. Angiotensin-(1–7) counterregulates angiotensin II signaling in human endothelial cells. *Hypertension* 2007;**50**:1093–8.

63. McCollum LT, Gallagher PE, Ann TE. Angiotensin-(1–7) attenuates angiotensin II-induced cardiac remodeling associated with upregulation of dual-specificity phosphatase 1. *Am J Physiol Heart Circ Physiol* 2012;**302**:H801–10.

64. Zhong J, Basu R, Guo D, Chow FL, Byrns S, Schuster M, et al. Angiotensin-converting enzyme 2 suppresses pathological hypertrophy, myocardial fibrosis, and cardiac dysfunction. *Circulation* 2010;**122**:717–28.

65. Moore ED, Kooshki M, Metheny-Barlow LJ, Gallagher PE, Robbins ME. Angiotensin-(1–7) prevents radiation-induced inflammation in rat primary astrocytes through regulation of MAP kinase signaling. *Free Radic Biol Med* 2013;**65**:1060–8.

66. Santos SH, Andrade JM, Fernandes LR, Sinisterra RD, Sousa FB, Feltenberger JD, et al. Oral angiotensin-(1–7) prevented obesity and hepatic inflammation by inhibition of resistin/TLR4/MAPK/NF-kappaB in rats fed with high-fat diet. *Peptides* 2013;**46**:47–52.

67. Santos RA, Castro CH, Gava E, Pinheiro SV, Almeida AP, Paula RD, et al. Impairment of in vitro and in vivo heart function in angiotensin-(1–7) receptor MAS knockout mice. *Hypertension* 2006;**47**:996–1002.

68. Pinheiro SV, Ferreira AJ, Kitten GT, da Silveira KD, da Silva DA, Santos SH, et al. Genetic deletion of the angiotensin-(1–7) receptor Mas leads to glomerular hyperfiltration and microalbuminuria. *Kidney Int* 2009;**75**:1184–93.

Chapter 31

ACE2 and Glycemic Control

Harshita Chodavarapu, Eric Lazartigues

Department of Pharmacology, Louisiana State University Health Sciences Center, New Orleans, Louisiana, USA

ABBREVIATIONS

T1DM	type 1 diabetes mellitus
T2DM	type 2 diabetes mellitus
IDDM	insulin-dependent diabetes mellitus
NIDDM	non-insulin-dependent diabetes mellitus
RAS	renin–angiotensin system
Ang-II	angiotensin II
ACE	angiotensin-converting enzyme
ACE2	angiotensin-converting enzyme type 2
Ang-(1-7)	angiotensin-(1-7)
ADAM17	a disintegrin and metalloprotease 17
HbA1C	hemoglobin A1C
ACEI	ACE inhibitor
AT$_1$R	angiotensin II type 1 receptor and
ARB	angiotensin II type 1 receptor blocker

One of the important physiological pathways that play a pivotal role in the pathogenesis of type 2 diabetes mellitus (T2DM) is the renin–angiotensin system (RAS). Identification of a local pancreatic RAS has led to a better understanding of the role of the RAS in diabetic pathophysiology. Angiotensin II (Ang-II)-induced impairment in insulin sensitivity, insulin secretion, and β-cell function was reversed by various RAS blockers.[1–4] However, the efficacy of RAS blockers has been under scrutiny and is an ongoing topic of debate because of the reports showing the inefficiency of RAS blockers in controlling hyperglycemic symptoms.[5,6]

ACE2/ANG-(1-7)/MAS: ROLE IN GLYCEMIC CONTROL

Angiotensin-converting enzyme type 2 (ACE2) has garnered significant attention over the past several years and is a promising target for its beneficial role in glycemic control. ACE2 was discovered in 2000 and shares 42% sequence homology with angiotensin-converting enzyme (ACE) but cannot be inhibited by ACE inhibitors (ACEI).[7,8] ACE2 is located on the X chromosome[8,9] and cleaves various substrates including an octapeptide Ang-II, angiotensin I (Ang-I), apelin, neurotensin, and des-Arg bradykinin with the highest catalytic efficiency towards Ang-II.[7,8,10] Angiotensin-(1-7) (Ang-(1-7)) generated from Ang-II, on the other hand, opposes the actions of the octapeptide. Of late, the protective axis of RAS "ACE2/Ang-(1-7)/Mas" has gained momentum because of its beneficial role in the pathophysiology of diabetes[11,12] and its related complications.[13–16]

ACE2 overexpression has been shown to reverse the detrimental phenotypes in cardiovascular disease,[15,17] diabetes,[11,18] and its related complications, an effect known to occur by suppressing the overactive Ang-II levels.[19] The plausible beneficial effects of ACE2 can be attributed to its capacity to enhance Ang-(1-7) signaling. Ang-(1-7) improved insulin sensitivity and glucose tolerance in experimental animal models, possibly by improving the insulin signaling via its receptor Mas.[20,21] While deficiency of Mas impairs glucose and lipid metabolism,[22] an increase in circulating Ang-(1-7) levels ameliorates it.[23] Under nonpathological conditions, loss of ACE2 results in progressive decline in first-phase insulin secretion and glucose tolerance as the mice age.[24] However, in pathological conditions, a high-fat high-sucrose diet resulted in a greater degree of glucose intolerance in mice lacking ACE2 compared with controls.[25] This effect was attributed to the reduced skeletal muscle levels of GLUT4 and myocyte enhancer factor 2A expression and was reversed upon administration of Ang-(1-7).[25] These suggest the importance of Ang-(1-7) signaling in maintaining glucose tolerance and insulin sensitivity and hence the requirement of ACE2. ACE2 levels decrease as the disease progresses, leading to an uninhibited rise in RAS activity.[11,26] Thus, it can be speculated that, as the Ang-II levels increase in hyperglycemic state, ACE2 is also upregulated as a compensatory mechanism and aids in the degradation and reduction of Ang-II. The importance of the alternative "ACE2/Ang-(1-7)/Mas" axis in maintaining β-cell function has been investigated *in vitro*, and the data clearly highlight an upregulation in the expression of ACE2 and Mas in association with an increase in insulin secretion at high glucose concentrations.[27] Moreover, ACE2 gene therapy and ACE2 activators, such as xanthenone and diminazene aceturate,[28] employed to increase ACE2 activity, proved beneficial in the face of diabetes[11,29] and its complications.[14,30]

Various mechanisms have been proposed via which ACE2 elicits its opposing effects on Ang-II signaling. Oxidative stress has been reported to be one of the predisposing factors of β-cell dysfunction during the hyperglycemic state[31,32] and is activated by Ang-II.[33] ACE2 overexpression reduced oxidative stress and corrected Ang-II-induced imbalance in AT$_1$R/ACE2 expression, thus improving glycemic control.[18] Pharmacological inhibition of ACE2 and Ang-(1-7) worsened Ang-II-mediated formation of ROS,[34] further supporting the fact that ACE2 by converting Ang-II to Ang-(1-7) decreases the capacity of Ang-II to form ROS.

Endoplasmic reticulum (ER) stress is another contributing factor towards hyperglycemia as evidenced by apoptotic β-cell death.[35] In addition to oxidative stress and ER stress, other potential mechanisms such as fibrosis and inflammation also play a role in impaired glucose homeostasis.[36,37] While loss of ACE2 has been reported to exacerbate fibrosis and inflammation in the kidney[38,39] and heart,[40] overexpression of ACE2 decreased fibrosis in the heart,[41] lungs[42], and pancreas.[43] Moreover, Ang-(1-7) proved beneficial in improving insulin resistance via its anti-inflammatory properties in the liver.[44] Surprisingly, this was also associated with an upregulation in ACE2 expression in the liver. However, there are not many reports demonstrating the role of ACE2/Ang-(1-7)/Mas in preventing fibrosis in islets.[43] Altogether, a role for ACE2 in modulating ER stress, fibrosis, and inflammation in the islets warrants further investigation. Despite having beneficial effects, ACE2 expression and activity goes down over time as the disease progresses.[11,18,45] This imbalance in ACE/Ang-II/AT$_1$R to ACE2/Ang(1-7)/Mas ratio results in loss of the compensatory mechanism offered by the protective axis of the RAS in diabetic disease pathophysiology (Figure 1).

ADAM17: A POTENTIAL MECHANISM FOR ACE2 SHEDDING?

The mechanism behind ACE2 depletion is "ACE2 shedding" that is a result of cleaving of ectodomain of ACE2 by proteases such as ADAM17 (a disintegrin and metalloprotease 17)[46,47] and TMPRSS2[48] into the extracellular milieu. ADAM17 upregulation[13,49,50] is accompanied by ACE2 shedding in several diseases,[13,50] and inhibition of ADAM17 prevented shedding of its substrates, including ACE2.[51,52] ADAM17, also called TNFα-converting enzyme (TACE), is a sheddase that is localized to islets of pancreas.[53] This coupled with available evidence on ACE2 localization[54] to islets further supports the notion that ADAM17 is the sheddase responsible for ACE2 shedding in pancreatic islets (Figure 2).

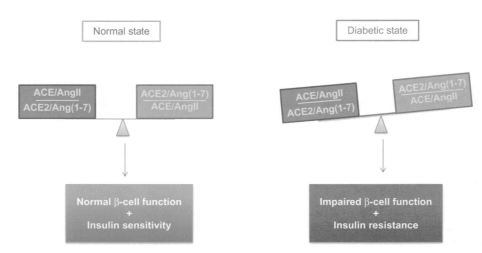

FIGURE 1   Schematic representing the significance of ACE2 in maintaining β-cell function and insulin sensitivity.

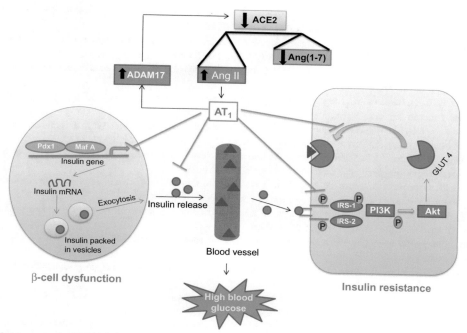

FIGURE 2   Schematic of Ang-II-induced ACE2 depletion and resultant impaired glucose homeostasis.

Ang-II has been shown to be a key player in initiating the phosphorylation and activation of TACE in several tissues[55,56] including the pancreas.[18] Based on current knowledge, it is tempting to speculate that hyperglycemia induces upregulation of Ang-II and Ang-II via the AT_1R phosphorylates and activates ADAM17, which then elicits ACE2 shedding. Thus, ADAM17-mediated ACE2 shedding could be one possible mechanism responsible for the loss of ACE2 and the accompanying deleterious effects observed during the late phase of T2DM (Figure 2).

PERSPECTIVE

Based on the current state of knowledge, it can be concluded that coordinated and regulated interplay between ACE2, Ang-(1-7), and Mas signaling is required to keep the Ang-II-induced harmful prodiabetic symptoms in check. It will also be interesting to know whether the experimental findings can actually be translated from bench to bed side. Hence, further investigation is required to check the efficacy and potency of ACE2 and Mas receptor activators and agonists in a clinical setting. Lastly, the ACE2/Ang-(1-7)/Mas axis could potentially revolutionize and replace years-long usage of traditional ACEI and ARB in the treatment of metabolic syndrome owing to its plethora of beneficial effects.

REFERENCES

1. Lau T, Carlsson PO, Leung PS. Evidence for a local angiotensin-generating system and dose-dependent inhibition of glucose-stimulated insulin release by angiotensin II in isolated pancreatic islets. *Diabetologia* 2004;**47**:240–8.
2. Wei Y, Sowers JR, Clark SE, Li W, Ferrario CM, Stump CS. Angiotensin II-induced skeletal muscle insulin resistance mediated by NF-kappaB activation via NADPH oxidase. *Am J Physiol Endocrinol Metab* 2008;**294**:E345–51.
3. Madec AM, Cassel R, Dubois S, Ducreux S, Vial G, Chauvin MA, et al. Losartan, an angiotensin II type 1 receptor blocker, protects human islets from glucotoxicity through the phospholipase C pathway. *FASEB J* 2013;**27**:5122–30.
4. Wang HW, Mizuta M, Saitoh Y, Noma K, Ueno H, Nakazato M. Glucagon-like peptide-1 and candesartan additively improve glucolipotoxicity in pancreatic beta-cells. *Metabolism* 2011;**60**:1081–9.
5. Bokhari S, Israelian Z, Schmidt J, Brinton E, Meyer C. Effects of angiotensin II type 1 receptor blockade on beta-cell function in humans. *Diabetes Care* 2007;**30**:181.
6. Dahlof B, Devereux RB, Kjeldsen SE, Julius S, Beevers G, de Faire U, et al. Cardiovascular morbidity and mortality in the Losartan Intervention For Endpoint reduction in hypertension study (LIFE): a randomised trial against atenolol. *Lancet* 2002;**359**:995–1003.
7. Donoghue M, Hsieh F, Baronas E, Godbout K, Gosselin M, Stagliano N, et al. A novel angiotensin-converting enzyme-related carboxypeptidase (ACE2) converts angiotensin I to angiotensin 1-9. *Circ Res* 2000;**87**:E1–9.

8. Tipnis SR, Hooper NM, Hyde R, Karran E, Christie G, Turner AJ. A human homolog of angiotensin-converting enzyme. Cloning and functional expression as a captopril-insensitive carboxypeptidase. *J Biol Chem* 2000;**275**:33238–43.

9. Komatsu T, Suzuki Y, Imai J, Sugano S, Hida M, Tanigami A, et al. Molecular cloning, mRNA expression and chromosomal localization of mouse angiotensin-converting enzyme-related carboxypeptidase (mACE2). *DNA Seq* 2002;**13**:217–20.

10. Vickers C, Hales P, Kaushik V, Dick L, Gavin J, Tang J, et al. Hydrolysis of biological peptides by human angiotensin-converting enzyme-related carboxypeptidase. *J Biol Chem* 2002;**277**:14838–43.

11. Bindom SM, Hans CP, Xia H, Boulares AH, Lazartigues E. Angiotensin-I converting enzyme type 2 (ACE2) gene therapy improves glycemic control in diabetic mice. *Diabetes* 2010;**59**:2540–8.

12. Chhabra KH, Chodavarapu H, Lazartigues E. Angiotensin converting enzyme 2: a new important player in the regulation of glycemia. *IUBMB Life* 2013;**65**:731–8.

13. Chodavarapu H, Grobe N, Somineni HK, Salem ES, Madhu M, Elased KM. Rosiglitazone treatment of type 2 diabetic db/db mice attenuates urinary albumin and angiotensin converting enzyme 2 excretion. *PLoS ONE* 2013;**8**, e62833.

14. Murca TM, Moraes PL, Capuruco CA, Santos SH, Melo MB, Santos RA, et al. Oral administration of an angiotensin-converting enzyme 2 activator ameliorates diabetes-induced cardiac dysfunction. *Regul Pept* 2012;**177**:107–15.

15. Zhao YX, Yin HQ, Yu QT, Qiao Y, Dai HY, Zhang MX, et al. ACE2 overexpression ameliorates left ventricular remodeling and dysfunction in a rat model of myocardial infarction. *Hum Gene Ther* 2010;**21**:1545–54.

16. Oudit GY, Liu GC, Zhong J, Basu R, Chow FL, Zhou J, et al. Human recombinant ACE2 reduces the progression of diabetic nephropathy. *Diabetes* 2010;**59**:529–38.

17. Huentelman MJ, Grobe JL, Vazquez J, Stewart JM, Mecca AP, Katovich MJ, et al. Protection from angiotensin II-induced cardiac hypertrophy and fibrosis by systemic lentiviral delivery of ACE2 in rats. *Exp Physiol* 2005;**90**:783–90.

18. Chhabra KH, Xia H, Pedersen KB, Speth RC, Lazartigues E. Pancreatic angiotensin-converting enzyme 2 improves glycemia in angiotensin II-infused mice. *Am J Physiol Endocrinol Metab* 2013;**304**:E874–84.

19. Feng Y, Yue X, Xia H, Bindom SM, Hickman PJ, Filipeanu CM, et al. Angiotensin-converting enzyme 2 overexpression in the subfornical organ prevents the angiotensin II-mediated pressor and drinking responses and is associated with angiotensin II type 1 receptor downregulation. *Circ Res* 2008;**102**:729–36.

20. Giani JF, Mayer MA, Munoz MC, Silberman EA, Hocht C, Taira CA, et al. Chronic infusion of angiotensin-(1-7) improves insulin resistance and hypertension induced by a high-fructose diet in rats. *Am J Physiol Endocrinol Metab* 2009;**296**:E262–71.

21. Prasannarong M, Santos FR, Henriksen EJ. ANG-(1-7) reduces ANG II-induced insulin resistance by enhancing Akt phosphorylation via a Mas receptor-dependent mechanism in rat skeletal muscle. *Biochem Biophys Res Commun* 2012;**426**:369–73.

22. Santos SH, Fernandes LR, Mario EG, Ferreira AV, Porto LC, Alvarez-Leite JI, et al. Mas deficiency in FVB/N mice produces marked changes in lipid and glycemic metabolism. *Diabetes* 2008;**57**:340–7.

23. Santos SH, Braga JF, Mario EG, Porto LC, Rodrigues-Machado Mda G, Murari A, et al. Improved lipid and glucose metabolism in transgenic rats with increased circulating angiotensin-(1-7). *Arterioscler Thromb Vasc Biol* 2010;**30**:953–61.

24. Niu M-J, Yang J-K, Lin S-S, Ji X-J, Guo L-M. Loss of angiotensin-converting enzyme 2 leads to impaired glucose homeostasis in mice. *Endocrine* 2008;**34**:56–61.

25. Takeda M, Yamamoto K, Takemura Y, Takeshita H, Hongyo K, Kawai T, et al. Loss of ACE2 exaggerates high-calorie diet-induced insulin resistance by reduction of GLUT4 in mice. *Diabetes* 2012;**62**:223–33.

26. Tikellis C, Johnston CI, Forbes JM, Burns WC, Burrell LM, Risvanis J, et al. Characterization of renal angiotensin-converting enzyme 2 in diabetic nephropathy. *Hypertension* 2003;**41**:392–7.

27. Hardtner C, Morke C, Walther R, Wolke C, Lendeckel U. High glucose activates the alternative ACE2/Ang-(1-7)/Mas and APN/Ang IV/IRAP RAS axes in pancreatic beta-cells. *Int J Mol Med* 2013;**32**:795–804.

28. Mecca AP, O'Connor TE, Dooies KA, Katovich MJ, Sumners C. Cerebroprotective action of angiotensin 1-7 in a rat model of ischemic stroke. *FASEB J* 2009;**23**, 947.1.

29. Bindom SM, Lazartigues E. The sweeter side of ACE2: physiological evidence for a role in diabetes. *Mol Cell Endocrinol* 2008;**302**:193–202.

30. Jarajapu YP, Bhatwadekar AD, Caballero S, Hazra S, Shenoy V, Medina R, et al. Activation of the ACE2/angiotensin-(1-7)/Mas receptor axis enhances the reparative function of dysfunctional diabetic endothelial progenitors. *Diabetes* 2012;**62**:1258–69.

31. Harmon JS, Stein R, Robertson RP. Oxidative stress-mediated, post-translational loss of MafA protein as a contributing mechanism to loss of insulin gene expression in glucotoxic beta cells. *J Biol Chem* 2005;**280**:11107–13.

32. Ihara Y, Toyokuni S, Uchida K, Odaka H, Tanaka T, Ikeda H, et al. Hyperglycemia causes oxidative stress in pancreatic beta-cells of GK rats, a model of type 2 diabetes. *Diabetes* 1999;**48**:927–32.

33. Hitomi H, Kiyomoto H, Nishiyama A. Angiotensin II and oxidative stress. *Curr Opin Cardiol* 2007;**22**:311–15.

34. Gwathmey TM, Pendergrass KD, Reid SD, Rose JC, Diz DI, Chappell MC. Angiotensin-(1-7)-angiotensin-converting enzyme 2 attenuates reactive oxygen species formation to angiotensin II within the cell nucleus. *Hypertension* 2010;**55**:166–71.

35. Negi S, Park SH, Jetha A, Aikin R, Tremblay M, Paraskevas S. Evidence of endoplasmic reticulum stress mediating cell death in transplanted human islets. *Cell Transplant* 2012;**21**:889–900.

36. Hayden M, Sowers J. Isletopathy in type 2 diabetes mellitus: implications of islet RAS, islet fibrosis, islet amyloid, remodeling and oxidative stress. *Antioxid Redox Signal* 2007;**9**:891–910.

37. Donath MY, Schumann DM, Faulenbach M, Ellingsgaard H, Perren A, Ehses JA. Islet inflammation in type 2 diabetes: from metabolic stress to therapy. *Diabetes Care* 2008;**31**:S161–4.

38. Liu Z, Huang XR, Chen HY, Penninger JM, Lan HY. Loss of angiotensin-converting enzyme 2 enhances TGF-beta/Smad-mediated renal fibrosis and NF-kappaB-driven renal inflammation in a mouse model of obstructive nephropathy. *Lab Invest* 2012;**92**:650–61.
39. Zhong J, Guo D, Chen CB, Wang W, Schuster M, Loibner H, et al. Prevention of angiotensin II-mediated renal oxidative stress, inflammation, and fibrosis by angiotensin-converting enzyme 2. *Hypertension* 2011;**57**:314–22.
40. Alghamri MS, Weir NM, Anstadt MP, Elased KM, Gurley SB, Morris M. Enhanced angiotensin II-induced cardiac and aortic remodeling in ACE2 knockout mice. *J Cardiovasc Pharmacol Ther* 2013;**18**:138–51.
41. Feng Y, Hans C, McIlwain E, Varner KJ, Lazartigues E. Angiotensin-converting enzyme 2 over-expression in the central nervous system reduces angiotensin-II-mediated cardiac hypertrophy. *PLoS ONE* 2012;**7**, e48910.
42. Shenoy V, Ferreira AJ, Qi Y, Fraga-Silva RA, Diez-Freire C, Dooies A, et al. The angiotensin-converting enzyme 2/angiogenesis-(1-7)/Mas axis confers cardiopulmonary protection against lung fibrosis and pulmonary hypertension. *Am J Respir Crit Care Med* 2010;**182**:1065–72.
43. Chodavarapu H, Chhabra K, Shenoy V, Raizada MK, Yue X, Lazartigues E. ACE2 gene therapy decreases fibrosis in the pancreas of high fat diet mice. *FASEB J* 2013;**27**, 1154.7.
44. Santos SH, Giani JF, Burghi V, Miquet JG, Qadri F, Braga JF, et al. Oral administration of angiotensin-(1-7) ameliorates type 2 diabetes in rats. *J Mol Med (Berl)* 2014;**92**:255–65.
45. Reich HN, Oudit GY, Penninger JM, Scholey JW, Herzenberg AM. Decreased glomerular and tubular expression of ACE2 in patients with type 2 diabetes and kidney disease. *Kidney Int* 2008;**74**:1610–16.
46. Lambert DW, Yarski M, Warner FJ, Thornhill P, Parkin ET, Smith AI, et al. Tumor necrosis factor-alpha convertase (ADAM17) mediates regulated ectodomain shedding of the severe-acute respiratory syndrome-coronavirus (SARS-CoV) receptor, angiotensin-converting enzyme-2 (ACE2). *J Biol Chem* 2005;**280**:30113–19.
47. Iwata M, Silva Enciso JE, Greenberg BH. Selective and specific regulation of ectodomain shedding of angiotensin-converting enzyme 2 by tumor necrosis factor alpha-converting enzyme. *Am J Physiol Cell Physiol* 2009;**297**:C1318–29.
48. Heurich A, Hofmann-Winkler H, Gierer S, Liepold T, Jahn O, Pohlmann S. TMPRSS2 and ADAM17 cleave ACE2 differentially and only proteolysis by TMPRSS2 augments entry driven by the SARS-coronavirus spike-protein. *J Virol* 2014;**88**:1293–307.
49. Cardellini M, Menghini R, Martelli E, Casagrande V, Marino A, Rizza S, et al. TIMP3 is reduced in atherosclerotic plaques from subjects with type 2 diabetes and increased by SirT1. *Diabetes* 2009;**58**:2396–401.
50. Xia H, Sriramula S, Chhabra KH, Lazartigues E. Brain angiotensin-converting enzyme type 2 shedding contributes to the development of neurogenic hypertension. *Circ Res* 2013;**113**:1087–96.
51. Uhal BD, Nguyen H, Dang M, Gopallawa I, Jiang J, Dang V, et al. Abrogation of ER stress-induced apoptosis of alveolar epithelial cells by angiotensin 1-7. *Am J Physiol Lung Cell Mol Physiol* 2013;**305**:L33–41.
52. Reddy AB, Ramana KV, Srivastava S, Bhatnagar A, Srivastava SK. Aldose reductase regulates high glucose-induced ectodomain shedding of tumor necrosis factor (TNF)-alpha via protein kinase C-delta and TNF-alpha converting enzyme in vascular smooth muscle cells. *Endocrinology* 2009;**150**:63–74.
53. Asayesh A, Alanentalo T, Khoo NK, Ahlgren U. Developmental expression of metalloproteases ADAM 9, 10, and 17 becomes restricted to divergent pancreatic compartments. *Dev Dyn* 2005;**232**:1105–14.
54. Fang HJ, Yang JK. Tissue-specific pattern of angiotensin-converting enzyme 2 expression in rat pancreas. *J Int Med Res* 2010;**38**:558–69.
55. Zhang H, Wada J, Hida K, Tsuchiyama Y, Hiragushi K, Shikata K, et al. Collectrin, a collecting duct-specific transmembrane glycoprotein, is a novel homolog of ACE2 and is developmentally regulated in embryonic kidneys. *J Biol Chem* 2001;**276**:17132–9.
56. Fukui K, Yang Q, Cao Y, Takahashi N, Hatakeyama H, Wang H, et al. The HNF-1 target collectrin controls insulin exocytosis by SNARE complex formation. *Cell Metab* 2005;**2**:373–84.

Chapter 32

Angiotensin-(1-7) and Cancer

E. Ann Tallant, Patricia E. Gallagher

Hypertension and Vascular Research Center, Wake Forest School of Medicine, Winston-Salem, North Carolina, USA

HISTORICAL PERSPECTIVE

Angiotensin-(1-7) (Ang-(1-7)) is a seven-amino-acid heptapeptide hormone of the renin–angiotensin system. Although the peptide was considered a breakdown product of the vasoactive hormone angiotensin II (Ang-II), numerous studies identified biological roles for Ang-(1-7), in the regulation of prostaglandin metabolism, in the reduction in blood pressure in hypertensive animals, and in the inhibition of cell growth. We showed that Ang-(1-7) inhibited the proliferation of rat aortic vascular smooth muscle cells (VSMCs), in contrast to the growth-stimulatory effects of Ang-II.[1] VSMC growth stimulated by Ang-II or platelet-derived growth factor (PDGF) was inhibited by Ang-(1-7) with an increase in prostacyclin and cAMP production, activation of protein kinase A, and a reduction in mitogen-activated protein (MAP) kinase activities.[2–4] Ang-(1-7) reduced neointimal formation following injury to the rat carotid artery[5,6] or abdominal aorta.[7] We observed that Ang-(1-7) inhibited the growth of isolated cardiac myocytes[8] and fibroblasts, through an increase in the dual-specificity protein phosphatase DUSP1 and a decrease in cyclooxygenase-2 (COX-2), prostaglandin E synthase (PGES), and the MAP kinases ERK1/2,[4] in agreement with the heptapeptide hormone-mediated reduction in cardiac hypertrophy and fibrosis in Ang-II-induced hypertension.[3]

ACE inhibitors, currently in widespread use for hypertension, cause a significant elevation in both tissue and circulating Ang-(1-7),[9–12] and hypertensive patients administered with ACE inhibitors have a decreased risk of cancers.[13–15] Since ACE inhibitors increase Ang-(1-7) and the heptapeptide hormone inhibits the growth of VSMCs, cardiomyocytes, and fibroblasts, these results suggested that Ang-(1-7) may also reduce the proliferation of cancer cells and tumors.

ANG-(1-7) IN LUNG CANCER: ROLE IN PROLIFERATION AND ANGIOGENESIS

It was estimated that in the United States, in 2014, 224,210 new cases of lung cancer would be diagnosed and 159,260 people would die due to this devastating disease, making lung cancer the leading cause of cancer mortality. Human adenocarcinoma SK-LU-1 and A549 cells and nonsmall cell lung cancer SK-MES-1 cells were incubated with Ang-(1-7) to determine the effect of the heptapeptide hormone on lung cancer cell growth. Ang-(1-7) inhibited the growth of all three lung cancer cell lines,[16] in agreement with studies in VSMCs and cardiomyocytes. The IC_{50} for the reduction in cell proliferation was subnanomolar in all three cell lines, demonstrating the potency of the heptapeptide hormone in reducing cell growth. Inhibition of SK-LU-1 lung cancer cell growth by Ang-(1-7) was unaffected by AT_1 or AT_2 receptor antagonists but was blocked by the Ang-(1-7) receptor antagonist [D-Ala^7]-Ang-(1-7) (Dala), suggesting that the growth-inhibitory effects were mediated by a unique Ang-(1-7) receptor. SK-LU-1, A549, and SK-MES-1 cells all contain *Mas*, the receptor for Ang-(1-7), suggesting that the antiproliferative response to Ang-(1-7) is mediated by *Mas* in lung cancer cells. Other angiotensin peptides—Ang-I, Ang-II, Ang-(2-8), Ang-(3-8), and Ang-(3-7)—were ineffective in reducing mitogen-stimulated DNA synthesis in SK-LU-1 cells, demonstrating the specificity of the response to the N-terminal heptapeptide. Pretreatment of SK-LU-1 cells with Ang-(1-7) reduced serum-stimulated phosphorylation of the MAP kinases ERK1/2, suggesting that the antiproliferative effects may occur through inhibition of MAP kinase signaling. The results of this initial study demonstrated that Ang-(1-7) inhibited the growth of human lung cancer cells, through activation of a unique Ang-(1-7) receptor and associated with a reduction in MAP kinases.

Mice with A549 human lung tumors were treated with Ang-(1-7) to determine the effect of the heptapeptide hormone on lung tumor growth.[17] Osmotic minipumps were implanted subcutaneously to deliver Ang-(1-7) intravenously into the jugular vein. Treatment with Ang-(1-7) was well tolerated by the mice, with no weight loss, change in eating or drinking, and loss in motor function nor were any gross pathologies observed at the time the mice were sacrificed. Tumor size in the control animals increased 2.5-fold over the 28-day treatment period. In contrast, treatment with Ang-(1-7) reduced tumor volume by 30% compared to the size prior to treatment. These results correlate with a reduction in the proliferation marker Ki67 in the Ang-(1-7)-infused tumors when compared with the control tumors, in agreement with the reduction in MAP kinase signaling in cultured A549 cells[16] and suggest that the heptapeptide hormone reduces lung cancer cell proliferation.

COX-2 is overexpressed in 70–90% of adenocarcinomas[18], and clinical trials with nonselective inhibitors demonstrated that attenuation of COX activity reduces the risk for lung cancer, suggesting that a reduction in COX-2 is associated with decreased lung tumor growth.[19,20] Ang-(1-7) significantly reduced COX-2 protein in both A549 tumor xenografts and A549 cells in culture, with no change in COX-1. COX-2 mRNA was also reduced by Ang-(1-7) in A549 cells and tumors, suggesting that the reduction in protein was due to transcription regulation or mRNA stabilization. While the selective inhibition of COX-2 was considered a promising treatment for lung cancer, the usefulness of COX-2 inhibitors was short-lived based on the increased risk for cardiovascular events (myocardial infarction, angina, and stroke) associated with their use;[21] in contrast, Ang-(1-7) has antithrombotic effects.[22,23] These results suggest that Ang-(1-7) has a significant advantage over a COX-2 inhibitor as a treatment for lung cancer as the heptapeptide hormone-mediated reduction in COX-2 is associated with antithrombotic and anti-inflammatory activities.

Blood vessel density in lung tumors from mice treated with Ang-(1-7) was quantified to determine whether the heptapeptide hormone regulates angiogenesis. Athymic mice with A549 flank tumors were injected daily with $1000\,\mu g/kg/day$ for 5 days, followed by a 2-day drug holiday, and sacrificed after 42 days.[24] Ang-(1-7) markedly decreased vessel density in association with reduced A549 lung tumor growth. Ang-(1-7) reduced vascular endothelial growth factor (VEGF) protein and mRNA in tumors from mice treated with the heptapeptide hormone as compared to controls and in the parent A549 human lung cancer cells, suggesting that Ang-(1-7) may attenuate tumor angiogenesis by reducing VEGF, a primary proangiogenic protein. The heptapeptide hormone significantly reduced tubule formation by endothelial cells in Matrigel and had similar antiangiogenic effects in the chick chorioallantoic membrane, causing a greater than 50% decrease in neovascularization; tubule formation and neovascularization were both blocked by a specific Ang-(1-7) receptor antagonist. These results are in agreement with previous studies showing that Ang-(1-7) reduces wound healing, an effect mediated by a selective Ang-(1-7) receptor.[25,26] These studies demonstrated that the heptapeptide hormone exhibits significant antiangiogenic activity and may be a novel therapeutic agent for lung cancer treatment by targeting a specific Ang-(1-7) receptor to reduce angiogenesis.

ANG-(1-7) IN BREAST CANCER: ROLE IN TUMOR FIBROSIS

It was estimated that there would be 235,030 new cases of breast cancer and 40,430 deaths due to breast cancer in the United States in 2014 as the second leading cause of cancer mortality in women. The three major types of breast cancer include estrogen receptor-positive (ER+) breast cancer, which can be treated with selective estrogen receptor modifiers (SERMs) or aromatase inhibitors to block estrogen production; HER2-overexpressing breast cancer where the human epidermal growth factor receptor 2 is overexpressed, which can be treated with an antibody to the HER2 receptor; and triple-negative breast cancers, which lack estrogen and progesterone receptors and do not overexpress HER2, significantly reducing the treatments that are available for women with triple-negative breast cancer. In spite of targeted treatments for both ER+ and HER2-overexpressing breast cancers, there is a continued need for additional therapies for women with breast cancer, for both the primary and the metastatic diseases.

Human ZR-75-1 ER+ and BT474 ER+/HER2-overexpressing breast tumors growing in the mammary fat pad of athymic mice were treated with Ang-(1-7) to determine the effect of the heptapeptide hormone on breast tumor growth. Ang-(1-7) caused a marked reduction in tumor volume and weight in both ZR-75-1 and BT474 breast tumors,[27] suggesting that Ang-(1-7) may represent a novel therapy for breast cancer. Increasing evidence demonstrates a major role for the tumor microenvironment in the regulation of tumor growth—in regulating both the growth of primary tumors and tumor metastases. Cancer-associated fibroblasts (CAFs) play a major role in various types of cancer including breast cancer, by stimulating fibrosis and the deposition of extracellular matrix (ECM) proteins including collagens and fibronectin. Increased ECM remodeling and stiffening by collagen and fibronectin and the secretion of growth factors, cytokines, and proangiogenic peptides by tumor-associated myofibroblasts promote breast tumor growth.[28] Human ZR-75-1 and BT474 breast tumors from athymic mice treated with Ang-(1-7) were stained with Picrosirius red, to visualize collagen deposition within the tumors and determine the amount of tumor-associated fibrosis.[27] Collagen deposition was present throughout both ZR-75-1 and BT474 breast tumors; however, treatment with Ang-(1-7) significantly reduced interstitial fibrosis within the tumors and perivascular fibrosis surrounding tumor vessels, in association with a decrease in immunoreactive collagen I deposition. This is in agreement with studies showing that Ang-(1-7) reduces fibrosis in the heart and coronary vessels of rats with hypertension[3,29] and the proliferation of isolated cardiac fibroblasts.[4,30]

CAFs were isolated from ZR-75-1 ER+ breast tumors, by proteolytic digestion and differential plating, and were identified as myofibroblasts by immunohistochemistry.[27] Ang-(1-7) treatment of the isolated CAFs significantly reduced serum-stimulated proliferation, suggesting that the heptapeptide hormone reduces fibroblast growth to decrease tumor fibrosis. The reduction in fibroblast proliferation by Ang-(1-7) was associated with an increase in the MAP kinase phosphatase DUSP1 and a corresponding decrease in phosphorylated ERK1/2. The upregulation of DUSP1 was blocked by a selective Ang-(1-7) receptor antagonist, indicating that DUSP1 induction by the heptapeptide hormone was an Ang-(1-7) receptor-mediated process. Ang-(1-7) also reduced production of transforming growth factor-beta (TGF-β), which regulates the growth of activated myofibroblasts and their subsequent production of growth factors, cytokines, and ECM proteins. The heptapeptide hormone reduced the TGF-β-mediated synthesis of fibronectin in both isolated CAFs and ZR-75-1 orthotopic tumors. The ability of Ang-(1-7) to prevent production of TGF-β and decrease TGF-β-stimulated ERK1/2 activation and fibronectin synthesis indicates that the heptapeptide is an antagonist of TGF-β-mediated fibrosis in breast cancer. Our findings indicate that Ang-(1-7) targets the tumor microenvironment to inhibit CAF growth and tumor fibrosis in breast cancer.

ANG-(1-7) IN PROSTATE CANCER: ROLE IN REGULATION OF ANGIOGENIC FACTORS AND METASTASIS

The estimated numbers of men with prostate cancer in the United State in 2014 includes 233,000 new cases and 29,480 deaths as the second leading cause of cancer deaths in men. Treatment options for localized prostate cancer include surgery, radiation therapy, and hormone ablation therapy. Although treatment is promising for men with primary prostate cancer, metastatic prostate cancer, predominantly to the bone, is often fatal. Human LNCaP prostate cancer cells were injected into the flank of athymic mice, and tumors were treated with Ang-(1-7) for 54 days. The heptapeptide hormone markedly reduced the volume and weight of LNCaP xenograft tumors.[31] Histological analysis of tumor sections indicated that treatment with Ang-(1-7) reduced Ki67 immunoreactivity and ERK1/2 activities, suggesting that the heptapeptide hormone reduces primary prostate cancer by decreasing cell proliferation. Vessel density in prostate tumors was decreased in Ang-(1-7)-treated mice compared to controls, demonstrating reduced angiogenesis.[31] The inhibition in angiogenesis was associated with a decrease in both VEGF and PlGF, suggesting that the heptapeptide hormone attenuates vascularization by reducing proangiogenic factors. Treatment of mice with Ang-(1-7) markedly increased the soluble fraction of VEGF receptor 1 (sFlt-1), with a concomitant reduction in both VEGF receptors 1(Flt-1) and 2 (Flk-1). sFlt-1 serves as a decoy receptor that traps VEGF and PlGF, making the ligands unavailable to membrane-bound VEGF receptors and preventing activation of proangiogenic signaling. sFlt-1, produced by alternative splicing of Flt-1, consists of the extracellular ligand-binding domain but lacks the transmembrane and intracellular signaling domains and is secreted into the circulation where it binds VEGF and PlGF. In addition, sFlt-1 binds to the extracellular regions of Flk-1 and Flt-1, forming heterodimers that perturb downstream signaling cascades. These results suggest that Ang-(1-7) reduces tumor angiogenesis in prostate tumors by multiple mechanisms—a decrease in production of the angiogenic factors VEGF and PlGF and an increase in the sFlt-1 decoy receptor.

Human PC3 prostate cancer cells were injected into the aortic arch of SCID mice pretreated with Ang-(1-7) or into the tibia of athymic mice, administered with Ang-(1-7) for 5 weeks beginning 2 weeks postinjection, to investigate the effect of the heptapeptide hormone on metastatic prostate cancer to the bone.[32] When PC3 cells were injected into the circulation of SCID mice, the mice developed tumors in the submandibular bone, the spinal column, or the long bone of the leg. In contrast, pretreatment with Ang-(1-7) prevented metastatic tumor formation; no tumors were detected in Ang-(1-7)-treated mice by bioluminescent imaging and MRI. Circulating VEGF was significantly higher in control mice compared with mice treated with Ang-(1-7), in agreement with previous studies showing that Ang-(1-7) reduced VEGF.[24] Tumors developed in all control mice that received intraossal injection of PC3 cells; however, the heptapeptide hormone attenuated intratibial tumor growth, suggesting that Ang-(1-7) reduced the formation of the niche for engraftment of prostate cancer cells into the bone. Bone marrow cells were extruded from the long bone of untreated mice, differentiated to induce osteoclastogenesis and treated with Ang-(1-7). Osteoclastogenesis was reduced by 50% in bone marrow cells differentiated in the presence of Ang-(1-7), suggesting that the heptapeptide hormone prevents the formation of osteolytic lesions to reduce tumor survival in the bone microenvironment.

ANG-(1-7) AND CLINICAL TRIALS

Four clinical trials with Ang-(1-7) are completed or ongoing and have demonstrated that the heptapeptide hormone has limited toxicity and clinical efficacy in cancer patients. Ang-(1-7) was administered to patients with breast cancer in the adjuvant setting to determine the effect of the heptapeptide hormone on cytopenia.[33] No dose-limiting toxicities were reported, and Ang-(1-7) reduced thrombocytopenia and lymphopenia. Patients with ovarian, Fallopian tube, or peritoneal carcinoma receiving gemcitabine and carboplatin or cisplatin were also treated with Ang-(1-7); the heptapeptide hormone reduced thrombocytopenia following gemcitabine and platinum chemotherapy, suggesting that Ang-(1-7) stimulated thrombogenesis following marrow-toxic chemotherapy.[34]

A phase I clinical trial was conducted at Wake Forest University Comprehensive Cancer Center to determine toxicity and efficacy in patients with advanced solid tumors that were refractory to standard therapy.[35] Escalating doses of Ang-(1-7) were administered for 5 days in 21 day cycles. Four of 15 evaluable patients had clinical benefit; one patient developed a 19% reduction in tumor measurements, while three patients had stabilization of disease for more than 3 months. Dose-limiting toxicities were encountered at the highest dose studied and included stroke (grade 4) and reversible cranial neuropathy (grade 3); other toxicities were generally mild. Clinical benefit was associated with a decrease in circulating PlGF, in agreement with preclinical studies showing a reduction in angiogenesis in mice with human lung and prostate tumors.[24,31] In a follow-up study of individual cancer types, PlGF was reduced in patients with sarcoma that was not detected in patients with other types of cancer,[36] leading to the initiation of a phase II clinical trial to evaluate clinical activity and PlGF in patients with metastatic sarcomas (NCT01553539), which is ongoing.

CONCLUSION

Preclinical studies demonstrating that Ang-(1-7) has antiproliferative, antiangiogenic, anti-inflammatory, antifibrotic, and antimetastatic effects in lung, breast, and prostate cancer and clinical trials demonstrating safety and efficacy suggest that the heptapeptide hormone may represent a novel therapy for the treatment of cancer.

REFERENCES

1. Freeman EJ, Chisolm GM, Ferrario CM, Tallant EA. Angiotensin-(1-7) inhibits vascular smooth muscle cell growth. *Hypertension* 1996;**28**:104–8.
2. Tallant EA, Clark MA. Molecular mechanisms of inhibition of vascular growth by angiotensin-(1-7). *Hypertension* 2003;**42**:574–9.
3. McCollum LT, Gallagher PE, Tallant EA. Angiotensin-(1-7) attenuates angiotensin II-dependent cardiac remodeling associated with up-regulation of dual specificity phosphatase 1. *Am J Physiol Heart Circ Physiol* 2012;**302**:H801–10.
4. McCollum LT, Gallagher PE, Tallant EA. Angiotensin-(1-7) abrogates mitogen-stimulated proliferation of cardiac fibroblasts. *Peptides* 2012;**34**:380–8.
5. Strawn WB, Ferrario CM, Tallant EA. Angiotensin-(1-7) reduces smooth muscle growth after vascular injury. *Hypertension* 1999;**33[part II]**:207–11.
6. Tallant EA, Diz DI, Ferrario CM. State-of-the-Art lecture. Antiproliferative actions of angiotensin-(1-7) in vascular smooth muscle. *Hypertension* 1999;**34**:950–7.
7. Langeveld B, Van Gilst WH, Gio RA, Zijlstra F, Roks AJ. Angiotensin-(1-7) attenuates neointimal formation after stent implantation in the rat. *Hypertension* 2005;**45**:138–41.
8. Tallant EA, Ferrario CM, Gallagher PE. Angiotensin-(1-7) inhibits growth of cardiac myocytes through activation of the *mas* receptor. *Am J Physiol Heart Circ Physiol* 2005;**289**:1560–6.
9. Campbell DJ, Kladis A, Duncan A-M. Nephrectomy, converting enzyme inhibition, and angiotensin peptides. *Hypertension* 1993;**22**:513–22.
10. Kohara K, Brosnihan KB, Chappell MC, Khosla MC, Ferrario CM. Angiotensin-(1-7). A member of circulating angiotensin peptides. *Hypertension* 1991;**17**:131–8.
11. Luque M, Martin P, Martell N, Fernandez C, Brosnihan KB, Ferrario CM. Effects of captopril related to increased levels of prostacyclin and angiotensin-(1-7) in essential hypertension. *J Hypertens* 1996;**14**:799–805.
12. Lawrence AC, Evin G, Kladis A, Campbell DJ. An alternative strategy for the radioimmunoassay of angiotensin peptides using amino-terminal-directed antisera: measurement of eight angiotensin peptides in human plasma. *J Hypertens* 1990;**8**:715–24.
13. Pahor M, Guralnik JM, Salive ME, Corti M-C, Carbonin P, Havlik RJ. Do calcium channel blockers increase the risk of cancer? *Am J Hypertens* 1996;**9**:695–9.
14. Jick H, Jick S, Derby LE, Vasilakis C, Myers MW, Meier CR. Calcium-channel blockers and risk of cancer. *Lancet* 1997;**349**:525–8.
15. Lever AF, Hole DJ, Gillis CR, McCallum IRMGT, MacKinnon PL, Meredith PA, et al. Do inhibitors of angiotensin-I-converting enzyme protect against risk of cancer? *Lancet* 1998;**352**:179–84.
16. Gallagher PE, Tallant EA. Inhibition of lung cancer cell growth by angiotensin-(1-7). *Carcinogenesis* 2004;**25**:2045–52.
17. Menon J, Soto-Pantoja DR, Callahan MF, Cline JM, Ferrario CM, Tallant EA, et al. Angiotensin-(1-7) inhibits growth of human lung adenocarcinoma xenografts in nude mice through a reduction in cyclooxygenase-2. *Cancer Res* 2007;**67**:2809–15.
18. Hida T, Yatabe Y, Achiwa H, Muramatsu H, Kozaki K, Nakamura S, et al. Increased expression of cyclooxygenase 2 occurs frequently in human lung cancers, especially in adenocarcinomas. *Cancer Res* 1998;**58**:3761–4.
19. Harris RE, Beebe-Donk J, Schuller HM. Chemoprevention of lung cancer by non-steroidal anti-inflammatory drugs among cigarette smokers. *Oncol Rep* 2002;**9**:693–5.

20. Lee EO, Lee HJ, Hwang HS, Ahn KS, Chae C, Kang KS, et al. Potent inhibition of Lewis lung cancer growth by heyneanol A from the roots of *Vitis amurensis* through apoptotic and anti-angiogenic activities. *Carcinogenesis* 2006;**27**:2059–69.

21. Mukherjee D, Topol EJ. Cox-2: where are we in 2003?—Cardiovascular risk and Cox-2 inhibitors. *Arthritis Res Ther* 2003;**5**:8–11.

22. Kucharewicz I, Chabielska E, Pawlak D, Matys T, Rolkowski R, Buczko W. The antithrombotic effect of angiotensin-(1-7) closely resembles that of losartan. *J Renin Angiotensin Aldosterone Syst* 2000;**1**:268–72.

23. Yoshida M, Naito Y, Urano T, Takada A, Takada Y. L-158,809 and (D-Ala(7))-angiotensin I/II (1-7) decrease PAI-1 release from human umbilical vein endothelial cells. *Thromb Res* 2002;**105**:531–6.

24. Soto-Pantoja DR, Menon J, Gallagher PE, Tallant EA. Angiotensin-(1-7) inhibits tumor angiogenesis in human lung cancer xenografts with a reduction in vascular endothelial growth factor. *Mol Cancer Ther* 2009;**8**:1676–83.

25. Machado RD, Santos RA, Andrade SP. Opposing actions of angiotensins on angiogenesis. *Life Sci* 2000;**66**:67–76.

26. Machado RDP, Santos RAS, Andrade SP. Mechanisms of angiotensin-(1-7)-induced inhibition of angiogenesis. *Am J Physiol Regul Integr Comp Physiol* 2001;**280**:994–1000.

27. Cook KL, Metheny-Barlow LJ, Tallant EA, Gallagher PE. Angiotensin-(1-7) reduces fibrosis in orthotopic breast tumors. *Cancer Res* 2010;**70**:8319–28.

28. Kalluri R, Zeisberg M. Fibroblasts in cancer. *Nat Rev Cancer* 2006;**6**:392–401.

29. Grobe JL, Mecca AP, Mao H, Katovich MJ. Chronic angiotensin-(1-7) prevents cardiac fibrosis in the DOCA-salt model of hypertension. *Am J Physiol Heart Circ Physiol* 2006;**290**(6):H2417–23.

30. Iwata M, Cowling RT, Gurantz D, Moore C, Zhang S, Yuan X-J, et al. Angiotensin-(1-7) binds to specific receptors on cardiac fibroblasts to initiate antifibrotic and antitrophic effects. *Am J Physiol Heart Circ Physiol* 2005;**289**:2356–63.

31. Krishnan B, Torti FM, Gallagher PE, Tallant EA. Angiotensin-(1-7) reduces proliferation and angiogenesis of human prostate cancer xenografts with a decrease in angiogenic factors and an increase in sFlt-1. *Prostate* 2013;**73**:60–70.

32. Krishnan B, Smith TL, Dubey P, Zapadka ME, Torti FM, Willingham MC, et al. Angiotensin-(1-7) attenuates metastatic prostate cancer and reduces osteoclastogenesis. *Prostate* 2013;**73**:71–82.

33. Rodgers KE, Oliver J. diZerega GS. Phase I/II dose escalation study of angiotensin 1-7 [A(1-7)] administered before and after chemotherapy in patients with newly diagnosed breast cancer. *Cancer Chemother Pharmacol* 2006;**57**:559–68.

34. Pham H, Schwartz BM, Delmore JE, Reed E, Cruickshank S, Drummond L, et al. Pharmacodynamic stimulation of thrombogenesis by angiotensin (1-7) in recurrent ovarian cancer patients receiving gemcitabine and platinum-based chemotherapy. *Cancer Chemother Pharmacol* 2013;**71**:965–72.

35. Petty WJ, Miller AA, McCoy TP, Gallagher PE, Tallant EA, Torti FM. Phase I and pharmacokinetic study of angiotensin-(1-7), an endogenous antiangiogenic hormone. *Clin Cancer Res* 2009;**15**:7398–404.

36. Petty WJ, Aklilu M, Varela VA, Lovato J, Savage PD, Miller AA. Reverse translation of phase I biomarker findings links the activity of angiotensin-(1-7) to repression of hypoxia inducible factor-1alpha in vascular sarcomas. *BMC Cancer* 2012;**12**:404.

Chapter 33

Mas and the Central Nervous System

Mariela M. Gironacci,* Maria J. Campagnole-Santos[†]

*Department of Biological Chemistry, IQUIFIB-CONICET, University of Buenos Aires, Buenos Aires, Argentina, [†]Department of Physiology and Biophysics, INCT-Nanobiofar, Institute of Biological Sciences, Federal University of Minas Gerais-UFMG, Brazil

MAS RECEPTOR DISTRIBUTION IN THE BRAIN

Mas receptor has been demonstrated to be present in the hippocampus, amygdala, cortex, and hypoglossal nucleus and in cardiovascular-related areas of medulla and forebrain of adult Wistar male rats.[1] A strong Mas immunostaining was observed in the nucleus tractus solitarii (NTS), caudal ventrolateral medulla (CVLM), rostral ventrolateral medulla (RVLM; Figure 1a–c), inferior olive, paraventricular nucleus (PVN), and supraoptic nucleus.[1] Mas receptor showed colocalization with neurofilaments in primary neuronal cultures from the hypothalami and brain stem of newborn rats, demonstrating the presence of Mas receptors in the neurons at birth.[2] Mecca et al.[3] have shown Mas immunostaining mainly in the soma of the neurons of the adult rat cerebral cortex and not in astroglia and to exist in both nonnuclear and nuclear compartments. Further immunostaining experiments revealed the presence of Mas receptor on macrophages/microglia of rat cerebral cortex, as illustrated by colocalization with a specific marker of these cells.[3] Within the murine brain, the strongest Mas protein expression was detected in the dentate gyrus of the hippocampus and within the pyriform cortex.[4] However, Mas protein expression is not restricted to these areas, since Mas immunopositive neurons were also seen in different parts of the cortex, hippocampus, amygdala, basal ganglia, thalamus, and hypothalamus.[4]

In a pathological situation, such us in hypertension or after ischemic stroke, Mas receptors are upregulated. Mas receptor expressions were higher in neurons from the hypothalami and brain stem of spontaneously hypertensive rats (SHR) than from normotensive Wistar-Kyoto rats.[5] PVN of renovascular hypertensive rats showed increased Mas receptor protein expression compared with Sham rats.[6] Furthermore, Mas receptors were enhanced after acute cerebral ischemic stroke in rats.[7] The Mas immunopositive neurons were also seen to have stronger expression in the ischemic cortex.[7] Altogether these results suggest that Mas receptors would play a pivotal role in neuronal function in cerebrovascular diseases.

MAS RECEPTOR-TRIGGERED ACTIONS IN THE CENTRAL NERVOUS SYSTEM

Ang-(1-7) acts as an important neuromodulator, especially in those areas related to tonic and reflex control of arterial pressure, in the hypothalamus, and in the dorsomedial and ventrolateral medulla. At these sites, the cardiovascular effects induced by Ang-(1-7) are blocked by A-779[8,9] that is recognized as the selective Mas receptor antagonist, suggesting that in the brain, Ang-(1-7) actions are mediated by the interaction with Mas. More consistently, Ang-(1-7) induces an important facilitation of the baroreflex, especially improving the bradycardic component of the baroreflex control of heart rate in normotensive[10,11] or hypertensive animals.[12–14] This effect was investigated with short-term infusion of Ang-(1-7) into the lateral ventricle. No alteration in blood pressure or drinking behavior was observed after acute infusion of Ang-(1-7), which contrasts with the classical effect mediated by AT_1 stimulation in the brain.[15] However, when long-term infusion (14 days) was performed, the chronic increase in Ang-(1-7) levels in the central nervous system strongly attenuated the increase in arterial pressure that is observed in DOCA salt[16] or TGR(mREN2)L-27 hypertensive rats.[17,18] Similar data were observed after the delivery of an Ang-(1-7) fusion protein at the cisterna magna of TGR(mREN2)L-27 hypertensive rats.[19] Further, the lowering effect of Ang-(1-7) in DOCA-salt hypertensive rats was related to an improvement in baroreflex bradycardia, restoration of the baroreflex control of renal sympathetic activity, and regaining of the balance of cardiac autonomic tone (Figure 1d–f).[16] These effects were blocked by A-779, suggesting again that they are mediated by Mas. Enhancement of the baroreflex control of the RSNA was also observed after 4 days of ICV infusion of Ang-(1-7) in rabbits subjected to chronic heart failure.[20] Similar effects were also observed in transgenic mice overexpressing human ACE2, which is the main enzyme responsible for Ang-(1-7) formation, selectively in the brain that were subjected to cardiac heart failure[21] or DOCA-salt hypertension.[22,23] Additionally, it was shown that mice lacking Mas receptor presented important imbalance

The Protective Arm of the Renin–Angiotensin System (RAS). http://dx.doi.org/10.1016/B978-0-12-801364-9.00033-X

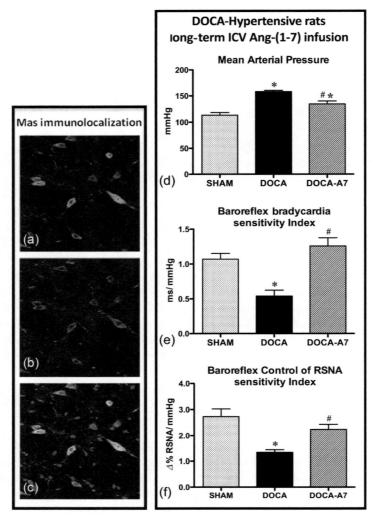

FIGURE 1   Mas receptor immunofluorescence and actions in the central nervous system. Panels (a–c) present colocalization of receptor Mas in neuronal cells of the ventrolateral medulla. The neuron marker anti-Neu-N is shown in green (a), the anti-Mas is shown in red (b), and the nuclear marker Draq-5 is shown in blue (c). Image showed in (c) represents the overlay. The effect of Mas stimulation through ICV infusion of Ang-(1-7) (A7) on blood pressure (d, g), baroreflex control (e, f, h), sympathetic tonus (i), and glucose (j), HOMA-IR (k), and HDL (l) in the blood. Panels (d–f) are from hypertensive DOCA-salt rats (DOCA), and

Continued

in the neural control of blood pressure, with blunted sensitivity of not only the baroreflex but also the chemoreflexes and Bezold-Jarisch reflexes.[24]

More recently, studying rats that develop metabolic syndrome after chronic fructose intake, it was shown that fructose-fed rats receiving Ang-(1-7) infusion into the lateral ventricle had normalized baseline MAP, baroreflex control of HR, and reduced cardiac sympathetic tone (Figure 1g–i).[25] More striking, along with these cardiovascular improvements, these rats presented normalized glucose tolerance, glycemia, insulinemia, and HOMA score (Figure 1j–l).[25] Moreover, fructose-fed rats treated with Ang-(1-7) had increased HDL and normalized hepatic and muscle glycogen content.[25] These data suggest that activation of the Ang-(1-7)/Mas pathway in the brain may play an important beneficial role against cardiovascular and metabolic disorders.

Molecular analysis of brain areas of fructose-fed rats treated with Ang-(1-7) ICV revealed a reduced mRNA expression of nNOS and NR1/NMDAr in the hypothalamus and medullary areas of these animals.[25] Indeed, these regions contain the major candidates to mediate the central actions of Ang Ang-(1-7)/Mas central actions—the PVN, the RVLM, and the NTS—crucial to sympathetic drive to the cardiovascular system and baroreflex control. In agreement, the overexpression of ACE2 selectively in the RVLM[26,27] or PVN[28] induces a significant decrease in blood pressure of SHR[26,27], attenuation of the sympathetic activity, and improvement of the baroreflex function in animals with congestive heart failure.[28] However, the stimulation of Mas receptor at these sites does not always present a straightforward response. It was shown in the first

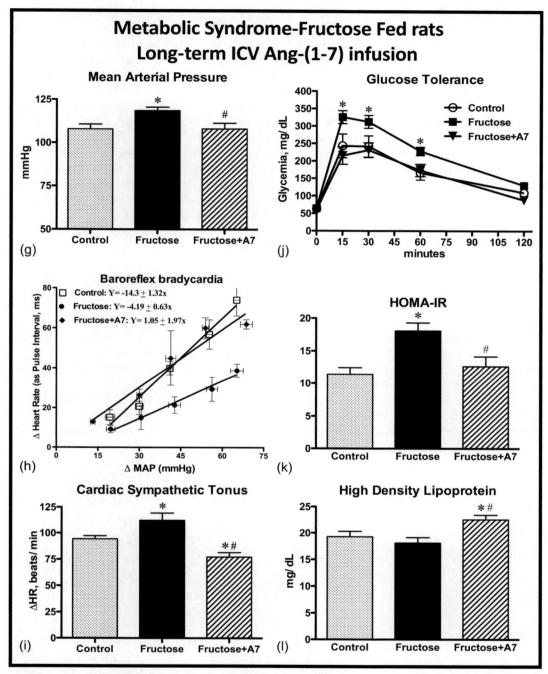

Metabolic Syndrome-Fructose Fed rats
Long-term ICV Ang-(1-7) infusion

FIGURE 1—CONT'D panels (g–l) are from chronically fructose-fed rats (FF). Abbreviations: DOCA—deoxycorticosterone; HOMA-IR—homeostasis model assessment of insulin resistance; HR—heart rate; MAP—mean arterial pressure; RSNA—renal sympathetic nerve activity. *(From Becker et al.[1] and Guimaraes et al.,[2,3] with permission.)*

studies that acute microinjections of Ang-(1-7) in the RVLM, NTS, and CVLM induce similar responses to those induced by Ang II, that is, increase or decrease of blood pressure depending on the selective area.[8,29–33] Importantly, however, the cardiovascular effects induced by Ang II and Ang-(1-7) were mediated by different receptor subtypes, AT_1 and Mas, respectively.[8,9] The only opposite effects induced by these peptides after acute injections were related to the modulation of the sensitivity of the baroreflex at the NTS.[34–36] NTS microinjection of Ang-(1-7) induces facilitation, while injection of Ang II produces attenuation of the baroreflex bradycardia, whereas A-779 and losartan, the selective Mas and AT_1 receptor antagonists, block the effect of the respective peptide in normotensive or hypertensive rats.[34,37]

Very interestingly, although the effects triggered by Mas and AT_1 receptors at the VLM are similar, they seem to be mediated by different peripheral mechanisms. Ang II effects at the VLM are mainly due to alteration in sympathetic activity,[38] whereas Ang-(1-7) includes vasopressin release and decrease of a vasodilator parasympathetic-related vascular tonus at the RVLM[39] and involves a nitrergic peripheral mechanism at the CVLM.[40,41] Distinct mechanisms of action seem to be also true for the effects of Ang-(1-7) at the hypothalamus, where it can elicit vasopressin release[8] and induce sympathoexcitation.[8] Mas receptors in the PVN[42] and in the RVLM[27] contribute to the enhanced sympathetic outflow and cardiac sympathetic afferent reflex through a cAMP-PKA pathway in renovascular hypertension. In contrast, it was recently reported that Ang-(1-7) produced a pronounced inhibition of β-adrenergic-mediated heart rate and renal sympathetic nerve activity response to air jet stress or disinhibition of the dorsomedial hypothalamus.[43]

Ang-(1-7)/Mas in the brain also triggers nonrelated cardiovascular actions. It was shown that Mas receptor stimulation of the dorsolateral-periaqueductal gray, which is importantly related to modulation of pain and certain behavioral activities, inhibits neuronal activity through a nitric-oxide-dependent pathway.[44] More recently, it was observed that ICV administration of Ang-(1-7), similar to the effect of the selective serotonin reuptake inhibitor fluoxetine, reversed the anxiety- and depressive-like behavior presented by transgenic rats with low brain angiotensinogen.[45]

MAS RECEPTOR AND CENTRAL NEUROTRANSMITTERS

Catecholamines are key neuromodulators in the mammalian brain. They are involved in a great number of physiological functions and in human pathologies associated with malfunction of them, including depression, appetite disorders, chronic pain, sleep disorders, essential hypertension, and some others.[46] Essential hypertension is characterized by an increased central sympathetic activity that results in augmented neurotransmitter levels in the synaptic cleft. AT_1 receptors have been implicated in the central Ang II-induced sympathetic overactivity and hypertension.[47] On the opposite side, Mas receptors and AT_2 receptors are coupled to an inhibitory neuromodulatory role on the brain NE system. Ang-(1-7) acting through Mas and AT_2 receptors induces a decrease in presynaptic NE release in the hypothalamus isolated from normotensive rats[48] and SHR.[49] Ang-(1-7) induces a decrease in NE release not only by itself but also by blockade of the central Ang II-stimulated NE release,[50] thus balancing the stimulatory action of Ang II. The inhibitory effect of Ang-(1-7) on NE release is coupled to Mas and AT_2 receptor stimulation, which in turn leads to bradykinin generation. Bradykinin, through stimulation of its B_2 receptor, enhances NO release activating the cGMP/PKG pathway signaling[48,49] (Figure 2). PKG may phosphorylate either

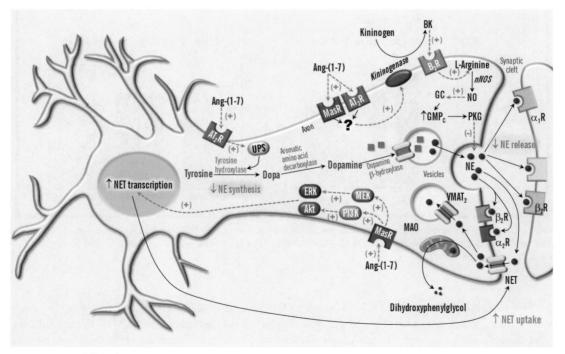

FIGURE 2 Mas receptor and the brain noradrenergic system. Abbreviations: NE—norepinephrine; NET—norepinephrine transporter; AT_2—angiotensin type 2 receptor; Mas R—Mas receptor; B2 R—bradykinin B2 receptor; NO—nitric oxide; nNOS—neuronal nitric oxide synthase; BK—bradykinin; UPS—ubiquitin-proteasome system; GC—guanylate cyclase; PKG—cGMP-dependent protein kinase; PI3K—phosphatidyl inositol 3 kinase.

voltage-dependent calcium channels, resulting in their inhibition, or synaptic vesicle proteins linked to neurotransmitter release, thus leading to a decrease in neurotransmitter levels in the synaptic cleft.

Tyrosine hydroxylase (TH), the rate-limiting enzyme in catecholamine biosynthesis, has been correlated with hypertension development in animals and humans.[51,52] While Ang II upregulates TH,[53] Ang-(1-7) downregulates this enzyme centrally.[54] Ang-(1-7) induces a decrease in TH enzymatic activity and expression in neurons from the hypothalami and brain stem of WKY and SHR.[54] Surprisingly, Mas receptors are not involved in this Ang-(1-7) action. Ang-(1-7), through AT_2 receptor activation, stimulates the ubiquitin-proteasome system, thus increasing TH degradation. Ang-(1-7) stimulates TH ubiquitination, which is further degraded by the proteasome, leading in this way to a reduced enzyme expression[54] (Figure 1). It is the first time that the ubiquitin-proteasome pathway stimulation is reported for Ang-(1-7), demonstrating a novel mechanism of action to downregulate a protein involved in hypertension development.

Intercellular communication in the central nervous system requires the precise control of the duration and intensity of neurotransmitter action at specific molecular targets. After they have been released at the synapse, neurotransmitters activate pre- and/or postsynaptic receptors. To terminate synaptic transmission, neurotransmitters are, in turn, inactivated by either enzymatic degradation or active transport in neuronal and/or glial cells by neurotransmitter transporters.[55] Thus, the concentrations of neurotransmitter in the synaptic cleft are tightly controlled by membrane transport proteins that are present in the cell membrane of neurons and/or glial cells. NE transporter (NET) regulates noradrenergic signaling in the central and peripheral nervous systems by mediating the clearance of NE.[55] Thus, changes in the activity of NET may alter the amount and duration of NE present in the synaptic cleft and consequently NE action. Thus, NET is essential for fine-tuned control of sympathetic activity. Ang-(1-7) does not modify acute NE neuronal uptake in the hypothalami from SHR. Conversely, Mas receptor activation by Ang-(1-7) induces an increase in NET expression. By activating the Mas receptor, Ang-(1-7) leads to both MEK1/2-ERK1/2 pathway stimulation and phosphatidyl PI3-kinase/Akt pathway stimulation resulting in an increase in NET transcription and transduction[5] (Figure 2). So, Ang-(1-7) through Mas receptors induces a long-term stimulatory effect on NE uptake by augmenting the amount of the transporter. This result was further confirmed by using an anti-Ang-(1-7) antibody. Neuronal NE uptake and NET expression were decreased after incubation with an anti-Ang-(1-7) antibody,[5] reinforcing the role of endogenous Ang-(1-7) through Mas receptor stimulation in the regulation of NE neuronal uptake. Since Mas receptors were demonstrated to be present in the neurons and not in the astroglia,[3] another cell assisting in the termination of signaling molecules that diffuse away from the synapse, the Ang-(1-7)-stimulated NET expression is the result of Mas receptor stimulation solely on neurons.

Concerning other central neurotransmitters, it has been shown that Mas receptors mediate the Ang-(1-7)-stimulated GABA release but not that of dopamine.[56] Ang-(1-7) caused a significant increase in dopamine and GABA release in the rat striatum but had no effect on glutamate release.[56] In contrast, Mas receptor stimulation at the CVLM induces an increase in glutamate and a decrease in taurine release, and in this way, it elicits a depressor effect.[57] Thus, by selective synaptic neurotransmitter level regulation, Ang-(1-7) through Mas receptors may contribute to the overall central effects.

CONCLUDING REMARKS

Inappropriate overactivity of the RAS plays an important role in the pathogenesis of cardiovascular diseases. On the other hand, several evidences show that Ang-(1-7)/Mas actions may oppose the deleterious effects induced by accumulation of Ang II during pathophysiological conditions, more importantly at the tissue level. In this chapter, we presented data showing that despite several local similarities in action, Ang II and Ang-(1-7) produce opposite effects on tonic and reflex control of blood pressure. Further, long-term stimulation of Mas may induce beneficial effects that, regardless of its complexity and yet not completely understood mechanisms of action, may counterbalance the deleterious effects of chronic Ang II on AT_1 receptors. Thus, new pharmacological strategies that increase tissue stimulation of Mas receptor, also in the brain, should be pursued in order to improve treatment of cardiovascular diseases.

REFERENCES

1. Becker LK, Etelvino GM, Walther T, Santos RA, Campagnole-Santos MJ. Immunofluorescence localization of the receptor Mas in cardiovascular-related areas of the rat brain. *Am J Physiol Heart Circ Physiol* 2007;**293**:H1416–24.
2. Gironacci MM, Cerniello FM, Longo Carbajosa N, Goldstein J, Cerrato BD. The protective axis of the renin angiotensin system at the central level. *Clin Sci* 2014;**127**(5):295–306.
3. Mecca AP, Regenhardt RW, O'Connor TE, Joseph JP, Raizada MK, Katovich MJ, et al. Cerebroprotection by angiotensin-(1-7) in endothelin-1-induced ischaemic stroke. *Exp Physiol* 2011;**96**:1084–96.
4. Freund M, Walther T, von Bohlen und Halbach O. Immunohistochemical localization of the angiotensin-(1-7) receptor Mas in the murine forebrain. *Cell Tissue Res* 2012;**348**:29–35.

5. Lopez Verrilli MA, Rodriguez Fermepin M, Longo Carbajosa N, Landa S, Cerrato BD, García S, et al. Angiotensin-(1-7) through Mas receptor up-regulates neuronal norepinephrine transporter via Akt and Erk1/2-dependent pathways. *J Neurochem* 2012;**120**:46–55.

6. Sun HJ, Li P, Chen WW, Xiong XQ, Han Y. Angiotensin II and angiotensin-(1-7) in paraventricular nucleus modulate cardiac sympathetic afferent reflex in renovascular hypertensive rats. *PLoS ONE* 2012;**7**, e52557.

7. Lu J, Jiang T, Wu L, Gao L, Wang Y, Zhou F, et al. The expression of angiotensin-converting enzyme 2-angiotensin-(1-7)-Mas receptor axis are upregulated after acute cerebral ischemic stroke in rats. *Neuropeptides* 2013;**47**:289–95.

8. Santos RA, Ferreira AJ, Pinheiro SV, Sampaio WO, Touyz R, Campagnole-Santos MJ. Angiotensin-(1-7) and its receptor as a potential targets for new cardiovascular drugs. *Expert Opin Investig Drugs* 2005;**14**(8):1019–31.

9. Diz DI, Arnold AC, Nautiyal M, Isa K, Shaltout HA, Tallant EA. Angiotensin peptides and central autonomic regulation. *Curr Opin Pharmacol* 2011;**11**:131–7.

10. Campagnole-Santos MJ, Heringer SB, Batista EN, Khosla MC, Santos RA. Differential baroreceptor reflex modulation by centrally infused angiotensin peptides. *Am J Physiol* 1992;**263**(1 Pt 2):R89–94.

11. Bomtempo CA, Santos GF, Santos RA, Campagnole-Santos MJ. Interaction of bradykinin and angiotensin-(1-7) in the central modulation of the baroreflex control of the heart rate. *J Hypertens* 1998;**16**(12 Pt 1):1797–804.

12. Oliveira DR, Santos RA, Santos GF, Khosla M, Campagnole-Santos MJ. Changes in the baroreflex control of heart rate produced by central infusion of selective angiotensin antagonists in hypertensive rats. *Hypertension* 1996;**27**:1284–90.

13. Britto RR, Santos RA, Fagundes-Moura CR, Khosla MC, Campagnole-Santos MJ. Role of angiotensin-(1-7) in the modulation of the baroreflex in renovascular hypertensive rats. *Hypertension* 1997;**30**:549–56.

14. Heringer-Walther S, Batista EN, Walther T, Khosla MC, Santos RA, Campagnole-Santos MJ. Baroreflex improvement in SHR after ACE inhibition involves angiotensin-(1-7). *Hypertension* 2001;**37**:1309–14.

15. Mahon JM, Allen M, Herbert J, Fitzsimons JT. The association of thirst, sodium appetite and vasopressin release with c-fos expression in the forebrain of the rat after intracerebroventricular injection of angiotensin II, angiotensin-(1-7) or carbachol. *Neuroscience* 1995;**69**:199–208.

16. Guimaraes PS, Santiago NM, Xavier CH, Velloso EP, Fontes MA, Santos RA, et al. Chronic infusion of angiotensin-(1-7) into the lateral ventricle of the brain attenuates hypertension in DOCA-salt rats. *Am J Physiol Heart Circ Physiol* 2012;**303**:H393–400.

17. Kangussu LM, Melo MB, Guimarães PS, Nadu AP, Bader M, Santos RA, et al. ICV Angiotensin-(1-7) improves cardiac function and attenuates remodeling in (mRen2)27 transgenic hypertensive rats. *Hypertension* 2013;**62**:A388 [abstract].

18. Dobruch J, Paczwa P, Łoń S, Khosla MC, Szczepańska-Sadowska E. Hypotensive function of the brain angiotensin-(1-7) in Sprague Dawley and renin transgenic rats. *J Physiol Pharmacol* 2003;**54**:371–81.

19. Garcia-Espinosa MA, Shaltout HA, Gallagher PE, Chappell MC, Diz DI. In vivo expression of angiotensin-(1-7) lowers blood pressure and improves baroreflex function in transgenic (mRen2)27 rats. *J Cardiovasc Pharmacol* 2012;**60**:150–7.

20. Kar S, Gao L, Belatti DA, Curry PL, Zucker IH. Central angiotensin-(1-7) enhances baroreflex gain in conscious rabbits with heart failure. *Hypertension* 2011;**58**:627–34.

21. Xiao L, Gao L, Lazartigues E, Zucker IH. Brain-selective overexpression of angiotensin-converting enzyme 2 attenuates sympathetic nerve activity and enhances baroreflex function in chronic heart failure. *Hypertension* 2011;**58**(6):1057–65.

22. Feng Y, Xia H, Cai Y, Halabi CM, Becker LK, Santos RA, et al. Brain-selective overexpression of human angiotensin-converting enzyme type 2 attenuates neurogenic hypertension. *Circ Res* 2010;**106**:373–82.

23. Xia H, Sriramula S, Chhabra KH, Lazartigues E. Brain angiotensin-converting enzyme type 2 shedding contributes to the development of neurogenic hypertension. *Circ Res* 2013;**113**(9):1087–96.

24. de Moura MM, Santos RA, Campagnole-Santos MJ, Todiras M, Bader M, Alenina N, et al. Altered cardiovascular reflexes responses in conscious angiotensin-(1-7) receptor Mas-knockout mice. *Peptides* 2010;**31**(10):1934–9.

25. Guimaraes PS, Oliveira MF, Braga JF, Nadu AP, Schreihofer A, Santos RA, et al. Increasing angiotensin-(1-7) levels in the brain attenuates metabolic syndrome-related risks in fructose-fed rats. *Hypertension* 2014;**63**(5):1078–85.

26. Yamazato M, Yamazato Y, Sun C, Diez-Freire C, Raizada MK. Overexpression of angiotensin-converting enzyme 2 in the rostral ventrolateral medulla causes long-term decrease in blood pressure in the spontaneously hypertensive rats. *Hypertension* 2007;**49**:926–31.

27. Li P, Sun HJ, Cui BP, Zhou YB, Han Y. Angiotensin-(1-7) in the rostral ventrolateral medulla modulates enhanced cardiac sympathetic afferent reflex and sympathetic activation in renovascular hypertensive rats. *Hypertension* 2013;**61**(4):820–7.

28. Sriramula S, Cardinale JP, Lazartigues E, Francis J. ACE2 overexpression in the paraventricular nucleus attenuates angiotensin II-induced hypertension. *Cardiovasc Res* 2011;**92**:401–8.

29. Campagnole-Santos MJ, Diz DI, Santos RA, Khosla MC, Brosnihan KB, Ferrario CM. Cardiovascular effects of angiotensin-(1-7) injected into the dorsal medulla of rats. *Am J Physiol* 1989;**257**(1 Pt 2):H324–9.

30. Silva LC, Fontes MA, Campagnole-Santos MJ, Khosla MC, Campos Jr RC, Guertzenstein PG, et al. Cardiovascular effects produced by microinjection of angiotensin-(1-7) on vasopressor and vasodepressor sites of the ventrolateral medulla. *Brain Res* 1993;**613**:321–5.

31. Fontes MA, Silva LC, Campagnole-Santos MJ, Khosla MC, Guertzenstein PG, Santos RA. Evidence that angiotensin-(1-7) plays a role in the central control of blood pressure at the ventrolateral medulla acting through specific receptors. *Brain Res* 1994;**665**:175–80.

32. Lima DX, Campagnole-Santos MJ, Fontes MA, Khosla MC, Santos RA. Haemorrhage increases the pressor effect of angiotensin-(1-7) but not of angiotensin II at the rat rostral ventrolateral medulla. *J Hypertens* 1999;**17**:1145–52.

33. Nakagaki T, Hirooka Y, Ito K, Kishi T, Hoka S, Sunagawa K. Role of angiotensin-(1-7) in rostral ventrolateral medulla in blood pressure regulation via sympathetic nerve activity in Wistar-Kyoto and spontaneous hypertensive rats. *Clin Exp Hypertens* 2011;**33**:223–30.

34. Chaves GZ, Caligiorne SM, Santos RA, Khosla MC, Campagnole-Santos MJ. Modulation of the baroreflex control of heart rate by angiotensin-(1-7) at the nucleus tractus solitarii of normotensive and spontaneously hypertensive rats. *J Hypertens* 2000;**18**:1841–8.

35. Couto AS, Baltatu O, Santos RA, Ganten D, Bader M, Campagnole-Santos MJ. Differential effects of angiotensin II and angiotensin-(1-7) at the nucleus tractus solitarii of transgenic rats with low brain angiotensinogen. *J Hypertens* 2002;**20**:919–25.

36. Sakima A, Averill DB, Kasper SO, Jackson L, Ganten D, Ferrario CM, et al. Baroreceptor reflex regulation in anesthetized transgenic rats with low glia-derived angiotensinogen. *Am J Physiol Heart Circ Physiol* 2007;**292**(3):H1412–19.

37. Diz DI, Garcia-Espinosa MA, Gallagher PE, Ganten D, Ferrario CM, Averill DB. Angiotensin-(1-7) and baroreflex function in nucleus tractus solitarii of (mRen2)27 transgenic rats. *J Cardiovasc Pharmacol* 2008;**51**(6):542–8.

38. Phillips MI, Sumners C. Angiotensin II in central nervous system physiology. *Regul Pept* 1998;**78**:1–11.

39. Oliveira RC, Campagnole-Santos MJ, Santos RA. The pressor effect of angiotensin-(1-7) in the rat rostral ventrolateral medulla involves multiple peripheral mechanisms. *Clinics (Sao Paulo)* 2013;**68**:245–52.

40. Alzamora AC, Santos RA, Campagnole-Santos MJ. Hypotensive effect of ANG II and ANG-(1-7) at the caudal ventrolateral medulla involves different mechanisms. *Am J Physiol Regul Integr Comp Physiol* 2002;**283**:R1187–95.

41. Ferreira PM, Alzamora AC, Santos RA, Campagnole-Santos MJ. Hemodynamic effect produced by microinjection of angiotensins at the caudal ventrolateral medulla of spontaneously hypertensive rats. *Neuroscience* 2008;**151**:1208–16.

42. Han Y, Sun HJ, Li P, Gao Q, Zhou YB, Zhang F, et al. Angiotensin-(1-7) in paraventricular nucleus modulates sympathetic activity and cardiac sympathetic afferent reflex in renovascular hypertensive rats. *PLoS ONE* 2012;**7**(11), e48966.

43. Martins Lima A, Xavier CH, Ferreira AJ, Raizada MK, Wallukat G, Velloso EP, et al. Activation of angiotensin-converting enzyme 2/angiotensin-(1-7)/Mas axis attenuates the cardiac reactivity to acute emotional stress. *Am J Physiol Heart Circ Physiol* 2013;**305**:H1057–67.

44. Xing J, Kong J, Lu J, Li J. Angiotensin-(1-7) inhibits neuronal activity of dorsolateral periaqueductal gray via a nitric oxide pathway. *Neurosci Lett* 2012;**522**:156–61.

45. Kangussu LM, Almeida-Santos AF, Bader M, Alenina N, Fontes MA, Santos RA, et al. Angiotensin-(1-7) attenuates the anxiety and depression-like behaviors in transgenic rats with low brain angiotensinogen. *Behav Brain Res* 2013;**257**:25–30.

46. Kasparov S, Teschemacher AG. Altered central catecholaminergic transmission and cardiovascular disease. *Exp Physiol* 2008;**93**:725–40.

47. Gelband CH, Sumners C, Raizada MK. Angiotensin receptors and norepinephrine neuromodulation: implications of functional coupling. *Regul Pept* 1997;**72**:139–45.

48. Gironacci MM, Vatta M, Rodriguez-Fermepin M, Fernandez BE, Peña C. Angiotensin-(1-7) reduces norepinephrine release through a nitric oxide mechanism in rat hypothalamus. *Hypertension* 2000;**35**:1248–52.

49. Gironacci MM, Valera MS, Yujnovsky I, Peña C. Angiotensin-(1-7) inhibitory mechanism of norepinephrine release in hypertensive rats. *Hypertension* 2004;**44**:783–7.

50. Gironacci MM, Yujnovsky I, Gorzalczany S, Taira C, Peña C. Angiotensin-(1-7) inhibits the angiotensin II-enhanced norepinephrine release in coarcted hypertensive rats. *Regul Pept* 2004;**118**:45–9.

51. Rao F, Zhang L, Wessel J, Zhang K, Wen G, Kennedy BP, et al. Tyrosine hydroxylase, the rate-limiting enzyme in catecholamine biosynthesis: discovery of common human genetic variants governing transcription, autonomic activity, and blood pressure in vivo. *Circulation* 2007;**116**:993–1006.

52. Reja V, Goodchild AK, Pilowski PM. Catecholamine-related gene expression correlates with blood pressures in SHR. *Hypertension* 2002;**40**:342–7.

53. Veerasingham SJ, Raizada MK. Brain renin-angiotensin system dysfunction in hypertension: recent advances and perspectives. *Br J Pharmacol* 2003;**139**(2):191–202.

54. Lopez Verrilli MA, Pirola CJ, Pascual MM, Dominici FP, Turyn D, Gironacci MM. Angiotensin-(1-7) through AT receptors mediates tyrosine hydroxylase degradation via the ubiquitin-proteasome pathway. *J Neurochem* 2009;**109**:326–35.

55. Bönisch H, Brüss M. The norepinephrine transporter in physiology and disease. *Handb Exp Pharmacol* 2006;**175**:485–524.

56. Stragier B, Hristova I, Sarre S, Ebinger G, Michotte Y. In vivo characterization of the angiotensin-(1-7)-induced dopamine and gamma-aminobutyric acid release in the striatum of the rat. *Eur J Neurosci* 2005;**22**:658–64.

57. Wang J, Peng YJ, Zhu DN. Amino acids modulate the hypotensive effect of angiotensin-(1-7) at the caudal ventrolateral medulla in rats. *Regul Pept* 2005;**129**:1–7.

Chapter 34

Role of the Alternate RAS in Liver Disease and the GI Tract

Chandana B. Herath, Kai Y. Mak, Peter W. Angus
Department of Medicine, The University of Melbourne, Austin Health, Heidelberg, Victoria, Australia

INTRODUCTION

The systemic renin–angiotensin system (RAS) is activated in patients with advanced liver disease in response to the circulatory changes that accompany the development of cirrhosis and portal hypertension in an attempt to maintain blood pressure and renal perfusion.[1] However, it is now recognized that in many tissues including the liver, there is local expression of the RAS, which functions in an autocrine and/or paracrine fashion to regulate the local response to disease or injury.[2,3] Indeed, there is now considerable evidence that the RAS contributes to fibrosis and tissue repair in diabetes and cardiovascular and renal diseases and that RAS blockers have beneficial effects in these diseases independent of their effects on blood pressure.[4,5] Many studies have shown that in the liver, as in other organs, the local RAS is upregulated in response to chronic diseases and contributes to the pathogenesis of tissue inflammation and fibrosis. Recently, it has been shown that the alternate RAS comprising ACE2, angiotensin-(1-7) (Ang-(1-7)), and its putative receptor Mas is also upregulated in liver disease[6,7] and that it can attenuate liver fibrosis and modulate portal pressure. This chapter will outline recent advances in understanding the aspects of RAS physiology in the gastrointestinal (GI) tract, focusing in particular on evidence of the role of both arms of the RAS in liver fibrosis and the circulatory abnormalities that underlie portal hypertension. Recent evidence suggesting possible roles for the RAS in the pathophysiology of other gastrointestinal disorders will also be presented.

THE CLASSIC RAS AND LIVER FIBROSIS

Hepatic fibrosis occurs when liver tissue is exposed to chronic damage from insults such as chronic viral hepatitis, alcohol abuse, and nonalcoholic steatohepatitis (NASH). It is characterized by the excessive accumulation of extracellular matrix (ECM) proteins and loss of functional parenchymal cells and eventually leads to distortion of the hepatic architecture with the formation of the dense fibrous scars and regenerative nodules that define cirrhosis.[8] In response to chronic liver injury,[9] hepatic stellate cells (HSCs), resident perisinusoidal cells, which are quiescent in the healthy liver, undergo phenotypic activation into ECM-secreting myofibroblasts. In the healthy liver, quiescent HSCs have very low levels of expression of components of the RAS and are unable to secrete angiotensin II (Ang II).[10] Following liver injury, activated HSCs and other hepatic myofibroblasts express many of the components of the RAS, including angiotensin-converting enzyme (ACE) and the Ang II type 1 receptor (AT_1R), and acquire the ability to synthesize Ang II. Ang II induces contraction and proliferation of HSCs via the AT_1R[10,11] and promotes activation and differentiation of quiescent HSCs into myofibroblasts.[8] Oxidative stress due to reactive oxygen species (ROS) generation by activated NADPH oxidase is likely to be a key driver of Ang II-induced hepatic fibrosis, since mice lacking $p47^{phox}$, a regulatory subunit of NADPH oxidase, have attenuated liver injury and fibrosis in response to bile duct ligation (BDL) compared with wild-type counterparts.[12] In keeping with this finding, incubation with Ang II increases collagen expression and the secretion of transforming growth factor-β1 (TGF-β1) and inflammatory cytokines by human HSCs, and these effects are attenuated by NADPH oxidase inhibition.[12] Furthermore, systemic infusion of Ang II induces liver fibrosis, whereas blockade of the AT_1R ameliorates liver fibrosis in several rat models of liver injury.[13,14] These findings collectively suggest that, acting via the AT_1R receptor in the liver, Ang II stimulates the activation of quiescent HSCs, induces oxidative stress, promotes the release of inflammatory cytokines, and increases ECM deposition.

The Protective Arm of the Renin–Angiotensin System (RAS). http://dx.doi.org/10.1016/B978-0-12-801364-9.00034-1

ACE INHIBITORS AND AT₁R BLOCKERS IN THE TREATMENT OF LIVER FIBROSIS

ACE inhibitors and Ang II type 1 receptor blockers (ARBs) block Ang II synthesis and action, respectively, and therefore might be expected to ameliorate liver fibrosis. In addition, ACE inhibition prevents the degradation of Ang-(1-7), thus increasing its plasma and local tissue levels. Numerous studies using a variety of experimental models of liver fibrosis have demonstrated antifibrotic effects of both ACE inhibitors and ARBs, although there are some conflicting findings, which may reflect the heterogeneity of disease models, drugs, and drug doses used.[15–17] There has been particular interest in the therapeutic use of RAS blockade in experimental NASH, given the lack of specific medical treatments for this condition and its association with hypertension and other features of the metabolic syndrome. Promising findings have come from work in several experimental models of NASH. For example, olmesartan, an ARB, is effective in reducing fibrosis in experimental NASH, while a study in obese Zucker rats found that both perindopril and irbesartan prevent hepatic steatosis and improve fibrosis with a reduction in the hepatic expression of the potent profibrotic cytokines tumor necrosis factor-α (TNF-α) and TGF-β1.[13,18]

Despite a large number of animal studies, at present, there is a relative paucity of clinical data to support the use of these drugs in human liver disease. A number of small prospective human studies have suggested beneficial effects of ACE inhibitors or ARBs in the treatment of liver fibrosis due to chronic hepatitis C.[19–21] Initial human studies examining the antifibrotic effects of RAS blockade in NASH have been limited by small patient numbers and the lack of a proper randomized control group, which make interpretation of results difficult. These initial studies suggested a possible beneficial effect of ARBs on liver inflammation and fibrosis due to NASH.[22,23]

THE ALTERNATE RAS AND LIVER FIBROSIS

As outlined above, the classic arm of the RAS, ACE-Ang II-AT₁R, is thought to be a primary pathway responsible for pathogenesis of liver fibrosis. However, there is recent evidence to indicate that the alternate (protective) arm of the RAS (ACE2-Ang-(1-7)-Mas receptor) may also be activated in chronic liver disease, potentially opposing some of the deleterious actions of Ang II. Indeed, several studies have provided a number of lines of evidence supporting the concept that ACE2 and Ang-(1-7) are part of a counterregulatory response of the RAS to liver injury and have beneficial effects in liver fibrosis. In experimental cholestatic liver disease induced by BDL, there is early upregulation of most components of the classic RAS including angiotensinogen, ACE, and AT₁R. This occurs within 1 week of bile duct ligation. Subsequently, from weeks 2 to 4 post-BDL, hepatic expression of the alternate RAS components ACE2 and Mas receptor (MasR) is increased and eventually parallels the increase in classic RAS components.[6,7,24] This is accompanied by a major increase in circulating levels of Ang-(1-7) (Figure 1a),[6] and has also been observed in patients with hepatitis C (Figure 1b and c).[25] It is tempting to speculate that this increased expression of the alternate RAS represents activation of a tissue-protective response aimed at modulating the harmful effects of Ang II.

These findings in experimental liver disease are supported by human data. In noncirrhotic patients with chronic hepatitis C, plasma Ang-(1-7) concentrations are elevated compared to healthy controls, while plasma Ang II concentrations are unchanged (Figure 1b and c).[25] In contrast, in cirrhotic patients, both Ang II and Ang-(1-7) concentrations are significantly increased (Figure 1b and c).[25] While ACE2 protein, which generates hepatic Ang-(1-7) from Ang II,[26] is expressed at very low levels in the healthy liver, in cirrhotic rat liver and in patients with hepatitis C cirrhosis, there is widespread ACE2 protein expression in the parenchymal tissues of diseased livers.[6,24] Organ perfusion experiments demonstrate that fibrotic livers, in which ACE2 expression is increased, produce increased amounts of Ang-(1-7) from Ang II, confirming the presence of functional ACE2 in the diseased liver, which opposes generation of Ang II by locally upregulated ACE.[26]

To date, there have been few studies investigating the effects of Ang-(1-7) in liver fibrosis. In the BDL model, Ang-(1-7) infusion reduces liver fibrosis as evidenced by reductions in picrosirius red staining (Figure 2a) and hydroxyproline content, representing a reduced collagen formation in association with reduced expression of α-smooth muscle actin (α-SMA) protein expression (Figure 2b), the surrogate marker for activated HSCs.[25] Furthermore, in isolated HSCs, Ang-(1-7) and MasR agonist AVE 0991 were shown to decrease α-SMA protein secretion, an effect that was blocked by MasR antagonism (Figure 2c).[25] To circumvent the short half-life of parenterally administered Ang-(1-7), an oral formulation has recently been developed, which has been shown to have cardioprotective effects following myocardial infarction in rats,[27] and the use of this formulation has been discussed elsewhere in this book. However, there are no published studies of its effects in experimental liver fibrosis.

Additional evidence for the potential beneficial effects of the alternate RAS in liver disease comes from a study that demonstrated exacerbation of experimental chronic liver injury, with increased expression of collagen and α-SMA protein in ACE2 gene-deleted mice compared with wild-type littermates following BDL or carbon tetrachloride (CCl_4)

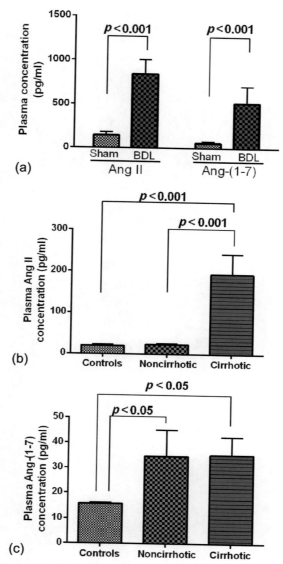

FIGURE 1   Circulating Ang II and Ang-(1-7) levels in rats that underwent bile duct ligation (BDL) and sham operation 4 weeks post-surgery (a) and Ang II (b) and Ang-(1-7) (c) levels in hepatitis C patients with and without cirrhosis. *Adapted from Refs. 6,25.*

administration.[28] These findings collectively suggest that, as in other organs,[28] in the liver, the Ang-(1-7)-ACE2-MasR axis protects against chronic tissue injury and fibrosis.

Given that the ACE2-Ang-(1-7)-MasR axis has effects that counterbalance those of Ang II and the AT_1R, it is tempting to speculate that stimulation of this protective arm might also be beneficial in the treatment of chronic liver disease. One strategy would be to enhance hepatic ACE2 expression and activity through pharmacological or genetic intervention since increasing hepatic ACE2 activity should produce dual beneficial effects, that is, degradation of the profibrotic peptide Ang II and generation of antifibrotic Ang-(1-7). Human recombinant ACE2 has been shown to have therapeutic benefits in cardiovascular disease and renal disease in animals[29,30] and had no adverse effects in healthy volunteers in a phase 1 clinical trial.[31] A single study using recombinant ACE2 to treat liver injury due to the toxin CCl_4 or cholestatic experimental liver injury induced by BDL in rats demonstrated its ability to attenuate fibrosis when given as "prophylactic" therapy for up to 2 weeks, commencing at the time of disease induction.[28] These results indicate that ACE2 treatment holds promise as a novel antifibrotic treatment in liver disease. However, it is unknown whether it is effective if given over longer periods in the treatment of established liver disease. Furthermore, the administration of recombinant ACE2 therapy as regular injections[28] would be cumbersome in clinical practice and it may be preferable to selectively increase ACE2 activity in the liver to avoid

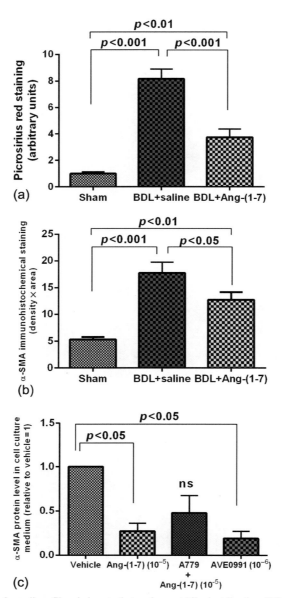

FIGURE 2 Ang-(1-7) infusion *in vivo* reduces liver fibrosis in rats that underwent bile duct ligation (BDL) (a) and inhibits α-SMA protein expression in the liver (b) and in cultured HSCs exposed to Ang-(1-7) (c). Specific MasR agonist AVE 0991 inhibits α-SMA protein secretion in cultured HSCs (c). *Adapted from Ref. 25.*

the likely side effects of systemic ACE2 administration. This may be particularly important in the treatment of chronic liver disease, since systemic RAS blockade is poorly tolerated in cirrhotic patients.[32] Thus, there is a need to develop and examine the effectiveness of other more convenient therapeutic strategies to increase intrahepatic ACE2 activity, which could potentially be used in the long-term treatment of patients with established chronic liver disease.

The use of viral vectors to carry transgenes into diseased target organs is being actively investigated as a means for correcting a range of inherited and metabolic disorders.[33] This approach is therapeutically attractive since long-term gene expression and protein expression can be achieved following a single injection of the vector with perhaps a booster at a later stage. Of the many viral vectors studied to date, adeno-associated viruses (AAVs) have been one of the most widely used for clinical trials because they are nonpathogenic in both animal models and man and have a broad tissue tropism.[33] In a recent study, we have shown that in mice, a single injection of recombinant AAV vector with a liver-specific serotype (AAV2/8) and promoter expressing ACE2 gene (rACE2-AAV) could result in sustained high-level expression of ACE2 gene and protein and increased ACE2 activity selectively in the liver for up to 6 months.[34] This treatment markedly reduced experimental liver fibrosis for up to 8 weeks when administered therapeutically, after the establishment of liver injury, in three experimental models such as BDL, toxic injury-induced hepatic fibrosis (CCl_4), and steatosis-associated hepatic fibrosis.[34] Importantly, in these studies, it was possible to show that the likely mechanism of action was via a selective increase in Ang-(1-7) levels and commensurate reduction of Ang II levels within the liver.[34]

ACE2 activator drugs offer another possible therapeutic approach since recent studies in animal models suggest these agents can reduce tissue fibrosis.[35] These drugs include small synthetic molecules such as diminazene aceturate (DIZE), which are discussed elsewhere in this book.[36] Whether such drugs, which appear to produce relatively small increases in systemic ACE2 activity,[35] might have efficacy in liver disease is not known. Another method of targeting the alternate RAS is via stimulation of the MasR, since pharmacological blockade of this receptor aggravates experimental liver fibrosis.[37] However, there are no published studies that have examined the effects of MasR agonists in experimental liver disease.

ROLE OF THE RAS IN HEPATIC RESISTANCE AND PORTAL HYPERTENSION

Portal hypertension contributes to many of the serious complications of cirrhosis, including variceal bleeding, ascites, spontaneous bacterial peritonitis, and hepatorenal syndrome. The development of portal hypertension results from both an increase in the intrahepatic resistance to portal flow and an associated vasodilation of the mesenteric vascular bed, which leads to an increase in mesenteric blood flow, much of which bypasses the liver through portosystemic collaterals (Figure 3). There is considerable experimental evidence that the RAS is involved in the pathogenesis of increased hepatic resistance in cirrhosis. This resistance has a fixed component that results from architectural distortion and obliteration of the normal sinusoidal vascular bed. However, approximately 30% of the increase is mediated by contraction of the sinusoidal vascular bed by activated HSCs and vascular smooth muscle cells (VSMCs) in response to potent local hepatic vasoconstrictors including Ang II.[26,38,39] As outlined earlier in this chapter, the phenotypic transformation of HSCs into ECM-secreting myofibroblasts following liver injury is promoted by Ang II and is associated with increased expression of RAS components and the acquisition of a contractile response to Ang II via the AT_1R.[11] In addition, ACE expression and activity are increased in the cirrhotic liver, providing a source of increased local Ang II synthesis.[6,7] Thus, not only are there elevated systemic Ang II levels in cirrhosis, but also there is likely to be elevated local Ang II production in the liver, which promotes both liver fibrosis and activation and contraction of myofibroblasts. This concept is supported by the observation that perfused cirrhotic rat liver is hyperresponsive to Ang II infusion[6] and that both AT_1R blockers and ACE inhibitors lower portal pressure in experimental cirrhosis. Unfortunately, the use of these drugs is contraindicated in patients with advanced cirrhosis who are dependent on the activation of the systemic RAS to maintain blood pressure and renal perfusion.[39]

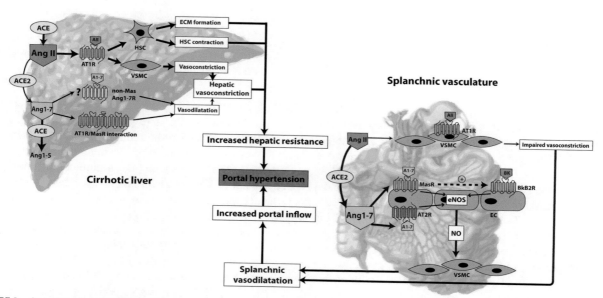

FIGURE 3 Overview of putative RAS-mediated pathophysiological changes in portal hypertension. In cirrhosis, the effects of the classic arm of the RAS and the vasoconstrictor Ang II within the hepatic vasculature predominate over those of the alternate RAS. This is primarily mediated via the AT_1R on myofibroblasts in the liver (activated HSCs) and VSMCs. In contrast, the effects of the protective arm of the RAS, mediated by ACE2 and its vasodilator peptide product (Ang-(1-7)), are predominant within the splanchnic vasculature, resulting in increased nitric oxide (NO) production by vascular endothelial cells (ECs). This exacerbates portal hypertension by increasing inflow to the portal circulation.

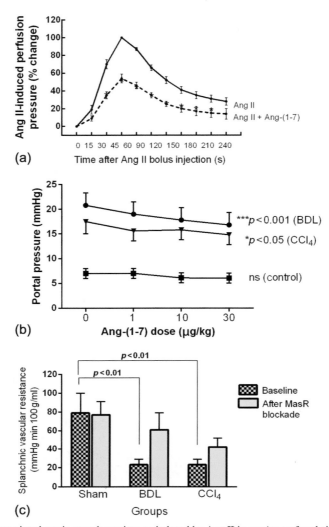

FIGURE 4 Ang-(1-7) infusion reduces intrahepatic vascular resistance induced by Ang II in *ex vivo* perfused cirrhotic rat liver (a). *In vivo* infusion of Ang-(1-7) causes a dose-dependent reduction in portal pressure in bile duct ligation (BDL) and CCl$_4$-induced cirrhotic rats (b). *In vivo*, MasR blocker A779 increases splanchnic vascular resistance in BDL and CCl$_4$-induced cirrhotic rats (c). *Adapted from Refs. 40,41.*

However, recent studies have suggested that the alternate RAS may provide other therapeutic targets for the treatment of portal hypertension. In this regard, it is of interest that the increased vasoconstrictive response evoked by both Ang II and the alpha-adrenergic agonist methoxamine in the perfused cirrhotic rat liver can be reduced by Ang-(1-7) infusion (Figure 4a).[40] This endothelium-dependent vasodilatory response to Ang-(1-7) was abolished by the nitric oxide synthase (NOS) inhibitor L-NAME or the guanylate cyclase inhibitor ODQ (1H-[1,2,4]oxadiazolo[4,3,-a]quinoxalin-1-one), suggesting the response is mediated via NO production and a NO-dependent guanylate cyclase pathway. Interestingly, while the specific MasR blocker D-alanine-Ang-(1-7) (A779) did not influence this vasodilatory effect, it was completely blocked by another Ang-(1-7) antagonist, D-Pro-Ang-(1-7).[40] This suggests that either D-Pro-Ang-(1-7) has a greater capacity to block the MasR or there is a subpopulation of other Ang-(1-7) receptors in the liver that are sensitive to D-Pro-Ang-(1-7). This latter argument is supported by the finding that the nonpeptidal MasR agonist AVE 0991 failed to inhibit the vasoconstrictive response evoked by Ang II.[40] Furthermore, we recently showed that *in vivo* infusion of Ang-(1-7) causes a dose-dependent reduction in portal pressure in rats with cirrhosis and portal hypertension (Figure 4b).[41] Taken together, these findings suggest that as in other vascular beds[42–45] in the cirrhotic liver, Ang-(1-7) causes a vasodilatory response and that by antagonizing the increase in portal pressure mediated by Ang II and other local vasoconstrictors, it could potentially produce a therapeutically useful lowering of portal pressure.

Splanchnic vasodilation, which increases portal inflow, is the other major contributor to the development and maintenance of portal hypertension. We have recently shown that both ACE2 and MasR, key components of the alternate RAS, are present in the mesenteric vascular bed and that their expression is increased in cirrhosis.[41] Similar to findings in the cirrhotic rat liver, in the mesenteric vascular bed, there is a major increase in Ang-(1-7) production from Ang II, which is completely abolished by ACE2 inhibition,[41] suggesting that an increase in mesenteric Ang-(1-7) production due to local ACE2 upregulation could contribute to mesenteric vasodilation in cirrhosis. This hypothesis is supported by the finding that Ang-(1-7) infusion attenuates methoxamine-induced vasoconstriction in the perfused mesenteric vasculature of cirrhotic rats but not in controls and that this response is mediated via increased endothelial NOS (eNOS) phosphorylation and can be blocked by inhibition of the MasR.[41] Furthermore, *in vivo* infusion of the MasR blocker A779 markedly reduced splanchnic vasodilation in animal models of cirrhosis (Figure 4c) without increasing hepatic resistance,[41] suggesting that blockade of mesenteric MasR could be used to counteract pathological splanchnic vasodilation in cirrhosis.

ROLE OF THE ALTERNATE RAS IN THE GUT

ACE2 is highly expressed throughout the GI tract in rodents.[46] The ileum appears to be the main site of intestinal ACE2 expression and activity but its role in this region of the gut is unclear. As outlined above, apart from generating local Ang-(1-7), it may be possible that ACE2 in the GI tract cleaves circulating Ang II in the mesenteric arterial blood into Ang-(1-7), which is destined to the portal circulation and liver. There is a high level of ACE2 activity in the stomach,[46] which generates local Ang-(1-7) and plays an important role in protecting gastric integrity against noxious stimuli such as stress, ethanol, and ischemia by inhibiting the generation of ROS and stimulating the production of NO and activating the free radical scavenging enzyme, catalase.[47] Components of the alternate RAS are usually upregulated in tissue injury, in response to or in concert with activation of the classic RAS. However, in patients with inflammatory bowel diseases (IBDs), Crohn's disease, and ulcerative colitis, circulating ACE2 activity and Ang-(1-7) levels are increased without changes in the levels of circulating classic RAS components.[48] Whether the alternate arm of the RAS plays a protective role in this setting or there are links between these diseases and ACE2 activity in the gut (particularly in the terminal ileum) is unknown.[46]

Interestingly, it appears that ACE2 has other functions in the GI tract independent of its role in RAS regulation. Recent evidence suggests that ACE2 deficiency causes epithelial damage leading to intestinal inflammation and that this enzyme has a key role in intestinal amino acid absorption. There is also evidence that it affects gut innate immunity and microbial ecology.[49,50]

CONCLUSION

There have been major recent advances in our understanding of the complexities of the RAS and its role in the pathogenesis of chronic liver disease, portal hypertension, and GI tract pathophysiology. It is apparent that the RAS has multiple roles in the pathogenesis of liver fibrosis and portal hypertension. Recent findings suggesting anti-inflammatory, antifibrotic, and vasodilatory functions of the alternate arm of the RAS in liver disease open avenues for the development of drugs selectively targeting the key enzymes and/or receptors of this protective arm. There is also new evidence showing that there is increased regional production of Ang-(1-7) in the liver and mesenteric vascular bed in cirrhosis and that long-term ACE2 therapy or oral formulations of Ang-(1-7) hold promise for the treatment of patients with chronic liver disease. The protective arm of the RAS driven by ACE2 may also play an important protective role in the GI tract including providing gastro-protection against noxious stimuli and in the pathophysiology of inflammatory bowel disease. ACE2 may also function independent of its role in angiotensin peptide metabolism to regulate gut amino acid homeostasis and to protect from intestinal inflammation. These findings suggest that ACE2-Ang-(1-7)-MasR axis provides a range of novel potential targets for the management of both liver and GI tract disorders.

REFERENCES

1. Lubel JS, Herath CB, Burrell LM, Angus PW. Liver disease and the renin-angiotensin system: Recent discoveries and clinical implications. *J Gastroenterol Hepatol* 2008;**23**:1327–38.
2. Paul M, Poyan Mehr A, Kreutz R. Physiology of local renin-angiotensin systems. *Physiol Rev* 2006;**86**:747–803.
3. Grace JA, Herath CB, Mak KY, Burrell LM, Angus PW. Update on new aspects of the renin-angiotensin system in liver disease: clinical implications and new therapeutic options. *Clin Sci (Lond)* 2012;**123**:225–39.
4. Dickstein K, Kjekshus J. Effects of losartan and captopril on mortality and morbidity in high-risk patients after acute myocardial infarction: the OPTIMAAL randomised trial. Optimal Trial in Myocardial Infarction with Angiotensin II Antagonist Losartan. *Lancet* 2002;**360**:752–60.
5. Parving HH, Persson F, Lewis JB, Lewis EJ, Hollenberg NK. Aliskiren combined with losartan in type 2 diabetes and nephropathy. *N Engl J Med* 2008;**358**:2433–46.

6. Herath CB, Warner FJ, Lubel JS, Dean RG, Jia Z, Lew RA, et al. Upregulation of hepatic angiotensin-converting enzyme 2 (ACE2) and angiotensin-(1-7) levels in experimental biliary fibrosis. *J Hepatol* 2007;**47**:387–95.

7. Paizis G, Cooper ME, Schembri JM, Tikellis C, Burrell LM, Angus PW. Up-regulation of components of the renin-angiotensin system in the bile duct-ligated rat liver. *Gastroenterology* 2002;**123**:1667–76.

8. Bataller R, Brenner DA. Liver fibrosis. *J Clin Invest* 2005;**115**:209–18.

9. Friedman SL. Mechanisms of hepatic fibrogenesis. *Gastroenterology* 2008;**134**:1655–69.

10. Bataller R, Sancho-Bru P, Gines P, Lora JM, Al-Garawi A, Sole M, et al. Activated human hepatic stellate cells express the renin-angiotensin system and synthesize angiotensin II. *Gastroenterology* 2003;**125**:117–25.

11. Bataller R, Gines P, Nicolas JM, Gorbig MN, Garcia-Ramallo E, Gasull X, et al. Angiotensin II induces contraction and proliferation of human hepatic stellate cells. *Gastroenterology* 2000;**118**:1149–56.

12. Bataller R, Schwabe RF, Choi YH, Yang L, Paik YH, Lindquist J, et al. NADPH oxidase signal transduces angiotensin II in hepatic stellate cells and is critical in hepatic fibrosis. *J Clin Invest* 2003;**112**:1383–94.

13. Hirose A, Ono M, Saibara T, Nozaki Y, Masuda K, Yoshioka A, et al. Angiotensin II type 1 receptor blocker inhibits fibrosis in rat nonalcoholic steatohepatitis. *Hepatology* 2007;**45**:1375–81.

14. Ueki M, Koda M, Yamamoto S, Matsunaga Y, Murawaki Y. Preventive and therapeutic effects of angiotensin II type 1 receptor blocker on hepatic fibrosis induced by bile duct ligation in rats. *J Gastroenterol* 2006;**41**:996–1004.

15. Iwata M, Cowling RT, Gurantz D, Moore C, Zhang S, Yuan JX, et al. Angiotensin-(1-7) binds to specific receptors on cardiac fibroblasts to initiate antifibrotic and antitrophic effects. *Am J Physiol Heart Circ Physiol* 2005;**289**:H2356–63.

16. Jonsson JR, Clouston AD, Ando Y, Kelemen LI, Horn MJ, Adamson MD, et al. Angiotensin-converting enzyme inhibition attenuates the progression of rat hepatic fibrosis. *Gastroenterology* 2001;**121**:148–55.

17. Paizis G, Gilbert RE, Cooper ME, Murthi P, Schembri JM, Wu LL, et al. Effect of angiotensin II type 1 receptor blockade on experimental hepatic fibrogenesis. *J Hepatol* 2001;**35**:376–85.

18. Toblli JE, Munoz MC, Cao G, Mella J, Pereyra L, Mastai R. ACE inhibition and AT1 receptor blockade prevent fatty liver and fibrosis in obese Zucker rats. *Obesity (Silver Spring)* 2008;**16**:770–6.

19. Colmenero J, Bataller R, Sancho-Bru P, Dominguez M, Moreno M, Forns X, et al. Effects of losartan on hepatic expression of nonphagocytic NADPH oxidase and fibrogenic genes in patients with chronic hepatitis C. *Am J Physiol Gastrointest Liver Physiol* 2009;**297**:G726–34.

20. Terui Y, Saito T, Watanabe H, Togashi H, Kawata S, Kamada Y, et al. Effect of angiotensin receptor antagonist on liver fibrosis in early stages of chronic hepatitis C. *Hepatology* 2002;**36**:1022.

21. Yoshiji H, Noguchi R, Fukui H. Combined effect of an ACE inhibitor, perindopril, and interferon on liver fibrosis markers in patients with chronic hepatitis C. *J Gastroenterol* 2005;**40**:215–16.

22. Yokohama S, Yoneda M, Haneda M, Okamoto S, Okada M, Aso K, et al. Therapeutic efficacy of an angiotensin II receptor antagonist in patients with nonalcoholic steatohepatitis. *Hepatology* 2004;**40**:1222–5.

23. Georgescu EF. Angiotensin receptor blockers in the treatment of NASH/NAFLD: could they be a first-class option? *Adv Ther* 2008;**25**:1141–74.

24. Paizis G, Tikellis C, Cooper ME, Schembri JM, Lew RA, Smith AI, et al. Chronic liver injury in rats and humans upregulates the novel enzyme angiotensin converting enzyme 2. *Gut* 2005;**54**:1790–6.

25. Lubel JS, Herath CB, Tchongue J, Grace J, Jia Z, Spencer K, et al. Angiotensin-(1-7), an alternative metabolite of the renin-angiotensin system, is up-regulated in human liver disease and has antifibrotic activity in the bile-duct-ligated rat. *Clin Sci (Lond)* 2009;**117**:375–86.

26. Herath CB, Lubel JS, Jia Z, Velkoska E, Casley D, Brown L, et al. Portal pressure responses and angiotensin peptide production in rat liver are determined by relative activity of ACE and ACE2. *Am J Physiol Gastrointest Liver Physiol* 2009;**297**:G98–106.

27. Marques FD, Ferreira AJ, Sinisterra RD, Jacoby BA, Sousa FB, Caliari MV, et al. An oral formulation of angiotensin-(1-7) produces cardioprotective effects in infarcted and isoproterenol-treated rats. *Hypertension* 2011;**57**:477–83.

28. Osterreicher CH, Taura K, De Minicis S, Seki E, Penz-Osterreicher M, Kodama Y, et al. Angiotensin-converting-enzyme 2 inhibits liver fibrosis in mice. *Hepatology* 2009;**50**:929–38.

29. Oudit GY, Liu GC, Zhong J, Basu R, Chow FL, Zhou J, et al. Human recombinant ACE2 reduces the progression of diabetic nephropathy. *Diabetes* 2010;**59**:529–38.

30. Wysocki J, Ye M, Rodriguez E, Gonzalez-Pacheco FR, Barrios C, Evora K, et al. Targeting the degradation of angiotensin II with recombinant angiotensin-converting enzyme 2: prevention of angiotensin II-dependent hypertension. *Hypertension* 2010;**55**:90–8.

31. Schuster M, Loibner H, Poglitsch M, Janzek E, Stranner S, Haschke M, et al. Pharmacology of recombinant human angiotensinconverting enzyme 2 in healthy volunteers. *Arteriosclerosis, thrombosis, and vascular biology 2011 scientific sessions, Chicago*; 2011. p. 450.

32. Tandon P, Abraldes JG, Berzigotti A, Garcia-Pagan JC, Bosch J. Renin-angiotensin-aldosterone inhibitors in the reduction of portal pressure: a systematic review and meta-analysis. *J Hepatol* 2010;**53**:273–82.

33. Alexander IE, Cunningham SC, Logan GJ, Christodoulou J. Potential of AAV vectors in the treatment of metabolic disease. *Gene Ther* 2008;**15**:831–9.

34. Mak KY, Chin R, Toressi J, Cunningham S, Alexander IE, Angus PW, et al. Angiotensin converting enzyme 2 (ACE2) gene therapy using high efficiency and liver specific adeno-associated viral vector attenuates experimental liver fibrosis in mice. *J Hepatol* 2014;**60**:598.

35. Hernandez Prada JA, Ferreira AJ, Katovich MJ, Shenoy V, Qi Y, Santos RA, et al. Structure-based identification of small-molecule angiotensin-converting enzyme 2 activators as novel antihypertensive agents. *Hypertension* 2008;**51**:1312–17.

36. Qi Y, Zhang J, Cole-Jeffrey CT, Shenoy V, Espejo A, Hanna M, et al. Diminazene aceturate enhances angiotensin-converting enzyme 2 activity and attenuates ischemia-induced cardiac pathophysiology. *Hypertension* 2013;**62**:746–52.

37. Pereira RM, Dos Santos RA, Teixeira MM, Leite VH, Costa LP, da Costa Dias FL, et al. The renin-angiotensin system in a rat model of hepatic fibrosis: evidence for a protective role of Angiotensin-(1-7). *J Hepatol* 2007;**46**:674–81.

38. Rockey D. The cellular pathogenesis of portal hypertension: stellate cell contractility, endothelin, and nitric oxide. *Hepatology* 1997;**25**:2–5.

39. Herath CB, Grace JA, Angus PW. Therapeutic potential of targeting the renin angiotensin system in portal hypertension. *World J Gastrointest Pathophysiol* 2013;**4**:1–11.

40. Herath CB, Mak K, Burrell LM, Angus PW. Angiotensin-(1-7) reduces the perfusion pressure response to angiotensin II and methoxamine via an endothelial nitric oxide-mediated pathway in cirrhotic rat liver. *Am J Physiol Gastrointest Liver Physiol* 2013;**304**:G99–108.

41. Grace JA, Klein S, Herath CB, Granzow M, Schierwagen R, Masing N, et al. Activation of the MAS receptor by angiotensin-(1-7) in the renin-angiotensin system mediates mesenteric vasodilatation in cirrhosis. *Gastroenterology* 2013;**145**:874–84.

42. Castro CH, Santos RA, Ferreira AJ, Bader M, Alenina N, Almeida AP. Evidence for a functional interaction of the angiotensin-(1-7) receptor Mas with AT1 and AT2 receptors in the mouse heart. *Hypertension* 2005;**46**:937–42.

43. Lemos VS, Silva DM, Walther T, Alenina N, Bader M, Santos RA. The endothelium-dependent vasodilator effect of the nonpeptide Ang(1-7) mimic AVE 0991 is abolished in the aorta of mas-knockout mice. *J Cardiovasc Pharmacol* 2005;**46**:274–9.

44. Santos RA, Simoes e Silva AC, Maric C, Silva DM, Machado RP, de Buhr I, et al. Angiotensin-(1-7) is an endogenous ligand for the G protein-coupled receptor Mas. *Proc Natl Acad Sci U S A* 2003;**100**:8258–63.

45. Silva DM, Vianna HR, Cortes SF, Campagnole-Santos MJ, Santos RA, Lemos VS. Evidence for a new angiotensin-(1-7) receptor subtype in the aorta of Sprague-Dawley rats. *Peptides* 2007;**28**:702–7.

46. Gembardt F, Sterner-Kock A, Imboden H, Spalteholz M, Reibitz F, Schultheiss HP, et al. Organ-specific distribution of ACE2 mRNA and correlating peptidase activity in rodents. *Peptides* 2005;**26**:1270–7.

47. Brzozowski T, Ptak-Belowska A, Kwiecien S, Krzysiek-Maczka G, Strzalka M, Drozdowicz D, et al. Novel concept in the mechanism of injury and protection of gastric mucosa: role of renin-angiotensin system and active metabolites of angiotensin. *Curr Med Chem* 2012;**19**:55–62.

48. Garg M, Burrell LM, Velkoska E, Griggs K, Angus PW, Gibson PR, et al. Upregulation of circulating components of the alternative renin-angiotensin system in inflammatory bowel disease: A pilot study. *J Renin Angiotensin Aldosterone Syst* 2014; February 6 [Epub ahead of print].

49. Hashimoto T, Perlot T, Rehman A, Trichereau J, Ishiguro H, Paolino M, et al. ACE2 links amino acid malnutrition to microbial ecology and intestinal inflammation. *Nature* 2012;**487**:477–81.

50. Perlot T, Penninger JM. ACE2 - from the renin-angiotensin system to gut microbiota and malnutrition. *Microbes infect* 2013;**15**:866–73.

Chapter 35

Metabolic Role of Angiotensin-(1-7)/Mas Axis

Sérgio Henrique Sousa Santos,* Fernando Pablo Dominici†

*Department of Pharmacology, Universidade Federal de Minas Gerais, Belo Horizonte, Minas Gerais, Brazil, †Department of Biological Chemistry, IQUIFIB, University of Buenos Aires, Buenos Aires, Argentina

ABBREVIATIONS

ACE	angiotensin-converting enzyme
AGT	angiotensinogen
Ang II	angiotensin II
Ang-(1-7)	angiotensin-(1-7)
ARB	type 1 angiotensin receptor blockers
AS160	Akt substrate of 160 kDa
AT_1R	type 1 Ang II receptor
AT_2R	type 2 Ang II receptor
FOXO1	forkhead box O1
GLUT4	(glucose transporter 4)
GSK3	glycogen synthase kinase-3
IRS	insulin receptor substrate
JAK2	Janus kinase 2
IL-6	interleukin 6; IR, insulin receptor
IRS	insulin receptor substrate
MAPK	mitogen-activated protein kinase
NOS	nitric oxide synthase
PI3K	phosphatidylinositol-3 kinase
PPARγ	peroxisome proliferator-activated receptor γ
RAS	renin–angiotensin system
Ser	serine
STZ	streptozotocin
Thr	threonine
TLR4	Toll-like receptor 4
TNF-α	tumor necrosis factor-α
Tyr	tyrosine
VSMC	vascular smooth muscle cells

INTRODUCTION

The body's metabolic homeostasis is a complex integrative tissue and systemic signaling responsible for the maintenance of energy balance (calorie intake vs. consumption). However, energy storage is mainly accomplished by the liver, adipose tissue, and muscle.[1–6] These three organs are also mainly responsible for the stimulation of the pancreas hormones: insulin and glucagon.[7,8]

The Protective Arm of the Renin–Angiotensin System (RAS). http://dx.doi.org/10.1016/B978-0-12-801364-9.00035-3

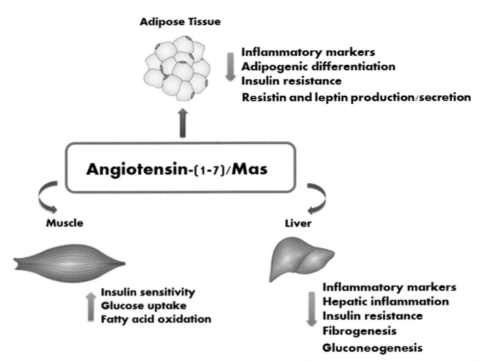

FIGURE 1 Effects of Ang-(1-7)/Mas axis activation in metabolic tissues: Modulation of adipokines in adipose tissue; reduction of inflammatory process and gluconeogenesis in the liver and metabolic regulation in the skeletal muscle.

Several endocrine components are involved in the regulation of body catabolism and anabolism. In recent years, the renin–angiotensin system gained an important role with the characterization of angiotensin II (Ang II) and angiotensin-(1-7) (Ang-(1-7)) metabolic effects.[9,10]

The renin–angiotensin system (RAS) is initiated by the expression of angiotensinogen (AGT) at different tissues and cells including hepatocytes, myocytes, and adipocytes. However, the liver is regarded as the primary source of AGT in the nondiseased state.[11] Renin enzyme cleaves AGT to produce Ang I, a peptide that is rapidly hydrolyzed by the angiotensin-converting enzyme (ACE) to the octapeptide, Ang II. Besides Ang II, several other angiotensin peptides formed from AGT have biological activity; nevertheless, the main one is Ang-(1-7), which broadly opposes Ang II actions. Ang-(1-7) can also be produced as a product from the degradation of Ang II through the action of the ACE homologue enzyme, the so-called ACE2.[11]

Ang-(1-7) and Ang II have been implicated in insulin resistance and associated with metabolic syndrome, within antagonistic actions.[12–19] Previous studies demonstrated that Ang II by acting through type 1 (AT$_1$) receptors produces endothelial dysfunction and metabolic changes leading to hypertension and metabolic syndrome.[18] More recent studies have demonstrated that Ang-(1-7) by acting at the Mas receptor opposes several metabolic effects of Ang II, which in turn improves insulin sensitivity, glucose tolerance, and type 2 diabetes; reduces body fat; increases adiponectin production; and reverses hyperleptinemia.[12–17,19,20] Indeed, Ang-(1-7) decreases the release of inflammatory mediators by the adipose tissue in obese animal models[15,21] (Figure 1). More recently, Andrade et al. demonstrated that Ang-(1-7) modulates sirtuins (enzymes that regulate metabolism)[12] and improves type 2 diabetes.[22,23] Besides, Muñoz et al. showed that the Mas receptor is an essential component of the insulin receptor signaling pathway.[24] Taken together, all these facts indicate that blunting Ang II actions and stimulating Ang-(1-7) signaling appears an attractive therapeutic approach for chronic metabolic disorders.

ANG-(1-7)/MAS MODULATING INSULIN EFFECTS AND SIGNALING

A cross talk between the insulin signaling system and Ang-(1-7) has been demonstrated (Figure 2). Given that Ang-(1-7) can improve insulin sensitivity, this signaling overlap could be implicated in the metabolic role of Ang-(1-7). Ang-(1-7) is capable of stimulating the phosphorylation of IRS-1 in rat heart.[1] We have demonstrated that this stimulating effect of

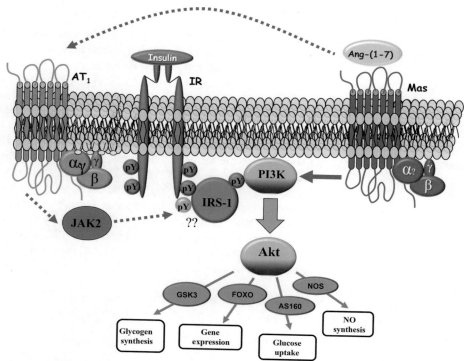

FIGURE 2 Signaling molecules engaged by Ang-(1-7) related to insulin action. Ang-(1-7) stimulates IRS-1 phosphorylation through an AT_1R-JAK2-mediated mechanism, but the role of this pathway is unknown (??). Via Mas receptor, Ang-(1-7) stimulates the PI3K/Akt pathway leading to an enhancement of insulin actions. pY: phosphotyrosine.

Ang-(1-7) involves the AT_1R through a mechanism that requires participation of JAK2.[25] Both the time and concentration necessary to attain the maximal stimulating response were similar to those previously described for Ang II.[26,27] The demonstration that IRS-1 is used by Ang-(1-7) reinforces the previously postulated concept that the IRS proteins serve as a convergence site for the signal transduction of several hormones and cytokines.[28]

The Ser/Thr kinase Akt has a crucial role in Ang-(1-7) signaling. Phosphorylation of Akt at regulatory sites Ser^{473} and Thr^{308} is stimulated by Ang-(1-7) in rat heart,[25] liver, skeletal muscle, and adipose tissue[19] *in vivo* as well as in human endothelial cells *in vitro*.[29] Several studies show that PI3K participates in Ang-(1-7)-induced Akt activation.[25,29,30] Time-resolved quantitative phosphoproteomic analysis corroborated these findings, situating Akt in a central place in Ang-(1-7) actions.[31] The PI3K/Akt pathway participates in several of the Ang-(1-7) actions including stimulation of glycerol release in adipocytes, promotion of the formation of small insulin-sensitive newly differentiated adipocytes,[32,33] enhancement of survival and proliferation of CD34(+) cells in diabetic patients,[34] maintenance of cardiomyocyte function,[35] and stimulation of atrial natriuretic peptide secretion.[36] Ang-(1-7) activates endothelial NOS leading to vasodilation. This process is triggered by the Mas receptor and requires activation of Akt.[29] The Ser/Thr kinase GSK-3β regulates a wide range of cellular processes including glycogen metabolism, gene transcription, protein translation, and cell apoptosis being one of the principal downstream targets of Akt.[37] Under basal conditions, GSK-3β is highly active and inhibits glycogen synthesis by phosphorylation of glycogen synthase.[37] Insulin-activated Akt phosphorylates GSK-3β leading to stimulation of glycogen synthesis and protein synthesis.[37] AS160 is a Rab GTPase-activating protein and important target of Akt. It regulates insulin-stimulated GLUT4 trafficking[38] and glucose uptake.[39] The importance of AS160 in insulin-mediated glucose transport has been highlighted recently by the phenotype of AS160-knockout mice that display whole-body insulin resistance.[40] Ang-(1-7) induces the phosphorylation of GSK-3β and AS160 in metabolic tissues of the rat *in vivo*.[19,24] As a corollary of this discovery, it has recently been shown that Ang-(1-7) facilitates glucose transport both in adipocytes and in neonatal cardiomyocytes.[22,41] Interestingly, in rats *in vivo*, the euglycemic clamp exhibited that Ang-(1-7) by itself did not promote glucose transport, but it increased insulin-stimulated glucose transport.[42] This was corroborated in a study published almost in parallel where it was shown that Ang-(1-7) recruits muscle microvasculature and enhances insulin's metabolic action via the Mas receptor.[43] In all, current knowledge suggests that endogenous (systemic and/or local) Ang-(1-7) could regulate insulin-mediated glucose transport *in vivo*.

LIVER METABOLIC EFFECTS OF ANG-(1-7)/MAS AXIS

As described previously, the liver is considered the primary source of circulating angiotensinogen (AGT) in normal physiology and modulates local and systemic RAS.[44] The RAS plays a pivotal role during several hepatic diseases. One of its main effectors, angiotensin II, is a potent proinflammatory, prooxidant, and profibrotic agent that interferes in several steps of intracellular insulin signaling.[11,45] On the other hand, ACE2 also plays a pivotal protective role in liver disorders: A relative ACE2 deficiency may lead to Ang II impaired degradation and accumulation contributing to increased inflammatory process and fibrosis.[46,47] A recent study by Bilman et al. showed a decreased gluconeogenesis in rats with increased circulating Ang-(1-7).[48]

Hepatic steatosis is characterized as an abnormal accumulation of lipids in hepatocytes and is associated with diseases such as type 2 diabetes mellitus, obesity, and some metabolic disorders. In addition, the increased fat accumulation in the liver is a marker of hepatic insulin resistance.[49,50] Rats treated with Ang-(1-7) have reduced hepatic fat accumulation, lowered fibrosis, and lobular inflammation. Ang-(1-7) can prevent in rats induced hepatic fibrosis by high-fructose diet.[20] Angiotensin (1-7) also reduces hepatic steatosis, liver weight, and fat mass, decreasing proinflammatory cytokines expression as IL-6 and TNF-α in high-fat-fed mice.[13] These results also were accompanied by regulation of RAS gene expression.[13] In the hypothalamus, the resistin-TLR4 signaling leads to the activation of MAPK induced by the inflammatory process.[51] Interestingly, transgenic rats overexpressing Ang-(1-7) showed diminished liver resistin levels and reduced TLR4, TNF-α, and IL-6 mRNA expressions accompanied by increased ACE2 and low phosphorylation of MAPK, reducing inflammation when challenged with a high-fat diet.[14]

In conclusion, the RAS appears to have a pivotal role on liver metabolism modulation. Thus, pharmacological approaches aiming to increase Ang-(1-7)/Mas effects and reduce those of Ang II could be a key tool for the treatment and prevention of liver disorders.

NEW METABOLIC REMARKS AND PERSPECTIVES

Promising preclinical studies suggest that modulation of the ACE2/Ang-(1-7)/Mas receptor axis could improve hemodynamic and metabolic diseases in humans. Several meta-analyses have underscored the positive effects of ARB and ACE inhibitors on insulin sensitivity and the progression to type 2 diabetes.[52,53] Since the ACE2/Ang-(1-7)/Mas receptor axis naturally counterbalances the effects of classical RAS components, it is reasonable to believe that part of the ARB and ACE inhibitors' positive effects on metabolic diseases could be mediated by overactivation of the Ang-(1-7) pathway. Indeed, it was recently suggested that the beneficial effects of the ARB olmesartan on vascular remodeling are mediated via activation of the ACE2/Ang-(1-7)/Mas axis.[54] However, ACE inhibitors fail to completely block Ang II formation since many other alternative proteases can still produce Ang II and generate hyperglycemia in the presence of ACE inhibitors.[55] ACE2, on the other hand, acts directly on Ang II, irrespective of the pathways involved in the production of this octapeptidic hormone. Thus, by ensuring depletion of Ang II in the body, ACE2 therapy also increases Ang-(1-7), emerging as a more efficient treatment than ACE inhibition in combating Ang II-mediated hyperglycemia. ACE2 is a novel target for the prevention of β-cell dysfunction and apoptosis, both hallmarks of type 2 diabetes onset. Overexpression of ACE2 in the pancreas of db/db mice significantly improved fasting glycemia, enhanced glucose tolerance, and increased islet insulin content and β-cell proliferation in these animals.[56]

Ang-(1-7) exerts beneficial effects on pathologies associated with type 2 diabetes. Chronic administration of Ang-(1-7) or of its synthetic analog, AVE 0991, prevented the diabetes-induced abnormal vascular responsiveness of the mesenteric bed and isolated carotid and renal arteries in diabetic rats.[57] Further studies demonstrated that the treatment with Ang-(1-7) ameliorated cardiomyopathy in diabetic animals.[58] Also, treatment with AVE 0991 was shown to rescue cardiac function under diabetic conditions as indicated by a normalization of blood pressure and contractility parameters.[59] Lately, it was shown that chronic infusion of the ACE2 activator XNT to STZ-induced diabetic rats improved endothelial-dependent vasorelaxation of aortic rings through a Mas receptor-dependent mechanism.[60] In general, these studies imply a cardioprotective role for Ang-(1-7) under hyperglycemic conditions and point to new therapeutic strategies using Ang-(1-7) agonists to treat cardiovascular complications associated with type 2 diabetes mellitus. Ang-(1-7) also exerts renoprotective effects on experimental models of diabetic nephropathy, associated with reduction of inflammation, fibrosis, oxidative stress, and lipotoxicity.[61–64] In view of the great therapeutic impact of the modulation of the ACE2/Ang-(1-7)/Mas receptor axis, a formulation of Ang-(1-7) included into the oligosaccharide hydroxypropyl β-cyclodextrin (HPβCD)[22] cavity that protects the peptide during the passage through the gastrointestinal tract when orally administered has been developed.[65] This oral formulation of Ang-(1-7) represents a new alternative in the treatment of the metabolic syndrome and diabetic complications.[13,14,22]

REFERENCES

1. Postic C, Dentin R, Girard J. Role of the liver in the control of carbohydrate and lipid homeostasis. *Diabetes Metab* 2004;**30**:398–408.
2. Mizgier ML, Casas M, Contreras-Ferrat A, Llanos P, Galgani JE. Potential role of skeletal muscle glucose metabolism on the regulation of insulin secretion. *Obes Rev* 2014;**15**(7):587–97. doi:10.1111/obr.12166.
3. Braun T, Gautel M. Transcriptional mechanisms regulating skeletal muscle differentiation, growth and homeostasis. *Nat Rev Mol Cell Biol* 2011;**12**:349–61.
4. Oosterveer MH, Schoonjans K. Hepatic glucose sensing and integrative pathways in the liver. *Cell Mol Life Sci* 2014;**71**:1453–67.
5. Klaus S. Adipose tissue as a regulator of energy balance. *Curr Drug Targets* 2004;**5**:241–50.
6. Rosen ED, Spiegelman BM. Adipocytes as regulators of energy balance and glucose homeostasis. *Nature* 2006;**444**:847–53.
7. Romacho T, Elsen M, Rohrborn D, Eckel J. Adipose tissue and its role in organ crosstalk. *Acta Physiol* 2014;**210**:733–53.
8. Leto D, Saltiel AR. Regulation of glucose transport by insulin: traffic control of GLUT4. *Nat Rev Mol Cell Biol* 2012;**13**:383–96.
9. Santos RA, Campagnole-Santos MJ, Andrade SP. Angiotensin-(1–7): an update. *Regul Pept* 2000;**91**:45–62.
10. Kackstein K, Teren A, Matsumoto Y, Mangner N, Mobius-Winkler S, Linke A, et al. Impact of angiotensin II on skeletal muscle metabolism and function in mice: contribution of IGF-1, Sirtuin-1 and PGC-1alpha. *Acta Histochem* 2013;**115**:363–70.
11. Santos SH, Simoes e Silva AC. The therapeutic role of renin-angiotensin system blockers in obesity- related renal disorders. *Curr Clin Pharmacol* 2014;**9**:2–9.
12. Oliveira Andrade JM, Paraiso AF, Garcia ZM, Ferreira AV, Sinisterra RD, Sousa FB, et al. Cross talk between angiotensin-(1–7)/Mas axis and sirtuins in adipose tissue and metabolism of high-fat feed mice. *Peptides* 2014;**55C**:158–65.
13. Feltenberger JD, Andrade JM, Paraiso A, Barros LO, Filho AB, Sinisterra RD, et al. Oral formulation of angiotensin-(1–7) improves lipid metabolism and prevents high-fat diet-induced hepatic steatosis and inflammation in mice. *Hypertension* 2013;**62**:324–30.
14. Santos SH, Andrade JM, Fernandes LR, Sinisterra RD, Sousa FB, Feltenberger JD, et al. Oral angiotensin-(1–7) prevented obesity and hepatic inflammation by inhibition of resistin/TLR4/MAPK/NF-kappaB in rats fed with high-fat diet. *Peptides* 2013;**46**:47–52.
15. Santos SH, Fernandes LR, Pereira CS, Guimaraes AL, de Paula AM, Campagnole-Santos MJ, et al. Increased circulating angiotensin-(1–7) protects white adipose tissue against development of a proinflammatory state stimulated by a high-fat diet. *Regul Pept* 2012;**178**:64–70.
16. Santos SH, Fernandes LR, Mario EG, Ferreira AV, Porto LC, Alvarez-Leite JI, et al. Mas deficiency in FVB/N mice produces marked changes in lipid and glycemic metabolism. *Diabetes* 2008;**57**:340–7.
17. Santos SH, Braga JF, Mario EG, Pôrto LC, Rodrigues-Machado Mda G, Murari A, et al. Improved lipid and glucose metabolism in transgenic rats with increased circulating angiotensin-(1–7). *Arterioscler Thromb Vasc Biol* 2010;**30**:953–61.
18. Olivares-Reyes JA, Arellano-Plancarte A, Castillo-Hernandez JR. Angiotensin II and the development of insulin resistance: implications for diabetes. *Mol Cell Endocrinol* 2009;**302**:128–39.
19. Giani JF, Mayer MA, Muñoz MC, Silberman EA, Höcht C, Taira CA, et al. Chronic infusion of angiotensin-(1–7) improves insulin resistance and hypertension induced by a high-fructose diet in rats. *Am J Physiol Endocrinol Metab* 2009;**296**:E262–71.
20. Marcus Y, Shefer G, Sasson K, Kohen F, Limor R, Pappo O, et al. Angiotensin 1–7 as means to prevent the metabolic syndrome: lessons from the fructose-fed rat model. *Diabetes* 2013;**62**:1121–30.
21. Souza LL, Costa-Neto CM. Angiotensin-(1–7) decreases LPS-induced inflammatory response in macrophages. *J Cell Physiol* 2012;**227**:2117–22.
22. Santos SH, Giani JF, Burghi V, Miquet JG, Qadri F, Braga JF, et al. Oral administration of angiotensin-(1–7) ameliorates type 2 diabetes in rats. *J Mol Med* 2014;**92**:255–65.
23. Bertagnolli M, Casali KR, De Sousa FB, Rigatto K, Becker L, Santos SH, et al. An orally active angiotensin-(1–7) inclusion compound and exercise training produce similar cardiovascular effects in spontaneously hypertensive rats. *Peptides* 2014;**51**:65–73.
24. Muñoz MC, Giani JF, Burghi V, Mayer MA, Carranza A, Taira CA, et al. The Mas receptor mediates modulation of insulin signaling by angiotensin-(1–7). *Regul Pept* 2012;**177**:1–11.
25. Giani JF, Gironacci MM, Muñoz MC, Peña C, Turyn D, Dominici FP. Angiotensin-(1–7) stimulates the phosphorylation of JAK2, IRS-1 and Akt in rat heart in vivo: role of the AT1 and Mas receptors. *Am J Physiol Heart Circ Physiol* 2007;**293**:1154–63.
26. Saad MJ, Velloso LA, Carvalho CR. Angiotensin II induces tyrosine phosphorylation of insulin receptor substrate 1 and its association with phosphatidylinositol 3-kinase in rat heart. *Biochem J* 2005;**310**:741–4.
27. Ali MS, Schieffer B, Delafontaine P, Bernstein KE, Ling BN, Marrero MB. Angiotensin II stimulates tyrosine phosphorylation and activation of insulin receptor substrate 1 and protein-tyrosine phosphatase 1D in vascular smooth muscle cells. *J Biol Chem* 1997;**272**:12373–9.
28. Shaw LM. The insulin receptor substrate (IRS) proteins: at the intersection of metabolism and cancer. *Cell Cycle* 2011;**10**:1750.
29. Sampaio WO, Souza dos Santos RA, Faria-Silva R, Mata Machado LT, Schiffrin EL, Touyz RM. Angiotensin-(1–7) through receptor Mas mediates endothelial nitric oxide synthase activation via Akt-dependent pathways. *Hypertension* 2007;**49**:185–92.
30. Tassone EJ, Sciacqua A, Andreozzi F, Presta I, Perticone M, Carnevale D, et al. Angiotensin (1–7) counteracts the negative effect of angiotensin II on insulin signalling in HUVECs. *Cardiovasc Res* 2013;**99**:129–36.
31. Verano-Braga T, Schwämmle V, Sylvester M, Passos-Silva DG, Peluso AAB, Etelvino Gisele M, et al. Time-resolved quantitative phosphoproteomics: new insights into angiotensin-(1–7) signaling networks in human endothelial cells. *J Proteome Res* 2012;**11**:3370–81.
32. Oh YB, Kim JH, Park BM, Park BH, Kim SH. Captopril intake decreases body weight gain via angiotensin-(1–7). *Peptides* 2012;**37**:79–85.
33. Than A, Leow MK, Chen P. Control of adipogenesis by the autocrine interplays between angiotensin (1–7)/Mas receptor and angiotensin II/AT1 receptor signaling pathways. *J Biol Chem* 2013;**288**:15520–31.
34. Jarajapu YP, Bhatwadekar AD, Caballero S, Hazra S, Shenoy V, Medina R, et al. Activation of the ACE2/angiotensin-(1–7)/Mas receptor axis enhances the reparative function of dysfunctional diabetic endothelial progenitors. *Diabetes* 2013;**62**:1258–69.

35. Dias-Peixoto MF, Santos RA, Gomes ER, Alves MN, Almeida PW, Greco L, et al. Molecular mechanisms involved in the angiotensin-(1–7)/Mas signaling pathway in cardiomyocytes. *Hypertension* 2008;**52**:542–8.

36. Shah A, Gul R, Yuan K, Gao S, Oh YB, Kim UH, et al. Angiotensin-(1–7) stimulates high atrial pacing-induced ANP secretion via Mas/PI3-kinase/Akt axis and Na+/H+ exchanger. *Am J Physiol Heart Circ Physiol* 2010;**298**:H1365–74.

37. Rayasam GV, Tulasi VK, Sodhi R, Davis JA, Ray A. Glycogen synthase kinase 3: more than a namesake. *Br J Pharmacol* 2009;**156**:885–98.

38. Sano H, Kane S, Sano E, Mîinea CP, Asara JM, Lane WS, et al. Insulin-stimulated phosphorylation of a Rab GTPase-activating protein regulates GLUT4 translocation. *J Biol Chem* 2003;**278**:14599–602.

39. Kramer HF, Witczak CA, Taylor EB, Fujii N, Hirshman MF, Goodyear LJ. AS160 regulates insulin- and contraction-stimulated glucose uptake in mouse skeletal muscle. *J Biol Chem* 2006;**281**:31478–85.

40. Wang HY, Ducommun S, Quan C, Xie B, Li M, Wasserman DH, et al. AS160 deficiency causes whole-body insulin resistance via composite effects in multiple tissues. *Biochem J* 2013;**449**:479–89.

41. Liu C, Lv XH, Li HX, Cao X, Zhang F, Wang L, et al. Angiotensin-(1–7) suppresses oxidative stress and improves glucose uptake via Mas receptor in adipocytes. *Acta Diabetol* 2012;**49**:291–9.

42. Echeverría-Rodríguez O, Del Valle-Mondragón L, Hong E. Angiotensin 1–7 improves insulin sensitivity by increasing skeletal muscle glucose uptake in vivo. *Peptides* 2014;**51**:26–30.

43. Fu Z, Zhao L, Aylor KW, Carey RM, Barrett EJ, Liu Z. Angiotensin-(1–7) recruits muscle microvasculature and enhances insulin's metabolic action via Mas receptor. *Hypertension* 2014;**63**(6):1219–27.

44. Dzau VJ, Herrmann HC. Hormonal control of angiotensinogen production. *Life Sci* 1982;**15–22**(30):577–84.

45. Velloso LA, Folli F, Perego L, Saad MJ. The multi-faceted cross-talk between the insulin and angiotensin II signaling systems. *Diabetes Metab Res Rev* 2006;**22**:98–107.

46. Thomas MC, Pickering RJ, Tsorotes D, Koitka A, Sheehy K, Bernardi S, et al. Genetic Ace2 deficiency accentuates vascular inflammation and atherosclerosis in the ApoE knockout mouse. *Circ Res* 2010;**107**:888–97.

47. Tikellis C, Bialkowski K, Pete J, Sheehy K, Su Q, Johnston C, et al. ACE2 deficiency modifies renoprotection afforded by ACE inhibition in experimental diabetes. *Diabetes* 2008;**57**:1018–25.

48. Bilman V, Mares-Guia L, Nadu AP, Bader M, Campagnole-Santos MJ, Santos RA, et al. Decreased hepatic gluconeogenesis in transgenic rats with increased circulating angiotensin-(1–7). *Peptides* 2012;**37**:247–51.

49. Cortez-Pinto H, de Moura MC, Day CP. Non-alcoholic steatohepatitis: from cell biology to clinical practice. *J Hepatol* 2006;**44**:197–208.

50. Dowman JK, Armstrong MJ, Tomlinson JW, Newsome PN. Current therapeutic strategies in non-alcoholic fatty liver disease. *Diabetes Obes Metab* 2011;**13**:692–702.

51. Benomar Y, Gertler A, De Lacy P, Crepin D, Ould Hamouda H, Riffault L, et al. Central resistin overexposure induces insulin resistance through Toll-like receptor 4. *Diabetes* 2013;**62**:102–14.

52. Abuissa H, Jones PG, Marso SP, O'Keefe Jr. JH. Angiotensin-converting enzyme inhibitors or angiotensin receptor blockers for prevention of type 2 diabetes. A meta-analysis of randomized clinical trials. *J Am Coll Cardiol* 2005;**46**:821–6.

53. Padwal R, Laupacis A. Antihypertensive therapy and incidence of type 2 diabetes. A systematic review. *Diabetes Care* 2004;**27**:247–55.

54. Iwai M, Nakaoka H, Senba I, Kanno H, Moritani T, Horiuchi M. Possible involvement of angiotensin-converting enzyme 2 and Mas activation in inhibitory effects of angiotensin II type 1 receptor blockade on vascular remodeling. *Hypertension* 2012;**60**:137–44.

55. Park S, Bivona BJ, Ford Jr SM, Hiroyuki Kobori SX, de Garavilla L, Harrison-Bernard LM. Direct evidence for intrarenal chymase-dependent angiotensin II formation on the diabetic renal microvasculature. *Hypertension* 2013;**61**:465–71.

56. Bindom SM, Hans CP, Xia H, Boulares AH, Lazartigues E. Angiotensin I-converting enzyme type 2 (ACE2) gene therapy improves glycemic control in diabetic mice. *Diabetes* 2010;**59**:2540–8.

57. Benter IF, Yousif MHM, Cojocel C, Al-Maghrebi M, Diz DI. Angiotensin-(1–7) prevents diabetes-induced cardiovascular dysfunction. *Am J Physiol Heart Circ Physiol* 2007;**292**:H666–72.

58. Singh K, Singh T, Sharma PL. Beneficial effects of angiotensin-(1–7) in diabetic rats with cardiomyopathy. *Ther Adv Cardiovasc Dis* 2011;**5**:159–67.

59. Ebermann L, Spillmann F, Sidiropoulos M, Escher F, Heringer-Walther S, Schultheiss HP, et al. The angiotensin-(1–7) receptor agonist AVE0991 is cardioprotective in diabetic rats. *Eur J Pharmacol* 2008;**590**:276–80.

60. Fraga-Silva RA, Costa-Fraga FP, Murça TM, Moraes PL, Martins Lima A, Lautner RQ, et al. Angiotensin-converting enzyme 2 activation improves endothelial function. *Hypertension* 2013;**61**:1233–8.

61. Benter IF, Yousif MH, Dhaunsi GS, Kaur J, Chappell MC, Diz DI. Angiotensin-(1–7) prevents activation of NADPH oxidase and renal vascular dysfunction in diabetic hypertensive rats. *Am J Nephrol* 2008;**28**:25–33.

62. Singh T, Singh K, Sharma PL. Ameliorative potential of angiotensin-1-7/Mas receptor axis in streptozotocin-induced diabetic nephropathy in rats. *Methods Find Exp Clin Pharmacol* 2010;**32**:19–25.

63. Giani JF, Burghi V, Veiras LC, Tomat A, Muñoz MC, Cao G, et al. Angiotensin-(1–7) attenuates diabetic nephropathy in Zucker diabetic fatty rats. *Am J Physiol Renal Physiol* 2012;**302**:F1606–15.

64. Mori J, Patel VB, Ramprasath T, Alrob OA, Desaulniers J, Scholey JW, et al. Angiotensin 1–7 mediates renoprotection against diabetic nephropathy by reducing oxidative stress, inflammation, and lipotoxicity. *Am J Physiol Renal Physiol* 2014;**306**:F812–21.

65. Lula I, Denadai AL, Resende JM, Sousa FB, Lima GF, Pilo-Veloso D, et al. Study of angiotensin-(1–7) vasoactive peptide and its β-cyclodextrin inclusion complexes: complete sequence-specific NMR assignments and structural studies. *Peptides* 2007;**28**:2199–210.

66. Schling P, Loffler G. Cross talk between adipose tissue cells: impact on pathophysiology. *News Physiol Sci* 2002;**17**:99–104.

67. Muñoz MC, Giani JF, Dominici FP. Angiotensin-(1–7) stimulates the phosphorylation of Akt in rat extracardiac tissues in vivo via receptor Mas. *Regul Pept* 2010;**161**:1–7.

Chapter 36

Mas and the Reproductive System

Adelina M. dos Reis,* Fernando M. Reis†

Department of Physiology & Biophysics, Universidade Federal de Minas Gerais, Belo Horizonte, Brazil, †Department of Obstetrics & Gynecology and Division of Human Reproduction, Universidade Federal de Minas Gerais, Belo Horizonte, Brazil

INTRODUCTION

The main components of the classical renin–angiotensin system (RAS) are present in the female and male reproductive organs of several mammal species, from rodents to human. The ovary contains renin and prorenin, angiotensinogen, angiotensin-converting enzyme (ACE), and angiotensin II (Ang II) and its receptors AT_1 and AT_2. These components have been implicated in folliculogenesis, steroidogenesis, oocyte maturation, and ovulation.[1] However, conflicting observations about the roles of ACE and Ang II in steroidogenesis and ovulation[2,3] have pointed to the possibility that other active RAS peptides generated from Ang II could act in the ovary independently from the classical Ang II receptors, such as angiotensin-(1-7) (Ang-(1-7)).

Most components of the RAS have also been identified in the uterus. The uterine mucosa contains renin[4] and produces Ang II through local ACE activity. The expression pattern of Ang II in the human endometrium suggests that Ang II promotes the growth and vascularization of the endometrial stroma.[5] However, the effects of Ang II are probably counterbalanced by the other arm of the RAS, the ACE2/Ang-(1-7)/Mas axis, which is also fully expressed in the endometrium.[6] Mas receptor signaling continues to be active in intrauterine tissues during pregnancy, possibly regulating placental blood flow and influencing fetal development.[7]

The male reproductive tract and the testes express classical members of the RAS that modulate steroidogenesis in Leydig cells and have effects on epididymal contractility and sperm cell function.[8] Further studies revealed that Mas is present in rat,[9] mouse,[10] and human[11] testes. In addition, the Mas receptor pathway is active in the corpus cavernosum, with vasodilator and antifibrotic effects.[12]

Thus, a consistent body of evidence produced in recent years suggests that Mas signaling is active and plays important physiological roles in both female and male reproductive systems.

MAS AND THE FEMALE REPRODUCTIVE SYSTEM

Ovary

Using high-pressure liquid chromatography and radioimmunoassay, Ang-(1-7) was detected in rat ovary in all phases of the estrous cycle.[13] The peptide is present in the ovarian stroma in diestrus and in mural and cumulus granulosa cells of preovulatory follicles and in oocytes in proestrus (Figure 1). The ovarian concentrations of Ang-(1-7) in proestrus and estrus are significantly higher than those in metestrus and diestrus,[13] suggesting that the peptide might be regulated by gonadotropins.

This regulation has been confirmed in rats, rabbits, cattle, and humans. In immature rats stimulated with exogenous gonadotropins, the ovarian Ang-(1-7) concentrations rise from virtually undetectable to the peak levels observed in adult animals during spontaneous estrous cycles.[13] After gonadotropin administration, immature rats also acquired strong immunoreactivity for Ang-(1-7) in the ovarian theca-interstitial cells, along with increased ovarian expression of Mas and ACE2.[14] In immature rats, stimulation with human chorionic gonadotropin (hCG), a potent luteinizing hormone (LH) receptor agonist, increased ACE2 mRNA expression in the ovaries. Immunohistochemistry showed that LH increased both Ang-(1-7) and ACE2 staining in isolated and cultured preovulatory follicles. These findings suggest that LH upregulates the ACE2/Ang-(1-7)/Mas axis in rat ovary.[15]

In immature rabbit ovaries, immunoreactivity for Ang-(1-7) has been found in the stroma and oocytes from primordial, primary, and secondary follicles, but not in granulosa cells. Upon gonadotropin stimulation, Ang-(1-7) became detectable

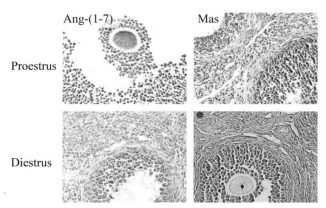

FIGURE 1   Immunohistochemical localization of Ang-(1-7) (left images) and Mas (right images) in rat ovaries in proestrus and diestrus. The brown color indicates specific immunostaining for Ang-(1-7) in the oocyte and granulosa cells of a preovulatory follicle in proestrus and in ovarian stromal cells in diestrus. Mas was found in ovarian theca-interstitial cells in proestrus and in diestrus.

in interstitial cells, inner theca cells, and granulosa cells of antral and preovulatory follicles.[16] In addition, the ovarian ACE2 and Ang-(1-7) levels increased after gonadal stimulation with intramuscular administration of GnRH in cattle.[17]

We have demonstrated that Ang-(1-7), the receptor Mas, and ACE2 are present in human ovary.[18] Ang-(1-7) and Mas were localized to the primordial, primary, secondary, and antral follicles; stroma; and corpora lutea of reproductive-age ovaries. Both Mas and ACE2 mRNAs were expressed in the ovarian tissue, and ovarian binding sites for Ang-(1-7) were identified by autoradiography.[18] In women preparing for *in vitro* fertilization (IVF), there is a significant increase in plasma Ang-(1-7) after ovulation induction with injectable gonadotropins.[19] Following gonadotropin stimulation, Ang-(1-7) is found in ovarian follicular fluid at concentrations as high as in the plasma. The follicular fluid Ang-(1-7) levels in these women do not correlate with plasma estradiol levels at the moment of follicle aspiration, suggesting that ovarian Ang-(1-7) release is directly stimulated by gonadotropins and is not a simple consequence of follicular growth and granulosa cell proliferation.[19]

Despite all the evidence that the ovaries release Ang-(1-7) in response to gonadotropins, it remains to be determined whether plasma Ang-(1-7) levels have circadian fluctuations and/or menstrual cycle changes. In nine volunteers, it was observed that urinary Ang-(1-7) excretion correlated with urinary progesterone excretion, which increased in the luteal phase of menstrual cycle.[20]

Mas mRNA and protein are detectable in granulosa-lutein cells of preovulatory ovarian follicles.[18,19] Mas immunoreactivity has also been localized to ovarian theca-interstitial cells of rats during the estrous cycle (Figure 1) and after gonadotropin administration,[14] suggesting that all follicular compartments are potential targets for Mas ligands, whose effects may encompass the modulation of steroidogenesis, oocyte maturation, and ovulation.

Ang-(1-7) induced a significant increase in estradiol and progesterone production in the ovaries of immature rats pretreated with gonadotropin and perfused in a closed-circuit system. This effect was blocked by the specific Mas antagonist A-779.[13]

Ang-(1-7) stimulated the resumption of meiosis in oocytes of rat preovulatory follicles. The specific antagonist of the Mas receptor, A-779, inhibited the germinal vesicle breakdown induced by Ang-(1-7) and reduced the oocyte maturation stimulated by LH, suggesting that Ang-(1-7) promotes meiotic resumption in rats, possibly as a gonadotropin intermediate.[15] In rabbits, Ang-(1-7) stimulated estradiol production and enhanced ovulatory efficiency, which was blocked by A-779.[16] Ovulation induced by hCG was antagonized by A-779 in this animal model, indicating that endogenous Ang-(1-7) plays an important role in rabbit ovulation, possibly as a mediator of gonadotropin action.[16]

During spontaneous ovulation, Mas-deficient female mice release less oocytes than their wild-type counterparts.[21] Thereby, female fecundity appears to be reduced in Mas-knockout mice (Figure 2). The possible involvement of ovarian Mas signaling in the mechanisms of ovulation is further suggested by indirect evidence in women receiving gonadotropin stimulation for IVF. The low responders, that is, those with less than five mature oocytes retrieved, had lower plasma Ang-(1-7) levels on the day of oocyte pickup, while the normal responders had a positive correlation between the number of mature oocytes retrieved and the ACE2 mRNA expression in the granulosa-lutein cell pool.[19]

Nonpregnant Uterus

Ang-(1-7) staining has been detected in the luminal and glandular epithelial cells of the endometrium in a pseudopregnant ovariectomized rat model.[22] Ang-(1-7) was significantly reduced in the decidualized horn and was irregular in the luminal epithelium but persisted in the glandular epithelium.

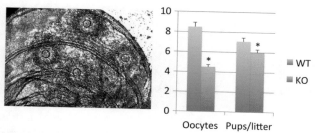

FIGURE 2   Left: 10× optical magnification of a mouse fallopian tube containing spontaneously released oocytes. Right: Mas-deficient mice (KO) release less oocytes and have less pups per litter compared with wild-type (WT) animals ($p < 0.05$, t-test).

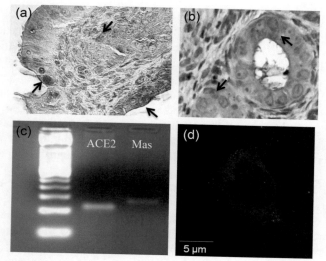

FIGURE 3   Immunolocalization of Ang-(1-7) (a) and Mas (b) in rat uterus. The arrows indicate examples of immunopositive cells in the lumen and glandular epithelia and in the stroma. (c) Gel electrophoresis with specific RT-PCR products corresponding to ACE2 and Mas mRNAs extracted from human endometrium. (d) Immunofluorescence for Mas in a human endometrial epithelial cell cultured *in vitro*.

We have evaluated the effect of treatment with estradiol alone and estradiol plus medroxyprogesterone acetate on the immunolocalization of Ang-(1-7) and Mas in the uterus of ovariectomized rats.[23] Ang-(1-7) was detected in all uterine tissues and in all compartments of the endometrium (Figure 3a and b). The intensity of the immunostaining decreased in the glandular epithelium of hormonally treated animals when compared with controls. In estrogen-treated rats, Ang-(1-7) labeling was scattered and sometimes included the nuclei of glandular cells. We also detected Ang-(1-7) expression in longitudinal myometrium and uterine serosa. Mas receptor was present in all tissues in the control and estrogen plus progestin groups. In the estrogen group, Mas staining was stronger in the luminal and glandular epithelium when compared with stroma or circular myometrium.

We have investigated whether the human endometrium expresses these components during the menstrual cycle.[6] By radioimmunoassay, Ang-(1-7) was detected in endometrial wash fluid at picomolar concentrations. Using immunofluorescence, both the peptide and its receptor were identified in cultured endometrial epithelial and stromal cells (Figure 3d). By immunohistochemistry, Ang-(1-7) was localized in the endometrium throughout the menstrual cycle, being more concentrated in the glandular epithelium of mid- and late secretory phase. This pattern corresponded to the ACE2 mRNA (Figure 3c), which was more abundant in epithelial cells than in stromal cells and in the secretory versus proliferative phase. The receptor Mas was equally distributed between epithelial and stromal cells and did not change during the menstrual cycle.

The physiological role of Mas in human endometrium remains unknown, and the current evidence is insufficient to determine whether and how it modulates endometrial growth and differentiation. Among women with hypertension, a multivariate analysis suggested that using ACE inhibitors and Ang receptor blockers increased the risk of endometrial polyps.[24] Women with polycystic ovary syndrome, a condition associated with anovulation and thereby with uterine exposure to estrogen unopposed by progesterone, were found to have higher mRNA expression not only of ACE2 and Mas but also of Ang II type 1 and type 2 receptors (AT_1 and AT_2) in the endometrium, compared with normal controls.[25] These preliminary data from clinical studies are inconclusive, and further studies are clearly needed to verify the role of Mas in physiological and pathological regulation in the endometrium.

Pregnant Uterus and Placenta

In the human placental bed, Ang-(1-7) and ACE2 are expressed at similar levels as in nonpregnant uterine fundus, while Mas receptor mRNA expression at this site is remarkably reduced in pregnancy.[26] Marques et al.[27] described RAS components in fetal membranes, decidua, and placenta collected at elective cesarean section (nonlaboring), after spontaneous labor and delivery, and in myometrium from elective (nonlaboring) or emergency cesarean (laboring) deliveries. ACE2 was largely expressed in the placenta, decidua, amnion, and chorion, but no change was associated with labor. Interestingly, ACE2 and Mas mRNA expression in explanted decidua appears to change according to the fetal gender.[28] Urinary Ang-(1-7) and Ang II increase throughout human gestation and fall during lactation and correlate with 17β-estradiol and progesterone.[20]

The physiological roles of placental Ang-(1-7) during pregnancy are both local and systemic. Plasma Ang-(1-7) levels are higher in pregnant women compared with nonpregnant controls.[29,30] A number of experimental models have enhanced our understanding of how this heptapeptide can influence the maternal adaptation to pregnancy, including the body fluid expansion required to irrigate the placenta and the uterine vascular remodeling that provides blood supply to face the fetal needs. One of these experiments tested the effects of the Mas antagonist A-779 on water balance in pregnant rats.[31] In 19-day pregnant rats, A-779 significantly decreased water intake and urine volume, increased urinary osmolality and kidney aquaporin-1 expression, and decreased the water balance. A-779 treatment increased plasma vasopressin in late gestation. It is suggested that Ang-(1-7) produces diuresis in late gestation with the downregulation of aquaporin-1 and that Ang-(1-7) stimulates water intake during pregnancy, thus contributing the maintenance of the physiological state of body fluid volume expansion.[31]

Pregnancy increases kidney expression of Ang-(1-7) and ACE2 in rats, as suggested by immunohistochemistry[32] and confirmed by a later study from the same group using a functional assay.[33] They showed that ACE2 activity measured using a fluorescent substrate was increased by almost twofold in the rat kidney at late gestation compared with nonpregnant animals.[33]

Plasma Ang-(1-7) levels are reduced in preeclamptic women compared with normal pregnant subjects.[30,34] In contrast, plasma Ang II levels were unchanged in preeclamptic patients.[34] In women with gestational diabetes, the plasma Ang-(1-7) levels and the Ang-(1-7)/Ang I ratio were lower than in healthy pregnancy.[29] Although these clinical observations do not establish a cause–effect relationship between low Ang-(1-7) levels and gestational complications, such a relationship is biologically plausible, considering the mechanisms observed in animal models of pathological gestation.

Genest et al.[35] found that the Mas receptor was decreased in the placenta and aorta of pregnant transgenic mice overexpressing human angiotensinogen and human renin, a model of preeclampsia superimposed on chronic hypertension. In another model of pregnancy-induced hypertension in rats, Ang-(1-7) concentration is significantly reduced in the uterus and placenta, as well as ACE2 mRNA in the uterus, compared with normal pregnancy.[36,37]

At late pregnancy, spontaneously hypertensive rats are more sensitive than normal rats to the vasodilator effect of Ang-(1-7) on isolated mesenteric arterial beds. ACE2 expression in mesenteric arterial beds is higher in pregnant than in nonpregnant rats regardless of hypertension. It is possible that both mechanisms are involved in the decrease of blood pressure seen at late pregnancy in normal and in hypertensive rats.[38] In a model of pregnancy-induced hypertension, Ang-(1-7) immunostaining and concentration in the kidney were significantly decreased compared with normal pregnant rats.[33] The expression of ACE2 in rat placenta is also reduced by maternal protein restriction in late pregnancy.[7] Because ACE2 protein was found predominantly in fetal mesenchymal tissue and fetal capillaries, it is possible to speculate that a reduced Ang-(1-7) production at this site, as a consequence of maternal protein restriction, may concur to placental insufficiency and fetal growth restriction.[7]

Breast

Preliminary data suggest that Ang-(1-7) is present in normal human breast, in benign breast lesions, and in breast cancer. Ang-(1-7) was localized to the cytoplasm and predominated in the glandular epithelium of normal mammary gland and benign lesions. In breast cancer, Ang-(1-7) was detected in the tumoral epithelial cells, vascular endothelium, and periglandular and perivascular stroma.[39]

Mas receptor mRNA has been identified in normal human breast tissue, benign tumor, and breast cancer.[40] Direct sequencing of the gene products in a small number of cases indicated that only one parental allele of Mas was expressed in breast cancer. The absence of an expression product from the other allele was due presumably to the presence of a functional imprint at this locus in human breast tissue.[40] However, the possible relevance of this phenomenon for breast cancer pathogenesis remains unexplored.

An important effect of Ang-(1-7) has been observed in athymic mice bearing orthotopic human breast tumors. Ang-(1-7) administration through subcutaneous osmotic minipumps during 18 days induced a marked decrease in tumor volume and weight and inhibited interstitial fibrosis, collagen deposition, and perivascular fibrosis.[41] These first experimental observations indicate that Ang-(1-7) may be a promising therapeutic agent to breast diseases.

MAS AND THE MALE REPRODUCTIVE SYSTEM

Testis

The receptor Mas has been identified in rat[9] and mouse[10] testis since early development, with increase around puberty and maximal expression during reproductive life. By *in situ* hybridization, Mas mRNA has been detected not only mainly in Leydig cells but also in Sertoli cells and primary spermatocytes.[10] In human, Mas was found in the interstitial compartment and throughout the seminiferous epithelium, from Sertoli cells and spermatogonia to sperm.[11] The human testis also has Ang-(1-7), which is more abundant in the cytoplasm of Leydig cells (Figure 4). In the seminiferous tubules, the expression of Ang-(1-7) is less intense compared to the interstitial compartment and predominates at the external layers, particularly in the cytoplasm of Sertoli cells and primary spermatocytes.[11]

Although Mas-deficient mice are still fertile,[42] their testes show aberrant expression of genes involved in mitochondrial function and steroidogenesis.[43] In addition, they have reduced testis weight, disturbed spermatogenesis, and an increased proportion of apoptotic cells, giant cells, and vacuoles in the seminiferous epithelium,[44] indicating that Mas signaling is an important physiological regulator of normal spermatogenesis.

This hypothesis is corroborated by observations in infertile men with distinct types of azoospermia: obstructive with normal spermatogenesis and nonobstructive with severely impaired spermatogenesis. In men with preserved spermatogenesis, we detected abundant distribution of Ang-(1-7) and Mas at both the interstitial and tubular compartments.[11] In contrast, men with nonobstructive azoospermia had a faint interstitial expression of the receptor Mas, while neither the peptide nor the receptor could be detected at the seminiferous tubules. In addition, the testicular samples of infertile men with impaired spermatogenesis expressed Mas and ACE2 mRNA at lower concentrations than samples with full spermatogenesis.[11]

Prostate

Ang-(1-7) prevents metastatic dissemination of prostate cancer cells in mice[45] and inhibits prostate malignant tumor growth in a human to mouse xenograft model.[46] This antineoplastic effect seems to involve mechanisms that inhibit both cell proliferation and angiogenesis, such as a marked increase of the soluble fraction of vascular endothelial growth factor (VEGF) receptor 1, with a concomitant reduction in VEGF receptors 1 and 2.[46]

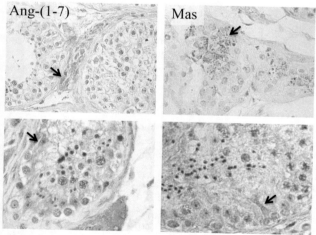

FIGURE 4   Immunohistochemical localization of Ang-(1-7) (left images) and Mas (right images) in rat ovaries and in human testis. Both the peptide and its receptor are present in Leydig cells (top images) and in the seminiferous epithelium (bottom images). The arrows indicate examples of immunopositive cells.

Corpus Cavernosum

The localization and function of Ang-(1-7) in the erectile tissue were firstly reported in rats and mice.[47] The receptor Mas was localized in rat corpus cavernosum by immunofluorescence and confocal microscopy. Mas-deficient knockout mice had penile fibrosis and severely depressed responses to electrical stimulation of the major pelvic ganglion. Intracavernosal infusion of Ang-(1-7) potentiated the erectile response to electrical stimulation of the major pelvic ganglion in rats, and this effect was reversed by A-779. Incubation of rat and mouse corpus cavernosum strips with Ang-(1-7) stimulated nitric oxide release.[47]

Moreover, the erectile function of hypertensive rats was improved by Ang-(1-7) administration.[47] In diabetic rats, the intraperitoneal injection of Ang-(1-7) reversed the oxidative DNA damage in the corpus cavernosum.[48] Ang-(1-7) produced nitric-oxide-dependent relaxation of the rabbit corpus cavernosum.[49] Older and diabetic rabbits showed impaired Ang-(1-7)-mediated relaxation, suggesting that aging- and diabetes-related erectile dysfunction may be partly due to decreased Ang-(1-7)-mediated relaxation of the corpus cavernosum.

Oral Ang-(1-7)-CyD treatment for 3 weeks reduced collagen content and attenuated ROS production in mouse corpus cavernosum. Ang-(1-7)-CyD treatment stimulated vascular nitric oxide production and improved cavernosal endothelial function.[50] Intracavernosal injection of the synthetic nonpeptide Mas agonist AVE 0991 potentiated the erectile response of rats; the effect of AVE 0991 was dose-dependent and was blocked by the nitric oxide synthase inhibitor L-NAME and by the specific Mas receptor blocker A-779.[51]

SUMMARY AND PERSPECTIVES

The evidence summarized here supports the concept that female and male reproductive organs synthesize, release, and are important targets of Ang-(1-7). Mas receptor activation promotes ovarian estradiol and progesterone release and improves ovulation and oocyte maturation. In the uterus, Ang-(1-7) is particularly found in the glandular epithelium of secretory endometrium and continues to be expressed in the decidua during pregnancy, as well as in the placenta and membranes. Ang-(1-7) modulates water balance in pregnant rats and its plasma levels are reduced in preeclamptic women. In males, Mas is relevant for gametogenesis and for the prevention of penile fibrosis, as suggested by the mouse knockout model. Ang-(1-7) has shown promising capability to restore erectile function in animal models of erectile dysfunction induced by hypertension and diabetes.

There are several open research questions and innovation perspectives in this field. These include understanding the dynamic regulation of Ang-(1-7) production throughout the process of ovarian follicle growth and oocyte maturation and to explore the potential pharmacological use of this peptide to induce ovulation. The role of Ang-(1-7)/Mas axis in the control of cell proliferation in the uterus, breast, and prostate remains to be clarified, as well as the potential use of this peptide to improve placental circulation and to treat erectile dysfunction.

REFERENCES

1. Yoshimura Y. The ovarian renin-angiotensin system in reproductive physiology. *Front Neuroendocrinol* 1997;**18**:247–91.
2. Kuji N, Sueoka K, Miyazaki T, Tanaka M, Oda T, Kobayashi T, et al. Involvement of angiotensin II in the process of gonadotropin-induced ovulation in rabbits. *Biol Reprod* 1996;**55**:984–91.
3. Daud AI, Bumpus FM, Husain A. Characterization of angiotensin I-converting enzyme (ACE)-containing follicles in the rat ovary during the estrous cycle and effects of ACE inhibitor on ovulation. *Endocrinology* 1990;**126**:2927–35.
4. Shah DM, Higuchi K, Inagami T, Osteen KG. Effect of progesterone on renin secretion in endometrial stromal, chorionic trophoblast, and mesenchymal monolayer cultures. *Am J Obstet Gynecol* 1991;**164**:1145–50.
5. Li XF, Ahmed A. Compartmentalization and cyclic variation of immunoreactivity of renin and angiotensin converting enzyme in human endometrium throughout the menstrual cycle. *Hum Reprod* 1997;**12**:2804–9.
6. Vaz-Silva J, Carneiro MM, Ferreira MC, Pinheiro SV, Silva DA, Silva-Filho AL, et al. The vasoactive peptide angiotensin-(1–7), its receptor Mas and the angiotensin-converting enzyme type 2 are expressed in the human endometrium. *Reprod Sci* 2009;**16**:247–56.
7. Gao H, Yallampalli U, Yallampalli C. Maternal protein restriction reduces expression of angiotensin I-converting enzyme 2 in rat placental labyrinth zone in late pregnancy. *Biol Reprod* 2012;**86**:31.
8. Leung PS, Sernia C. The renin-angiotensin system and male reproduction: new functions for old hormones. *J Mol Endocrinol* 2003;**30**:263–70.
9. Metzger R, Bader M, Ludwig T, Berberich C, Bunnemann B, Ganten D. Expression of the mouse and rat Mas proto-oncogene in the brain and peripheral tissues. *FEBS Lett* 1995;**357**:27–32.
10. Alenina N, Baranova T, Smirnow E, Bader M, Lippoldt A, Patkin E, et al. Cell type-specific expression of the Mas proto-oncogene in testis. *J Histochem Cytochem* 2002;**50**:691–6.
11. Reis AB, Araujo FC, Pereira VM, Dos Reis AM, Santos RA, Reis FM. Angiotensin (1–7) and its receptor Mas are expressed in the human testis: implications for male infertility. *J Mol Histol* 2010;**41**:75–80.

12. Fraga-Silva RA, Montecucco F, Mach F, Santos RA, Stergiopulos N. Pathophysiological role of the renin-angiotensin system on erectile dysfunction. *Eur J Clin Invest* 2013;**43**:978–85.
13. Costa AP, Fagundes-Moura CR, Pereira VM, Silva LF, Vieira MA, Santos RA, et al. Angiotensin-(1–7): a novel peptide in the ovary. *Endocrinology* 2003;**144**:1942–8.
14. Pereira VM, Reis FM, Santos RA, Cassali GD, Santos SH, Honorato-Sampaio K, et al. Gonadotropin stimulation increases the expression of angiotensin-(1–7) and MAS receptor in the rat ovary. *Reprod Sci* 2009;**16**:1165–74.
15. Honorato-Sampaio K, Pereira VM, Santos RA, Reis AM. Evidence that angiotensin-(1–7) is an intermediate of gonadotrophin-induced oocyte maturation in the rat preovulatory follicle. *Exp Physiol* 2012;**97**:642–50.
16. Viana GE, Pereira VM, Honorato-Sampaio K, Oliveira CA, Santos RA, Reis AM. Angiotensin-(1–7) induces ovulation and steroidogenesis in perfused rabbit ovaries. *Exp Physiol* 2011;**96**:957–65.
17. Tonellotto dos Santos J, Ferreira R, Gasperin BG, Siqueira LC, de Oliveira JF, Santos RA, et al. Molecular characterization and regulation of the angiotensin-converting enzyme type 2/angiotensin-(1–7)/MAS receptor axis during the ovulation process in cattle. *J Renin Angiotensin Aldosterone Syst* 2012;**13**:91–8.
18. Reis FM, Bouissou DR, Pereira VM, Camargos AF, dos Reis AM, Santos RA. Angiotensin-(1–7), its receptor Mas, and the angiotensin-converting enzyme type 2 are expressed in the human ovary. *Fertil Steril* 2011;**95**:176–81.
19. Cavallo IK, Oliveira ML, Del Puerto HL, Lobach V, Camargos MG, Dela Cruz C, et al. Angiotensin-(1–7)/Mas/ACE2 and ovulation induction in IVF women. *Reprod Sci* 2014;**21**:233A.
20. Valdes G, Germain AM, Corthorn J, Berrios C, Foradori AC, Ferrario CM, et al. Urinary vasodilator and vasoconstrictor angiotensins during menstrual cycle, pregnancy, and lactation. *Endocrine* 2001;**16**:117–22.
21. Honorato-Sampaio K. Angiotensina-(1-7) no processo ovulatório: da maturação ovocitária à ovulação. PhD Thesis, Universidade Federal de Minas Gerais, 2010.
22. Brosnihan KB, Bharadwaj MS, Yamaleyeva LM, Neves LA. Decidualized pseudopregnant rat uterus shows marked reduction in Ang II and Ang-(1–7) levels. *Placenta* 2012;**33**:17–23.
23. Vaz-Silva J, Tavares RL, Ferreira MC, Honorato-Sampaio K, Cavallo IK, Santos RA, et al. Tissue specific localization of angiotensin-(1–7) and its receptor Mas in the uterus of ovariectomized rats. *J Mol Histol* 2012;**43**:597–602.
24. Indraccolo U, Matteo M, Greco P, Barbieri F, Mignini F. ACE inhibitors and angiotensin receptor blockers could promote the onset of endometrial polyps in hypertensive women. *Int J Clin Pharmacol Ther* 2013;**51**:963–8.
25. Qin S, Yan X, Tang HJ, Zhou YJ, Liu Y, Shen HM, et al. Expression and significance of ACE2-Ang-(1–7)-Mas axis in endometrium of patients with polycystic ovary syndrome. *Nat Med J China* 2013;**93**:1989–92.
26. Anton L, Merrill DC, Neves LA, Diz DI, Corthorn J, Valdes G, et al. The uterine placental bed renin-angiotensin system in normal and preeclamptic pregnancy. *Endocrinology* 2009;**150**:4316–25.
27. Marques FZ, Pringle KG, Conquest A, Hirst JJ, Markus MA, Sarris M, et al. Molecular characterization of renin-angiotensin system components in human intrauterine tissues and fetal membranes from vaginal delivery and cesarean section. *Placenta* 2011;**32**:214–21.
28. Wang Y, Pringle KG, Sykes SD, Marques FZ, Morris BJ, Zakar T, et al. Fetal sex affects expression of renin-angiotensin system components in term human decidua. *Endocrinology* 2012;**153**:462–8.
29. Nogueira AI, Souza Santos RA, Simoes ESAC, Cabral AC, Vieira RL, Drumond TC, et al. The pregnancy-induced increase of plasma angiotensin-(1–7) is blunted in gestational diabetes. *Regul Pept* 2007;**141**:55–60.
30. Merrill DC, Karoly M, Chen K, Ferrario CM, Brosnihan KB. Angiotensin-(1–7) in normal and preeclamptic pregnancy. *Endocrine* 2002;**18**:239–45.
31. Joyner J, Neves LA, Stovall K, Ferrario CM, Brosnihan KB. Angiotensin-(1–7) serves as an aquaretic by increasing water intake and diuresis in association with downregulation of aquaporin-1 during pregnancy in rats. *Am J Physiol Regul Integr Comp Physiol* 2008;**294**:R1073–80.
32. Brosnihan KB, Neves LA, Joyner J, Averill DB, Chappell MC, Sarao R, et al. Enhanced renal immunocytochemical expression of ANG-(1–7) and ACE2 during pregnancy. *Hypertension* 2003;**42**:749–53.
33. Joyner J, Neves LA, Granger JP, Alexander BT, Merrill DC, Chappell MC, et al. Temporal-spatial expression of ANG-(1–7) and angiotensin-converting enzyme 2 in the kidney of normal and hypertensive pregnant rats. *Am J Physiol Regul Integr Comp Physiol* 2007;**293**:R169–77.
34. Velloso EP, Vieira R, Cabral AC, Kalapothakis E, Santos RA. Reduced plasma levels of angiotensin-(1–7) and renin activity in preeclamptic patients are associated with the angiotensin I-converting enzyme deletion/deletion genotype. *Braz J Med Biol Res* 2007;**40**:583–90.
35. Genest DS, Falcao S, Michel C, Kajla S, Germano MF, Lacasse AA, et al. Novel role of the renin-angiotensin system in preeclampsia superimposed on chronic hypertension and the effects of exercise in a mouse model. *Hypertension* 2013;**62**:1055–61.
36. Neves LA, Stovall K, Joyner J, Valdes G, Gallagher PE, Ferrario CM, et al. ACE2 and ANG-(1–7) in the rat uterus during early and late gestation. *Am J Physiol Regul Integr Comp Physiol* 2008;**294**:R151–61.
37. Yamaleyeva LM, Neves LA, Coveleskie K, Diz DI, Gallagher PE, Brosnihan KB. AT1, AT2, and AT(1–7) receptor expression in the uteroplacental unit of normotensive and hypertensive rats during early and late pregnancy. *Placenta* 2013;**34**:497–502.
38. Ognibene DT, Carvalho LC, Costa CA, Rocha AP, de Moura RS, Castro RA. Role of renin-angiotensin system and oxidative status on the maternal cardiovascular regulation in spontaneously hypertensive rats. *Am J Hypertens* 2012;**25**:498–504.
39. Ribeiro AG, Bacon J, Amaral VF, Cassali GD, Santos RA, Reis FM. Expressão da angiotensina(1–7) na glândula mamária humana: correlações fisiopatológicas. In: Livro de Resumos da FeSBE; 2008. p. 1871.
40. Miller N, McCann AH, O'Connell D, Pedersen IS, Spiers V, Gorey T, et al. The MAS proto-oncogene is imprinted in human breast tissue. *Genomics* 1997;**46**:509–12.

41. Cook KL, Metheny-Barlow LJ, Tallant EA, Gallagher PE. Angiotensin-(1–7) reduces fibrosis in orthotopic breast tumors. *Cancer Res* 2010;**70**:8319–28.
42. Walther T, Balschun D, Voigt JP, Fink H, Zuschratter W, Birchmeier C, et al. Sustained long term potentiation and anxiety in mice lacking the Mas protooncogene. *J Biol Chem* 1998;**273**:11867–73.
43. Xu P, Santos RA, Bader M, Alenina N. Alterations in gene expression in the testis of angiotensin-(1–7)-receptor Mas-deficient mice. *Regul Pept* 2007;**138**:51–5.
44. Leal MC, Pinheiro SV, Ferreira AJ, Santos RA, Bordoni LS, Alenina N, et al. The role of angiotensin-(1–7) receptor Mas in spermatogenesis in mice and rats. *J Anat* 2009;**214**:736–43.
45. Krishnan B, Smith TL, Dubey P, Zapadka ME, Torti FM, Willingham MC, et al. Angiotensin-(1–7) attenuates metastatic prostate cancer and reduces osteoclastogenesis. *Prostate* 2013;**73**:71–82.
46. Krishnan B, Torti FM, Gallagher PE, Tallant EA. Angiotensin-(1–7) reduces proliferation and angiogenesis of human prostate cancer xenografts with a decrease in angiogenic factors and an increase in sFlt-1. *Prostate* 2013;**73**:60–70.
47. da Costa Goncalves AC, Leite R, Fraga-Silva RA, Pinheiro SV, Reis AB, Reis FM, et al. Evidence that the vasodilator angiotensin-(1–7)-Mas axis plays an important role in erectile function. *Am J Physiol Heart Circ Physiol* 2007;**293**:H2588–96.
48. Kilarkaje N, Yousif MH, El-Hashim AZ, Makki B, Akhtar S, Benter IF. Role of angiotensin II and angiotensin-(1–7) in diabetes-induced oxidative DNA damage in the corpus cavernosum. *Fertil Steril* 2013;**100**:226–33.
49. Yousif MH, Kehinde EO, Benter IF. Different responses to angiotensin-(1–7) in young, aged and diabetic rabbit corpus cavernosum. *Pharmacol Res* 2007;**56**:209–16.
50. Fraga-Silva RA, Costa-Fraga FP, Savergnini SQ, De Sousa FB, Montecucco F, da Silva D, et al. An oral formulation of angiotensin-(1–7) reverses corpus cavernosum damages induced by hypercholesterolemia. *J Sex Med* 2013;**10**:2430–42.
51. da Costa Goncalves AC, Fraga-Silva RA, Leite R, Santos RA. AVE 0991, a non-peptide Mas-receptor agonist, facilitates penile erection. *Exp Physiol* 2013;**98**:850–5.

Chapter 37

ACE2/Ang-(1-7)/Mas Axis and Physical Exercise

Daisy Motta-Santos,* Lenice Becker†

*Department of Physiology and Biophysics, Federal University of Minas Gerais, Belo Horizonte, Minas Gerais, Brazil, †Sport Center of UFOP, Federal University of Ouro Preto, Ouro Preto, Minas Gerais, Brazil

PROTECTIVE EFFECTS OF PHYSICAL EXERCISE

Physical exercise has been widely recommended as an essential nonpharmacological strategy for the treatment and prevention of various diseases. The most significant cardiovascular benefits of exercise include left ventricular remodeling, improvement of myocardial contractile function, improvement of endothelial and autonomic function, increased oxygen uptake, and reduced blood pressure.[1–3] Exercise training also promotes anti-inflammatory effects mediated in part by body fat decrease which induces changes in adipokines release.[4] Thus, the physical exercise also acts on metabolic disorders such as obesity and diabetes. In addition, voluntary running enhanced neurogenesis in the adult mouse dentate gyrus[5] and the capacity of exercise to stimulate the neuroplasticity explains in part why it is recommended to treat neurological disorders.[6]

Adaptations Promoted by Physical Exercise: Cardiac and Musculoskeletal Effects

One of the first studies that identified regular physical activity as cardioprotective was published in 1953.[7] The authors investigated in Routemaster workers the number of deaths from coronary heart disease in drivers, who spent the most part of the day sitting versus conductors, who went up and down the stairs several times a day. The results showed that heart disease mortality was higher in workers who had low levels of physical activity. The "sedentary hearts" seemed to be 10–15 years older than "physical activity hearts."

In hypertensive subjects, studies have demonstrated that acute and chronic exercise can reduce blood pressure.[8] Postexercise hypotension (PEH) is characterized by an acute reduction in blood pressure during the postexercise period to levels below the baseline values preworkout. [9] The effect of aerobic training on blood pressure is directly associated with changes in endothelial function and autonomic system.[10] Exercise is able to increase the bioavailability of nitric oxide to promote endothelium-dependent vasodilation,[11] similar effects as those induced by angiotensin-(1-7).[12]

An important adaptation of chronic exercise is cardiac hypertrophy, and it can occur physiologically or pathologically.[1] The pathological remodeling is associated with higher levels of vasoconstriction inducing dilated cardiomyopathy, heart failure (HF), and hypertension. Despite this pathological condition, the "athlete's heart" occurs due to cardiac remodeling characterized by increases in both muscle mass and the size of cardiomyocytes and improved cardiac function.[13]

Beyond the myocardium, the skeletal muscle is another important organ that has a great ability to adapt to physical training. PGC1α appears to be the main mediator of these actions.[14,15] The PGC1α is expressed in tissues that have high oxidative capacity including the brown adipose tissue, kidney, heart, skeletal muscle, and brain.[16] Various stimuli can lead to an increase in the expression of PGC1α such as cold exposure and decrease in the ATP levels in the skeletal muscle.[17] Recently, it was demonstrated that the increased expression of PGC1α stimulates the expression of FNDC5, a membrane protein that when cleaved releases the hormone irisin.[18] The irisin acts on white adipose tissue by stimulating the expression of UCP1 inducing the brown adipose tissue and increasing caloric expenditure.[19] There is evidence that irisin is related to angiotensin II levels in children.[20]

Renin–Angiotensin System and Physical Exercise

The first evidence of the relationship between the renin–angiotensin system (RAS) and physical exercise was found between angiotensin-converting enzyme (ACE) polymorphism and performance.[21] The ACEI/D polymorphism is a possible genetic

The Protective Arm of the Renin–Angiotensin System (RAS). http://dx.doi.org/10.1016/B978-0-12-801364-9.00037-7

263

marker for physical performance with allele I associated with endurance and higher oxygen uptake and allele D with strength and power.[19] In addition to its systemic effect, the RAS has a significant local effect on performance. Evidence showed that the RAS plays an important modulation of gene expression and physiological responses induced by physical exercise.[22]

Regarding the activity of the RAS in the muscle, the major findings involve the activity of ACE. Most studies were conducted in animal models. Animal experiments demonstrated that muscular ACE activity in the muscle could act independently from that in the plasma. Cultures of skeletal muscle cells (myoblasts) have shown that ACE degraded bradykinin.[23] Studies using tissue perfusion without RAS components showed that vasoconstriction produced by topical application of angiotensin I was preserved in the skeletal muscle.[24]

The RAS modulates cardiovascular responses and electrolyte balance and its participation in exercise is important since several changes in the cardiovascular system and homeostasis occur during acute or chronic exercise. During an exercise bout, there is an increase in plasma angiotensin II levels that contributes to the pressor effect of exercise through a facilitation of norepinephrine release by postganglionic sympathetic neurons.[25]

Hayashi et al. showed that physical training significantly decreases the plasma renin activity.[26] These responses were related to improvement in cardiopulmonary capacity. Braith et al. demonstrated that, after physical training, the plasmatic levels of angiotensin II, aldosterone, vasopressin, and ANP decrease during the rest and exercise in patients with HF. However, at the peak of exercise, no differences in these neurohormonal levels were found. These data indicate that the exercise training alters only the neuroendocrine hyperactivity associated with HF, without changing the physiological responses during intense exercise.[27]

The association of exercise training with the administration of AT_1 receptor antagonists (TCV-116 and losartan) prevents excessive hypotension observed during the use of these antagonists. This effect is due to a better stabilization of arterial pressure. In addition, physical exercise increases the reduction in systolic blood pressure at rest produced by the administration of TCV-116.[28]

Exercise training promotes endothelial growth and angiogenesis, and these effects are mediated by angiotensin II. The participation of this peptide was confirmed using the ACE inhibitor captopril and AT_1 antagonist losartan. Treatment with these drugs in trained animals did not result in cell growth and angiogenesis as observed in trained control animals.[29]

Coirault et al., in a study with rabbits, found that preventive therapy with Perindopril has a beneficial effect on the intrinsic performance of the skeletal muscle during moderate cardiac hypertrophy.[30] Another study concluded that in SHR, Perindopril mediated the exercise capacity improvement.[31] These effects mentioned above appear to be similar to those presented in pathological conditions in spontaneously hypertensive rats (SHRs). Perindopril associated with physical training can improve the physical ability, the skeletal muscle fiber type and the capillarity in SHR. It was suggested that Perindopril promotes adaptive changes in the skeletal muscle in response to exercise including increases in capillary density and percentage of type I fibers in SHR.

Other studies have reported the effects of exercise on the RAS. Pereira et al. evaluated the effect of exercise training in transgenic mice with HF induced by sympathetic hyperactivity through deleting alfa2a/alfa2c adrenergic receptor.[32] Physical training over 8 weeks reduced the levels of Ang II and ACE activity in these transgenic mice, which had an increase in both components of the RAS. These data provide evidence that the reduced cardiac RAS explains, in part, the beneficial effects of training on cardiac function in a genetic model of HF.

During graded exercise considering low exercise intensities, the central AT_1 blockade inhibits the increase in glucose levels and free fatty acid mobilization from the adipose tissue. These data indicated that the central angiotensinergic system is involved in metabolic adjustments during exercise.[33] In the central nervous system, the physical exercise attenuated the amino acid pressor response in the rostral ventrolateral medulla (RVLM). The central nucleus is an important area involved in the modulation of the sympathetic outflow to the cardiovascular system.[34] Becker et al. showed that the central RAS system in RVLM has a different pressor responsiveness to Ang II and Ang-(1-7) in trained rats. In these animals, the RVLM pressor effect induced by Ang II was higher while Ang-(1-7) pressor effect was smaller. These results were corroborated by the selective antagonists, indicating that the RVLM is a site in the central nervous system involved in the adaptive mechanisms triggered during exercise training.[35]

The Angiotensin-(1-7) Effects Related to Physical Exercise: Role of Mas Receptor

The association between Ang-(1-7) and physical exercise could take part in the amelioration of arterial hypertension. Treatment with Ang-(1-7) attenuated increases in myocyte diameter and cardiac fibrosis induced by hypertension in trained 2K1C rats. This effect is probably mediated by the Ang-(1-7) receptor, because the receptor is upregulated in ventricles of the trained 2K1C rats. The association of exercise and Ang-(1-7) also increases the protein expression of the angiotensin II type 2 receptor and endothelial nitric oxide synthase phosphorylation in this hypertension model.[36]

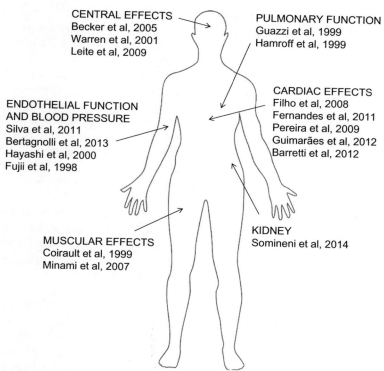

CENTRAL EFFECTS
Becker et al, 2005
Warren et al, 2001
Leite et al, 2009

PULMONARY FUNCTION
Guazzi et al, 1999
Hamroff et al, 1999

CARDIAC EFFECTS
Filho et al, 2008
Fernandes et al, 2011
Pereira et al, 2009
Guimarães et al, 2012
Barretti et al, 2012

ENDOTHELIAL FUNCTION
AND BLOOD PRESSURE
Silva et al, 2011
Bertagnolli et al, 2013
Hayashi et al, 2000
Fujii et al, 1998

MUSCULAR EFFECTS
Coirault et al, 1999
Minami et al, 2007

KIDNEY
Somineni et al, 2014

FIGURE 1 Some studies demonstrated the interaction between the ACE2/Ang-(1-7)/Mas axis and physical exercise, most of which have been conducted in rodents.

In SHR, the oral treatment with Ang-(1-7) and the chronic physical exercise induced similar responses in autonomic control. The heart rate variability components have improved through the activation of parasympathetic modulation after Ang-(1-7) treatment and physical exercise. The sympathovagal balance LF/HF index in the heart was also improved by both interventions. Physical exercise and the Ang-(1-7) oral treatment induced morphometric improvements as evidenced by cardiomyocyte diameter reduction in SHRs. The association between physical exercise and Ang-(1-7) was effective in reducing myocyte hypertrophy in SHR. Chronic Ang-(1-7) treatment and physical exercise also improved LV function, as well as LV maximum and minimum derivatives (dP/dT).[37]

After physical training, SHR had shown improvement of their endothelial function. Interestingly, the vasodilatory response of Ang-(1-7) was augmented in trained SHRs, indicating a synergic response between Ang-(1-7) and physical exercise. This vasodilatory response was mediated by Mas once the authors observed that the Ang-(1-7) effect was abolished when aortic rings were preincubated with two Ang-(1-7)/Mas receptor antagonists A-779 and D-Pro7-Ang-(1-7).[38]

Studies using physiological and pathological responses could help explain the interaction between Mas and physical exercise (Figure 1). The exercise training is well recognized as a model to promote a physiological left ventricular hypertrophy (LVH). LVH induced by physical exercise is an important physiological compensatory mechanism in response to chronic hemodynamic overload increase. Recent studies suggested that the Mas receptor participates in this effect.[22,39] However, it seems that in cardiac physiological hypertrophy, the Mas involvement is present only when an imbalance of local RAS exists. This hypothesis was raised based on the increase in Mas cardiac expression promoted by physical exercise mainly in diseased states. Filho et al.[22] demonstrated that Mas expression increases only in the left ventricle of trained SHR and not in normotensive rats after swimming training. This evidence suggests that the Mas receptor is essential to promote physiological benefits especially in disease conditions.

Is the ACE2/Ang-(1-7)/Mas Axis Required for Physical Exercise Benefits?

The pharmacological treatment with RAS blockers, ACE inhibitors, or angiotensin II type 1 (AT_1) receptor is known to be effective in reducing blood pressure.[40] The nonpharmacological treatment through regular physical exercise can prevent and treat hypertension and other cardiovascular diseases. First, several authors investigated the effect of RAS blockade on physical performance.[41,42] Later, some studies have found that exercise was able to reduce the levels of angiotensin II.[43,44] The exercise can activate ACE2/Ang-(1-7)/MAS and reduce the activation of the classic RAS axis.[32,45]

The activation of the ACE2/Ang-(1-7)/Mas axis was observed after physical exercise in the cardiac muscle.[22,39,45] The positive balance between ACE2/Ang-(1-7)/Mas and ACE/Ang II/AT$_1$ receptor was demonstrated through ACE2, AT$_2$, and Ang-(1-7) increase in the left ventricular levels and simultaneous decrease in angiotensinogen, ACE, and Ang II levels in the left ventricle.[45] In the Mas-knockout mouse (Mas-KO), it was observed that the ACE2/Ang-(1-7)/Mas axis is crucial for physical exercise benefits. The physical exercise associated with Mas deficiency induced a paradoxical deleterious cardiac effect as the deposition of extracellular matrix proteins and increased expression of collagens I and III. These data suggest that Ang-(1-7) through Mas may exert a compensatory mechanism counteracting an increase in the extracellular matrix after chronic exercise. In obese Zucker rats (OZRs), the exercise induced changes in local and systemic RAS. In the obese Zucker rats, the physical exercise reduced systemic Ang II levels and downregulated ACE, Ang II, and AT$_2$ in the left ventricular tissue. In addition, exercise training increased cardiac ACE2 expression in the obese Zucker rats.[46]

Another significant association between physical exercise and the Mas receptor was evidenced in a recent study involving preeclamptic mice that overexpress human angiotensinogen and human renin. Genest et al. showed that voluntary physical exercise during pregnancy induced prevention in pathological features associated with preeclampsia disorder.[47] The AT$_1$ levels were increased whereas the Mas receptor was decreased in the placenta and the aorta of pregnant preeclampsia mice.[47] Physical exercise also decreases the placental angiotensin II type 1 receptor; on the other hand, the Mas receptors are increased in the aorta. This effect contributes to a better balance of Ang-(1-7) and Ang II minimizing the vasoconstriction and helping to control blood pressure. Physical exercise controls the blood pressure during pregnancy not only through RAS balance but also by improving glucose metabolism and endothelium function.

Another organ that could benefit from physical exercise involving ACE2 is the kidney. ACE2 is highly expressed in the kidney and has been shown to be renoprotective by degrading Ang II to Ang-(1-7). In diabetic mice, a decrease in ACE2 is observed and the physical exercise corrected the ACE2 shedding. Lower blood glucose and urinary albumin excretion were demonstrated. The results with physical exercise were more efficient than observed with metformin treatment.[48]

In conclusion, angiotensin-(1-7) and physical exercise are effective for the treatment of cardiovascular and metabolic disease. Mas deficiency leads to pathological responses induced by exercise training evidenced in Mas-KO animals. In diseases such as hypertension and HF, the exercise can balance the RAS components. The Ang-(1-7)/ACE2/Mas axis is activated by physical exercise and is apparently necessary to promote countless exercise benefits.

REFERENCES

1. Weeks KL, McMullen JR. The athlete's heart vs. the failing heart: can signaling explain the two distinct outcomes? *Physiology (Bethesda)* 2011;**26**:97–105.
2. Joyner MJ, Green DJ. Exercise protects the cardiovascular system: effects beyond traditional risk factors. *J Physiol* 2009;**587**:5551–8.
3. Pescatello L, Franklin B, Fagard R, Farquhar W, Kelley G, Ray C. American College of Sports Medicine position stand exercise and hypertension. *Med Sci Sports Exerc* 2004;**36**:533–53.
4. Walsh NP, Gleeson M, Shephard RJ, Woods JA, Bishop NC, Fleshner M, et al. Position statement. Part one: immune function and exercise. *Exerc Immunol Rev* 2011;**17**:6–63.
5. van Praag H, Kempermann G, Gage FH. Running increases cell proliferation and neurogenesis in the adult mouse dentate gyrus. *Nat Neurosci* 1999;**2**:266–70.
6. Gómez-Pinilla F, Ying Z, Roy RR, Molteni R, Edgerton VR. Voluntary exercise induces a BDNF-mediated mechanism that promotes neuroplasticity. *J Neurophysiol* 2002;**88**:2187–95.
7. Morris JN, Heady JA, Raffle PA, Roberts CG, Parks JW. Coronary heart-disease and physical activity of work. *Lancet* 1953;**265**:1111–20, concl.
8. Hamer M. The anti-hypertensive effects of exercise: integrating acute and chronic mechanisms. *Sports Med* 2006;**36**:109–16.
9. MacDonald J, Hogben C, Tarnopolsky M, MacDougall J. Post exercise hypotension is sustained during subsequent bouts of mild exercise and simulated activities of daily living. *J Hum Hypertens* 2001;**15**:567–71.
10. Thijssen DH, Cable NT, Green DJ. Impact of exercise training on arterial wall thickness in humans. *Clin Sci (Lond)* 2012;**122**:311–22.
11. Goto C, Nishioka K, Umemura T, Jitsuiki D, Sakagutchi A, Kawamura M, et al. Acute moderate-intensity exercise induces vasodilation through an increase in nitric oxide bioavailability in humans. *Am J Hypertens* 2007;**20**:825–30.
12. Sampaio WO, Nascimento AA, Santos RA. Systemic and regional hemodynamic effects of angiotensin-(1–7) in rats. *Am J Physiol Heart Circ Physiol* 2003;**284**:H1985–94.
13. Bernardo BC, Weeks KL, Pretorius L, McMullen JR. Molecular distinction between physiological and pathological cardiac hypertrophy: experimental findings and therapeutic strategies. *Pharmacol Ther* 2010;**128**:191–227.
14. Holloszy JO. Regulation by exercise of skeletal muscle content of mitochondria and GLUT4. *J Physiol Pharmacol* 2008;**59**(Suppl. 7):5–18.
15. Ventura-Clapier R, Garnier A, Veksler V. Transcriptional control of mitochondrial biogenesis: the central role of PGC-1alpha. *Cardiovasc Res* 2008;**79**:208–17.
16. Puigserver P. Tissue-specific regulation of metabolic pathways through the transcriptional coactivator PGC1-alpha. *Int J Obes (Lond)* 2005;**29**(Suppl. 1):S5–9.
17. Handschin C, Spiegelman BM. Peroxisome proliferator-activated receptor gamma coactivator 1 coactivators, energy homeostasis, and metabolism. *Endocr Rev* 2006;**27**:728–35.

18. Boström P, Wu J, Jedrychowski MP, Korde A, Ye L, Lo JC, et al. A PGC1-α-dependent myokine that drives brown-fat-like development of white fat and thermogenesis. *Nature* 2012;**481**:463–8.

19. Puthucheary Z, Skipworth JR, Rawal J, Loosemore M, Van Someren K, Montgomery HE. The ACE gene and human performance: 12 years on. *Sports Med* 2011;**41**:433–48.

20. Al-Daghri NM, Al-Attas OS, Alokail MS, Alkharfy KM, Yousef M, Vinodson B, et al. Maternal inheritance of circulating irisin in humans. *Clin Sci (Lond)* 2014;**126**:837–44.

21. Rigat B, Hubert C, Alhenc-Gelas F, Cambien F, Corvol P, Soubrier F. An insertion/deletion polymorphism in the angiotensin I-converting enzyme gene accounting for half the variance of serum enzyme levels. *J Clin Invest* 1990;**86**:1343–6.

22. Filho AG, Ferreira AJ, Santos SH, Neves SR, Silva Camargos ER, Becker LK, et al. Selective increase of angiotensin(1–7) and its receptor in hearts of spontaneously hypertensive rats subjected to physical training. *Exp Physiol* 2008;**93**:589–98.

23. Ward PE, Russell JS, Vaghy PL. Angiotensin and bradykinin metabolism by peptidases identified in skeletal muscle. *Peptides* 1995;**16**:1073–8.

24. Vicaut E, Hou X. Arteriolar constriction and local renin-angiotensin system in rat microcirculation. *Hypertension* 1993;**21**:491–7.

25. Warren JH, Lewis W, Wraa CE, Stebbins CL. Central and peripheral effects of angiotensin II on the cardiovascular response to exercise. *J Cardiovasc Pharmacol* 2001;**38**:693–705.

26. Hayashi A, Kobayashi A, Takahashi R, Suzuki F, Nakagawa T, Kimotro K. Effects of voluntary running exercise on blood pressure and renin-angiotensin system in spontaneously hypertensive rats and normotensive Wistar-Kyoto rats. *J Nutr Sci Vitaminol (Tokyo)* 2000;**46**:165–70.

27. Braith RW, Welsch MA, Feigenbaum MS, Kluess HA, Pepine CJ. Neuroendocrine activation in heart failure is modified by endurance exercise training. *J Am Coll Cardiol* 1999;**34**:1170–5.

28. Fujii N, Nagashima S, Yukawa K, Miyauchi T, Maki S, Sakai S, et al. Hypotensive effects of an angiotensin II type 1 receptor antagonist differ between exercised and sedentary rats aged from 4 to 19 weeks. *Jpn J Physiol* 1998;**48**:215–18.

29. Amaral SL, Papanek PE, Greene AS. Angiotensin II and VEGF are involved in angiogenesis induced by short-term exercise training. *Am J Physiol Heart Circ Physiol* 2001;**281**:H1163–9.

30. Coirault C, Langeron O, Lambert F, Blanc F, Lerebours G, Claude N, et al. Impaired skeletal muscle performance in the early stage of cardiac pressure overload in rabbits: beneficial effects of angiotensin-converting enzyme inhibition. *J Pharmacol Exp Ther* 1999;**291**:70–5.

31. Minami N, Li Y, Guo Q, Kawamura T, Mori N, Nagasaka M, et al. Effects of angiotensin-converting enzyme inhibitor and exercise training on exercise capacity and skeletal muscle. *J Hypertens* 2007;**25**:1241–8.

32. Pereira M, Ferreira J, Bueno CJ, Mattos K, Rosa K, Irigoyen M, et al. Exercise training reduces cardiac angiotensin II levels and prevents cardiac dysfunction in a genetic model of sympathetic hyperactivity-induced heart failure in mice. *Eur J Appl Physiol* 2009;**105**:843–50.

33. Leite LH, Lacerda AC, Balthazar CH, Marubayashi U, Coimbra CC. Central angiotensin AT1 receptors are involved in metabolic adjustments in response to graded exercise in rats. *Peptides* 2009;**30**:1931–5.

34. Martins-Pinge MC, Becker LK, Garcia MR, Zoccal DB, Neto RV, Basso LS, et al. Attenuated pressor responses to amino acids in the rostral ventrolateral medulla after swimming training in conscious rats. *Auton Neurosci* 2005;**122**:21–8.

35. Becker LK, Santos RA, Campagnole-Santos MJ. Cardiovascular effects of angiotensin II and angiotensin-(1–7) at the RVLM of trained normotensive rats. *Brain Res* 2005;**1040**:121–8.

36. Shah A, Oh YB, Lee SH, Lim JM, Kim SH. Angiotensin-(1–7) attenuates hypertension in exercise-trained renal hypertensive rats. *Am J Physiol Heart Circ Physiol* 2012;**302**:H2372–80.

37. Bertagnolli M, Casali KR, De Sousa FB, Rigatto K, Becker L, Santos SH, et al. An orally active angiotensin-(1–7) inclusion compound and exercise training produce similar cardiovascular effects in spontaneously hypertensive rats. *Peptides* 2014;**51**:65–73.

38. Silva DM, Gomes-Filho A, Olivon VC, Santos TM, Becker LK, Santos RA, et al. Swimming training improves the vasodilator effect of angiotensin-(1–7) in the aorta of spontaneously hypertensive rat. *J Appl Physiol (1985)* 2011;**111**:1272–7.

39. Guimarães GG, Santos SH, Oliveira ML, Pimenta-Velloso EP, Motta DF, Martins AS, et al. Exercise induces renin-angiotensin system unbalance and high collagen expression in the heart of Mas-deficient mice. *Peptides* 2012;**38**:54–61.

40. Bader M. Tissue renin-angiotensin-aldosterone systems: targets for pharmacological therapy. *Annu Rev Pharmacol Toxicol* 2010;**50**:439–65.

41. Guazzi M, Palermo P, Pontone G, Susini F, Agostoni P. Synergistic efficacy of enalapril and losartan on exercise performance and oxygen consumption at peak exercise in congestive heart failure. *Am J Cardiol* 1999;**84**:1038–43.

42. Hamroff G, Katz SD, Mancini D, Blaufarb I, Bijou R, Patel R, et al. Addition of angiotensin II receptor blockade to maximal angiotensin-converting enzyme inhibition improves exercise capacity in patients with severe congestive heart failure. *Circulation* 1999;**99**:990–2.

43. Liu JL, Irvine S, Reid IA, Patel KP, Zucker IH. Chronic exercise reduces sympathetic nerve activity in rabbits with pacing-induced heart failure: a role for angiotensin II. *Circulation* 2000;**102**:1854–62.

44. Negrão CE, Middlekauff HR. Exercise training in heart failure: reduction in angiotensin II, sympathetic nerve activity, and baroreflex control. *J Appl Physiol* 2008;**104**:577–8.

45. Fernandes T, Hashimoto NY, Magalhães FC, Fernandes FB, Casarini DE, Carmona AK, et al. Aerobic exercise training-induced left ventricular hypertrophy involves regulatory MicroRNAs, decreased angiotensin-converting enzyme-angiotensin ii, and synergistic regulation of angiotensin-converting enzyme 2-angiotensin (1–7). *Hypertension* 2011;**58**:182–9.

46. Barretti DL, Magalhaes Fde C, Fernandes T, do Carmo EC, Rosa KT, Irigoyen MC, et al. Effects of aerobic exercise training on cardiac renin-angiotensin system in an obese Zucker rat strain. *PLoS ONE* 2012;**7**, e46114.

47. Genest DS, Falcao S, Michel C, Kajla S, Germano MF, Lacasse AA, et al. Novel role of the renin-angiotensin system in preeclampsia superimposed on chronic hypertension and the effects of exercise in a mouse model. *Hypertension* 2013;**62**:1055–61.

48. Somineni HK, Boivin GP, Elased KM. Daily exercise training protects against albuminuria and angiotensin converting enzyme 2 shedding in db/db diabetic mice. *J Endocrinol* 2014;**221**:243–59.

Chapter 38

Angiotensin-Converting Enzyme 2/ Angiotensin-(1-7)/Mas Receptor Axis: Emerging Pharmacological Target for Pulmonary Diseases

Vinayak Shenoy,* Anderson J. Ferreira,† Michael Katovich,* Mohan K. Raizada‡

*Department of Pharmacodynamics, College of Pharmacy, University of Florida, Gainesville, FL 32610, USA, †Department of Morphology, Federal University of Minas Gerais, Belo Horizonte, MG 31270-901, Brazil, ‡Department of Physiology and Functional Genomics, College of Medicine, University of Florida, Gainesville, FL 32610, USA

INTRODUCTION

The renin–angiotensin system (RAS) has long been recognized as an important player in the pathogenesis of several lung diseases, including pulmonary arterial hypertension (PAH), pulmonary fibrosis (PF), chronic obstructive pulmonary disease (COPD), and acute respiratory distress syndrome (ARDS). Evidence for this stems from the following observations: (a) PAH and PF are associated with higher circulating levels of angiotensin II (Ang II)[1,2]; (b) increased concentrations of angiotensinogen (the precursor for Ang II peptide) and angiotensin-converting enzyme (ACE), the major generator of Ang II, have been observed in the lungs of fibrotic and pulmonary hypertensive subjects[3,4]; (c) patients carrying the ACE ID/DD genotype, which confers increased ACE levels, are susceptible to COPD and ARDS[5,6]; (d) human lung fibroblasts obtained from patients with PF, but not from normal lungs, generate Ang II,[2] suggesting a causative role for this peptide in disease pathogenesis; (e) Ang II induces apoptosis of the alveolar epithelial cells, a key initiating factor for lung fibrogenesis[7]; (f) Ang II is a potent pulmonary vasoconstrictor with mitogenic properties, that produces migratory, hypertrophic, and proliferative effects on the lung smooth muscles to cause PAH[8]; (g) Ang II mediates oxidative stress and cytokine signaling,[9] factors that contribute to the pathophysiology of lung diseases; and (h) pharmacological blockade of the RAS using ACE inhibitors (ACEi) or angiotensin receptor blockers (ARB) offers protection against animal models of COPD, PAH, and lung fibrosis.[5,10,11] Collectively, these findings indicate an active involvement of the RAS in lung diseases. However, clinical studies with ACEi or ARBs have yielded mixed results, failing to reach a consensus opinion on the use of these agents for treating pulmonary disorders.

Discovery of several new members has led to the emergence of a new arm of the RAS, which has forced a reevaluation of the role of this hormonal system in pulmonary physiology and pathobiology. These new components include the monocarboxypeptidase angiotensin-converting enzyme 2 (ACE2), the vasoactive peptide angiotensin-(1-7) (Ang-(1-7)), and its G protein-coupled receptor Mas, which together form a protective arm, the ACE2/Ang-(1-7)/Mas axis of the RAS.[12] This axis counteracts the vasoconstrictive, proliferative, inflammatory, and fibrotic actions of Ang II in several target organs, including the lungs. Our objective in this chapter will be to (i) summarize the importance of the ACE2/Ang-(1-7)/Mas axis in lung diseases and (ii) explore novel strategies to upregulate this axis that hold great potential for clinical development in the treatment of pulmonary disorders.

EVIDENCE FOR THE EXISTENCE OF VASODELETERIOUS AND VASOPROTECTIVE AXES OF THE RAS IN THE LUNGS

The classical RAS comprises a cascade of proteolytic enzymes and peptides located primarily within the circulatory system. However, in addition to the circulating RAS, a tissue-specific RAS has been identified in several organs, including the heart, kidneys, and lungs. Biochemical and molecular studies have revealed that all the necessary components required for the active biosynthesis of Ang II, namely, renin, angiotensinogen, and ACE, are present within the lung tissue.[13] Furthermore,

The Protective Arm of the Renin–Angiotensin System (RAS). http://dx.doi.org/10.1016/B978-0-12-801364-9.00038-9

expression of AT_1R and AT_2R at both mRNA and protein levels has been detected in the pulmonary system. Since AT_1R gene expression is more abundant than AT_2R in the adult lungs, this receptor mediates most of the biological effects of Ang II. The intrapulmonary ACE-Ang II-AT_1R axis has been shown to promote vasoconstriction, smooth muscle cell proliferation, and fibrosis, along with induction of oxidative stress and inflammation. Thus, this axis forms the vasodeleterious arm of the RAS. On the other hand, identification of ACE2 and Mas receptor within the lung parenchymal cells has lent credence to the existence of a vasoprotective arm of the RAS.[14] We believe that an imbalance between vasodeleterious and vasoprotective axes of the RAS plays a crucial role in the pathophysiology of lung diseases.

THE ACE2/ANG-(1-7)/MAS AXIS IN PULMONARY DISEASES

ACE2 is a zinc-dependent metalloproteinase that is widely expressed on the pulmonary endothelium, type II alveolar epithelial and smooth muscle cells of the lungs.[15] Its primary physiological function is to metabolize Ang II to Ang-(1-7), a biologically active heptapeptide that binds to the Mas receptor to exert vasodilatory, antiproliferative, and antifibrotic actions in the pulmonary system. It is becoming increasingly evident that downregulation of the ACE2/Ang-(1-7)/Mas axis contributes to the development and progression of lung diseases. The importance of ACE2 in pulmonary physiology was appreciated when this enzyme was identified as a functional receptor for the SARS coronavirus (SARS-CoV), the causative agent of severe acquired respiratory syndrome (SARS).[16] *In vivo* experimental studies revealed that lung ACE2 expression is markedly decreased upon SARS-CoV infection, which consequently leads to respiratory failure and death.[17] The protective role of ACE2 against lung injury was further underscored by studies carried out on ACE2-deficient animals. Knockout mice for ACE2 displayed enhanced vascular permeability, increased lung edema and neutrophil accumulation, which worsened pulmonary function in experimental models of ARDS.[18]

With respect to fibrotic lung diseases, Li et al. were the first to demonstrate severe downregulation of ACE2 and its catalytic activity in the lungs of animals, as well as patients with PF.[19] Also, inhibition/silencing of pulmonary ACE2 in rodents was associated with excessive lung collagen accumulation, suggesting that decreased ACE2 levels play a causal role in lung fibrogenesis. Consistent with this interpretation, mice deficient in ACE2 gene were more susceptible to bleomycin-induced injury, exhibiting excessive deposition of extracellular matrix proteins in the lungs as compared with wild-type animals.[20]

A growing body of studies has focused on the relevance of the ACE2/Ang-(1-7)/Mas axis in PAH. PAH is a fatal lung disease that is associated with increased blood pressure in the pulmonary arteries. Decreased expression of ACE2 and enzymatic activity has been observed in animal models of PAH and pulmonary hypertensive patients, along with a substantial increase in circulating Ang II levels.[1,12] Thus, it is evident that restoring the balance between the vasodeleterious and the vasoprotective axes of the RAS could be therapeutically beneficial in attenuating disease pathogenesis.

THE ACE2/ANG-(1-7)/MAS AXIS PROTECTS AGAINST ARDS AND PAH

ARDS is the most severe form of acute lung injury that is triggered by sepsis, gastric juice aspiration, or pathogenic infections such as SARS-CoV and H5N1 avian influenza virus. In experimental models of ARDS, treatment of wild-type mice and ACE2 knockout mutants with catalytically active recombinant ACE2 protein rescued them from lung failure.[18] Likewise, administration of Ang-(1-7) conferred beneficial effects against acute lung injury, an effect that was abolished by blocking the Mas receptor.[21]

With regard to PAH, the enzymatic activity of ACE2 offers a dual beneficial effect. On one hand, it decreases the levels of Ang II, which are significantly elevated during PAH, and on the other, it increases the concentration of the vasoprotective heptapeptide, Ang-(1-7). We were the first to demonstrate that pulmonary overexpression of ACE2 attenuates monocrotaline (MCT)-induced PAH and associated pathophysiology.[22] Similarly, genetic overexpression of Ang-(1-7) rendered beneficial effects against this disease.[23] Interestingly, coadministration of A-779, a Mas antagonist, abolished the cardiopulmonary protection offered by Ang-(1-7), suggesting the involvement of Mas-mediated signaling pathway. In a separate set of experiments, Kleinsasser et al. demonstrated that administration of recombinant human ACE2 protein (rhACE2) decreased hypoxia-induced pulmonary vasoconstriction.[24] On the clinical front, it was observed that the serum ACE2 activity was significantly reduced in patients with PAH.[25] Apparently, elevating circulating levels of ACE2 in these pulmonary hypertensive patients could yield beneficial results.

THE ACE2/ANG-(1-7)/MAS AXIS IN PF

PF is a debilitating disease marked by scarring in the lungs. Treatment of ACE2 knockout animals with rhACE2 attenuated bleomycin-induced PF and improved lung function.[20] Furthermore, studies from our group have shown that pulmonary overexpression of either ACE2 or Ang-(1-7) by lentiviral-mediated gene transfer prevents lung fibrosis in the bleomycin model.[23] One of the hallmarks of PF is the loss of alveolar epithelial cells (AECs). Mechanistic studies carried out in cultured AECs revealed that ACE2 mediates cell survival by balancing the proapoptotic Ang II and its antiapoptotic metabolic product Ang-(1-7). This is evidenced by the fact that silencing of the ACE2 gene in AECs or its pharmacological inhibition resulted in increased Ang II and decreased Ang-(1-7) levels, which was associated with apoptotic cell death.[14] On the contrary, treatment of AECs with Ang-(1-7) attenuated Ang II and bleomycin-induced apoptosis. This cell survival effect of Ang-(1-7) was mediated through inhibition of caspases (caspases 3 and 9) and downregulation of the Jun N-terminal kinase (JNK) pathways. Also, these Ang-(1-7)-mediated effects were reversed by blocking the Mas receptor or with antisense nucleotides against the Mas mRNA. Taken together, these findings implicate a potent antifibrotic role for the ACE2/Ang-(1-7)/Mas axis against PF.

THERAPEUTIC STRATEGIES TO ACTIVATE THE ACE2/ANG-(1-7)/MAS AXIS

Recombinant Human ACE2 (rhACE2)

Administration of rhACE2 has been shown to exert promising effects in a variety of disease models with pathologically elevated Ang II levels or a dysregulated RAS. These experimental findings have provided compelling evidence for initiating clinical trials with rhACE2. In this regard, GlaxoSmithKline is currently evaluating the effects of GSK2586881 (formerly APN01), a soluble form of rhACE2, in a multicenter phase II clinical trial for the treatment of acute lung injury. Previous phase I trials have established that administration of single and multiple doses of rhACE2 is well tolerated by healthy humans, with no serious adverse events or dose-limiting toxicities.[26] Furthermore, in a separate clinical study, the efficacy of rhACE2 is being assessed for the treatment of PAH (ClinicalTrials.gov; NCT01884051). Although clinical trials are currently under way, the use of protein therapy faces several challenges. Among these, the cost of manufacturing the recombinant protein, its stability, repetitive intravenous dosing, and patient compliance pose major obstacles in realizing the full therapeutic potential of ACE2 therapy.

ACE2 ACTIVATORS

An ideal approach to overcome the aforementioned limitations of recombinant protein therapy would be to identify small synthetic molecules that can favorably activate endogenous ACE2. Using a structure-based drug design, we have identified synthetic enhancers of ACE2 such as resorcinolnaphthalein, 1-[[2-(dimethylamino)ethyl]amino]-4-(hydroxymethyl)-7-[[(4-methylphenyl)sulfonyl]oxy]-9H-xanthone9 (XNT), and diminazene aceturate (DIZE).[27–29] *In silico* docking studies revealed that binding of these small molecules to a novel pocket in the hinge-bending region of the ACE2 protein stabilizes the enzyme in an open conformation. This is important because only the open conformation favors/accelerates enzyme activity. Among the three compounds, DIZE was found to be the most potent, followed by XNT, and finally resorcinolnaphthalein. Pharmacologically, all these ACE2 activators were found to be effective in attenuating lung injury. In addition, XNT exhibited strong antithrombotic actions along with decreased platelet attachment,[30] whereas DIZE therapy was associated with enhanced endothelial progenitor cell (EPC) function in pulmonary hypertensive animals, factors that could potentiate the cardiopulmonary beneficial effects of these agents. However, there are some critical issues that impede the successful translation of these ACE2 activators into the clinic. XNT has an undesirable pharmacokinetic profile. It is poorly soluble in water and requires acidic pH for solubilization, making it an unsuitable drug candidate. With regard to DIZE, it is associated with toxic side effects. Chronic treatment with DIZE has been shown to harm the liver, kidneys, and brain, which could potentially result in life-threatening situations. In fact, a preslaughter withdrawal period of 21–35 days has been recommended for DIZE when treated animals are intended for human consumption. Besides, DIZE was found to be mutagenic, but not teratogenic. Nonetheless, XNT and DIZE could serve as lead compounds that require structural modification to make them safe and druggable.

STRATEGIES TARGETING ANG-(1-7)

It is evident that Ang-(1-7) holds promise for the treatment of lung diseases. However, the use of Ang-(1-7) as a therapeutic agent is limited because of its short half-life *in vivo*. Also, being of peptide nature, Ang-(1-7) will be rapidly degraded in the gastrointestinal tract when given orally. To overcome these issues, formulations of Ang-(1-7) with extended half-life and improved oral bioavailability are currently being developed. Ang-(1-7) complexed with beta-cyclodextrin (HPβCD/Ang-(1-7)) enabled effective oral administration, with significant increases in plasma Ang-(1-7) levels.[31] In addition,

Kluskens and coworkers synthesized a cyclic analog of Ang-(1-7) (thioether-bridged Ang-(1-7)) that exhibited better *in vivo* pharmacokinetic profile.[32] Alternatively, liposomal delivery system represents an effective method to administer Ang-(1-7). Administration of liposomes containing Ang-(1-7) was also found to increase the half-life of this heptapeptide.[33] Importantly, all these modified forms of Ang-(1-7) were pharmacologically active when tested in experimental models. It is worth noting that TXA127, a synthetic analog of Ang-(1-7) from Tarix Pharmaceuticals, is currently in clinical trials for the treatment of PAH. TXA127 could also be a promising candidate for PF therapy.

MAS RECEPTOR AGONISTS

Given that stimulation of the Mas receptor is protective, it would be reasonable to evaluate the effects of Mas agonists for pulmonary therapeutics. AVE 0991 is the first nonpeptide agonist of the Mas receptor that was shown to promote beneficial effects against several cardiovascular diseases.[34] An important aspect of this compound is that it is orally active and mimics the effects of Ang-(1-7) in several organs by binding to the Mas receptor. Recently, two novel peptides (CGEN 856 and CGEN 857) with high specificity for the Mas receptor have been discovered.[35] All these Mas agonists have the potential to treat pulmonary diseases.

FUTURE PERSPECTIVES

Oral Delivery of ACE2 and Ang-(1-7)

The oral route is the most widely accepted means of drug administration. But oral delivery of therapeutic proteins like ACE2 and Ang-(1-7) is not feasible since they will be easily degraded in the gastrointestinal tract. However, we have developed a low-cost oral delivery system for administering ACE2 and Ang-(1-7) using transplastomic technology. The use of transplastomic technology allows chloroplasts to generate high levels of therapeutic proteins within plant leaves.[36] Since chloroplast expression of ACE2 or Ang-(1-7) enables the bioencapsulation of these therapeutic proteins within plant cells, it protects them against gastric enzymatic degradation upon oral delivery. We observed that oral feeding of rats with bioencapsulated ACE2 or Ang-(1-7) prevented the development of MCT-induced PH and was associated with improved right heart function.[37] It is possible to produce bulk quantities of therapeutically active ACE2 and Ang-(1-7) using transplastomic technology, which holds great potential for clinical development as an oral delivery system.

Genetic Modification of Stem Cells

Reduced number and impaired function of circulating/resident stem cells have been associated with poor cardiopulmonary outcomes in patients and experimental models of lung diseases. In this regard, exogenous administration of stem cells such as endothelial progenitor cells, bone marrow- or adipose-derived mesenchymal stromal cells (MSCs), and umbilical cord blood cells could offer protection. However, such cell-based therapy presents a number of unique challenges before they can be effectively employed in the clinical setting. The hostile environment characterized by inflammation and high oxidative stress within the diseased lungs could threaten survival and impair homing of the injected stem/progenitor cells. Genetic modification of these cells to overexpress ACE2 or Ang-(1-7) could improve survival and enhance their potential to effectively repair lung injury. In fact, we have shown that ACE2 priming of endothelial progenitor cells not only enhances their function but also increases their therapeutic efficacy to render protection against stroke.[38]

CONCLUSIONS

A dysregulated pulmonary angiotensin system contributes to the pathogenesis and progression of lung diseases. Discovery of the protective ACE2/Ang-(1-7)/Mas axis has provided a novel and more promising target for disease treatment than the classical approach of using ACEi or ARBs. Innovative pharmaceutical strategies that can effectively activate this beneficial ACE2/Ang-(1-7)/Mas axis hold great promise for treating pulmonary pathologies.

REFERENCES

1. de Man FS, Tu L, Handoko ML, Rain S, Ruiter G, François C, et al. Dysregulated renin–angiotensin–aldosterone system contributes to pulmonary arterial hypertension. *Am J Respir Crit Care Med* 2012;**186**:780–9.
2. Li X, Molina-Molina M, Abdul-Hafez A, Ramirez J, Serrano-Mollar A, Serrano-Mollar A, et al. Extravascular sources of lung angiotensin peptide synthesis in idiopathic pulmonary fibrosis. *Am J Physiol Lung Cell Mol Physiol* 2006;**291**:L887–995.

3. Uhal BD, Li X, Piasecki CC, Molina-Molina M. Angiotensin signalling in pulmonary fibrosis. *Int J Biochem Cell Biol* 2012;**44**:465–8.

4. Orte C, Polak JM, Haworth SG, Yacoub MH, Morrell NW. Expression of pulmonary vascular angiotensin-converting enzyme in primary and secondary plexiform pulmonary hypertension. *J Pathol* 2000;**192**:379–84.

5. Shrikrishna D, Astin R, Kemp PR, Hopkinson NS. Renin-angiotensin system blockade: a novel therapeutic approach in chronic obstructive pulmonary disease. *Clin Sci* 2012;**123**:487–98.

6. Marshall RP, Webb S, Bellingan GJ, Montgomery HE, Chaudhari B, McAnulty RJ, et al. Angiotensin converting enzyme insertion/deletion polymorphism is associated with susceptibility and outcome in acute respiratory distress syndrome. *Am J Respir Crit Care Med* 2002;**166**:646–50.

7. Wang R, Zagariya A, Ibarra-Sunga O, Gidea C, Ang E, Deshmukh S, et al. Angiotensin II induces apoptosis in human and rat alveolar epithelial cells. *Am J Physiol* 1999;**276**:L885–9.

8. Lipworth BJ, Dagg KD. Vasoconstrictor effects of angiotensin II on the pulmonary vascular bed. *Chest* 1994;**105**:1360–4.

9. Ruiz-Ortega M, Lorenzo O, Rupérez M, Esteban V, Suzuki Y, Mezzano S, et al. Role of the renin-angiotensin system in vascular diseases: expanding the field. *Hypertension* 2001;**38**:1382–7.

10. Morrell NW, Morris KG, Stenmark KR. Role of angiotensin-converting enzyme and angiotensin II in development of hypoxic pulmonary hypertension. *Am J Physiol* 1995;**269**:H1186–94.

11. Molina-Molina M, Serrano-Mollar A, Bulbena O, Fernandez-Zabalegui L, Closa D, Marin-Arguedas A, et al. Losartan attenuates bleomycin induced lung fibrosis by increasing prostaglandin E2 synthesis. *Thorax* 2006;**61**:604–10.

12. Bradford CN, Ely DR, Raizada MK. Targeting the vasoprotective axis of the renin-angiotensin system: a novel strategic approach to pulmonary hypertensive therapy. *Curr Hypertens Rep* 2010;**12**:212–19.

13. Marshall RP. The pulmonary renin–angiotensin system. *Curr Pharm Des* 2003;**9**:715–22.

14. Uhal BD, Li X, Xue A, Gao X, Abdul-Hafez A. Regulation of alveolar epithelial cell survival by the ACE-2/angiotensin 1-7/Mas axis. *Am J Physiol Lung Cell Mol Physiol* 2011;**301**:L269–74.

15. Hamming I, Timens W, Bulthuis ML, Lely AT, Navis G, van Goor H. Tissue distribution of ACE2 protein, the functional receptor for SARS coronavirus. A first step in understanding SARS pathogenesis. *J Pathol* 2004;**203**:631–7.

16. Li W, Moore MJ, Vasilieva N, Sui J, Wong SK, Berne MA, et al. Angiotensin-converting enzyme 2 is a functional receptor for the SARS coronavirus. *Nature* 2003;**426**:450–4.

17. Kuba K, Imai Y, Rao S, Gao H, Guo F, Guan B, et al. A crucial role of angiotensin converting enzyme 2 (ACE2) in SARS coronavirus-induced lung injury. *Nat Med* 2005;**11**:875–9.

18. Imai Y, Kuba K, Rao S, Huan Y, Guo F, Guan B, et al. Angiotensin-converting enzyme 2 protects from severe acute lung failure. *Nature* 2005;**436**:112–16.

19. Li X, Molina-Molina M, Abdul-Hafez A, Uhal V, Xaubet A, Uhal BD. Angiotensin converting enzyme-2 is protective but downregulated in human and experimental lung fibrosis. *Am J Physiol Lung Cell Mol Physiol* 2008;**295**:L178–85.

20. Rey-Parra GJ, Vadivel A, Coltan L, Hall A, Eaton F, Schuster M, et al. Angiotensin converting enzyme 2 abrogates bleomycin-induced lung injury. *J Mol Med (Berl)* 2012;**90**:637–47.

21. Klein N, Gembardt F, Supé S, Kaestle SM, Nickles H, Erfinanda L, et al. Angiotensin-(1–7) protects from experimental acute lung injury. *Crit Care Med* 2013;**41**:e334–43.

22. Yamazato Y, Ferreira AJ, Hong KH, Sriramula S, Francis J, Yamazato M, et al. Prevention of pulmonary hypertension by angiotensin-converting enzyme-2 gene transfer. *Hypertension* 2009;**54**:365–71.

23. Shenoy V, Ferreira AJ, Qi Y, Fraga-Silva RA, Díez-Freire C, Dooies A, et al. The angiotensin converting enzyme 2/angiotensin-(1–7)/Mas axis confers cardiopulmonary protection against lung fibrosis and pulmonary hypertension. *Am J Respir Crit Care Med* 2010;**182**:1065–72.

24. Kleinsasser A, Pircher I, Treml B, Schwienbacher M, Schuster M, Janzek E, et al. Recombinant angiotensin-converting enzyme 2 suppresses pulmonary vasoconstriction in acute hypoxia. *Wilderness Environ Med* 2012;**23**:24–30.

25. Dai HL, Guo Y, Guang XF, Xiao ZC, Zhang M, Yin XL. The changes of serum angiotensin-converting enzyme 2 in patients with pulmonary arterial hypertension due to congenital heart disease. *Cardiology* 2013;**12**:208–12.

26. Haschke M, Schuster M, Poglitsch M, Loibner H, Salzberg M, Bruggisser M, et al. Pharmacokinetics and pharmacodynamics of recombinant human angiotensin-converting enzyme 2 in healthy human subjects. *Clin Pharmacokinet* 2013;**52**(9):783–92.

27. Hernández Prada JA, Ferreira AJ, Katovich MJ, Shenoy V, Qi Y, Santos RA, et al. Structure-based identification of small-molecule angiotensin-converting enzyme 2 activators as novel antihypertensive agents. *Hypertension* 2008;**51**:1312–17.

28. Shenoy V, Gjymishka A, Jarajapu YP, Qi Y, Afzal A, Rigatto K, et al. Diminazene attenuates pulmonary hypertension and improves angiogenic progenitor cell functions in experimental models. *Am J Respir Crit Care Med* 2013;**187**:648–57.

29. Ferreira AJ, Shenoy V, Yamazato Y, Sriramula S, Francis J, Yuan L, et al. Evidence for angiotensin-converting enzyme 2 as a therapeutic target for the prevention of pulmonary hypertension. *Am J Respir Crit Care Med* 2009;**179**:1048–54.

30. Fraga-Silva RA, Sorg BS, Wankhede M, Dedeugd C, Jun JY, Baker MB, et al. ACE2 activation promotes antithrombotic activity. *Mol Med* 2010;**16**:210–15.

31. Fraga-Silva RA, Costa-Fraga FP, De Sousa FB, Alenina N, Bader M, Sinisterra RD, et al. An orally active formulation of angiotensin-(1–7) produces an antithrombotic effect. *Clinics* 2011;**66**:837–41.

32. Kluskens LD, Nelemans SA, Rink R, de Vries L, Meter-Arkema A, Wang Y, et al. Angiotensin-(1–7) with thioether bridge: an angiotensin-converting enzyme-resistant, potent angiotensin-(1–7) analog. *J Pharmacol Exp Ther* 2009;**328**:849–55.

33. Silva-Barcellos NM, Frézard F, Caligiorne S, Santos RA. Long-lasting cardiovascular effects of liposome entrapped angiotensin-(1–7) at the rostral ventrolateral medulla. *Hypertension* 2001;**38**:1266–71.

34. Ferreira AJ, Jacoby BA, Araujo CA, Macedo FA, Silva GA, Almeida AP, et al. The nonpeptide angiotensin-(1–7) receptor Mas agonist AVE-0991 attenuates heart failure induced by myocardial infarction. *Am J Physiol Heart Circ Physiol* 2007;**292**:H1113–19.

35. Savergnini SQ, Beiman M, Lautner RQ, de Paula-Carvalho V, Allahdadi K, Pessoa DC, et al. Vascular relaxation, antihypertensive effect, and cardioprotection of a novel peptide agonist of the Mas receptor. *Hypertension* 2010;**56**:112–20.

36. Kwon KC, Verma D, Singh ND, Herzog RW, Daniell H. Oral delivery of human biopharmaceuticals, autoantigens and vaccine antigens bioencapsulated in plant cells. *Adv Drug Deliv Rev* 2013;**65**:782–99.

37. Shenoy V, Kwon KC, Rathinasabapathy A, Lin S, Jin G, Song C, et al. Oral delivery of Angiotensin-converting enzyme 2 and Angiotensin-(1-7) bioencapsulated in plant cells attenuates pulmonary hypertension. *Hypertension* 2014;**64**:1248–59.

37. Chen J, Xiao X, Chen S, Zhang C, Chen J, Yi D, et al. Angiotensin-converting enzyme 2 priming enhances the function of endothelial progenitor cells and their therapeutic efficacy. *Hypertension* 2013;**61**:681–9.

Chapter 39

Nanocarriers for Improved Delivery of Angiotensin-(1-7)

Frédéric Frézard, Robson Augusto Souza dos Santos, Ana Paula Correa Oliveira Bahia
Departamento de Fisiologia e Biofísica—ICB, Universidade Federal de Minas Gerais, Belo Horizonte, Minas Gerais, Brazil

INTRODUCTION

The heptapeptide angiotensin-(1-7) (Ang-(1-7)) is an important component of the renin–angiotensin system (RAS) with pleiotropic actions, including vasodilator, antiproliferative, antifibrotic, antiarrhythmic, antihypertrophic, and antithrombotic effects. Because of these biological actions, this peptide serves as a basis for the development of new pharmacotherapeutic drugs for the treatment of hypertension, heart failure, cardiac hypertrophy, diabetic complications, metabolic syndrome, atherosclerosis, cancer, muscular dystrophy, glaucoma, erectile dysfunction, and alopecia.[1,2]

As an endogenous peptide, Ang-(1-7) has the great advantage of exhibiting a low-toxicity profile. Mordwinkin et al.[3] showed that subcutaneous administration of Ang-(1-7) at 10 mg/(kg day) for 28 days did not lead to any detectable toxicities in either rats or dogs. No toxicity was reported in human subjects submitted to anticancer chemotherapy, after receiving the peptide subcutaneously at 300 μg/kg.[4] On the other hand, the rapid *in vivo* metabolism of the peptide through inactivation by proteolytic enzymes results in a very short plasma half-life, typically about 10 min in rodents and 30 min in humans,[5,6] and short biological actions, limiting its therapeutic potential. Furthermore, the high molecular weight and hydrophilic and peptidergic character of Ang-(1-7) are physicochemical factors that limit its absorption across biological barriers, such as the skin and gastrointestinal tract.

In this context, different nanocarriers have been investigated to circumvent these limitations and improve the delivery of Ang-(1-7). Liposomes have emerged as a promising vehicle of the peptide for its sustained and controlled release and for its dermal or transdermal delivery. On the other hand, cyclodextrins have been shown to improve the peptide bioavailability by the oral route.

This chapter will describe briefly the basic properties of the nanocarriers used so far for Ang-(1-7) and the investigated applications and discuss other potential, not yet explored, therapeutic applications.

LIPOSOME FORMULATIONS OF ANG-(1-7)

Basic Properties of Liposomes

Liposomes are artificial spherical vesicles composed of concentric lipid bilayers separating one or more internal aqueous compartments from the external environment. These characteristics are the consequence of the amphiphilic nature of the lipid(s): a polar head covalently attached to one or two hydrophobic chains. These vesicles are formed spontaneously when lipids are dispersed in aqueous solution. Hydrophilic substances, such as Ang-(1-7), can be encapsulated in the aqueous compartments, and lipophilic or amphiphilic substances may be incorporated into the lipid bilayer.[7]

A major advantage of liposomes over other drug carrier systems is their high biocompatibility, especially when those are composed of natural lipids such as phosphatidylcholine. Moreover, they are highly versatile systems, whose size, lamellarity surface characteristics, membrane fluidity, phase behavior, internal aqueous volume, and composition can be manipulated through selection of appropriate lipid composition and preparation method, as a function of pharmaceutical and pharmacological requirements.[7,8]

Liposomes can retain water-soluble substances and protect them from fast degradation or renal excretion. The slow drug-release properties of this carrier system result in the reduction of the free drug concentration and in the prolongation

The Protective Arm of the Renin–Angiotensin System (RAS). http://dx.doi.org/10.1016/B978-0-12-801364-9.00039-0

of its presence in the organism. Consequently, in many cases, drug action is prolonged, drug efficacy is enhanced, and/or its undesired side effects are reduced.

The release of water-soluble substances from liposomes in storage conditions is essentially mediated by drug diffusion across the membrane and depends on both the substance physicochemical characteristics and the membrane fluidity. Membrane in the gel state is much less permeable, and more stable in the serum, than those in the liquid-crystal state. The addition of cholesterol as a lipid component tends to reduce the membrane permeability and increases the vesicle stability in the serum.[7] After parenteral administration of liposomes, the release of encapsulated substance can also be mediated by endocytic or phagocytic cells, following vesicle capture and entrance in the lysosomal apparatus. There, liposomes are disrupted within the lysosomes, the solute is released locally and, depending on its nature, can be either transferred into the cytoplasm or exocytosized. In this context, controlled drug release *in vivo* can be achieved through manipulation of liposome–cell interactions. In this respect, vesicle size and surface properties appear as critical parameters for the modulation of this interaction. Small-sized liposomes (diameter less than 100 nm) are captured more slowly than larger vesicles. The incorporation in liposome membrane of lipid(s) with polar head group conjugated to poly(ethylene) glycol (PEG) reduces the rate of opsonization of the vesicles and its capture by cells and promotes the sustained release of encapsulated drug.[7]

Liposome modification with PEG increases their field of usage by enhancing circulation time in blood, allowing for their passive accumulation in tumor, hypoxic, and inflammation sites. Active accumulation can be achieved through attachment of antibodies or different targeting moieties to their surface to target specific affected areas. Such modified liposomes and immunoliposomes can be considered for intravascular drug delivery, using cells and noncellular components as the targeted sites for treating important cardiovascular pathologies, such as myocardial infarction, coronary thrombosis, and atherosclerosis.[9]

Furthermore, liposomes can be designed to release their content specifically in acidic environments, such as tumor sites, as a result of pH-induced membrane destabilization. pH-sensitive liposomes are classically obtained by combining the hexagonal II phase-forming phospholipid, dioleoylphosphatidylethanolamine, with the pH-sensitive cholesteryl hemisuccinate. Steric stabilization, for instance, with PEG lipid, is also necessary to preserve the membrane stability and pH sensitivity and enhance the accumulation of liposomal drugs in the tumor.[10,11]

Lipid vesicles have been extensively studied to carry drugs and enhance their penetration across the skin. Conventional liposomes, mainly those made from lipids that form liquid-crystal state membranes such as lecithin, are generally reported to accumulate in the stratum corneum, in the upper skin layers, and in the appendages.[12,13]

Other vesicle compositions have been investigated to develop systems that are capable of carrying drugs and macromolecules to deeper tissues and/or systemic circulation. These belong to the category that is variously termed as deformable, elastic vesicles. Transfersomes® (IDEA AG) is an example of deformable vesicles, which were first introduced in the early 1990s by Cevc and Blume.[14] Elasticity is generated by the incorporation of an edge activator in the lipid bilayer structure. The original composition of these vesicles consisted of soybean phosphatidylcholine and sodium cholate. These vesicles were claimed to improve the penetration of a range of small hydrophilic and lipophilic molecules, peptides, macromolecules to deep peripheral tissues and/or to the systemic blood circulation.[15] Regarding the mechanism responsible for the transdermal drug absorption, there is still considerable debate as to whether deformable vesicles behave as a true carrier system by penetrating intact through the skin or act as a permeation enhancer.[16,17]

Liposomes as a Tool for Site-Specific Chronic Infusion of Ang-(1-7)

Several studies have indicated that peptides of the RAS may act as important neuromodulators, especially in the brain medullary areas related to the tonic and reflex control of arterial pressure. Microinjection of aqueous solutions containing 25–50 ng of Ang-(1-7) into the rostral ventrolateral medulla (RVLM) of normotensive rats elicited a significant increase of blood pressure.[18] However, the duration of these effects is very short, typically 8 min.

In the search of an appropriate methodology for assessing the chronic actions of the peptide, Silva-Barcellos et al. have investigated the ability of liposomes to act as a nanoreservoir and mimic the physiological conditions of its chronic endogenous production.[19,20]

The proposed method combined and took advantage of three different techniques: liposome encapsulation, site-specific microinjection, and telemetry. Ideally, liposomes may be designed to remain located at the injection site for a long period of time, where they protect encapsulated peptide from rapid degradation and act as a sustained release system. Secondly, microinjection allowed the administration of bioactive substances in some specific sites of the brain with minimal side effects. Finally, using telemetry, it was possible to register physiological parameters and their circadian variations in undisturbed, free-moving animals for days and even months.

Several liposome preparations differing in their vesicle size, membrane surface, and fluidity were evaluated. The most effective preparation consisted of vesicles of diameter close to 200 nm, gel-state membrane containing PEG lipid.[20]

Microinjection of these Ang-(1-7)-containing liposomes (35 ng/200 nL) into the RVLM of normotensive rats elicited a prolonged pressor effect for at least 5 days.[19] Moreover, microinjection of the same liposomes labeled with the lipophilic fluorescent dye, diI, showed that vesicles remained localized at the RVLM even after 7 days. This long-lasting action also led to unmask a new physiological role for Ang-(1-7): its modulation of the circadian rhythms of mean arterial blood pressure.

This work has established the use of liposomes as a tool for the sustained release of the short half-life heptapeptide in a specific site of the brain. It also stimulates investigation of the potential therapeutic applications of liposome-encapsulated Ang-(1-7) after systemic administration.

Topical Liposome Formulation of Ang-(1-7) for Treatment of Alopecia

The potential of Ang-(1-7) for the prevention of alopecia has been described in a previous patent application.[2] This was based on clinical trials in cancer patients submitted to chemotherapy, in association or not with Ang-(1-7) administered subcutaneously. Interestingly, the heptapeptide significantly decreased alopecia in chemotherapy patients relative to control. The mechanism involved was not elucidated. This effect could be related to the vasodilating action of the peptide or to a more specific effect on the hair follicle.

These results prompted our group to investigate topical formulation of Ang-(1-7) for the treatment of alopecia. A lipid vesicle-based formulation was chosen, because of the ability of lipid vesicles to act as a localized drug depot in the skin and skin appendages including the hair follicle[21] and to promote the dermal and transdermal drug absorption.[15] The formulation was based on deformable vesicles with liquid-crystal state membrane.[22]

A clinical trial was performed in 60 subjects with androgenetic alopecia to compare the efficacy of the liposomal Ang-(1-7) to that of the liposome vehicle (placebo). The topical application of liposomal Ang-(1-7) once daily for 90 days resulted in a significantly higher proportion of hairs in the anagen phase, when compared to placebo (Frézard F, Bahia APCO, Santos RA, unpublished data). Therefore, this study clearly established the potential of this formulation for preventing hair loss.

Potential Therapeutic Applications of Liposomal Ang-(1-7)

As illustrated in Table 1, from the actual knowledge of the biological actions of Ang-(1-7), the properties of liposomes, and the results achieved specifically with Ang-(1-7), several therapeutic applications of liposome formulations can be anticipated including cancer, myocardial infarction, atherosclerosis, muscular dystrophy, glaucoma, and erectile dysfunction.[1]

TABLE 1 Potential Therapeutic Applications of Liposome Formulations of Ang-(1-7), Preferred Liposome Characteristics, and Expected Benefits

Therapeutic Application	Explored Biological Action(s)	Route	Preferred Liposome Characteristics	Expected Benefit(s)
Cancer	Antiproliferative Antiangiogenic	IV	Pegylated, pH-sensitive liposomes	Improved targeting and delivery
Myocardial infarction	Cardioprotection	IV	Pegylated liposomes, coated or not with specific antibody (e.g., antimyosin)	Targeting of infarcted area and long-lasting action
Atherosclerosis	Atheroprotection	IV	Pegylated liposomes, coated or not with specific antibody (e.g., anti-ICAM$_1$)	Targeting of atherosclerotic plaque and long-lasting action
Muscular dystrophy	Antifibrotic Antiapoptotic	IM	Pegylated liposomes with 200 nm diameter	Localized and long-lasting action
Glaucoma	Reduction of intraocular pressure	Topical	Liposomes with liquid-crystal membrane	Enhanced absorption and long-lasting action
Erectile dysfunction	Vasodilation	Topical	Liposomes with liquid-crystal membrane	Enhanced transdermal delivery

CYCLODEXTRIN-BASED ORAL FORMULATION OF ANG-(1-7)

Cyclodextrins, which are cyclic oligosaccharides composed of glucose units joined through α-1,4 glucosidic bonds, are well known in recognition chemistry as molecular hosts capable of including, with a degree of selectivity, water-insoluble guest molecules by means of noncovalent interactions within their hydrophobic cavity. Thus, this carrier has been widely used to improve the oral bioavailability of water-insoluble drugs, owing to the enhancement of the drug apparent solubility and dissolution rate.[23] β-Cyclodextrin (βCD) is the most common natural cyclodextrin and has 21 hydroxyl groups, seven being primary and 14 secondary hydroxyls. These are available as starting points for structural modifications, and various functional groups have been introduced to modify the physicochemical properties of the parent host molecule.[23] A typical example of pharmaceutically useful βCD derivatives is 2-hydroxypropyl β-cyclodextrin (HPβCD), obtained through hydroxyalkylation of βCD hydroxyl groups, which exhibits enhanced water solubility and biocompatibility compared to βCD.

It has been recently demonstrated that Ang-(1-7) forms a complex with βCD through inclusion of nonpolar Tyr residue into its hydrophobic cavity.[24] In a subsequent work, Marques et al. demonstrated the efficacy by the oral route of the complex between Ang-(1-7) and HPβCD in reducing the deleterious effects induced by myocardial infarction in rats.[25] The plasma levels of the peptide were also evaluated following oral administration of HPβCD/Ang-(1-7) in rats. Plasma levels increased three-fold following 2 h of administration and peaked at 6 h (12-fold increase). No significant effect on plasma Ang-(1-7) levels was observed when vehicle or free peptide control was administered. Thus, the oral bioavailability of Ang-(1-7) was clearly enhanced through complexation with HPβCD. In subsequent works, the efficacy of the oral treatment with HPβCD/Ang-(1-7) was established in experimental models of thrombosis, metabolic syndrome, erectile dysfunction, type 2 diabetes, hypertension, muscular dystrophy, and atherosclerotic lesions.[26–32]

An interesting and innovative feature of the HPβCD/Ang-(1-7) composition is that it associates a highly hydrophilic peptide to a highly hydrophilic cyclodextrin. At the time it was described, it was the first report of a hydrophilic cyclodextrin being capable of enhancing the oral bioavailability of a hydrophilic peptide.[33] The mechanism by which HPβCD improves the bioavailability of the peptide is still unclear. Most probably, complexation of the heptapeptide to HPβCD may prevent its degradation in the gastrointestinal tract and the peptide may be released and absorbed in the intestine after dilution-induced dissociation of the complex. Considering that HPβCD is essentially excreted in the feces and that the active substance entering the circulation is the endogenous peptide Ang-(1-7), this complex appears much more promising from the point of view of drug development than the synthetic derivatives of Ang-(1-7). Next critical steps include the development of an oral pharmaceutical formulation of HPβCD/Ang-(1-7) and the demonstration of its efficacy in humans.

CONCLUSION

The cyclodextrin derivative HPβCD has been successfully used to improve the bioavailability of Ang-(1-7) by the oral route, and several potential therapeutic applications of the innovative peptide–cyclodextrin complex have been established in preclinical models. While this carrier system is suitable to elevate the systemic concentration of the peptide for chronic oral therapy, liposomes may allow the peptide delivery to specific body sites after parenteral or topical applications. Liposomes were also used with good results as a tool for the sustained release of the heptapeptide in a specific site of the brain.

REFERENCES

1. Santos RAS. Angiotensin-(1–7). *Hypertension* 2014;**63**:1138–47.
2. Rodgers KE, Dizerega GS. Methods for treating and preventing alopecia. *World Patent Information* WO 01/98325, 2001.
3. Mordwinkin NM, Russell JR, Burke AS, Dizerega GS, Louie SG, Rodgers KE. Toxicological and toxicokinetic analysis of angiotensin (1–7) in two species. *J Pharm Sci* 2012;**101**:373–80.
4. Pham H, Schwartz BM, Delmore JE, Reed E, Cruickshank S, Drummond L, et al. Pharmacodynamic stimulation of thrombogenesis by angiotensin (1–7) in recurrent ovarian cancer patients receiving gemcitabine and platinum-based chemotherapy. *Cancer Chemother Pharmacol* 2013;**71**:965–72.
5. Yamada K, Iyer SN, Chappell MC, Ganten D, Ferrario CM. Converting enzyme determines plasma clearance of angiotensin-(1–7). *Hypertension* 1998;**32**:496–502.
6. Rodgers KE, Ellefson DD, Espinoza T, Hsu YH, diZerega GS, Mehrian-Shai R. Expression of intracellular filament, collagen, and collagenase genes in diabetic and normal skin after injury. *Wound Repair Regen* 2006;**14**:298–305.
7. Frézard F. Liposomes: from biophysics to design of peptide vaccines. *Braz J Med Biol Res* 1999;**32**:181–9.

8. Frézard F, Schetini DA, Rocha OGF, Demicheli C. Lipossomas: propriedades físico-químicas e farmacológicas, aplicações na quimioterapia à base de antimônio. *Quím Nova* 2005;**29**:511–18.

9. Levchenko TS, Hartner WC, Torchilin VP. Liposomes for cardiovascular targeting. *Ther Deliv* 2012;**3**:501–14.

10. Momekova D, Rangelov S, Lambov N. Long-circulating, pH-sensitive liposomes. *Methods Mol Biol* 2010;**605**:527–44.

11. Slepushkin VA, Simões S, Dazin P. Sterically stabilized pH-sensitive liposomes. Intracellular delivery of aqueous contents and prolonged circulation in vivo. *J Biol Chem* 1997;**272**:2382–8.

12. Mezei M, Gulasekharam V. Liposomes: a selective drug delivery system for the topical route of administration. Lotion dosage form. *Life Sci* 1980;**26**:1473–7.

13. Li L, Hoffman RM. Topical liposome delivery of molecules to hair follicles in mice. *J Dermatol Sci* 1997;**14**:101–8.

14. Cevc G, Blume G. Lipid vesicles penetrate into intact skin owing to the transdermal osmotic gradients and hydration force. *Biochim Biophys Acta* 1992;**1104**:226–32.

15. Cevc G. Lipid vesicles and other colloids as drug carriers on the skin. *Adv Drug Deliv Rev* 2004;**56**:675–711.

16. El Maghraby GM, Barry BW, Williams AC. Liposomes and skin: from drug delivery to model membranes. *Eur J Pharm Sci* 2008;**34**:203–22.

17. Bahia APCO, Azevedo EG, Ferreira LAM, Frézard F. New insights into the mode of action of ultradeformable vesicles using calcein as hydrophilic fluorescent marker. *Eur J Pharm Sci* 2010;**39**:90–6.

18. Fontes MAP, Pinge MCM, Naves V, Campagnole-Santos MJ, Lopes OU, Khosla MC, et al. Cardiovascular effects produced by microinjection of angiotensins and angiotensin antagonists into the ventrolateral medulla of freely moving rats. *Brain Res* 1997;**750**:305–10.

19. Silva-Barcellos NM, Frézard F, Caligiorne S, Santos RA. Long-lasting cardiovascular effects of liposome-entrapped angiotensin-(1–7) at the rostral ventrolateral medulla. *Hypertension* 2001;**38**:1266–71.

20. Silva-Barcellos NM, Caligiorne S, Santos RA, Frézard F. Site-specific microinjection of liposomes into the brain for local infusion of a short-lived peptide. *J Control Release* 2004;**95**:301–7.

21. Lauer AC, Ramachandran C, Lieb LM, Niemiec S, Weiner ND. Targeted delivery to the pilosebaceous unit via liposomes. *Adv Drug Deliv Rev* 1996;**18**:311–24.

22 Ferreira AJ, Frezard FJG, Santos RAS, Fraga-Silva RA, Lautner RQ, Bahia APCO, et al. Formulações tópicas para a prevenção e tratamento da alopecia e para inibição do crescimento de pelos. *Brazil Patent* BR 102013030151-5, 2013.

23. Hirayama F, Uekama K. Cyclodextrin-based controlled drug release system. *Adv Drug Deliv Rev* 1999;**36**:125–41.

24. Lula I, Denadai AL, Resende JM, Sousa FB, Lima GF, Pilo-Veloso D, et al. Study of angiotensin-(1–7) vasoactive peptide and its β-cyclodextrin inclusion complexes: complete sequence-specific NMR assignments and structural studies. *Peptides* 2007;**28**:2199–210.

25. Marques FD, Ferreira AJ, Sinisterra RD, Jacoby BA, Sousa FB, Caliari MV, et al. An oral formulation of angiotensin-(1–7) produces cardioprotective effects in infarcted and isoproterenol-treated rats. *Hypertension* 2011;**57**:477–83.

26. Fraga-Silva RA, Costa-Fraga FP, De Sousa FB, Alenina N, Bader M, Sinisterra RD, et al. An orally active formulation of angiotensin-(1–7) produces an antithrombotic effect. *Clinics* 2011;**66**:837–41.

27. Fraga-Silva RA, Montecucco F, Mach F, Santos RA, Stergiopulos N. Pathophysiological role of the renin-angiotensin system on erectile dysfunction. *Eur J Clin Invest* 2013;**43**:978–85.

28. Feltenberger JD, Andrade JM, Paraíso A, Barros LO, Filho AB, Sinisterra RD, et al. Oral formulation of angiotensin-(1–7) improves lipid metabolism and prevents high-fat diet-induced hepatic steatosis and inflammation in mice. *Hypertension* 2003;**62**:324–30.

29. Santos SH, Giani JF, Burghi V, Miquet JG, Qadri F, Braga JF, et al. Oral administration of angiotensin-(1–7) ameliorates type 2 diabetes in rats. *J Mol Med* 2014;**92**:255–65.

30. Bertagnolli M, Casali KR, De Sousa FB, Rigatto K, Becker L, Santos SH, et al. An orally active angiotensin-(1–7) inclusion compound and exercise training produce similar cardiovascular effects in spontaneously hypertensive rats. *Peptides* 2014;**51**:65–73.

31. Sabharwal R, Cicha MZ, Sinisterra RD, De Sousa FB, Santos RA, Chapleau MW. Chronic oral administration of Ang-(1–7) improves skeletal muscle, autonomic and locomotor phenotypes in muscular dystrophy. *Clin Sci* 2014;**127**:101–9.

32. Fraga-Silva RA, Savergnini SQ, Montecucco F, Nencioni A, Caffa I, Soncini D, et al. Treatment with angiotensin-(1–7) reduces inflammation in carotid atherosclerotic plaques. *Thromb Haemost* 2014;**111**:736–47.

33 Sinisterra RD, Santos RA, Frezard FJG, Nadu AP. Process of preparation of formulations of the peptide angiotensin-(1–7) and its analogues, agonistic and antagonists using cyclodextrins, liposomes and biodegradable polymers and/or mixtures and products thereof. *World Patent Information* WO/2003/039434, 2003.

Chapter 40

Mas Agonists

Rodrigo A. Fraga-Silva,* Walkyria Oliveira Sampaio,†,‡ Robson Augusto Souza dos Santos§

*Institute of Bioengineering, Ecole Polytechnique Federale de Lausanne, Lausanne, Vaud, Switzerland, †Department of Biological Sciences, Universidade de Itaúna, Itaúna, Minas Gerais, Brazil, ‡Department of Physiology, Universidade Federal de Minas Gerais, Belo Horizonte, Minas Gerais, Brazil, §Departamento de Fisiologia e Biofísica—ICB, Universidade Federal de Minas Gerais, Belo Horizonte, Minas Gerais, Brazil

INTRODUCTION

The existence of the renin–angiotensin system (RAS) was first postulated by Tigerstedt and Bergman, in 1898[1,2]; nevertheless, its fundamental concept is still being altered, and several recent studies have made great advances into our understanding of the system, revealing new components and signaling pathways.[1] Among many achievements of the last decades, one of the most significant was the discovery by Santos et al. that Mas is a functional receptor for Ang-(1-7).[3,4] Subsequently, a number of studies confirmed that Mas mediates protective actions, often contrary to those produced by AT_1 stimulation.[1,4] Beyond its own stimulatory pathway, which often involves nitric oxide release, the Mas receptor may also inhibit AT_1-mediated signaling pathways, modulating the vasoconstrictor, proliferative, and oxidative effects of the Ang II.[4,5] Recently, Lautner and colleagues identified a new vasoactive peptide, alamandine, and its receptor, the Mas-related G-coupled receptor type D (MrgD),[6] suggesting a mutual reinforcement of alamandine/MrgD and Ang-(1-7)/Mas to modulate the deleterious actions of Ang II/AT_1 axis. Thus, considerable efforts have been made to develop novel pharmacological approaches to activate Mas receptor and its modulatory actions within the RAS[7] (Figure 1).

MAS AGONISTS

The Mas agonists are one of the important sets of tools to study the Ang-(1-7)/Mas receptor axis and a promising strategy for the development of new treatments for cardiovascular and other diseases.[8] The current pharmacological therapies based on the inhibition of Ang II formation or the blockage of AT_1 receptor (angiotensin-converting enzyme inhibitors (ACEi) and AT_1 receptor blockers (ARBs), respectively, (Figure 1)) represent important achievements in the treatment of cardiovascular diseases.[5,9] In addition to the inhibition of Ang II synthesis and receptor binding, ACEi and ARBs substantially increase Ang-(1-7) plasma levels,[7,8] leading to consequent Mas activation. Indeed, several studies have suggested that Ang-(1-7)/Mas contributes to the clinical effects of these drugs. For instance, A-779, a well-known Mas antagonist, can attenuate or even block, in some conditions, many effects of ACEi and ARBs.[10,11] Therefore, the beneficial effects of Ang-(1-7) and its involvement in the effects of ACEi and ARBs are evidence for the therapeutic potential of the Mas receptor agonist for treating cardiovascular disease. Besides its well-known actions in the cardiovascular system, the antiproliferative and antiangiogenic proprieties of Ang-(1-7)/Mas indicate that Mas agonists can also be used as chemotherapeutic agents.[12]

Ang-(1-7)

Ang-(1-7) is the natural agonist of Mas receptor[3]; hence, Ang-(1-7) itself represents an excellent choice for clinical purposes. However, the unfavorable pharmacokinetics[13] hampers the feasibility of Ang-(1-7) administration, and to overcome this barrier, new strategies such as the inclusion of Ang-(1-7) in hydroxypropyl β-cyclodextrin (Ang-(1-7)-CyD)[14–16] as well as a cyclized form of Ang-(1-7)[17] were developed.

Cyclodextrins (CyDs) are a family of pharmaceutical tools used to enhance the drug stability and absorption across biological barriers and offer gastric protection.[18] Typically, CyDs are composed of a number of cyclic oligosaccharides, ranging from six to eight units, that form a ring and create a conical shape. The structure generates an amphiphilic molecule with a hydrophilic outer surface and a hydrophobic inner surface, allowing the formation of a supramolecular inclusion complex, stabilized by noncovalent interactions with a guest molecule.[18] The supramolecular complex of Ang-(1-7) and hydroxypropyl β-cyclodextrin occurs between the Ang-(1-7) tyrosine amino acid residue and the inner surface of hydroxypropyl

The Protective Arm of the Renin–Angiotensin System (RAS). http://dx.doi.org/10.1016/B978-0-12-801364-9.00040-7

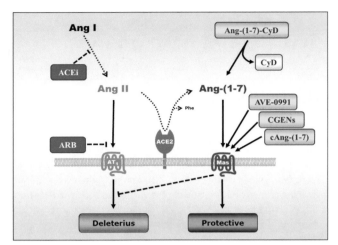

FIGURE 1 Schematic representation of the RAS cascade and the main known Mas agonists. Ang I, angiotensin I; Ang II, angiotensin II; Ang-(1-7), angiotensin-(1-7); AT$_1$, angiotensin II type 1 receptor; Mas, Mas receptor; ACEi, angiotensin-converting enzyme inhibitor; ARB, AT$_1$ receptor blockers; ACE2, angiotensin-converting enzyme 2; Phe, phenylalanine; Ang-(1-7)-CyD, angiotensin-(1-7) formulation, supramolecular inclusion complex of Ang-(1-7) in hydroxypropyl β-cyclodextrin; CyD, hydroxypropyl β-cyclodextrin; AVE-0991, synthetic Mas agonist (5-formyl-4-methoxy-2-phenyl-1-[[4-[2-ethyl-aminocarbonylsulfonamido-5-isobutyl-3-thienyl]-phenyl]-methyl]-imidazole); CGENs, synthetic Mas agonists CGEN-856 and CGEN-857; cAng-(1-7), cyclized angiotensin-(1-7), thioether-bridged angiotensin-(1-7).

β-cyclodextrin.[19] This complex protects Ang-(1-7) against proteolytic degradation within the gastrointestinal tract, permitting its absorption. When given orally to rats, Ang-(1-7)-CyD increased Ang-(1-7) plasma levels after 2 h of administration, which was maintained for at least 6 h.[12]

The inclusion compound (Ang-(1-7) included in hydroxypropyl β-cyclodextrin) is not entirely absorbed into the systemic circulation; however, only the free form of Ang-(1-7) is fully absorbed.[13,14] Apparently, the inclusion compound crosses through the stomach and small intestine remaining intact. In the colon microflora, the cyclodextrin is broken down into saccharide monomers, delivering the free form of Ang-(1-7) to the large intestine where it may be absorbed.[18] The formulation Ang-(1-7)-CyD acts as an Ang-(1-7) delivery system. For example, 76 µg/kg of Ang-(1-7)-CyD (relative to 30 µg/kg of Ang-(1-7)) increases Ang-(1-7) plasma level from 50 to 100 pg/mL after 2 h and to 400 pg/mL after 6 h, which is then normalized to baseline levels after 24 h.[12] Therefore, Ang-(1-7) is administered in the order of micrograms, but the sustained increase in plasma is in the order of picograms (10^6 less).

As expected, Ang-(1-7)-CyD produces similar actions as reported to Ang-(1-7). Chronic administration of Ang-(1-7)-CyD significantly attenuated the heart function impairment and cardiac remodeling induced by isoproterenol and myocardial infarction in rats.[12] Moreover, acute and chronic oral administration of Ang-(1-7)-CyD produced antithrombotic activity, which was abolished in *Mas* gene-deleted mice.[14] Additionally, Ang-(1-7)-CyD ameliorates diabetes type 2 symptoms,[15] stabilizes vulnerable atherosclerotic plaques,[16] improves lipid metabolism,[20,21] reverses corpus cavernosum damage induced by hypercholesterolemia,[22] and produces cardioprotective action in spontaneously hypertensive rats.[23]

Overall, Ang-(1-7)-CyD is a feasible formulation, which permits the oral administration and long-lasting action of Ang-(1-7). Since it is based on the action of the Ang-(1-7) peptide itself, the endogenous agonist, toxicological reactions are less expected. Therefore, Ang-(1-7)-CyD represents a great opportunity to treat cardiovascular disease (Figure 1).

Cyclized Ang-(1-7)

A cyclized form of Ang-(1-7) (cAng-(1-7)) has been proposed as an innovative approach to yield more resistance against proteolytic breakdown.[17] This compound has been developed based on the ability of prokaryotes to cyclize peptides. An intramolecular thioether-bridged was introduced to the Ang-(1-7) peptide sequence, increasing its stability against purified angiotensin-converting enzyme and homogenates of different organs.[17] Indeed, cAng-(1-7) has been shown to be stable at pH 2.0 and has enhanced resistance to pancreatic and liver proteases. Moreover, it has been demonstrated that cAng-(1-7) can be orally and pulmonary administered.[24]

Importantly, cAng-(1-7) was able to mimic Ang-(1-7) effects. This compound produced Mas-dependent vasorelaxation in aorta rings of rats with two-fold larger magnitude compared to the natural peptide.[17] Subcutaneous administration of cAng-(1-7) lowered left ventricular end-diastolic pressure and improved endothelial function in a rat model of myocardial infarction but did not change infarct size or area.[25] Moreover, cAng-(1-7) was able to attenuate cardiopulmonary injury by reducing pulmonary inflammation and prevent right ventricular hypertrophy.[26]

Generally, it appears that cAng-(1-7) is a proteolytic-resistant peptide compound that acts through the Mas receptor (Figure 1). However, only a few studies have reported its actions, and, therefore, further studies are required to better evaluate cAng-(1-7) potential.

The Nonpeptide Mas Agonist, AVE-0991

In 2002, prompted by the findings that demonstrated beneficial Ang-(1-7)-induced actions, Wiemer and colleagues[27] initiated studies directed toward the synthesis and identification of a nonpeptide mimetic of Ang-(1-7), which resulted in the discovery of the compound, AVE-0991, 5-formyl-4-methoxy-2-phenyl-1-[[4-[2-ethyl aminocarbonylsulfonamido-5-isobutyl-3-thienyl]-phenyl]-methyl]-imidiazole (Figure 1).

This nonpeptide synthetic compound competed for high-affinity binding with [^{125}I]-Ang-(1-7) in bovine aortic endothelial cells, demonstrating the same specific interaction for the molecular site of Ang-(1-7).[27] Moreover, AVE-0991 was able to mimic several Ang-(1-7) effects.[27–29] AVE-0991 was able to induce NO release in Mas-transfected CHO cells, which was blocked by A-779, a Mas antagonist,[30] but not by AT$_1$ (CV11974) or AT$_2$ (PD123319) antagonists.[30] Supporting this finding, in vitro receptor autoradiography in mouse kidney slices demonstrated that the specific binding of [^{125}I]-Ang-(1-7) was displaced by AVE-0991, whereas no interference were observed in the binding of [^{125}I]-angiotensin II or [^{125}I]-angiotensin IV.[31] Similarly, AVE-0991 displaced ^{125}I-Ang-(1-7) binding in *Mas*-transfected monkey kidney cells and the fluorescent rhodamine-Ang-(1-7) binding in *Mas*-transfected Chinese hamster ovary (CHO) cells.[30] In an elegant evaluation, Lemos and colleagues[32] demonstrated that AVE-0991 caused vasorelaxation of rat and mouse aortic rings and was abolished by two different Mas antagonists, A-779 and D-pro-Ang-(1-7); moreover, the vasorelaxant response of AVE-0991 was absent in *Mas*-deficient mice.[32]

AVE-0991 mimics various Ang-(1-7) effects. In the vasculature, this compound produced vasodilation[32] and improved endothelial function through NO release.[33] In a water-loaded mouse model, AVE-0991 produced antidiuretic actions, which were completely blocked by A-779.[30] In the heart, AVE-0991 also showed similar cardioprotective actions to Ang-(1-7).[31,34] This compound reduced myocardial infarct area and improves cardiac function in a rat model of myocardial infarction induced by left coronary artery ligation.[31] Furthermore, this compound attenuated isoproterenol-induced impairment of cardiac function and remodeling.[34] Interestingly, AVE-0991 potentiated the hypotensive effect of bradykinin in a Mas-dependent manner.[35] In a spontaneously hypertensive rat model chronically treated with N^G-nitro-L-arginine methyl ester (L-NAME), AVE-0991 attenuated the development of hypertension and end-organ damage.[36] Moreover, AVE-0991 dose-dependently decreased arterial blood pressure in DOCA-treated rats[37] and produced synergic blood-pressure-lowering effects in combination with aliskiren, a renin inhibitor.[37] Additionally, AVE-0991 produced protective actions against atherosclerosis,[38] acute lung injury,[39] pulmonary remodeling in chronic asthma,[40] and nephropathy.[41]

As expected and similar to Ang-(1-7), AVE-0991 actions have been shown to be dependent on NO production. Generally, coadministration of L-NAME completely blocks the vasodilatory actions evoked by AVE-0991,[31,32] the facilitatory effect on erectile function,[42] and the potentiation of bradykinin.[35]

Taken together, AVE-0991 is postulated to be a potential tool for the treatment of cardiovascular disease. Clinically, one important advantage of AVE-0991 is that it is a more stable compound in regard to oral administration and demonstrates a greater half-life compared to Ang-(1-7).[27] However, pharmacokinetics and toxicity of the compound are still unknown and further studies are required to address these issues.

CGEN-856S

Using a novel computational screening platform designed to predict and identify hypothetical natural ligands (essentially peptides encrypted in the human proteome) that may activate GPCR, Shemesh and colleagues identified two peptides, p33 and p61 (designed as CGEN-856 and CGEN-857), that were able to activate the Mas receptor (Figure 1).[43] Interestingly, these synthetic peptides have no correlation with Ang-(1-7) amino acid sequence but display high binding affinity to Mas.[43]

The CGEN-856S is a derivative peptide from CGEN-856. CGEN-856S has been shown to displace with high-affinity fluorescent-Ang-(1-7) binding in CHO *Mas*-transfected cells and elicits a vasorelaxant response that was abolished in *Mas* gene-deleted mice.[44] Moreover, this compound produced various biological actions similar to Ang-(1-7). For instance, CGEN-856S produced antiarrhythmogenic effects, decreased arterial pressure of conscious SHR,[44] elicited NO-dependent vasodilation, reduced the degree of hypertrophy and attenuated collagen deposition induced by isoproterenol, and improved cardiac function and decreased the infarct area lesion in a rat model of myocardial infarction.[45]

These findings suggest CGEN-856S could be used as a putative tool to activate the Mas receptor. However, additional preclinical studies with this compound and/or derivatives still need to be evaluated, as well as toxicological aspects.

Other Mas Agonists

Besides the more characterized Mas agonists described above, other low-affinity ligands have been described. Bikkavilli et al.,[46] using a phage-displayed random peptide library, identified the synthetic peptide MBP7 as a Mas agonist through the activation of phospholipase C, intracellular calcium mobilization, and Mas internalization in stable CHO cells overexpressing Mas. In HEK Mas-transfected cells, the RFamide neuropeptide NPFF also activated Mas receptor.[47] The compound described by Zhang et al.[48] as a Mas agonist, AR234960, produced opposite effects to those described for Ang-1-7 and other Mas agonists in isolated hearts. In contrast, treatment of hearts with the Mas inverse agonist AR244555 during reperfusion resulted in significantly elevated coronary flow during reperfusion.[48] Further studies are necessary to clarify these intriguing observations.

CONCLUSIONS

As discussed in this chapter, the Mas receptor is a key component of the RAS, acting in an opposite manner as AT_1 receptor, and "Mas agonists" may represent a novel class of drugs for the treatment of cardiovascular diseases. In clinical trials enrolling cancer patients, Ang-(1-7) was well tolerated with limited toxic side effects and showed beneficial actions.[49] However, further clinical investigations using the Mas agonist compounds are necessary to explore their safety and efficacy in the treatment of cardiovascular and other diseases.

REFERENCES

1. Bader M. Tissue renin-angiotensin-aldosterone systems: targets for pharmacological therapy. *Annu Rev Pharmacol Toxicol* 2010;**50**:439–65.
2. Basso N, Terragno NA. History about the discovery of the renin-angiotensin system. *Hypertension* 2001;**38**(6):1246–9.
3. Santos RA, Simoes e Silva AC, Maric C, Silva DM, Machado RP, de Buhr I, et al. Angiotensin-(1–7) is an endogenous ligand for the G protein-coupled receptor Mas. *Proc Natl Acad Sci U S A* 2003;**100**(14):8258–63.
4. Ferreira AJ, Santos RA, Bradford CN, Mecca AP, Sumners C, Katovich MJ, et al. Therapeutic implications of the vasoprotective axis of the renin-angiotensin system in cardiovascular diseases. *Hypertension* 2010;**55**(2):207–13.
5. Schiffrin EL. Vascular and cardiac benefits of angiotensin receptor blockers. *Am J Med* 2002;**113**(5):409–18.
6. Lautner RQ, Villela DC, Fraga-Silva RA, Silva N, Verano-Braga T, Costa-Fraga F, et al. Discovery and characterization of alamandine: a novel component of the renin-angiotensin system. *Circ Res* 2013;**112**(8):1104–11.
7. Iyer SN, Ferrario CM, Chappell MC. Angiotensin-(1–7) contributes to the antihypertensive effects of blockade of the renin-angiotensin system. *Hypertension* 1998;**31**(1 Pt 2):356–61.
8. Iyer SN, Chappell MC, Averill DB, Diz DI, Ferrario CM. Vasodepressor actions of angiotensin-(1–7) unmasked during combined treatment with lisinopril and losartan. *Hypertension* 1998;**31**(2):699–705.
9. Nicholls MG, Richards AM, Agarwal M. The importance of the renin-angiotensin system in cardiovascular disease. *J Hum Hypertens* 1998;**12**(5):295–9.
10. Collister JP, Hendel MD. The role of Ang (1–7) in mediating the chronic hypotensive effects of losartan in normal rats. *J Renin Angiotensin Aldosterone Syst* 2003;**4**(3):176–9.
11. Kucharewicz I, Pawlak R, Matys T, Pawlak D, Buczko W. Antithrombotic effect of captopril and losartan is mediated by angiotensin-(1–7). *Hypertension* 2002;**40**(5):774–9.
12. Marques FD, Ferreira AJ, Sinisterra RD, Jacoby BA, Sousa FB, Caliari MV, et al. An oral formulation of angiotensin-(1–7) produces cardioprotective effects in infarcted and isoproterenol-treated rats. *Hypertension* 2011;**57**(3):477–83.
13. Fraga-Silva RA, Ferreira AJ, Dos Santos RA. Opportunities for targeting the angiotensin-converting enzyme 2/angiotensin-(1–7)/mas receptor pathway in hypertension. *Curr Hypertens Rep* 2013;**15**(1):31–8.
14. Fraga-Silva RA, Costa-Fraga FP, De Sousa FB, Alenina N, Bader M, Sinisterra RD, et al. An orally active formulation of angiotensin-(1–7) produces an antithrombotic effect. *Clinics* 2011;**66**(5):837–41.
15. Santos SH, Giani JF, Burghi V, Miquet JG, Qadri F, Braga JF, et al. Oral administration of angiotensin-(1–7) ameliorates type 2 diabetes in rats. *J Mol Med* 2014;**92**(3):255–65.
16. Fraga-Silva RA, Savergnini SQ, Montecucco F, Nencioni A, Caffa I, Soncini D, et al. Treatment with angiotensin-(1–7) reduces inflammation in carotid atherosclerotic plaques. *Thromb Haemost* 2014;**111**(4):736–47.
17. Kluskens LD, Nelemans SA, Rink R, de Vries L, Meter-Arkema A, Wang Y, et al. Angiotensin-(1–7) with thioether bridge: an angiotensin-converting enzyme-resistant, potent angiotensin-(1–7) analog. *J Pharmacol Exp Ther* 2009;**328**(3):849–54.
18. Uekama K. Design and evaluation of cyclodextrin-based drug formulation. *Chem Pharm Bull* 2004;**52**(8):900–15.
19. Lula I, Denadai AL, Resende JM, de Sousa FB, de Lima GF, Pilo-Veloso D, et al. Study of angiotensin-(1–7) vasoactive peptide and its beta-cyclodextrin inclusion complexes: complete sequence-specific NMR assignments and structural studies. *Peptides* 2007;**28**(11):2199–210.
20. Feltenberger JD, Andrade JM, Paraiso A, Barros LO, Filho AB, Sinisterra RD, et al. Oral formulation of angiotensin-(1–7) improves lipid metabolism and prevents high-fat diet-induced hepatic steatosis and inflammation in mice. *Hypertension* 2013;**62**(2):324–30.

21. Santos CF, Santos SH, Ferreira AV, Botion LM, Santos RA, Campagnole-Santos MJ. Association of an oral formulation of angiotensin-(1–7) with atenolol improves lipid metabolism in hypertensive rats. *Peptides* 2013;**43**:155–9.

22. Fraga-Silva RA, Costa-Fraga FP, Savergnini SQ, De Sousa FB, Montecucco F, da Silva D, et al. An oral formulation of angiotensin-(1–7) reverses corpus cavernosum damages induced by hypercholesterolemia. *J Sex Med* 2013;**10**(10):2430–42.

23. Bertagnolli M, Casali KR, De Sousa FB, Rigatto K, Becker L, Santos SH, et al. An orally active angiotensin-(1–7) inclusion compound and exercise training produce similar cardiovascular effects in spontaneously hypertensive rats. *Peptides* 2014;**51**:65–73.

24. de Vries L, Reitzema-Klein CE, Meter-Arkema A, van Dam A, Rink R, Moll GN, et al. Oral and pulmonary delivery of thioether-bridged angiotensin-(1–7). *Peptides* 2010;**31**(5):893–8.

25. Durik M, van Veghel R, Kuipers A, Rink R, Haas Jimoh Akanbi M, Moll G, et al. The effect of the thioether-bridged, stabilized angiotensin-(1–7) analogue cyclic ang-(1–7) on cardiac remodeling and endothelial function in rats with myocardial infarction. *Int J Hypertens* 2012;**2012**:536426.

26. Wagenaar GT, Laghmani el H, Fidder M, Sengers RM, de Visser YP, de Vries L, et al. Agonists of MAS oncogene and angiotensin II type 2 receptors attenuate cardiopulmonary disease in rats with neonatal hyperoxia-induced lung injury. *Am J Physiol Lung Cell Mol Physiol* 2013;**305**(5):L341–51.

27. Wiemer G, Dobrucki LW, Louka FR, Malinski T, Heitsch H. AVE 0991, a nonpeptide mimic of the effects of angiotensin-(1–7) on the endothelium. *Hypertension* 2002;**40**(6):847–52.

28. Santos RA, Ferreira AJ. Pharmacological effects of AVE 0991, a nonpeptide angiotensin-(1–7) receptor agonist. *Cardiovasc Drug Rev* 2006;**24**(3–4):239–46.

29. da Costa-Goncalves AC, Fraga-Silva RA, Leite R, Santos RA. AVE 0991, a non-peptide Mas-receptor agonist facilitates penile erection. *Exp Physiol* 2012;**98**(3):850–5.

30. Pinheiro SV, Simoes e Silva AC, Sampaio WO, de Paula RD, Mendes EP, Bontempo ED, et al. Nonpeptide AVE 0991 is an angiotensin-(1–7) receptor Mas agonist in the mouse kidney. *Hypertension* 2004;**44**(4):490–6.

31. Ferreira AJ, Jacoby BA, Araujo CA, Macedo FA, Silva GA, Almeida AP, et al. The nonpeptide angiotensin-(1–7) receptor Mas agonist AVE-0991 attenuates heart failure induced by myocardial infarction. *Am J Physiol Heart Circ Physiol* 2007;**292**(2):H1113–19.

32. Lemos VS, Silva DM, Walther T, Alenina N, Bader M, Santos RA. The endothelium-dependent vasodilator effect of the nonpeptide Ang(1–7) mimic AVE 0991 is abolished in the aorta of mas-knockout mice. *J Cardiovasc Pharmacol* 2005;**46**(3):274–9.

33. Faria-Silva R, Duarte FV, Santos RA. Short-term angiotensin(1–7) receptor MAS stimulation improves endothelial function in normotensive rats. *Hypertension* 2005;**46**(4):948–52.

34. Ferreira AJ, Oliveira TL, Castro MC, Almeida AP, Castro CH, Caliari MV, et al. Isoproterenol-induced impairment of heart function and remodeling are attenuated by the nonpeptide angiotensin-(1–7) analogue AVE 0991. *Life Sci* 2007;**81**(11):916–23.

35. Carvalho MB, Duarte FV, Faria-Silva R, Fauler B, da Mata Machado LT, de Paula RD, et al. Evidence for Mas-mediated bradykinin potentiation by the angiotensin-(1–7) nonpeptide mimic AVE 0991 in normotensive rats. *Hypertension* 2007;**50**(4):762–7.

36. Benter IF, Yousif MH, Anim JT, Cojocel C, Diz DI. Angiotensin-(1–7) prevents development of severe hypertension and end-organ damage in spontaneously hypertensive rats treated with L-NAME. *Am J Physiol Heart Circ Physiol* 2006;**290**(2):H684–91.

37. Singh Y, Singh K, Sharma PL. Effect of combination of renin inhibitor and Mas-receptor agonist in DOCA-salt-induced hypertension in rats. *Mol Cell Biochem* 2013;**373**(1–2):189–94.

38. Jawien J, Toton-Zuranska J, Gajda M, Niepsuj A, Gebska A, Kus K, et al. Angiotensin-(1–7) receptor Mas agonist ameliorates progress of atherosclerosis in apoE-knockout mice. *J Physiol Pharmacol* 2012;**63**(1):77–85.

39. Klein N, Gembardt F, Supe S, Kaestle SM, Nickles H, Erfinanda L, et al. Angiotensin-(1–7) protects from experimental acute lung injury. *Crit Care Med* 2013;**41**(11):e334–43.

40. Rodrigues-Machado MG, Magalhaes GS, Cardoso JA, Kangussu LM, Murari A, Caliari MV, et al. AVE 0991, a non-peptide mimic of angiotensin-(1–7) effects, attenuates pulmonary remodelling in a model of chronic asthma. *Br J Pharmacol* 2013;**170**(4):835–46.

41. Silveira KD, Barroso LC, Vieira AT, Cisalpino D, Lima CX, Bader M, et al. Beneficial effects of the activation of the angiotensin-(1–7) MAS receptor in a murine model of adriamycin-induced nephropathy. *PLoS ONE* 2013;**8**(6), e66082.

42. da Costa Goncalves AC, Fraga-Silva RA, Leite R, Santos RA. AVE 0991, a non-peptide Mas-receptor agonist, facilitates penile erection. *Exp Physiol* 2013;**98**(3):850–5.

43. Shemesh R, Toporik A, Levine Z, Hecht I, Rotman G, Wool A, et al. Discovery and validation of novel peptide agonists for G-protein-coupled receptors. *J Biol Chem* 2008;**283**(50):34643–9.

44. Savergnini SQ, Beiman M, Lautner RQ, de Paula-Carvalho V, Allahdadi K, Pessoa DC, et al. Vascular relaxation, antihypertensive effect, and cardioprotection of a novel peptide agonist of the MAS receptor. *Hypertension* 2010;**56**(1):112–20.

45. Savergnini SQ, Ianzer D, Carvalho MB, Ferreira AJ, Silva GA, Marques FD, et al. The novel Mas agonist, CGEN-856S, attenuates isoproterenol-induced cardiac remodeling and myocardial infarction injury in rats. *PLoS ONE* 2013;**8**(3), e57757.

46. Bikkavilli RK, Tsang SY, Tang WM, Sun JX, Ngai SM, Lee SS, et al. Identification and characterization of surrogate peptide ligand for orphan G protein-coupled receptor mas using phage-displayed peptide library. *Biochem Pharmacol* 2006;**71**(3):319–37.

47. Dong X, Han S, Zylka MJ, Simon MI, Anderson DJ. A diverse family of GPCRs expressed in specific subsets of nociceptive sensory neurons. *Cell* 2001;**106**(5):619–32.

48. Zhang T, Li Z, Dang H, Chen R, Liaw C, Tran TA, et al. Inhibition of Mas G-protein signaling improves coronary blood flow, reduces myocardial infarct size, and provides long-term cardioprotection. *Am J Physiol Heart Circ Physiol* 2012;**302**(1):H299–311.

49. Rodgers KE, Oliver J, diZerega GS. Phase I/II dose escalation study of angiotensin 1–7 [A(1–7)] administered before and after chemotherapy in patients with newly diagnosed breast cancer. *Cancer Chemother Pharmacol* 2006;**57**(5):559–68.

Chapter 41

Preclinical and Clinical Development of Angiotensin Peptides (Mas/Ang(1-7)/ACE-2): Future Clinical Application

Kathleen E. Rodgers,* Gere S. diZerega†

*School of Pharmacy, University of Southern California, Los Angeles, California, USA, †US Biotest, San Luis Obispo, CA, USA; Keck School of Medicine, Los Angeles, CA, USA

INTRODUCTION

Multilineage suppression or ablation (requiring hematopoietic stem cell support) of bone marrow progenitors occurs following chemotherapy, as well as radiation (therapeutic, diagnostic, or environmental), resulting in cytopenias of their formed elements in the peripheral circulation. The manifestations of myelosuppression include anemia, thrombocytopenia, lymphopenia, and neutropenia, which are often managed with a delay and/or a dose reduction in the next scheduled cycle of chemotherapy in order to allow recovery of bone marrow and circulating formed elements. Modifications to the chemotherapy regimen result in lower relative dose intensity (delivered dose intensity/planned dose intensity). Studies show that long-term survival may be compromised if the total dose or relative dose intensity falls below a threshold value.[1–5] Prolonged or severe myelotoxic effects may reflect a diminished hematopoietic reserve, which occurs with aging, age-related comorbidity, intensive chemotherapy, combined radiation therapy/chemotherapy, or previous exposures to myelosuppressive therapies.[6]

The most widely utilized hematopoietic stimulants act on later-stage precursors and usually induce proliferation, differentiation, and mobilization of single cell lineages from the bone marrow into the peripheral circulation; therefore, they do not impact chronic and progressive multilineage cytopenias that occur after myelosuppression. Thus, additional therapies are needed to reduce clinically significant cytopenias.

Chemotherapy-induced thrombocytopenia (CIT) is a common side effect of treating solid tumors that occurs in up to 50% of treatment cycles and affects upward of 300,000 patients annually.[7] CIT results in increased risk of bleeding and the inability to deliver necessary treatment.[7,8] Current strategies to manage CIT utilize platelet transfusions along with chemotherapy dose reductions and delays, leading to compromised clinical outcomes. Transfusions are costly and increase the risk of allergic reactions and infection. It is estimated that CIT increases costs by approximately 40%.[9,10] Neumega® was shown to reduce platelet transfusions and was approved by FDA in 1997. However, Neumega® has substantial toxicity that limits its use for the routine treatment of CIT.[11]

THE RENIN–ANGIOTENSIN SYSTEM IN BONE MARROW AND HEMATOPOIESIS

Research over the last few decades revealed the presence of local, integrated renin–angiotensin system (RAS) within several tissues including the bone marrow.[12,13] Further, mice that are deficient in individual components of the RAS have been shown to have impairment of hematopoiesis. ACE knockout mice have normal to elevated levels of erythropoietin, but are anemic; administration of angiotensin II normalized hematocrit.[14] Further, ACE knockout mice were found to have myelopoietic abnormalities characterized by increased bone marrow myeloblasts and myelocytes, as well as extramedullary myelopoiesis.[15] Mas knockout mice have reduced numbers of myeloid and erythroid progenitors in their bone marrow, again showing the involvement of the endogenous RAS in optimal hematopoiesis.

The Protective Arm of the Renin–Angiotensin System (RAS). http://dx.doi.org/10.1016/B978-0-12-801364-9.00041-9

These observations point to a significant role for the RAS in the regulation of hematopoiesis and the development of hematopoietic progenitor cells.[16–18] The central role of the RAS in regulating early hematopoiesis was recently reviewed.[16,19–22] The earliest marker for the isolation of hemangioblasts (hematopoietic stem cells) is ACE_1 (CD143).[19,23,24] ACE_1 was shown to be expressed in all presumptive and developing blood-forming tissues of the human embryo and fetus.[25] Further, expression of CD143 by cells derived from human pluripotent cells is currently the earliest marker that identifies the hemangioblast, an early cell found in the fetus that results in the development of the hematopoietic and vascular systems.[24,26] Preferential induction of the AT_2 receptor directs the hemangioblast towards the hematopoietic lineage.[19]

Angiotensin 1-7 as a Therapeutic in Bone Marrow Recovery

Because of the increased sensitivity of immature cells to the effects of angiotensin 1-7 (A(1-7)), therapeutic opportunities exist to enhance tissue regeneration, particularly in the repair of injuries in the bone marrow associated with chemotherapy and radiation. Potential populations include cancer patients receiving antineoplastic or radiation therapy with myelosuppressive side effects, stem cell transplant patients after myeloablative conditioning, patients with conditions resulting in ineffective hematopoiesis, and individuals exposed to nuclear radiation. Peptides of the RAS are potent stimulators of progenitor cell proliferation.[17,27–30] Studies show that hematopoietic progenitor cells are sensitive to stimulation by A(1-7) and the multilineage effect is most robust after injury.[17,28–33] The effects of A(1-7) on bone marrow recovery are thought to occur through the Mas receptor. The expression of Mas in normal bone marrow is low, but expression is increased by myelosuppression.[31] Further, a Mas antagonist, A779, blocks the ability of A(1-7) to accelerate bone marrow recovery.[17]

After total body irradiation (TBI), an early increase and sustained expansion in early mixed, myeloid, erythroid, and megakaryocytic progenitors as well as platelet recovery were observed in A(1-7)-treated animals, as compared to controls. However, initiation of treatment must occur once the damage resulting from the myelotoxic exposure has ceased, as evidenced by improved efficacy at higher doses of A(1-7) when drug was initiated at 48 rather than 24 h post TBI. In fact, initiation of treatment could be delayed up to 5 days. The multilineage bone marrow response also resulted in platelet and white blood cell recovery after TBI.[17,28]

Nonhuman Primate Studies

Two studies sponsored and conducted by NIAID evaluated the effect of A(1-7) in nonhuman primates (NHP). In the first NHP (nonirradiated) study, platelet concentration was shown to increase following 10 consecutive days of subcutaneous (SC) dosing with 100 µg/kg/day A(1-7) (Figure 1). In the second study, A(1-7) was administered to irradiated NHP beginning 24 h post TBI. The average onset and duration of severe thrombocytopenia was delayed by approximately 1 day and shortened to one-third the time, respectively, in the 0.3 mg/kg A(1-7) group compared to placebo. The percent of animals with severe thrombocytopenia (<20,000 platelets/µL) by day 10 was 58% in control animals and 9–20% in A(1-7)-treated animals. The initiation of antibiotic treatment was triggered by the occurrence of severe neutropenia (ANC $<50 \times 10^9 \, L^{-1}$); this occurred on day 5 for the vehicle control and low-dose groups but was delayed until day 10 in the high-dose group. These results indicate hematopoietic benefit to NHP, which received lethal doses of radiation.

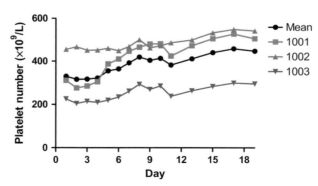

FIGURE 1   A(1-7) increases platelets in NHP.

A(1-7) Stimulates Mucosal/Gastrointestinal Recovery

After the administration of cytotoxic agents, either chemotherapy or radiation, lesions develop in rapidly dividing tissues. Studies conducted with A(1-7) in animals and confirmed in human phase 2a clinical trials show a reduction in the incidence and severity of mucosal lesions after chemotherapy.[32] In a model of chemotherapy-induced mucositis, A(1-7) led to a 20-fold reduction in the area of the intestine affected by lesions, and a reduction in lesion severity was observed in the crypts and epithelial lining.

Hematopoietic Stem Cell Transplant after TBI

A(1-7) was also shown in preclinical studies to increase survival and the concentration of WBCs, platelets, and bone marrow progenitors of both murine (syngeneic) and human origins, after transplantation of suboptimal levels of hematopoietic progenitors.

A(1-7) also stimulated engraftment of human hematopoietic progenitors into NOD/SCID mice.[30] Engraftment of total human and human CD34+ cells in the bone marrow and spleen as well as mature cells in the spleen was increased (up to 600-fold) by A(1-7) in irradiated NOD/SCID mice.[30]

Synergy and Multilineage Effects After Combining A(1-7) with Colony-Stimulating Factors

Combining A(1-7) with commonly used growth factors post chemotherapy increased the numbers of bone marrow progenitor cells and formed elements in the peripheral circulation.[22,31] A(1-7) administered in combination with suboptimal doses of Neupogen throughout the postmyelosuppressive interval increased the number of progenitors and circulating WBC concentration.[31]

Administration of A(1-7) with 3 days of filgrastim decreased ten-fold the amount of filgrastim needed for optimal recovery of bone marrow progenitors (CFU-GEMM, CFU-GM, CFU-Meg, and BFU-E) and circulating formed elements (WBC and platelets).[22] In addition to the synergistic effects of combining A(1-7) and Neupogen on white blood cell and neutrophil recovery, combining these therapies increased platelet concentrations above those observed with A(1-7) alone.

Further, A(1-7) or A(1-7) in combination with erythropoietin (Epogen) increased all erythroid progenitors, with the largest effect on early erythroid progenitors (immature BFU-E). As with Neupogen, combining A(1-7) with Epogen has hematologic effects outside of the erythroid lineage in that the concentration of circulating neutrophils was increased with this combination. These factors acted synergistically with A(1-7) to increase the concentration of myeloid, megakaryocytic, and erythroid progenitor cells in the bone marrow, suggesting that A(1-7)'s multilineage effect on early progenitors in the marrow facilitates proliferation in response to lineage-specific growth factors.

Progenitor mobilization in combination with mobilizing factors: Mobilization of progenitors from the bone marrow into the circulation occurs with injury to allow the progenitors to home to the site of injury and initiate repair. Further, progenitors are mobilized from the bone marrow into the periphery to allow collection via apheresis rather than bone marrow biopsy. In animal studies, A(1-7) increased bone marrow progenitors in the peripheral blood after chemotherapy.[29] In noninjured animals, combining exposure of animals to Neupogen or Mozobil with A(1-7) resulted in increased numbers of circulating progenitors (Figure 2a).

CLINICAL DEVELOPMENT OF A(1-7) FOR HEMATOPOIETIC INDICATIONS

Multilineage suppression of marrow precursors occurs following myelosuppressive radiotherapy and chemotherapy, resulting in cytopenias of one or more of the mature formed elements of blood. Preclinical studies of A(1-7) demonstrated an increase in multiple lineages of early hematopoietic precursors from the bone marrow and the peripheral blood of mice and after *in vitro* exposure of cells from human cord blood. IND-enabling safety studies were performed with no concerns raised.[34]

Phase 1 Evaluation of A(1-7) with and without Neupogen

In one phase 1 study in healthy, male volunteers, no adverse effect was seen at the injection site, in vital signs, or in routine clinical laboratory tests. A(1-7) was safe and well tolerated up to 300 µg/kg for 7 days.

In a second phase 1, double-blind study, the safety, tolerance, and effectiveness of five daily subcutaneous doses of A(1-7) at 300 µg/kg, with and without Neupogen at 10 µg/kg, to increase peripheral blood stem cell (CD34+) counts were evaluated in healthy volunteers. A(1-7) was safely administered in combination with Neupogen with no dose-limiting

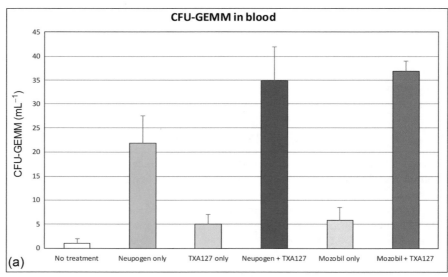

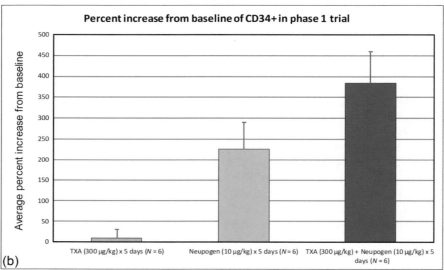

FIGURE 2   A(1-7) synergizes with mobilizing agents to increase progenitors in the blood (a) mouse; (b) phase 1.

toxicity (DLT) or significant adverse events (SAEs) observed. When combined with Neupogen, A(1-7) increased CD34+ mobilization and lymphocyte number as compared to levels achieved with either molecule alone. From baseline through day 6, mean CD34+ cell count increased 156 percent for the TXA127 plus Neupogen group as compared to the other treatment groups (Figure 2b).

Phase 2a Clinical Trial in Breast Cancer Patients

A phase 2a prospective, blinded, randomized, dose-escalation study of A(1-7) administered at up to 100 μg/kg was conducted in breast cancer subjects receiving three cycles of adjuvant chemotherapy.[32] Patients were treated for 7 days starting 14 days prior to chemotherapy, to evaluate potential toxicity. A(1-7) was found to be safe and was well tolerated with no DLT or SAE observed. No A(1-7)-treated patients experienced any NCI grade 1-4 platelet reductions, as compared to 60% of the controls. Further, patients had a lower incidence of lymphopenia, anemia, and mucositis. Following the completion of the third cycle of chemotherapy, recovery of hemoglobin, lymphocytes, leukocytes, neutrophils, and platelets was superior in A(1-7)-treated subjects. A(1-7) also reduced filgrastim use. Single-dose pharmacokinetics of A(1-7) showed a C^{max} of the 0.5 h sample, which correlated with dose. The plasma half-life of A(1-7) was 29 min, with levels being undetectable by 3–6 h post dosing.[32]

Phase 2b Clinical Trial in Recurrent Ovarian Cancer Patients

A phase 2b study was conducted to evaluate the safety and efficacy of A(1-7) in reducing the incidence and severity of thrombocytopenia in subjects receiving a combination of gemcitabine and platinum therapy for ovarian carcinoma. The primary end point of this study was severity and incidence of thrombocytopenia as determined by the number of chemotherapy cycles during which the platelet count was below 50,000 mm^{-13} (grade 3 or 4 thrombocytopenia). A significant reduction in grade 4 thrombocytopenia was seen in the 100 µg/kg group.[33] In addition, there was a significant maximal percent increase in platelet count and delivery of scheduled chemotherapy dose on time in subjects treated with 100 µg/kg/day versus placebo control. Maintenance of the scheduled chemotherapy at the desired dose has been linked with improved tumor control and long-term survival. No DLT was observed during the course of the study and no investigational product-related SAEs or deaths occurred.

CONCLUSIONS

These studies show that the effects of A(1-7) on hematopoietic progenitors in the bone marrow can be translated into peripheral effects and that synergy is seen when A(1-7) is combined with appropriate differentiating agents. Studies of A(1-7) have shown the ability to increase multiple lineages of early hematopoietic precursors cultured from the bone marrow and the peripheral blood of mice after myelosuppression due to chemotherapy, radiation, or hematopoietic stem cell transplantation. In addition, clinical data from multiple trials show decreased incidence of CIT and maintenance of chemotherapy dose density in A(1-7)-treated patients.[32,33] The consistent A(1-7)-mediated return to baseline hematopoietic values may be due to an increase in the number of hematopoietic progenitor cells, thereby pharmacologically replenishing the hematopoietic system.

It is anticipated that demonstration of the benefit of A(1-7) administration after TBI or improvement in hematopoietic stem cell transplantation will be translatable to humans. A favorable safety profile has been observed with A(1-7) in FDA-required toxicology as well as in all clinical studies.

In conclusion, the results from preclinical studies are supported by clinical trials showing that A(1-7) is a potent stimulator of hematopoietic recovery after injury.

REFERENCES

1. Bonadonna G, Valagussa P, Moliterni A, Zambetti M, Brambilla C. Adjuvant cyclophosphamide, methotrexate, and fluorouracil in node-positive breast cancer: the results of 20 years of follow-up. *N Engl J Med* 1995;**332**:901–6.
2. Budman DR, Berry DA, Cirrincione CT, Henderson IC, Wood WC, Weiss RB, et al. Dose and dose intensity as determinants of outcome in the adjuvant treatment of breast cancer. *J Natl Cancer Inst* 1998;**90**:1205–11.
3. Kwak LW, Halpern J, Olshen RA, Horning SJ. Prognostic significance of actual dose intensity in diffuse large-cell lymphoma: results of a tree-structured survival analysis. *J Clin Oncol* 1990;**8**:963–77.
4. Epelbaum R, Faraggi D, Ben-Arie Y, Ben-Shahar M, Haim N, Ron Y, et al. Survival of diffuse large cell lymphoma: a multivariate analysis including dose intensity variables. *Cancer* 1990;**66**:1124–9.
5. Lepage E, Gisselbrecht C, Haioun C, Sebban C, Tilly H, Bosly A, et al. Prognostic significance of received relative dose intensity in non-Hodgkin's lymphoma patients: application to LNH-87 protocol. *Ann Oncol* 1993;**4**:651–6.
6. Crivellari D, Bonetti M, Castiglione-Gertsch M, Gelber RD, Rudenstam CM, Thurlimann B, et al. Burdens and benefits of adjuvant cyclophosphamide, methotrexate, and fluorouracil and tamoxifen for elderly patients with breast cancer: the International Breast Cancer Study Group trial VII. *J Clin Oncol* 2000;**18**:1412–22.
7. Kaushansky K. The thrombocytopenia of cancer. Prospects for effective cytokine therapy. *Hematol Oncol Clin North Am* 1996;**10**(2):431–55.
8. Prow D, Vadhan-Raj S. Thrombopoietin: biology and potential clinical applications. *Oncology* 1998;**12**(11):1597–614.
9. Liou SY, Stephens JM, Carpiuc KT, Feng W, Botteman MF, Hay JW. Economic burden of haematological adverse effects in cancer patients. *Clin Drug Investig* 2007;**27**:381–96.
10. Elting LS, Cantor SB, Martin CG, Hamblin L, Kurtin D, Rivera E, et al. Cost of chemotherapy-induced thrombocytopenia among patients with lymphoma or solid tumors. *Cancer* 2003;**97**(6):1541–50.
11. Pfizer. Neumega package insert. http://www.pfizer.com/files/products/uspi_neumega.pdf; 2011. (Accessed 14 June, 2014.)
12. Haznedaroglu IC, Ozturk MA. Towards the understanding of the local hematopoietic bone marrow renin-angiotensin system. *Int J Biochem Cell Biol* 2003;**35**(6):867–80.
13. Strawn WB, Richmond RS, Ann TE, Gallagher PE, Ferrario CM. Renin-angiotensin system expression in rat bone marrow haematopoietic and stromal cells. *Br J Haematol* 2004;**126**:120–6.
14. Cole J, Ertoy D, Lin H, Sutliff RL, Ezan E, Guyene TT, et al. Lack of angiotensin II-facilitated erythropoiesis causes anemia in angiotensin-converting enzyme-deficient mice. *J Clin Invest* 2000;**106**:1391–8.

15. Lin C, Datta V, Okwan-Duodu D, Chen X, Fuchs S, Alsabeh R, et al. Angiotensin-converting enzyme is required for normal myelopoiesis. *FASEB J* 2011;**25**:1145–55.
16. Shen XZ, Bernstein KE. The peptide network regulated by angiotensin converting enzyme (ACE) in hematopoiesis. *Cell Cycle* 2011;**10**:1363–9.
17. Rodgers KE, Espinoza T, Roda N, Meeks CJ, Hill C, Louie SG, et al. Accelerated hematopoietic recovery with angiotensin-(1–7) after total body radiation. *Int J Radiat Biol* 2012;**88**(6):466–76.
18. Passos-Silva DG, Verano-Braga T, Santos RA. Angiotensin-(1–7): beyond the cardio-renal actions. *Clin Sci* 2013;**124**(7):443–56.
19. Zambidis ET, Park TS, Yu W, Tam A, Levine M, Yuan X, et al. Expression of angiotensin-converting enzyme (CD143) identifies and regulates primitive hemangioblasts derived from human pluripotent stem cells. *Blood* 2008;**112**:3601–14.
20. Durik M, Seva Pessoa B, Roks AJ. The renin-angiotensin system, bone marrow and progenitor cells. *Clin Sci* 2012;**123**(4):205–23.
21. Haznedaroglu IC, Beyazit Y. Local bone marrow renin-angiotensin system in primitive, definitive and neoplastic haematopoiesis. *Clin Sci (Lond)* 2013;**124**(5):307–23.
22. Rodgers K, diZerega GS. Contribution of the local RAS to hematopoietic function: a novel therapeutic target. *Front Endocrinol* 2013;**4**:157. doi:10.3389/fendo.2013.00157.
23. Jokubaitis VJ, Sinka L, Driessen R, Whitty G, Haylock DN, Bertoncello I, et al. Angiotensin-converting enzyme (CD143) marks hematopoietic stem cells in human embryonic, fetal, and adult hematopoietic tissues. *Blood* 2008;**111**:4055–63.
24. Tavian M, Biasch K, Sinka L, Vallet J, Péault B. Embryonic origin of human hematopoiesis. *Int J Dev Biol* 2010;**54**(6):1061.
25. Sinka L, Biasch K, Khazaal I, Péault B, Tavian M. Angiotensin-converting enzyme (CD143) specifies emerging lympho-hematopoietic progenitors in the human embryo. *Blood* 2012;**119**(16):3712–23.
26. Roks AJ, Rodgers K, Walther T. Effects of the renin angiotensin system on vasculogenesis-related progenitor cells. *Curr Opin Pharmacol* 2011;**11**(2):162–74.
27. Rodgers KE, Xiong S, Steer R, diZerega GS. Effect of angiotensin II on hematopoietic progenitor cell proliferation. *Stem Cells* 2000;**18**:287–94.
28. Rodgers KE, Xiong S, diZerega GS. Accelerated recovery from irradiation injury by angiotensin peptides. *Cancer Chemother Pharmacol* 2002;**49**:403–11.
29. Rodgers K, Xiong S, DiZerega GS. Effect of angiotensin II and angiotensin(1–7) on hematopoietic recovery after intravenous chemotherapy. *Cancer Chemother Pharmacol* 2003;**51**:97–106.
30. Heringer-Walther S, Eckert K, Schumacher SM, Uharek L, Wulf-Goldenberg A, Gembardt F, et al. Angiotensin-(1–7) stimulates hematopoietic progenitor cells in vitro and in vivo. *Haematologica* 2009;**94**:857–60.
31. Ellefson DD, diZerega GS, Espinoza T, Roda N, Maldonado S, Rodgers KE. Synergistic effects of co-administration of angiotensin 1–7 and Neupogen on hematopoietic recovery in mice. *Cancer Chemother Pharmacol* 2004;**53**:5–24.
32. Rodgers KE, Oliver J, diZerega GS. Phase I/II dose escalation study of angiotensin 1–7 [A(1–7)] administered before and after chemotherapy in patients with newly diagnosed breast cancer. *Cancer Chemother Pharmacol* 2006;**57**:559–68.
33. Pham H, Schwartz BM, Delmore JE, Davis KP, Reed E, Cruickshank S, et al. Pharmacodynamic stimulation of thrombogenesis by angiotensin 1–7 in recurrent ovarian cancer patients receiving gemcitabine and platin-based chemotherapy. *Cancer Chemother Pharmacol* 2013;**4**:965–72.
34. Mordwinkin NM, Russell JR, Burke AS, diZerega GS, Louie SG, Rodgers KE. Pharmacokinetics, pharmacodynamics and drug metabolism toxicological and toxicokinetic analysis of angiotensin (1–7) in two species. *J Pharm Sci* 2012;**101**:373–80.

Index

Note: Page numbers followed by *f* indicate figures, and *t* indicate tables.

Printed in the United States
By Bookmasters